INTRODUCTION TO The Uniform Geometrical Theory of Diffraction

The Artech House Microwave Library

Analysis, Design, and Applications of Fin Lines, Bharathi Bhat and Shiban K. Koul
E-Plane Integrated Circuits, P. Bhartia and P. Pramanick, eds.
Filters with Helical and Folded Helical Resonators, Peter Vizmuller
GaAs MESFET Circuit Design, Robert A. Soares, ed.
Gallium Arsenide Processing Techniques, Ralph Williams
Handbook of Microwave Integrated Circuits, Reinmut K. Hoffmann
Handbook for the Mechanical Tolerancing of Waveguide Components, W.B.W. Alison
Handbook of Microwave Testing, Thomas S. Laverghetta
High Power Microwave Sources, Victor Granatstien and Igor Alexeff, eds.
Introduction to Microwaves, Fred E. Gardiol
LOSLIN: Lossy Line Calculation Software and User's Manual, Fred. E. Gardiol
Lossy Transmission Lines, Fred E. Gardiol
Materials Handbook for Hybrid Microelectronics, J.A. King, ed.
Microstrip Antenna Design, K.C. Gupta and A. Benalla, eds.
Microstrip Lines and Slotlines, K.C. Gupta, R. Garg, and I.J. Bahl
Microwave Engineer's Handbook: 2 volume set, Theodore Saad, ed.
Microwave Filters, Impedance Matching Networks, and Coupling Structures, G.L. Matthaei, L. Young and E.M.T. Jones
Microwave Integrated Circuits, Jeffrey Frey and Kul Bhasin, eds.
Microwaves Made Simple: Principles and Applications, Stephen W. Cheung, Frederick H. Levien, et al.
Microwave and Millimeter Wave Heterostructure Transistors and Applications, F. Ali, ed.
Microwave Mixers, Stephen A. Maas
Microwave Transition Design, Jamal S. Izadian and Shahin M. Izadian
Microwave Transmission Line Filters, J.A.G. Malherbe
Microwave Transmission Line Couplers, J.A.G. Malherbe
Microwave Tubes, A.S. Gilmour, Jr.
MMIC Design: GaAs FETs and HEMTs, Peter H. Ladbrooke
Modern Spectrum Analyzer Theory and Applications, Morris Engelson
Monolithic Microwave Integrated Circuits: Technology and Design, Ravender Goyal, et al.
Nonlinear Microwave Circuits, Stephen A. Maas
Terrestrial Digital Microwave Communications, Ferdo Ivanek, et al.

INTRODUCTION TO The Uniform Geometrical Theory of Diffraction

D.A. McNamara
C.W.I. Pistorius
J.A.G. Malherbe
University of Pretoria

Artech House
Boston • London

Library of Congress Cataloging-in-Publication Data

McNamara, Derek, A., 1955-
Introduction to the geometrical theory of diffraction /
Derek A. McNamara, Carl W.I. Pistorius, J.A.G. Malherbe.
p. cm.
Includes bibliographical references .
ISBN 0-89006-301-X
1. Diffraction, Geometrical. 2. Optics, Geometrical.
I. Pistorius, Carl W.I., 1958- . II. Malherbe, J.A.G., 1940-
III. Title.
QC415.M34 1990 89-49004
535.4--dc20 CIP

British Library Cataloguing in Publication Data

McNamara, Derek A. (Derek Albert), 1955-
Introduction to the uniform geometrical theory of
diffraction.
1. Diffraction. Geometrical theory
I. Title II. Pistorius, Carl W. I., 1958- III. Malherbe,
J.A.G., 1940-
539.2

ISBN 0-89006-301-X

ARTECH HOUSE
685 Canton Street
Norwood, MA 02062

International Standard Book Number: 0-89006-301-X
Library of Congress Catalog Card Number: 89-49004
10 9 8 7 6 5 4 3 2 1

To our teachers at Ohio State University

CONTENTS

Preface

Knowledge of the geometrical theory of diffraction (GTD) is best accomplished in two steps. On first looking into the subject we should accept the principal results and concepts, properly understand their various parameters, and apply the methods to a wide variety of problems—this, roughly, is the way we first approach network theory. Once this has been accomplished, the asymptotic methods used to piece together the GTD coefficients can be studied through the many scholarly works available in engineering and mathematical journals. This text has been written to provide a means of achieving the first step.

Although material on the GTD and its uniform extension (UTD) has appeared in advanced handbooks and texts, as well as in collections of journal article reprints, no text suitable for an introductory course or self-study is currently available. We hope that this book will bridge the gap between specialist papers and the use of the GTD in the practical problems for which the method is eminently suitable. As such, the text is intended for senior undergraduate or beginning graduate students, as well as practicing engineers. After a brief introductory chapter, which discusses the basic concepts and ideas, reflection and various diffraction phenomena are discussed in turn. As with network theory, applying the GTD to problems of practical interest is possible without a complete understanding of all the minute details of its derivation, which indeed is a most attractive feature. But although the approach may be described as axiomatic, motivation is given before any results are postulated in their general form. Thereafter, special cases that provide increased insight are discussed and the theory immediately applied to worked problems.

We wished to include only those aspects that at the time of the writing, were well proven and established. We therefore deal with conducting objects only. Having worked through this introductory text, the reader will be able to proceed with increased confidence to additional problems discussed in the references. References also are given to the derivations behind the results quoted and used. Most of these assume a basic knowledge of the mathematics of asymptotic methods, and some guidance as to where this can be gathered is provided in Appendix D.

The uniform geometrical theory of diffraction developed at Ohio State University is not the only uniform theory available. However, it has been the one widely applied in the literature to specific problems on a systematic basis. It is the only form used in this book.

We have benefitted greatly from the instruction of our teachers at OSU, and to them this book is dedicated. In the preparation of this book we have made use of the class notes for the course at OSU written by Professors L. Peters, Jr., and W.D. Burnside. Use has also been made of the notes for the short course on the Modern Geometrical Theory of Diffraction presented by OSU, and we are grateful for the permission granted us to do so.

We also wish to thank our own graduate students who have spent many hours working through early drafts of the manuscript. The University of Pretoria provided us the opportunity to complete the text, and for this we express our appreciation. Finally, we thank our families and friends for their support.

DAM[c]
CWIP
JAGM
UNIVERSITY OF PRETORIA
AUGUST 1989

Chapter 1
The Nature of High-Frequency Methods

"Die Optika, soos die leer van die Lig dikwels genoem word, word gewoonlik onder twee hoofafdelinge bestudeer, n.l., die Geometriese en die Fiesiese Optika. Die twede afdeling handel oor die wese van die lig en sal vir die doel van hierdie boek maar min aandag kan geniet. Die verskynsels wat onder die geometriese optika begrepe is kan verklaar word uit 'n paar eenvoudege wette wat betrekking het op die voortplanting, die terugkaatsing en die breking van lig."

W.E. Malherbe,
Beknopte Leerboek Oor Die Natuurkunde,
Pro-Ecclesia, 1922

1.1 INTRODUCTION

Ever since 1864 when James Clerk Maxwell deduced the set of equations bearing his name, which describe all macroscopic electromagnetic phenomena, we have known that light is an electromagnetic wave of extremely high frequency. Thus, all results obtained from studies on light include models of high-frequency electromagnetic wave phenomena in general and not merely the visible portion of the electromagnetic spectrum. As Maxwell's theory superseded the older geometrical approaches on which classical optics had been based, we might presume that geometrical optics could be cast aside. Although this presumption is correct in principle, just the contrary is true in practice; geometrical concepts have been found, as might be expected, to provide increased understanding of certain electromagnetic phenomena and effects. Furthermore, the extension of classical geometrical optics (that is, modern geometrical optics and the geometrical theory of diffraction) that forms the subject of this text enables us to solve in an acceptable manner high-frequency electromagnetic problems for which exact solutions of Maxwell's equations as boundary value problems are not possible.

A proper understanding of a subject is helped greatly by a review of its history, even if this is brief. Such a review forms the content of Section 1.2. In Section 1.3 a number of important concepts are discussed qualitatively.

1.2 A BRIEF HISTORICAL OVERVIEW

Observations of electrostatic and magnetostatic phenomena were begun by the ancient Greeks at about the same time as their studies in optics. Probably, the Greek fascination with geometry enabled them to look at the properties of light from a geometric point of view. The first people on record as involved with the subject of optics were Pythagoras and Empedocles, but Euclid [1], the great Alexandrian mathematician (*circa* 330–275 BC), first stated the law of reflection: that when light is reflected from a smooth surface, the angle of incidence is equal to the angle of reflection. He also apparently deduced the properties of paraboloidal, spherical, and ellipsoidal reflectors.

Another Alexandrian by the name of Heronis (we know only that he lived sometime between 200 BC and 300 AD!) showed that the law of reflection can be put in the alternative form [1]: that the light follows the shortest path from the source point to the observation point, subject to the condition that it must strike the reflector at some point in its journey. Thus, if ABC in Figure 1.1 is the actual path, this is shorter than $AB'C$ or $AB''C$, or any similar path. Heronis appears not to have attached any special importance to this result [1] and could not have suspected that he was in fact introducing a new and far-reaching idea later to become known as *Fermat's principle*.

Ptolemy, also of Alexandria, *circa* 140 AD, performed experiments on the refraction of light by glass and water and formulated a law of refraction. His law,

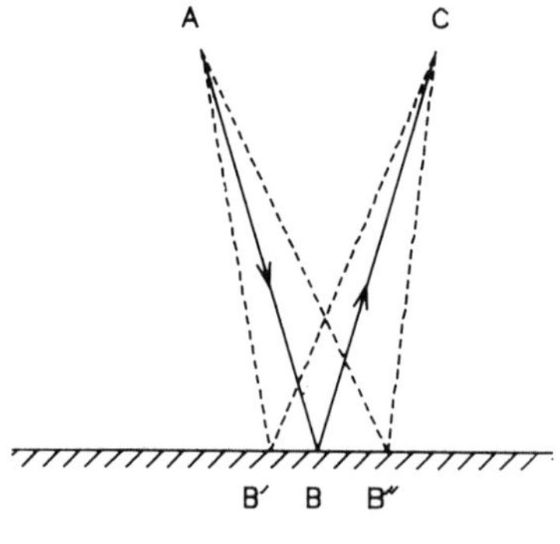

Figure 1.1 The Heronis-Fermat principle for reflection.

although not exact, is very nearly correct for small angles the sine of which is approximately equal to the angle itself. In medieval times, Alhazen (965–1038 AD), working in Cairo, wrote a treatise on optics [1]. He found Ptolemy's law of refraction valid only for small angles but did not discover the true law. He also studied the action of spherical and paraboloidal mirrors and the magnification produced by lenses and solved the problem of finding the relation between the positions of a source of light and its image formed by a lens. With Alhazen, classical geometrical optics was beginning to assume a more quantitative form.

Not until 1621 did Willebrord Snell experimentally determine the correct law of refraction: that for a ray impinging on the interface between two media, the ratio of the sines of the angles of incidence and refraction equals the ratio of the refractive indices of the two media. In 1654 the French mathematician Fermat postulated that regardless of what kind of reflection or refraction to which a ray is subjected, it travels from one point to another in such a way as to make the time taken (rather than the distance) a minimum. This so-called Fermat's principle was later put on a firmer mathematical basis by William Hamilton during the years 1824 to 1844. This completed the mathematical theory of what can be called *classical geometrical optics,* which was conceived quite independently of any reference to electromagnetic theory. This is the same geometrical optics that comprises the fundamentals taught in elementary physics courses, and that is concerned with the problem of ray-tracing through systems of lenses and mirrors. The design of most optical instruments has been and still is based on these principles. However, as a possible method of solving high-frequency electromagnetics problems in general, classical geometrical optics is limited because there is no mention of the concept of phase (and hence wavelength and interference phenomena), polarization (the vector properties of electromagnetic fields), intensity, or diffraction. The classical theory deals solely with geometrical curves.

Much of the latter phenomena, however, could be explained in terms of wave theory, rather than the geometrical approach. Using a wave theory approach and the correct assumption that the velocity of light is smaller in a denser medium, Christiaan Huygens in 1690 was able to give the first theoretical derivation of the law of refraction. Diffraction and interference had been observed by Robert Hooke and the Italian monk, F.M. Grimaldi, in 1665. In the first half of the nineteenth century new experimental work by Thomas Young, Augustin Fresnel, and others clarified that a wave theory of light was needed to account for diffraction and other phenomena not explained by classical geometrical optics. After light had been shown to be an electromagnetic wave governed by Maxwell's equations and the resultant wave equations, the subject of diffraction was studied by Lord Rayleigh, Lord Kelvin, Sir George Stokes, Kirchoff, Helmholtz, Mie, and many other famous physicists. The result most significant to the work in this text is the exact solution for plane wave diffraction by a conducting half-plane, derived by Arnold Sommerfeld in 1884. However, although the electromagnetic wave theory approach can explain what classical geometrical optics cannot, the description is by means of partial differential equations and, though mathematically elegant, is not without its practical disadvantages. First, the insight afforded by geometrical concepts is lost. Also, only a very limited number of problems formulated as boundary value problems actually can be solved exactly; in the majority of cases of physical interest we have to resort to approximate methods, and then geometrical insight is invaluable. To exploit classical geometrical optics and to retain the wave theory effects, modern geometrical optics and the geometrical theory of diffraction had to be devised.

To be able to account for interference effects (by retaining phase information), the polarization properties of electromagnetic fields, and to obtain quantitative results for the field amplitudes, in the latter part of the present century the classical geometrical optics has been extended to what may be called *modern geometrical optics,* but which in the remainder of this text will simply be called *geometrical optics* (GO). This has been achieved by establishing a more rigorous connection between Maxwell's electromagnetic field equations and geometrical concepts, provided chiefly through the work of Luneberg [2] and Kline [3], and is considered in Chapter 2. The GO theory of reflection from smooth surfaces is taken up in Chapter 3.

A method by which diffraction phenomena can be incorporated into a geometrical strategy and phrased in geometric terms was presented by Joseph B. Keller in 1953, though published only some time later [4]. This laid the basis for what has become known as the *geometrical theory of diffraction* (GTD). In studying diffraction by a wedge, Keller introduced diffracted rays that behave like the ordinary rays of GO once they leave the edge. These rays have paths determined by a generalization of Fermat's principle: an extremum path subject to the constraint that it must include a point somewhere on the edge. These diffracted rays are added to those of GO and together correctly account for high-frequency field behavior in the presence of diffracting obstacles with edges. Keller [5] later extended the principle to encompass diffraction by smooth surfaces (the phenomenon of creeping rays). In these cases, too, the ray paths are found from the generalized Fermat principle; this generalization "consists of a generalization of the concept of rays to include any extremal path satisfying constraints inferred by the environment" [6]. In the GTD context, diffraction today is regarded as any process whereby electromagnetic wave propagation differs from the predictions of GO [6]. The interference between different diffracted terms and direct and reflected rays gives rise to the maxima and minima usually found in physical field distributions.

The original form of the *geometrical theory of diffraction* (which is the form henceforth denoted as the GTD) suffers from a number of problems. The most serious of these is at shadow boundaries, where the GO fields fall abruptly and nonphysically to zero (more will be said about this in the next chapter) and the predicted diffracted fields become infinite. The reason for this is roughly as follows.

In the mathematics of asymptotic methods, a narrow region in which equation solutions change rapidly is called a *boundary layer*. For diffraction problems, the shadow boundaries fall inside such boundary layers. Keller based his original theory on asymptotic expressions valid outside such boundary layers, and they therefore are inapplicable in the vicinity of shadow boundaries. Asymptotic expressions valid even in boundary layers are known as *uniform solutions*. The *uniform theory of diffraction* (UTD), developed principally by Kouyoumjian's group [7–10] at Ohio State University, is presented and applied in Chapters 4 to 8. The UTD has the additional property that its solutions always reduce continuously to those of the GTD in the domains where the latter indeed are valid.

1.3 HIGH-FREQUENCY PHENOMENA

The geometrical theory of diffraction, which will be considered in this text to encompass modern geometrical optics, is applicable to high-frequency field problems. By *high-frequency phenomena,* we mean that fields are being considered in a system where the properties of the medium and scatterer size parameters vary little over an interval on the order of a wavelength. The problems in engineering to which these techniques are applied typically are of such dimensions that the methods are valid at microwave frequencies and above; hence the term *high-frequency*. It is important to realize that the method does not have an intrinsic frequency limitation however; what restricts its application is the fact that the size of a scatterer must be large in terms of the wavelength at the given frequency. Thus, geometrical optics ray-tracing procedures can be used for studying electromagnetic wave propagation in the ionosphere in the megahertz frequency region, for example.

The mathematical techniques germane to the analysis of high-frequency electromagnetic field phenomena are known as asymptotic methods [11]. In the literature, several asymptotic methods have been developed for the solution of Maxwell's equations. The application of these methods can be either direct or indirect, depending on the stage in the solution process where the asymptotic method is applied. In the indirect approach, the exact solution of the scattering problem at hand first must be determined. Then, the asymptotic method is applied to this exact solution, yielding an asymptotic expansion for the exact solution. In the direct approach, the asymptotic method is applied directly to each of Maxwell's equations at the beginning of the solution process, instead of to the result, and the alternative equations are solved. The geometrical optics theory considered in Chapters 2 and 3 use the direct approach, and these solutions are given a ray interpretation. An indirect approach to the reflection problem also is possible and mention is made of it in Chapter 3 and again in Chapter 8. The diffraction phenomena discussed in the remainder of the text are derived (though this text concentrates on their significance and application, not their complete derivation) through the application of the indirect approach to a number of key geometries (referred to as *canonical* or *etalon* problems) for which exact solutions are available. The essential step is to then cast these solutions into a ray picture. Thereafter, the fact that diffraction effects are dependent principally on the properties of the scatterer in the vicinity of certain crucial points permits the extension of the results for the canonical geometries to those of a more general nature. The practice of the UTD consists of the systematic piecing together of results gleaned from such canonical geometries to analyze more complex structures.

REFERENCES

[1] Sir James Jeans, *The Growth Of Physical Science,* Cambridge University Press, 1950.

[2] R.M. Luneberg, *Mathematical Theory Of Optics,* Brown University Press, 1944.

[3] M. Kline, "An Asymptotic Solution of Maxwell's Equations," *Commun. Pure Appl. Math.,* Vol. 4, 1951, pp. 225–262.

[4] J.B. Keller, "Geometrical Theory of Diffraction," *J. Opt. Soc. Am.,* Vol. 52, 1962, pp. 116–130.

[5] J.B. Keller, "Diffraction by a Convex Cylinder," *IRE Trans. Antennas and Propagation,* Vol. AP-24, 1956, pp. 312–321.

[6] J.B. Keller, "One Hundred Years of Diffraction Theory," *IEEE Trans. Antennas and Propagation,* Vol. AP-33, No. 2, February 1985, pp. 123–126.

[7] R.G. Kouyoumjian and P.H. Pathak, "A Uniform Geometrical Theory of Diffraction for an Edge in a Perfectly Conducting Surface," *Proc. IEEE,* Vol. 62, November 1974, pp. 1448–1461.

[8] P.H. Pathak, W.D. Burnside, and R.J. Marhefka, "A Uniform GTD Analysis of the Diffraction of Electromagnetic Waves by a Smooth Convex Surface," *IEEE Trans. Antennas and Propagation,* Vol. AP-28, No. 5, September 1980, pp. 631–642.

[9] P.H. Pathak, N. Wang, W.D. Burnside, and R.G. Kouyoumjian, "A Uniform GTD Solution for the Radiation from Sources on a Convex Surface," *IEEE Trans. Antennas and Propagation,* Vol. AP-29, No. 4, July 1981, pp. 609–621.

[10] P.H. Pathak and N. Wang, "Ray Analysis of Mutual Coupling between Antennas on a Convex Surface," *IEEE Trans. Antennas and Propagation,* Vol. AP-29, No. 6, November 1981, pp. 911–922.

[11] R.G. Kouyoumjian, "Asymptotic High-Frequency Methods," *Proc. IEEE,* Vol. 53, No. 8, August 1965, pp. 864–876.

Chapter 2
Geometrical Optics Fields

"By the rays of light I understand its least parts, and those as well successive in the same lines, as contemporary in several lines. For it is manifest that light consists of parts, both successive and contemporary; because in the same place you may stop that which comes one moment, and let pass that which comes presently after; and in the same time you may stop it in any one place, and let it pass in any other. For that part of light which is stopp'd cannot be the same with that which is let pass. The least light or part of light, which may be stopp'd alone without the rest of the light, or propagated alone, or do or suffer any thing alone, which the rest of the light doth not or suffers not, I call a ray of light."

Sir Isaac Newton,
Opticks, 1704

2.1 INTRODUCTION

The most significant property of the high-frequency field is that of its local plane wave nature and its consequences. This is one of the principal assets exploited methodically in the geometrical theory of diffraction. The incorporation of such local plane wave behavior of the field allows reducing the electromagnetic wave equations to the simpler equations for the polarization, amplitude, phase, and propagation path of the high-frequency field. In this chapter such ideas are given a quantitative basis. The chronological approach of Section 1.2 will not be pursued here. Instead, means will be adopted whereby the desired results can be reached as directly as possible. We will strive, however, to place the high-frequency field concepts in the correct context with the body of electromagnetic theory as a whole.

2.2 RAY OPTICAL CONSTRUCTION OF THE HIGH-FREQUENCY FIELD

2.2.1 Preliminary Remarks

Many useful solutions to physical problems develop through what can be considered a two-step process, which may take many years. The first step is that in which intuitive insight is built up. Only then is it possible to turn with any success to the second task of transforming that view into the formal quantitative language of calculation. When the results of the latter process are subsequently presented, any assumptions required (based on earlier intuitive or partial quantitative understanding) are then traditionally given the name *ansatz,* which as translated essentially means "anticipated solution." These may seem puzzling at first, having been based on a considerable amount of hindsight. However, the familiarity gained through use of the results soon supplies an insight similar to that of the originator—and all is well. Such is the case for the Luneberg-Kline ansatz to be discussed in Section 2.2.3, but we first set down some more conventional equations of electromagnetic theory. We note here that, because the necessary mathematical treatment is no simpler in the three-dimensional case than in the two-dimensional, the former will be treated directly and then later in the chapter reduced to two dimensions. The insight afforded in doing it this way will provide the motivation for sometimes beginning discussions of new phenomena in the reverse order in later chapters.

2.2.2 Some Conventional Electromagnetic Theory

Let the spatial coordinate be denoted by $\mathbf{r}$, and the radian frequency by ω. Then exterior to source regions (i.e., at any point $\mathbf{r}$ at which there is no source), in a medium of permittivity $\epsilon(\mathbf{r})$ and permeability $\mu(\mathbf{r})$, the time-harmonic ($e^{j\omega t}$ variation) electric and magnetic fields satisfy the Maxwell equations [1]:

$$\nabla \times \mathbf{E}(\mathbf{r}, \omega) + j\omega\mu(\mathbf{r})\mathbf{H}(\mathbf{r}, \omega) = 0 \tag{2.1}$$

$$\nabla \times \mathbf{H}(\mathbf{r}, \omega) - j\omega\epsilon(\mathbf{r})\mathbf{E}(\mathbf{r}, \omega) = 0 \tag{2.2}$$

$$\nabla \cdot [\epsilon(\mathbf{r})\mathbf{E}(\mathbf{r}, \omega)] = 0 \tag{2.3}$$

$$\nabla \cdot [\mu(\mathbf{r})\mathbf{H}(\mathbf{r}, \omega)] = 0 \tag{2.4}$$

The electric and magnetic field vectors, $\mathbf{E}$ and $\mathbf{H}$ respectively, are *root mean square* (rms) phasor quantities. For these differential forms of Maxwell's equations to be valid, $\epsilon(\mathbf{r})$ and $\mu(\mathbf{r})$ must either be smoothly varying or constant in the region of interest so that their spatial derivatives and those of the field vectors are well defined. In most cases of practical interest to which the GTD has been applied, and certainly in all problems to be considered in this text, the material media in which propagation and scattering occur will be one of two types.

The first is that of a homogeneous medium for which the permittivity and permeability are constants, ϵ and μ, respectively, in the region of interest. As a further yet not overly restrictive simplification we will assume that $\mu = \mu_0$. The wave impedance of the medium is then $Z = \sqrt{\mu_0/\epsilon}$, the wave admittance $Y = 1/Z$, and the wave velocity (speed of light in the medium) is $v = 1/\sqrt{\mu_0\epsilon}$.

The second category is one where the permittivity may be constant piecewise; that is, the overall region of interest may consist of subregions over which ϵ is constant but different, with $\mu = \mu_0$ in each. Such abrupt permittivity changes at interfaces between subregions will make the total field vectors discontinuous there. The earlier differential equations therefore are not valid at such interfaces. Rather, they have to be applied separately in each subregion and the solutions matched across the interfaces through application of the boundary conditions for electromagnetic fields. Thus the details of the solutions for the constant ϵ media are applicable in such a situation as well. We therefore will restrict our treatment to homogeneous media, which in addition will be assumed isotropic, linear, and lossless (permittivity and permeability are real). Any results presented should not simply be assumed to apply directly in media of smoothly varying electrical parameters.

Now, if the earlier assumption regarding the homogeneity of the medium is accepted, (2.1) and (2.2) become

$$\nabla \times \mathbf{E}(\mathbf{r}, \omega) + j\omega\mu_0\mathbf{H}(\mathbf{r}, \omega) = 0 \tag{2.5}$$

$$\nabla \times \mathbf{H}(\mathbf{r}, \omega) - j\omega\epsilon\mathbf{E}(\mathbf{r}, \omega) = 0 \tag{2.6}$$

and the zero divergence equations (2.3) and (2.4) are

$$\nabla \cdot \mathbf{E}(\mathbf{r}, \omega) = 0 \tag{2.7}$$

$$\nabla \cdot \mathbf{H}(\mathbf{r}, \omega) = 0 \tag{2.8}$$

One further equation that will be of use in Section 2.2.4 is the vector Helmholtz equation derived by eliminating **H** from (2.1) and (2.2) to obtain

$$\nabla^2\mathbf{E}(\mathbf{r}, \omega) + k^2\mathbf{E}(\mathbf{r}, \omega) = 0 \tag{2.9}$$

where $k^2 = \omega^2\mu_0\epsilon$ and $\nabla^2\mathbf{E} = \nabla\nabla \cdot \mathbf{E} - \nabla \times \nabla \times \mathbf{E}$. A similar equation holds for **H**.

To define here certain special electromagnetic wave types—namely *plane waves, cylindrical waves,* and *spherical waves*—will be beneficial for later sections of the book. Therefore consider an electromagnetic field existing in some region of space. This field is referred to as a (true) *uniform plane wave* or simply a (true) *plane wave* if its surfaces of constant phase *and* its surfaces of constant amplitude are planar everywhere and coincident. If the equiphase and equiamplitude surfaces are planar but do not coincide, it is a *nonuniform* or *inhomogeneous plane wave.* The fields of a plane wave have a number of interesting properties. If $\hat{\mathbf{s}}$ is the unit vector in the propagation direction, then $\mathbf{E}(\mathbf{r})$, $\mathbf{H}(\mathbf{r})$, and $\hat{\mathbf{s}}$ are mutually orthogonal and interelated through $\mathbf{H} = Y(\hat{\mathbf{s}} \times \mathbf{E})$. In other words, a plane wave is a *transverse electromagnetic* (TEM) wave. The qualifier *true* is added to contrast these definitions with those of *locally* plane waves, which follow and appear many times later in the text.

A (true) *cylindrical wave* is one that has equiphase surfaces that are concentric cylinders, whereas a (*true*) *spherical wave* is one with constant phase surfaces that are concentric spheres. For vector fields, it is not physically possible to obtain equiamplitude surfaces that are concentric spheres [2], and thus to speak of uniform and nonuniform spherical waves makes little sense. However, if we are dealing with scalar fields as in acoustics, true uniform spherical waves are not theoretically forbidden. At sufficiently large distances from cylindrical and spherical wave sources, the field components oriented in the observation direction $\hat{\mathbf{s}}$ usually have decayed so rapidly that the only significant components are those transverse to $\hat{\mathbf{s}}$ and related in precisely the same way as those of a true plane wave; that is, $\mathbf{E}(\mathbf{r})$, $\mathbf{H}(\mathbf{r})$, and $\hat{\mathbf{s}}$ are mutually orthogonal and connected through $\mathbf{H} = Y(\hat{\mathbf{s}} \times \mathbf{E})$. This is true even though the phasefronts are not planar; we say that such fields are *locally plane.*

2.2.3 The Luneberg-Kline Anticipated Solution (Ansatz)

Kline [3, 4] adopted the following ansatz for the form of a high-frequency electromagnetic field in a source-free region occupied by an isotropic medium with constitutive parameters ϵ and $\mu = \mu_0$:

$$\mathbf{E}(\mathbf{r}, \omega) \sim e^{-jk\Psi(\mathbf{r})} \sum_{n=0}^{\infty} \frac{\mathbf{E}_n(\mathbf{r})}{(j\omega)^n} \tag{2.10}$$

$$\mathbf{H}(\mathbf{r}, \omega) \sim e^{-jk\Psi(\mathbf{r})} \sum_{n=0}^{\infty} \frac{\mathbf{H}_n(\mathbf{r})}{(j\omega)^n} \tag{2.11}$$

where $k^2 = \omega^2\mu_0\epsilon$ and $\Psi(\mathbf{r})$ is the so-called phase function to which we will return later. This is based on a number of physical observations from classical geometrical optics, the form of electromagnetic fields sufficiently far from their sources, the success of the Sommerfeld-Runge ansatz [5], and the work of Luneberg [6]. The expansions (2.10) and (2.11), which are extensions of the Sommerfeld-Runge ansatz, are referred to as the Luneberg-Kline *asymptotic expansions,* the symbol ~ meaning "equal to, in the asymptotic sense." Kouyoumjian [7] has succinctly defined asymptotic techniques as follows:

> "From the mathematical viewpoint, asymptotic methods are methods for expanding functions, evaluating integrals, and solving differential equations, which become increasingly accurate as some parameter approaches a limiting value. Here we are interested in solutions to electromagnetic scattering problems which are valid as the free-space wavenumber k (or angular frequency ω) approaches infinity."

Thus, here we hope that, as ω and hence the largeness parameter k become larger, expansions (2.10) and (2.11) become increasingly accurate representations of the electromagnetic field. In fact, as the frequency tends to infinity (the high-frequency or zero-wavelength limit), only the first term ($n = 0$) remains, called the *geometrical optics* (GO) field because it will be seen to encompass the classical geometrical optics field characteristics in Section 2.2.6. Note that the terms $\mathbf{E}_n$, $\mathbf{H}_n$, and Ψ depend only on the spatial coordinates $\mathbf{r}$ and not on the frequency ω. If these are used, the mathematical solutions derived place on a firm quantitative foundation not only the observed "classical" characteristics of high-frequency electromagnetic fields, but in addition incorporate further properties. However, there is no *a priori* knowledge that the expressions in (2.10) and (2.11) are, according to Jones,

> ". . . valid whether as a convergent or an asymptotic series. Even in special circumstances it can be extremely difficult to prove the legitimacy of the expansion. Be that as it may, experience is overwhelmingly in favour of its introduction" [8, p.5]

Electromagnetic fields that can be written as a Luneberg-Kline expansion are called *ray-optic fields* or simply *ray fields* [9,10]. However, the designations *geometrical optics field* or *high-frequency field* are reserved in this text for the first term in the Luneberg-Kline expansion. If a field is not ray-optic, of course, it is not a geometrical optics field either, but a ray-optic field need not be a geometrical optics field.

2.2.4 The Eikonal Equation

In Section 2.2.2 the mathematical manipulations were exact and certainly not unique to the geometrical theory of diffraction in any way. We now wish to obtain high-frequency solutions for $\mathbf{E}$ and $\mathbf{H}$ of the form given by the Luneberg-Kline expansions (2.10) and (2.11). In the mathematical manipulations to follow, we should remember that the ∇-operator acts on the spatial coordinate $\mathbf{r}$, and thus on $\mathbf{E}(\mathbf{r})$, $\mathbf{H}(\mathbf{r})$, and $\Psi(\mathbf{r})$ only. The derivation is done in five steps.

Step 1

Equations (2.10) and (2.11) are substituted into (2.5), and differentiability assumed for the individual terms in the Luneberg-Kline expansion. In the vector identity:

$$\nabla \times (f\mathbf{F}) = \nabla f \times \mathbf{F} + f\nabla \times \mathbf{F}$$

the scalar function is set to

$$f = \mathrm{e}^{-\mathrm{j}k\Psi}$$

and the vector function to

$$\mathbf{F} = \sum_{n=0}^{\infty} \frac{\mathbf{E}_n}{(\mathrm{j}\omega)^n}$$

After differentiation,

$$\sum_{n=0}^{\infty} \left[\frac{\nabla \times \mathbf{E}_n \, \mathrm{e}^{-\mathrm{j}k\Psi}}{(\mathrm{j}\omega)^n} - \frac{\mathrm{j}k(\nabla\Psi \times \mathbf{E}_n)\, \mathrm{e}^{-\mathrm{j}k\Psi}}{(\mathrm{j}\omega)^n} \right] = -\mathrm{j}\omega\mu_0 \, \mathrm{e}^{-\mathrm{j}k\Psi} \sum_{n=0}^{\infty} \frac{\mathbf{H}_n}{(\mathrm{j}\omega)^n} \tag{2.12}$$

This can be rewritten in the form:

$$\sum_{n=0}^{\infty} \left[\frac{\nabla \times \mathbf{E}_n \, \mathrm{e}^{-\mathrm{j}k\Psi}}{(\mathrm{j}\omega)^n} - \frac{1/\nu \, (\nabla\Psi \times \mathbf{E}_n)\, \mathrm{e}^{-\mathrm{j}k\Psi}}{(\mathrm{j}\omega)^{n-1}} \right] = -\mu_0 \mathrm{e}^{-\mathrm{j}k\Psi} \sum_{n=0}^{\infty} \frac{\mathbf{H}_n}{(\mathrm{j}\omega)^{n-1}} \tag{2.13}$$

Extracting the $n = 0$ and $n = 1$ terms explicitly, for reasons we will see shortly, we obtain

$$\begin{aligned} &\left[-\frac{1/\nu \, (\nabla\Psi \times \mathbf{E}_0)}{(\mathrm{j}\omega)^{-1}} + \frac{\nabla \times \mathbf{E}_1}{(\mathrm{j}\omega)} - \frac{1/\nu \, (\nabla\Psi \times \mathbf{E}_1) - \nabla \times \mathbf{E}_0}{(\mathrm{j}\omega)^0} \right] \mathrm{e}^{-\mathrm{j}k\Psi} \\ &\quad + \sum_{n=2}^{\infty} \left[\frac{\nabla \times \mathbf{E}_n \, \mathrm{e}^{-\mathrm{j}k\Psi}}{(\mathrm{j}\omega)^n} - \frac{1/\nu \, (\nabla\Psi \times \mathbf{E}_n)\, \mathrm{e}^{-\mathrm{j}k\Psi}}{(\mathrm{j}\omega)^n} \right] \\ &\quad = -\frac{\mu_0 \mathbf{H}_0 \, \mathrm{e}^{-\mathrm{j}k\Psi}}{(\mathrm{j}\omega)^{-1}} - \frac{\mu_0 \mathbf{H}_1 \, \mathrm{e}^{-\mathrm{j}k\Psi}}{(\mathrm{j}\omega)^0} - \mu_0 \, \mathrm{e}^{-\mathrm{j}k\Psi} \sum_{n=2}^{\infty} \frac{\mathbf{H}_n}{(\mathrm{j}\omega)^{n-1}} \end{aligned} \tag{2.14}$$

Equating terms in (jω) gives us the relation:

$$1/\nu\ (\nabla\Psi \times \mathbf{E}_0) = \mu_0 \mathbf{H}_0 \tag{2.15}$$

Step 2

Equation (2.10) is substituted into the zero divergence condition (2.7) and differentiability assumed for the individual terms in the Luneberg-Kline expansion. The vector identity

$$\nabla \cdot (f\mathbf{F}) = \mathbf{F} \cdot \nabla f + f\nabla \cdot \mathbf{F}$$

is applied and we obtain

$$e^{-jk\Psi} \sum_{n=0}^{\infty} \left[\frac{1/\nu\ \nabla\Psi \cdot \mathbf{E}_n}{(j\omega)^{n-1}} - \frac{\nabla \cdot \mathbf{E}_n}{(j\omega)^n} \right] = 0 \tag{2.16}$$

Once again, extraction of the $n = 0$ and $n = 1$ terms separately gives

$$\begin{aligned} & e^{-jk\Psi} \left[\frac{1/\nu\ \nabla\Psi \cdot \mathbf{E}_0}{(j\omega)^{-1}} + \frac{1/\nu\ \nabla\Psi \cdot \mathbf{E}_1 - \nabla \cdot \mathbf{E}_0}{(j\omega)^0} - \frac{\nabla \cdot \mathbf{E}_1}{(j\omega)} \right] \\ & + e^{-jk\Psi} \sum_{n=2}^{\infty} \left[\frac{1/\nu\ \nabla\Psi \cdot \mathbf{E}_n}{(j\omega)^{n-1}} - \frac{\nabla \cdot \mathbf{E}_n}{(j\omega)^n} \right] = 0 \end{aligned} \tag{2.17}$$

Equating terms in powers (jω) yields the expression:

$$\nabla\Psi \cdot \mathbf{E}_0 = 0 \tag{2.18}$$

Step 3

If the procedure in Step 1 is repeated, but using equation (2.11) in (2.2), we have the following:

$$1/\nu\ (\nabla\Psi \times \mathbf{H}_0) = -\epsilon \mathbf{E}_0 \tag{2.19}$$

Step 4

Repeating the process of Step 2, but substituting (2.11) into (2.8), we obtain

$$\nabla\Psi \cdot \mathbf{H}_0 = 0 \tag{2.20}$$

Step 5

Rearranging (2.15) with $\mathbf{H}_0$ the subject of the expression we derive

$$\mathbf{H}_0 = Y(\nabla\Psi \times \mathbf{E}_0) \tag{2.21}$$

Figure 2.1 shows the spatial relationship between vectors $\mathbf{E}_0$, $\mathbf{H}_0$, and $\nabla\Psi$ that satisfies (2.18), (2.19), and (2.21). When this is inserted in (2.19), the latter becomes

$$-\nabla\Psi \times (\nabla\Psi \times \mathbf{E}_0) = \mathbf{E}_0 \tag{2.22}$$

Through application of the vector identity:

$$\mathbf{A} \times (\mathbf{B} \times \mathbf{C}) = \mathbf{B}(\mathbf{A} \cdot \mathbf{C}) - \mathbf{C}(\mathbf{A} \cdot \mathbf{B})$$

(2.22) can be rewritten as

$$(\nabla\Psi \cdot \nabla\Psi)\mathbf{E}_0 - (\nabla\Psi \cdot \mathbf{E}_0)\nabla\Psi = \mathbf{E}_0 \tag{2.23}$$

From (2.18) we know that the second term on the left-hand side of (2.23) is identically zero, so that (2.23) finally reduces to what is called the *eikonal equation*:

$$|\nabla\Psi|^2 = 1 \tag{2.24}$$

or equivalently $|\nabla\Psi| = 1$. The quantity $\Psi(\mathbf{r})$ is called the *eikonal* or *phase function,* from the Greek word $\epsilon\iota\kappa\omega\nu$ (eikon) for "image." We may note immediately that there are at least three, almost obvious solutions of the first-order nonlinear differential equation (2.24). Working in rectangular coordinates, with α, β,

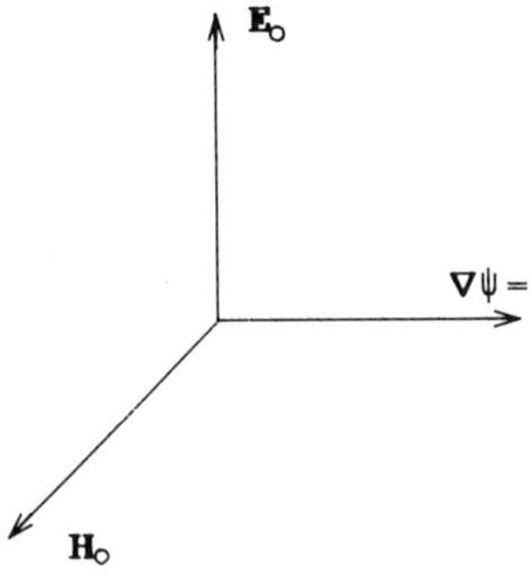

Figure 2.1 Spatial relationship between $\mathbf{E}_0$, $\mathbf{H}_0$, and $\nabla\psi$.

and γ the usual direction cosines, these solutions are $\Psi(x, y, z) = \alpha x + \beta y + \gamma z$, $\Psi(x, y, z) = \sqrt{x^2 + y^2 + z^2}$, and $\Psi(x, y, z) = \sqrt{x^2 + y^2}$, corresponding to planar, spherical, and cylindrical surfaces, respectively.

The eikonal equation later will be seen to provide a convenient description for the phase variation along the propagation path of high-frequency fields. At that stage we also will require quantitative information on the amplitude variation of such fields, for which the so-called transport equations are utilized. These are derived in the section that follows.

2.2.5 The Transport Equations

The derivation of the transport equations begins with the vector Helmholtz equation (2.9). The procedure followed is similar to that of Section 2.2.4, and a detailed derivation will not be presented. This is left as the proverbial exercise for the reader.

Substitution of the asymptotic expansion (2.10) into (2.9) yields, under the previous assumption of differentiability,

$$e^{-jk\Psi} \sum_{n=0}^{\infty} \left[\frac{1/\nu\,(1 - |\nabla\Psi|^2 \mathbf{E}_n)}{(j\omega)^{n-2}} + \frac{(1/\nu)\,[\mathbf{E}_n \nabla^2\Psi + 2(\nabla\Psi \cdot \nabla)\mathbf{E}_n]}{(j\omega)^{n-1}} + \frac{\nabla^2 \mathbf{E}_n}{(j\omega)^n} \right] = 0 \qquad (2.25)$$

We recall that the terms $\mathbf{E}_n$ and Ψ are not functions of ω, so to satisfy equation (2.25), the coefficients of the various powers of ω individually must be zero. Equating to zero the coefficients of $(j\omega)^2$ once more gives us the eikonal equation (2.24), whereas the same process applied to the coefficients of $(j\omega)^1$ obtrudes the *zero*th-order *transport equation* (termed such because it determines the flow of energy),

$$2(\nabla\Psi \cdot \nabla)\mathbf{E}_0 + (\nabla^2\Psi)\mathbf{E}_0 = 0 \qquad (2.26)$$

Also, the single coefficient of the $(j\omega)^0$ power yields the condition,

$$\nabla^2 \mathbf{E}_0 = 0 \qquad (2.27)$$

The *higher-order transport equations* ($n \geqslant 1$) also can be obtained [7, 8] from this as

$$2(\nabla\Psi \cdot \nabla)\mathbf{E}_n + (\nabla^2\Psi)\mathbf{E}_n = -\nu\,\nabla^2 \mathbf{E}_{n-1} \qquad (2.28)$$

but they will not be used in this text.

At this stage the reader might be wondering about the significance of these rather cumbersome equations. The remainder of the chapter is devoted to answering this question. We now have at our disposal all the results necessary to undertake an examination of the properties of high-frequency or geometrical optics fields. Before continuing, some remarks are in order on other forms that the reader may encounter. These forms will not be used in this text, but are included for the sake of completeness.

(a) It is often seen in journal articles dealing with regions of constant permittivity, that the Luneberg-Kline expansion is written in terms of the wavenumber k as

$$\mathbf{E}(\mathbf{r}, k) \sim e^{-jk\Psi(\mathbf{r})} \sum_{n=0}^{\infty} \frac{\mathbf{E}_n(\mathbf{r})}{(jk)^n}$$

That is, the series is expanded in terms of $k = \omega\sqrt{\mu_0\epsilon}$ instead of ω.

(b) An alternative form of the Luneberg-Kline expansion that is used in many research papers [7] is

$$\mathbf{E}(\mathbf{r}, \omega) \sim e^{-jk_0 S(\mathbf{r})} \sum_{n=0}^{\infty} \frac{\mathbf{E}_n(\mathbf{r})}{(j\omega)^n}$$

The only change is in the term containing the phase function $S(\mathbf{r})$, where $k_0 = \omega\sqrt{\mu_0\epsilon_0}$ has been used instead of the $k = \omega\sqrt{\mu_0\epsilon}$ of the medium through which the propagation takes place. The phase functions $\Psi(\mathbf{r})$ and $S(\mathbf{r})$ are simply related at any point $\mathbf{r}$ as $S(\mathbf{r}) = \sqrt{\epsilon_r}\,\Psi(\mathbf{r})$, and the eikonal equation becomes $|\nabla S(\mathbf{r})|^2 = \epsilon_r$. This form may be the more useful one when considering propagation in media with continuously varying $\epsilon(\mathbf{r})$. In such cases, it is customary to use the eikonal equation in the form $|\nabla S(\mathbf{r})|^2 = n^2(\mathbf{r})$, where the refractive index of the medium is $n(\mathbf{r}) = \sqrt{\epsilon_r(\mathbf{r})}$. For the homogeneous or piecewise-homogeneous materials of concern in this text, which form is actually used is immaterial. The forms (2.10) and (2.11) that we have selected are more consistent with those used in the seminal literature [10–14] on the geometrical theory of diffraction.

(c) The derivation of the eikonal and *zero*th-order transport equations also can be undertaken starting with the Sommerfeld-Runge ansatz [5]. This, in essence, is the first term in the Luneberg-Kline expansion, and is sometimes also called the *Wentzel-Kramers-Brillouin* (WKB) approximation:

$$U(\mathbf{r},\omega) \approx A(\mathbf{r})\, e^{-jk_0 S(\mathbf{r})}$$

where $U(\mathbf{r}, \omega)$ is a scalar component of the vector field, considered for simplicity. The resulting equation is [15, 16]

$$k_0^2[\epsilon_r - |\nabla S|^2] + jk_0\left[\nabla^2 S + 2\frac{\nabla S \cdot \nabla A}{A}\right] + \frac{\nabla^2 A}{A} = 0$$

To obtain the eikonal equation in this case, we must assume that the term $\nabla^2 A/A$ is negligible. Then the expression reduces to two equations; the eikonal equation $|\nabla S|^2 = \epsilon_r$, and the *zero*th-order transport equation $2\nabla S \cdot \nabla A + (\nabla^2 S)A = 0$.

2.2.6 The Geometrical Optics Terms and Their Interpretation

In the limit, as the radian frequency ω in the asymptotic expansions (2.10) and (2.11) tends to infinity, the only terms remaining are the lowest order ones $\mathbf{E}_0$ and $\mathbf{H}_0$. These are the GO field terms:

$$\mathbf{E}(\mathbf{r}) \underset{\omega\to\infty}{\sim} \mathbf{E}_0(\mathbf{r})\, e^{-jk\Psi(\mathbf{r})} \tag{2.29}$$

$$\mathbf{H}(\mathbf{r}) \underset{\omega\to\infty}{\sim} \mathbf{H}_0(\mathbf{r})\, e^{-jk\Psi(\mathbf{r})} \tag{2.30}$$

Fortunately, we already know something about the behavior of these *zero*th-order terms from expressions derived in Sections 2.2.4 and 2.2.5. To interpret these results in a physically meaningful way is important at this stage. For convenience, the relevant expressions from the earlier sections are repeated:

$$Y(\nabla\Psi \times \mathbf{E}_0) = \mathbf{H}_0 \tag{2.31}$$

$$\nabla\Psi \cdot \mathbf{E}_0 = 0 \tag{2.32}$$

$$Z(\nabla\Psi \times \mathbf{H}_0) = -\mathbf{E}_0 \tag{2.33}$$

$$\nabla\Psi \cdot \mathbf{H}_0 = 0 \tag{2.34}$$

$$|\nabla\Psi|^2 = 1 \tag{2.35}$$

$$2(\nabla\Psi \cdot \nabla)\mathbf{E}_0 + (\nabla^2\Psi)\mathbf{E}_0 = 0 \tag{2.36}$$

With the fields in the form of (2.29) and (2.30) the complex Poynting vector $\mathbf{E} \times \mathbf{H}^*$ is easily found to be $\mathbf{E}_0 \times \mathbf{H}_0^*$, which, from (2.31), is simply

$$\mathbf{E}_0 \times \mathbf{H}_0^* = Y[\mathbf{E}_0 \times (\nabla\Psi \times \mathbf{E}_0)^*]$$

Because, in lossless media, Ψ cannot be complex, $\nabla\Psi$ is its own complex conjugate. Using the vector identity:

$$\mathbf{A} \times (\mathbf{B} \times \mathbf{C}) = \mathbf{B}(\mathbf{A} \cdot \mathbf{C}) - \mathbf{C}(\mathbf{A} \cdot \mathbf{B})$$

the complex Poynting vector becomes

$$\mathbf{E} \times \mathbf{H}^* = Y[(\nabla\Psi)(\mathbf{E}_0 \cdot \mathbf{E}_0^*) - \mathbf{E}_0^*(\mathbf{E}_0 \cdot \nabla\Psi)]$$

However, from (2.32) the second term on the left-hand side of this expression is identically zero so that the complex Poynting vector becomes

$$\mathbf{E} \times \mathbf{H}^* = Y(\mathbf{E}_0 \cdot \mathbf{E}_0^*)(\nabla\Psi)$$

Because this expression is pure-real, the time-averaged Poynting vector is

$$\text{Re}[\mathbf{E} \times \mathbf{H}^*] = Y(\mathbf{E}_0 \cdot \mathbf{E}_0^*)(\nabla\Psi) = Y|\mathbf{E}_0|^2(\nabla\Psi)$$

Power flow thus is in the direction

$$\hat{\mathbf{s}} = (\nabla\Psi) \tag{2.37}$$

and this defines the GO ray direction. Remember that from (2.24) we know that $\hat{\mathbf{s}}$ is a unit vector. In other words, GO rays are defined as curves tangent to the $\hat{\mathbf{s}}$ direction (the ray vector), and energy transport occurs along these ray trajectories. There is no energy transport transverse to a ray. If we consider any bundle of rays adjacent to some axial ray, these form what is known as a *ray tube* or *ray pencil,* as shown in Figure 2.2. Because the sides of such ray tubes are formed by the trajectories of the adjacent rays, the power flow through any cross section of a given ray tube is the same insofar as no power flows across the sides of the tube.

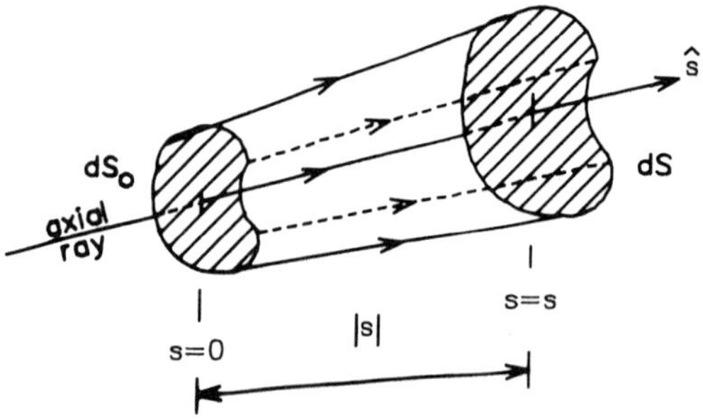

Figure 2.2 Narrow ray tube (ray pencil or flux tube).

The power conservation property in fact is inherent in the *zero*th-order transport equation (2.36). This will be shown at the end of the next section.

In practice, launching a single ray is not possible. Rather, to be able to perform quantitative analyses, we work with a selected axial ray plus an infinitesimally narrow ray tube surrounding it. (The rays in the immediate neighborhood of the axial ray are called *paraxial rays.*) This allows us to compute the field phase, amplitude, and polarization along any ray as it propagates, which will be discussed further in Section 2.2.7.

Because the vectors $\nabla\Psi = \hat{\mathbf{s}}$ are perpendicular to surfaces of constant Ψ (the equiphase surfaces), that the rays defined earlier are normal to these equiphase surfaces thus follows. As has been shown, $\mathbf{E}_0$, $\mathbf{H}_0$, and $\hat{\mathbf{s}}$ are mutually perpendicular at any point on a ray, and there are no field components in the propagation direction. Thus, we see that GO fields in homogeneous, lossless media satisfy the usual uniform plane wave relationships between the quantities $\mathbf{E}_0$, $\mathbf{H}_0$, and $\hat{\mathbf{s}}$ at any point along a ray; hence, we say that GO fields are *locally plane.* This should not be misunderstood to mean that all GO fields in fact are uniform plane waves. "Thouroughbred" uniform plane wave fields not only must have this relationship among $\mathbf{E}_0$, $\mathbf{H}_0$, and $\nabla\Psi$ (the locally plane property) but also equiphase surfaces that are planar and no amplitude variation along a ray or in the equiphase planes. So, although all GO fields are locally plane, they are certainly not all plane waves.

To provide a proper quantitative description of the GO fields in a homogeneous lossless medium we need at least four items of information: details of the ray trajectory, polarization information, and the form of the amplitude and phase of the field at certain reference planes, as well as their variation as the field propagates along the ray path.

2.2.7 Ray Paths, Amplitude Functions, and Phase Functions

Ray Trajectories

Thus far we have used the coordinate vector $\mathbf{r}$ to define position in space. It is expedient in what follows rather to use the ray coordinate s measured along a ray; the relationship between them is illustrated in Figure 2.3. Then the operator

$$(\nabla\Psi \cdot \nabla) = \hat{\mathbf{s}} \cdot \nabla = d/ds$$

is the directional derivative along the ray, where ds is the incremental distance along the ray curve orthogonal to the $\Psi(\mathbf{r})$ surfaces. In this notation the eikonal equation (2.35) can be written as

$$\frac{d\Psi}{ds} = 1 \tag{2.38}$$

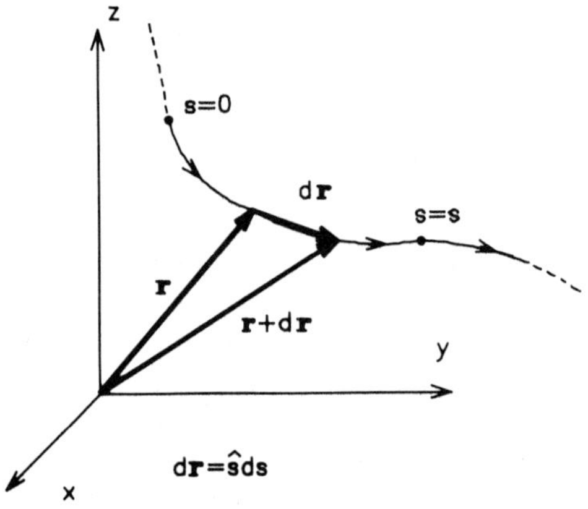

Figure 2.3 Ray coordinate system.

and the *zero*th-order transport equation (2.36) in the form:

$$\frac{d\mathbf{E}_0}{ds} + \frac{1}{2}(\boldsymbol{\nabla}^2\Psi)\mathbf{E}_0 = 0 \tag{2.39}$$

From basic differential geometry (see Appendix C) we can write

$$\hat{\mathbf{s}} = \frac{d\mathbf{r}}{ds} \tag{2.40}$$

Now,

$$\begin{aligned}\frac{d(\boldsymbol{\nabla}\Psi)}{ds} &= \frac{d\hat{s}}{ds}\\ &= \frac{d}{ds}(d\mathbf{r}/ds)\\ &= \frac{d^2\mathbf{r}}{ds^2}\end{aligned}$$

However, we can also write

$$\begin{aligned}\frac{d(\boldsymbol{\nabla}\Psi)}{ds} &= (\boldsymbol{\nabla}\Psi)\cdot\boldsymbol{\nabla}\Psi\\ &= \frac{1}{2}\boldsymbol{\nabla}[|\boldsymbol{\nabla}\Psi|^2]\\ &= 0\end{aligned}$$

and, as a consequence, the trajectory equation is

$$\frac{d^2\mathbf{r}}{ds^2} = 0 \tag{2.41}$$

which has as its solution:

$$\mathbf{r}(s) = \mathbf{A}s + \mathbf{B}$$

where **A** and **B** are constant vectors. Thus, GO rays in a homogeneous medium are straight lines. We caution the reader not to assume that the same can be said for rays in inhomogeneous media, where the ray trajectories usually are curved.

As a matter of interest, in differential geometry, a system of curves that fills a portion of space in such a way that a single curve generally passes through each point in the region is called a *congruence* [17, pp. 126–127]. We call it a *normal congruence* if a family of surfaces exists that cut each of the curves orthogonally. If each curve of the congruence is a straight line, the congruence also is said to be *rectilinear.* So what we have found in this derivation is that GO rays in a homogeneous medium represent a *normal rectilinear congruence.*

Polarization Properties

The polarization properties of GO rays next must be established. Polarization is defined as the orientation of the electric field, and a polarization unit vector $\hat{\mathbf{e}} = \mathbf{E}_0/|\mathbf{E}_0|$ parallel to $\mathbf{E}_0$ thus may be designated. Immediately following from the local plane wave property:

$$\hat{\mathbf{e}} \cdot \hat{\mathbf{s}} = 0$$

and the unit vector $\hat{\mathbf{h}} = \mathbf{H}_0/|\mathbf{H}_0|$, oriented in the direction of the magnetic field, is given by

$$\hat{\mathbf{h}} = \hat{\mathbf{s}} \times \hat{\mathbf{e}}$$

The changes in $\hat{\mathbf{e}}$ along the ray trajectory is described by the derivative $d\hat{\mathbf{e}}/ds$. Using the definition for $\hat{\mathbf{e}}$ just given,

$$\begin{aligned}\frac{d\hat{\mathbf{e}}}{ds} &= \frac{d}{ds}[\mathbf{E}_0/|\mathbf{E}_0|] \\ &= |\mathbf{E}_0|^{-1}\frac{d\mathbf{E}_0}{ds} - \hat{\mathbf{e}}|\mathbf{E}_0|^{-1}\frac{d|\mathbf{E}_0|}{ds}\end{aligned} \tag{2.42}$$

If we now take the dot product of both sides of equation (2.36) with $\mathbf{E}_0^*$, and add the resulting expression to its own complex conjugate, we are left with

$$\mathbf{E}_0 \cdot \mathbf{E}_0^*(\nabla^2\Psi) + \frac{d(\mathbf{E}_0 \cdot \mathbf{E}_0^*)}{ds} = 0 \tag{2.43}$$

Using (2.42) and (2.43) the *zero*th-order transport equation (2.36) can be reduced to the simple relation:

$$\frac{d\hat{\mathbf{e}}}{ds} = 0 \tag{2.44}$$

Thus, the polarization vector $\hat{\mathbf{e}}$ is constant in homogeneous regions. Once $\hat{\mathbf{e}}$ is known at some reference point s_0 on the ray, $\hat{\mathbf{e}}$ is known at any other because it does not change during propagation.

We will see in later chapters that this polarization direction may be altered during reflection or diffraction, remaining constant once more as soon as a ray has detached itself from the reflecting or diffracting obstacle. In the terminology of differential geometry, we say that the constancy of the polarization in homogeneous media is due to the zero torsion of the ray curves. In the chapter on curved surface diffraction, we will mention that rays traveling along the surface of a smooth scatterer may have nonzero torsion, and special attention is paid to polarization changes along such trajectories.

Finally, we comment that in the application of the GTD, circularly or elliptically polarized fields are accommodated by resolving them into their two linearly polarized components, treating these separately, and then recombining the resultant components. This naturally assumes that all the media present are linear so that superposition is valid.

Phase Continuation along Trajectories

For homogeneous isotropic media the rays were seen to be straight lines and normal to the wavefronts (equiphase surfaces). Thus [7],

$$d\Psi = |\nabla\Psi| ds = ds$$

where ds is an incremental distance along the ray trajectory, as already defined. Integration of this equation along this ray path from some phase reference point $s = s_0$ to some general point s a distance $(s - s_0)$ from this reference point on the ray yields

$$\Psi(s) - \Psi(s_0) = s - s_0 \tag{2.45}$$

To select the ray coordinates for a given problem such that the reference point is always at $s_0 = 0$ is convenient, and this always will be done in what follows. This convention having been adopted, (2.45) becomes

$$\Psi(s) = \Psi(0) + s \tag{2.46}$$

Hence, the description of the phase change along the ray path is given by

$$e^{-jk\Psi(s)} = e^{-jk\Psi(0)}\, e^{-jks} \tag{2.47}$$

Amplitude Continuation along Trajectories

The final item of information required before the GO field behavior is completely specified is the amplitude function. To obtain this function, a solution must be found for the *zero*th-order transport equation (2.36). Let the known field amplitude (and polarization) at the reference point $s = 0$ be denoted by $\mathbf{E}_0(0)$. Equation (2.36) is a simple differential equation with solution given by [7, 9]

$$\mathbf{E}_0(s) = \mathbf{E}_0(0)\exp\left[-\frac{1}{2}\int_{s_0}^{s} \nabla^2\Psi\, ds'\right] \tag{2.48}$$

In its present form, (2.48) is not yet of much use to us, as the precise form of $\nabla^2\Psi$ still must be evaluated. In the next few paragraphs, an expression for the exponent in (2.48) will be derived in terms of the principal radii of curvature of the (wavefront) surface.

Consider [2, pp. 169–171] an infinitesimally narrow ray tube positioned about some central ray with ray vector $\hat{\mathbf{s}}$ and along which the distance s is measured, with two closely spaced equiphase surfaces $\Psi(0)$ and $\Psi(s)$ a distance s apart, as shown in Figure 2.4. This is called an *astigmatic ray tube,* from the classical Greek words *a* for "not" and *stigma* for "a point." The distance ρ between the focal lines is called the *astigmatic difference.* The reference constant phase surface again is $\Psi(0)$, and it has principal radii of curvature ρ_1 and ρ_2 measured on the central ray. The $\Psi(s)$ surface has principal radii of curvature $(\rho_1 + s)$ and $(\rho_2 + s)$. Following Born and Wolf [17], Kouyoumjian [7], or Kline and Kay [4], we introduce the Gaussian curvatures:

$$G(0) = \frac{1}{\rho_1\rho_2}$$

and

$$G(s) = \frac{1}{(\rho_1 + s)(\rho_2 + s)}$$

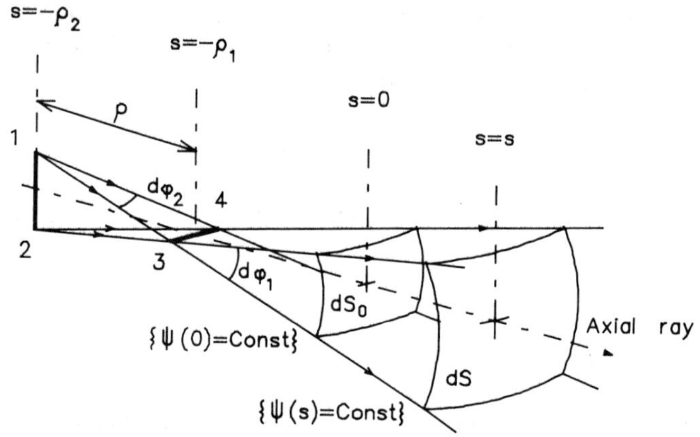

Figure 2.4 Infinitesimally narrow diverging astigmatic ray tube, for which both ρ_1 and ρ_2 are positive (ρ is the astigmatic difference).

for surfaces $\Psi(0)$ and $\Psi(s)$, respectively. Next we define a vector function of the ray coordinate:

$$\mathbf{F}(s) = G(s)\hat{\mathbf{s}}$$

and consider the integral of **F** over the closed surface consisting of the elemental surfaces dS_0 and dS on the ray tube and the sides of the tube between them. (Note that dS_0 and dS are elements of different constant Ψ surfaces.) This integral is

$$\begin{aligned} \iint \mathbf{F} \cdot \hat{\mathbf{n}} dS &= \mathbf{F}(s) \cdot \hat{\mathbf{s}}\, dS + \mathbf{F}(0) \cdot (-\hat{\mathbf{s}})\, dS_0 \\ &= G(s) dS - G(0) dS_0 \end{aligned} \tag{2.49}$$

However, from simple geometry, the incremental area at s is easily seen to be

$$\begin{aligned} dS &= (\rho_1 + s)(\rho_2 + s) d\phi_1 d\phi_2 \\ &= \frac{d\phi_1 d\phi_2}{G(s)} \end{aligned}$$

and the area at $s = 0$ is

$$\begin{aligned} dS_0 &= \rho_1 \rho_2 d\phi_1 d\phi_2 \\ &= \frac{d\phi_1 d\phi_2}{G(0)} \end{aligned}$$

Therefore, from (2.49),

$$\iint \mathbf{F} \cdot \hat{\mathbf{n}} dS = 0 \tag{2.50}$$

Invoking the divergence theorem of vector analysis, this result also implies that

$$\iiint \nabla \cdot \mathbf{F} \, dV = 0 \tag{2.51}$$

Because this condition holds for arbitrary $\mathbf{F}(s)$ we can conclude that $\nabla \cdot \mathbf{F} = 0$. From the original definition for $\mathbf{F}$ given earlier, this means that

$$\begin{aligned} \nabla \cdot [G(s)\hat{\mathbf{n}}] &= \hat{\mathbf{n}} \cdot \nabla G(s) + G(s)\nabla \cdot \hat{\mathbf{n}} \\ &= 0 \end{aligned}$$

Therefore,

$$\hat{\mathbf{n}} \cdot \nabla G(s) = -G(s)\nabla \cdot \hat{\mathbf{n}} \tag{2.52}$$

Now,

$$\hat{\mathbf{n}} \cdot \nabla G(s) = \frac{dG(s)}{ds}$$

and

$$\nabla \cdot \hat{\mathbf{n}} = \nabla^2 \Psi$$

so that (2.52) can be rearranged to read:

$$\frac{dG}{ds} = -G(s) \, \nabla^2 \Psi \tag{2.53}$$

Integrating this equation along the central ray from $\Psi(0)$ to $\Psi(s)$, we obtain [7]

$$G(s)/G(0) = \exp\left[-\int_0^s \nabla^2 \Psi \, ds'\right] \tag{2.54}$$

The exponent in (2.54) is immediately recognizable as the same as that in (2.48), and after substitution we find that

$$\mathbf{E}_0(s) = \mathbf{E}_0(0)\sqrt{G(s)/G(0)} \tag{2.55}$$

Incorporating the explicit forms of $G(s)$ and $G(0)$ given earlier, (2.55) can be rewritten as

$$\mathbf{E}_0(s) = \mathbf{E}_0(0)\sqrt{\rho_1\rho_2/(\rho_1 + s)(\rho_2 + s)} \tag{2.56}$$

We now have all the items of information necessary to construct a general expression for the form of the GO field along a ray trajectory. Incorporating (2.47) and (2.56), (2.29) takes the final form:

$$\mathbf{E}(s) = \mathbf{E}(0)\ \sqrt{\rho_1\rho_2/(\rho_1 + s)(\rho_2 + s)}\ \mathrm{e}^{-jks} \tag{2.57}$$

where

$$\mathbf{E}(0) = \mathbf{E}_0(0)\ \mathrm{e}^{-jk\Psi(0)}$$

The latter expression contains amplitude, phase, and polarization information at the reference position. We have written $\mathbf{E}(s)$ for $\mathbf{E}_0(s)$, the GO field at any observation point s; henceforth in this text, unless specifically stated to the contrary, the subscript "0" signifying the GO field will be suppressed. Thus, when we write $\mathbf{E}(s)$ we mean the GO field. By specifying the complex amplitude and polarization $\mathbf{E}(0)$ of the GO field at some reference point $s = 0$, and the principal radii of curvature ρ_1 and ρ_2 of the wavefront there, (2.57) allows us to compute the characteristics of the GO field at any other point a distance s away from the reference point. This process is known as *ray optical continuation* [15].

If many (yet a finite number of) rays pass through a given observation point s, linearity allows us to obtain the resultant GO field at position s as the sum of the individual ray fields. For instance, we will see in later chapters that the total ray field at some point may be the sum of a ray proceeding directly from some source, reflected rays, and diffracted rays. It is essential however that there not be an infinite number of rays passing through the point at which we wish to determine the field, as the amplitude predicted then of necessity also will be infinite. What is happening at such a point of course is not that the physical fields actually become infinite (usually, however, they become large), but that the GO description of the phenomenon breaks down, and alternative methods are required. Further comment on this topic will be made in Section 2.2.8.

Finally, recall that in Section 2.2.6 we were able to infer that power is conserved along any ray tube simply as a consequence of the definition of the ray direction and that of a ray tube. Here we wish to demonstrate that this property follows directly from the *zero*th-order transport equation (2.36). Earlier in this section we found that sheer mathematical manipulation of (2.36) provided us with equation (2.43). Recalling that

$$\mathbf{E}_0 \cdot \mathbf{E}_0^* = |\mathbf{E}_0|^2$$

and

$$\frac{d}{ds}(\mathbf{E}_0 \cdot \mathbf{E}_0^*) = \nabla|\mathbf{E}_0|^2 \cdot \nabla\Psi$$

we may rewrite (2.43) as

$$|\mathbf{E}_0|^2\, \nabla^2\Psi + \nabla|\mathbf{E}_0|^2 \cdot \nabla\Psi = 0$$

which from the product rule of differentiation simply is

$$\nabla \cdot (|\mathbf{E}_0|^2\, \nabla\Psi) = 0$$

However, because

$$|\mathbf{E}_0|^2\, \nabla\Psi = |\mathbf{E}_0|^2\, \hat{\mathbf{s}}$$

is the power density vector, this equation simply states that the divergence of the power density at any point is zero. As this statement applies for any GO field, it implies that the integral of $\nabla \cdot (|\mathbf{E}_0|^2\hat{\mathbf{s}})$ throughout any closed volume also is zero. Applying the divergence theorem, the integral of $|\mathbf{E}_0|^2\hat{\mathbf{s}}$ over any closed surface hence also is zero. If we now select this closed surface to be the section of ray tube shown in Figure 2.4, because $|\mathbf{E}_0|^2\hat{\mathbf{s}}$ everywhere is perpendicular to the unit vector normal to the side surfaces of the ray tube, only that portion of the surface integral over the ends will remain. This gives us

$$|\mathbf{E}_0(s)|^2\, dS = |\mathbf{E}_0(0)|^2\, dS_0 \tag{2.58}$$

which is a statement of power conservation along the ray tube, and in classical geometrical optics is called the *intensity law*, $|\mathbf{E}_0(s)|^2/Z$ being the intensity or power density at any point s along the ray. This relation more commonly is written in the form:

$$|\mathbf{E}_0(s)| = |\mathbf{E}_0(0)|\sqrt{dS_0/dS} \tag{2.59}$$

Because, from differential geometry,

$$\sqrt{dS_0/dS} = \sqrt{\frac{\rho_1\rho_2}{(\rho_1 + s)(\rho_2 + s)}}$$

(2.56) is seen to follow as a consequence of the intensity law (power conservation along the ray tube) and may have been derived along these lines; here, we consider it to be a physical interpretation of (2.56).

2.2.8 Sign Conventions and Caustics of the Geometrical Optics Fields

Sign Conventions

A very important convention to note is that, *at the selected reference front* $s = 0$, a *positive* (*negative*) radius of curvature implies *diverging* (*converging*) paraxial rays in the corresponding principal plane. Because the use of this convention is seldom explained in any great detail in the literature, here we will illustrate its correct application for three different cases. The first is shown in Figure 2.5(a). For this case, the paraxial rays are diverging in both principal planes at the reference front $s = 0$, and thus both $\rho_1 > 0$ and $\rho_2 > 0$. However, in Figure 2.5(b), at the selected reference front $s = 0$ the paraxial rays are converging in principal plane 1 and diverging in principal plane 2, so that $\rho_1 < 0$ and $\rho_2 > 0$. Third, for the situation in Figure 2.5(c), at $s = 0$ the wavefront is convergent in both principal planes, so that $\rho_1 < 0$ and $\rho_2 < 0$.

Caustics

Although (2.57) is adequate for many situations, it possesses an easily identified deficiency. Observe that, when $s = -\rho_1$ or $-\rho_2$ in (2.57), the field becomes infinite, and geometrical optics therefore is invalid at these positions, yielding an insufficient description of the actual field. The cause of the deficiency can be seen by considering

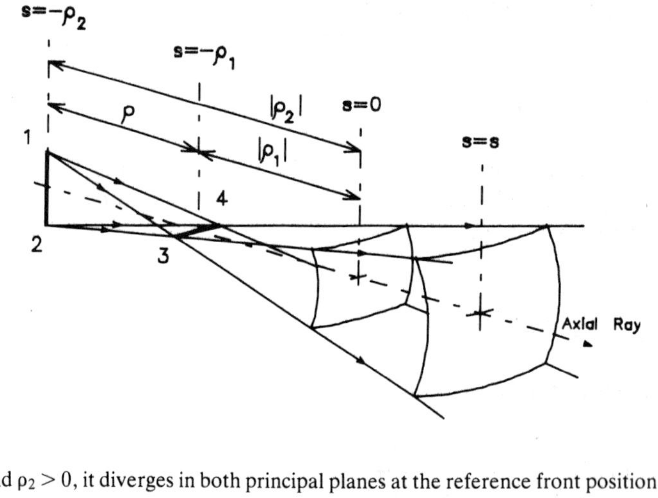

(a) With $\rho_1 > 0$ and $\rho_2 > 0$, it diverges in both principal planes at the reference front position.

Figure 2.5 Astigmatic ray tube.

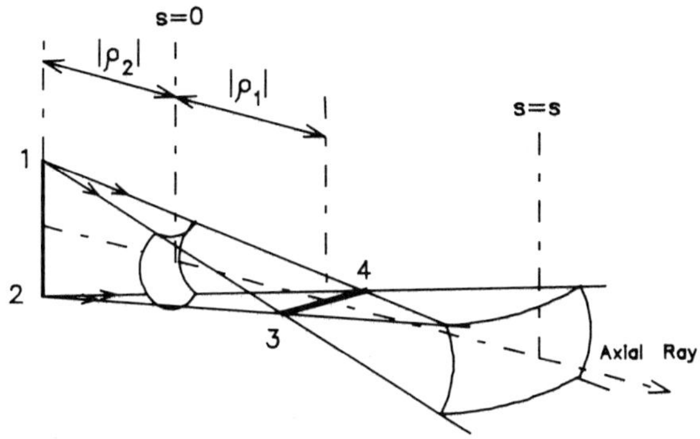

(b) With $\rho_1 < 0$ and $\rho_2 > 0$, it diverges in one of the principal planes at the reference front position, and converges in the other.

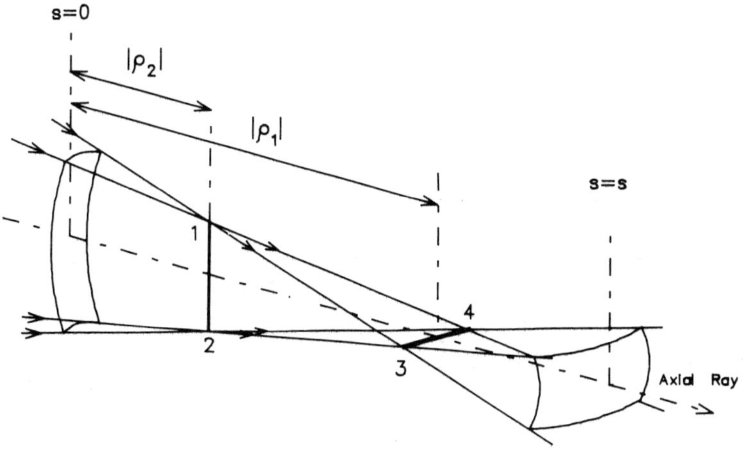

(c) With $\rho_1 < 0$ and $\rho_2 < 0$, it converges in both principal planes at the reference front position.

Figure 2.5 cont'd.

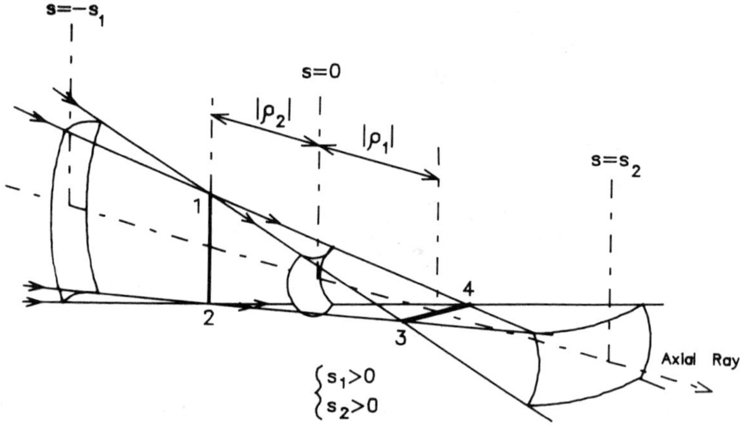

(d) With the reference position as in (b), it shows the effects of the caustics at various observation points along the axial ray; s_1 and s_2 are positive.

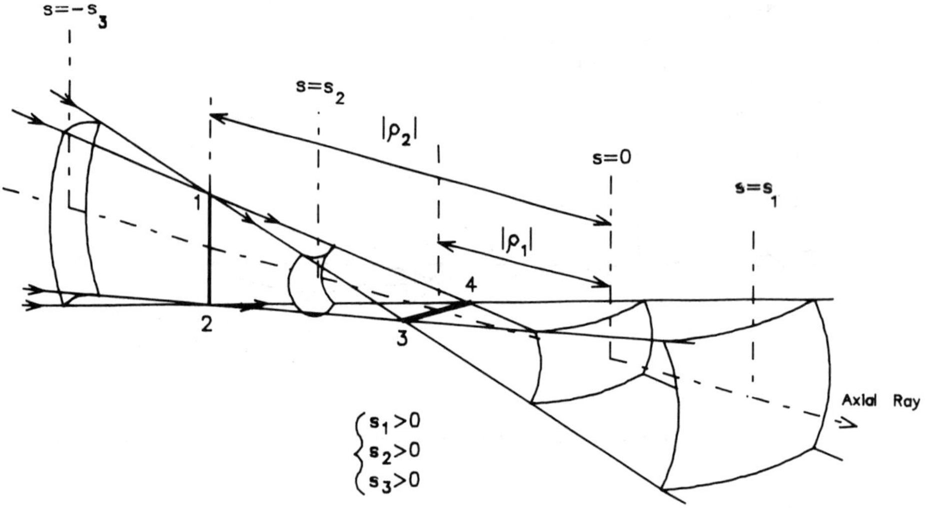

(e) With the reference position as in (a), it shows the effects of the caustics at various observation points along the axial ray; s_1, s_2, and s_3 are positive.

Figure 2.5 cont'd.

the intensity law in (2.59). If, as at lines 1–2 or 3–4 in Figure 2.5, the cross section dS of the ray tube vanishes, this implies that $|\mathbf{E}_0(s)|$ becomes singular. The congruence of rays at the lines 1–2 or 3–4 in Figure 2.5 are particular situations that occur when the rays forming a ray tube converge on a line, point, or surface. Any point at which the GO prediction of the field diverges is called a *caustic point*, and a collection of such points a *caustic point set*. These caustic point sets may take the form of a line (*caustic line*) or surface (*caustic surface*). In applications, the scattering of GO fields from obstacles gives rise to caustics, and the topic will arise many times in later sections of this text. It will be wise therefore to lay down a number of rules at this early stage.

Although in many instances we can live with the fact that we may not use a GO description to find the fields at and in the neighborhood of caustics, it is crucial that we nevertheless examine the effect on a ray of passing through a caustic. Passing through a caustic line, say 3–4 of Figure 2.5, from one side to the other, the sign of $(\rho_1 + s)$ changes abruptly. Because this factor is inside a square root sign in (2.57), such a sign change represents a jump in the phase of the ray field of either $+\pi/2$ or $-\pi/2$. This phenomenon has received the proper attention in the literature (e.g., [19]) and is properly taken into account by always selecting the correct square root in (2.57). Confusion is best avoided by stating and applying the following convention (with the phase reference, as usual, at the selected reference front $s = 0$):

(a) If we move along a ray through a caustic line, in the direction of propagation, we multiply the ray field expression by $e^{j\pi/2}$.

(b) If we move along a ray through a caustic line, in a direction opposite to that of propagation, we multiply the ray field expression by $e^{-j\pi/2}$.

In other words, accepting this convention, we may write (2.57) as

$$\mathbf{E}(s) = \mathbf{E}(0)\sqrt{|\rho_1\rho_2/(\rho_1 + s)(\rho_2 + s)|}\,e^{-jks}\,e^{j(n-m)\pi/2} \tag{2.60}$$

where n is the number of caustic lines crossed by the observer in moving from the references position $s = 0$ to the given observation point s in the direction of propagation, and m is the number crossed in the opposite direction. This form, together with the sign convention discussed at the start of this section, perhaps is the surest way of obtaining correct answers in all situations. The surd in (2.60) is termed the *spreading factor*:

$$A(s) = \sqrt{|\rho_1\rho_2/(\rho_1 + s)(\rho_2 + s)|} \tag{2.61}$$

Application of the principles are best illustrated through a number of examples.

Example 2.1

At the reference front shown in Figure 2.5(d), $\rho_1 < 0$ and $\rho_2 > 0$, and the complex

GO field is $\mathbf{E}(0)$. Moving from the reference position $s = 0$ to the observation position $s = s_2$, we pass through caustic line 3–4 in the direction of propagation, and thus add a factor $e^{j\pi/2}$ to the phase term. Then, the field at $s = s_2$ is given by

$$\mathbf{E}(s_2) = \mathbf{E}(0)A(s_2)\, e^{-jks_2}\, e^{j\pi/2}$$

However, moving from the reference position $s = 0$ to the observation position $s = -s_1$, we pass through caustic line 1–2 in the direction opposite to that of propagation, and thus a factor $e^{-j\pi/2}$ is added to the phase term. Then, the field at $s = -s_1$ is given by

$$\mathbf{E}(-s_1) = \mathbf{E}(0)A(-s_1)\, e^{jks_1}\, e^{-j\pi/2}$$

Example 2.2

Figure 2.5(e) shows the case where the principal radii of curvature at the reference position are such that $\rho_1 > 0$ and $\rho_2 > 0$, and the complex GO field once more is denoted $\mathbf{E}(0)$. Moving from the reference position $s = 0$ to the observation position $s = s_1$, the observer does not cross any caustic, the field at $s = s_1$ simply is given by

$$\mathbf{E}(s_1) = \mathbf{E}(0)\, A(s_1)\, e^{-jks_1}$$

By contrast, in going from the reference position $s = 0$ to the observation position $s = -s_2$, we pass through caustic line 3–4 in the direction opposite to that of propagation, and thus add a factor $e^{-j\pi/2}$ to the phase term. Therefore, the field at $s = -s_2$ is given by

$$\mathbf{E}(-s_2) = \mathbf{E}(0)\, A(-s_2)\, e^{jks_2}\, e^{-j\pi/2}$$

Finally, if we move all the way from $s = 0$ to the position $s = -s_3$, passing through both caustic lines 3–4 and 1–2 in a direction opposite to that of propagation, we add a factor $e^{-j\pi}$ to give

$$\mathbf{E}(-s_3) = \mathbf{E}(0)\, A(-s_3)\, e^{jks_3}\, e^{-j\pi}$$

Warning: In the preceding two examples, we must stress once more that the spreading factor (2.61) containing the absolute value signs has been assumed. Further examples of the application of this rule will be given in Section 2.5.

Another way of stating the convention regarding travel through a caustic simply is to say that the convention adopted for the square roots in (2.57) is that $\sqrt{\rho_{1,2}+s}$ takes only positive-real, positive-imaginary, or zero values. The sign convention on ρ_1 and ρ_2 at $s = 0$ still must be applied, of course.

Note that when we have a situation in which $\rho_1 = \rho_2$, the two caustic lines coalesce in a focal point through which all rays converge. Then the preceding convention still may be applied, but by considering the focal point to have the same effect on the phase as two caustic lines.

Further remarks on caustics will be reserved for the discussion on reflection in the next chapter because this reflection process more often than not is the cause of caustic formation.

2.2.9 The Geometrical Optics Field and Fermat's Principle

If a ray travels a distance s along its trajectory, the "optical path length" over which it has traveled is by definition the quantity:

$$F = \int_0^s \sqrt{\epsilon_r} ds' \tag{2.62}$$

where the integration is taken along the ray trajectory. Fermat's principle of classical geometrical optics states that the path followed by a ray is always such that the optical path length F is an extremum, but not necessarily a minimum. Following directly from this principle, in a homogeneous medium (i.e., constant ϵ_r throughout) the ray trajectories are straight lines, a result obtained earlier by using the eikonal equation. Indeed, through the calculus of variations [20], extremization of F and solution of the eikonal equation can be shown to be mathematically the same idea, different ways of stating the same law. In this particular case of a ray in a homogeneous medium, the straight line trajectory not only extremizes but in fact minimizes the optical path length functional (2.62).

The Fermat principle provides an alternative viewpoint that proves to be extremely useful in the study of phenomena such as reflection and diffraction considered later in this text. In those cases the ray trajectory is determined by requiring that (2.62) be extremized subject to certain constraints; namely, that the ray strike some surface (in the case of reflection) in its journey from one point in space to another, or strike some edge (in the case of edge diffraction) in its journey from one point to another. This then is referred to as the *generalized Fermat principle,* and obtrudes the laws of reflection and diffraction respectively discussed in Chapters 3 and 4. The principle is also invoked in the creeping wave diffraction discussion of Chapter 8.

2.3 SUMMARY OF THE PROPERTIES OF A HIGH-FREQUENCY FIELD AND SOME SPECIAL CASES

The expression describing the transmission of the geometrical optics (GO) field, for the general astigmatic ray tube shown in Figure 2.4 is given by,

$$\mathbf{E}(s) = \mathbf{E}(0) \sqrt{|\rho_1\rho_2/(\rho_1 + s)(\rho_2 + s)|}\ e^{-jks}\ e^{j(n-m)\pi/2} \tag{2.63}$$

We list the meaning of each term in the above expression:

(a) $\mathbf{E}(0)$ gives the field amplitude, phase, and polarization at the reference point $s = 0$.
(b) s is the distance along the ray path from the reference point $s = 0$, and hence
(c) e^{-jks} gives the phase shift along the ray path.

(d) The term $A(s) = \sqrt{|\rho_1\rho_2/(\rho_1 + s)(\rho_2 + s)|}$ (2.64)

is the divergence factor (or spreading factor), which governs the amplitude variation of the GO field along the ray path.
(e) ρ_1 and ρ_2 are the principal radii of curvature of the wavefront (which is a surface) at the reference point $s = 0$. The sign convention is that a *positive* (*negative*) radius of curvature implies *diverging* (*converging*) rays in the corresponding principal plane.
(f) n (m) is the number of caustic lines crossed by the observer in moving from the references position $s = 0$ to the given observation point s in a direction of (opposite to that of) propagation.

All of these considerations have been in terms of **E**, but the reader should remember that

$$\mathbf{H}(s) = \mathbf{H}(0) \sqrt{|\rho_1\rho_2/(\rho_1 + s)(\rho_2 + s)|}\ e^{-jks}\ e^{j(n-m)\pi/2} \tag{2.65}$$

is equally valid, with $\mathbf{H}(s)$ and $\mathbf{E}(s)$ related *locally* by the expression:

$$\mathbf{H}(s) = Y\,\hat{\mathbf{s}} \times \mathbf{E}(s) \tag{2.66}$$

with

$$Y = \sqrt{\epsilon_0\epsilon_r/\mu_0}$$
$$= \sqrt{\epsilon_r}\sqrt{\epsilon_0/\mu_0}$$

the characteristic admittance of the medium through which the ray is traveling. Expressions (2.63) and (2.65) are those used for calculation. Note that in this text we will be considering reflection and diffraction at conducting objects embedded in a homogeneous medium that always will be taken to be the free-space value ($\epsilon_r = 1$).

For both instruction and expedience, we consider three special cases of (2.63) or (2.65): plane wave ray tube; cylindrical wave ray tube; and spherical wave ray tube.

Plane Wave Ray Tube

Suppose we are interested in the case for which the surfaces of constant phase at the selected reference point are planar. This means that in (2.63) we must let the radii of curvature $\rho_1 \to \infty$ and $\rho_2 \to \infty$. As a consequence the spreading factor in (2.64) reduces to $A(s) = 1$, and there is no variation of the amplitude with s along the ray path and the wavefront remains planar. The expression for such a GO field simplifies to

$$\mathbf{E}(s) = \mathbf{E}(0)\, e^{-jks} \tag{2.67}$$

Cylindrical Wave Ray Tube

In a cylindrical wave ray tube one of the principal radii of curvature is infinite whereas the other is finite, at some reference point. Therefore, let $\rho_1 \to \infty$ and $\rho_2 = \rho$, with ρ being a finite value. The divergence (spreading) factor (2.64) then becomes

$$A(s) = \sqrt{\rho/(\rho + s)}$$

The general astigmatic ray tube expression (2.63) therefore simply is

$$\mathbf{E}(s) = \mathbf{E}(0)\sqrt{|\rho/(\rho + s)|}\, e^{-jks} \tag{2.68}$$

To refer the field to the caustic line at $s = -\rho$ (equivalent to letting $\rho \to 0$) often is convenient, although we cannot use (2.68) to evaluate the field there. To achieve this, the amplitude of the ray tube field is examined by taking the modulus of either side of (2.68):

$$|\mathbf{E}(s)| = |\mathbf{E}(0)|\, |\sqrt{\rho}|/|\sqrt{\rho + s}| \tag{2.69}$$

We then recognize that the field amplitude $|\mathbf{E}(s)|$ in (2.69), for fixed s, must be a constant irrespective of the reference position. Next, (2.69) is rewritten as

$$|\mathbf{E}(s)|\,|\sqrt{\rho + s}| = |\mathbf{E}(0)|\,|\sqrt{\rho}| \tag{2.70}$$

Clearly, as $\rho \to 0$ (i.e., the reference position is shifted to the caustic line) for fixed s the left-hand side of (2.70) becomes $|\mathbf{E}(s)|\,|\sqrt{s}|$, which also is a constant, e.g., A_o, and so will the right-hand side $|\mathbf{E}(0)|\,|\sqrt{\rho}|$. In other words, when $\rho \to 0$ expression (2.69) will become

$$|\mathbf{E}(s)| = \frac{A_o}{\sqrt{s}} \tag{2.71}$$

Incorporating the phase at the reference point, as well as the vector properties, into a constant vector quantity $\mathbf{A}_0$, for example, the complete expression for the cylindrical wave ray tube, with reference position at the caustic line and s the distance from this caustic line, simply is

$$\mathbf{E}(s) = \mathbf{A}_0 \frac{e^{-jks}}{\sqrt{s}} \tag{2.72}$$

The constant vector $\mathbf{A}_0$ essentially is an excitation factor for the cylindrical ray tube source. It often can be usefully employed in the simulation of real antennas to incorporate an amplitude pattern that is a function of angle about the cylindrical source.

Spherical Wave Ray Tube

If $\rho_1 = \rho_2 = \rho$, for example, we have the case of a spherical wave ray tube. The spreading factor $A(s)$ then reduces to the form $A(s) = \rho/(\rho + s)$, and the field expression is given by

$$\mathbf{E}(s) = \mathbf{E}(0)\frac{\rho}{(\rho + s)}\, e^{-jks} \tag{2.73}$$

If we wish to refer the field to the caustic at $s = -\rho$, then, following arguments similar to those for the cylindrical wave ray tube, we obtain the expression:

$$\mathbf{E}(s) = \mathbf{A}_o\, e^{-jks}/s \tag{2.74}$$

and the same comments regarding the term $\mathbf{A}_o$ can be made.

2.4 SPECIFIC EXAMPLES OF GEOMETRICAL OPTICS FIELDS

2.4.1 Initial Comments and Some Definitions

In situations of engineering importance, we must appreciate the relevance and applicability of fields of the GO type. To this end, a number of examples of fields used in "real-world" problems are presented in the following few sections. This material also will serve as a useful collection of expressions for source fields that are used in the example problems in the text.

2.4.2 Uniform Plane Wave Fields

From the definition of a uniform plane wave given in Section 2.2.2 we note that such a wave not only is locally plane (i.e., it has $\mathbf{E}_o$, $\mathbf{H}_o$, and $\hat{\mathbf{s}}$ everywhere spatially orthogonal to each other), but truly plane.

Uniform plane waves are the workhorse of the engineer interested in scattering problems. It therefore will be useful to examine various expressions for such plane wave fields. In a rectangular coordinate system a unit vector, pointing from the origin in the direction (θ_i, ϕ_i)—the latter being spherical coordinates—can be written as

$$\hat{\mathbf{r}} = \hat{\mathbf{x}} \sin\theta_i \cos\phi_i + \hat{\mathbf{y}} \sin\theta_i \sin\phi_i + \hat{\mathbf{z}} \cos\theta_i \tag{2.75}$$

Thus, a unit vector pointing toward the origin from this direction is given by

$$\hat{\mathbf{s}} = -(\hat{\mathbf{x}} \sin\theta_i \cos\phi_i + \hat{\mathbf{y}} \sin\theta_i \sin\phi_i + \hat{\mathbf{z}} \cos\theta_i) \tag{2.76}$$

The electric field of a uniform plane wave of unit amplitude traveling in direction $\hat{\mathbf{s}}$, with its phase referenced to the origin of the coordinate system, is given by

$$\mathbf{E}(\mathbf{r}) = \hat{\mathbf{e}}\, e^{-jk\hat{\mathbf{s}}\cdot\mathbf{r}} \tag{2.77}$$

where $\hat{\mathbf{e}}$ is the fixed polarization vector and $\mathbf{r} = x\hat{\mathbf{x}} + y\hat{\mathbf{y}} + z\hat{\mathbf{z}}$ is the position vector, as used earlier. More explicitly, (2.77) can be written in the form:

$$\mathbf{E}(\mathbf{r}) = \hat{\mathbf{e}}\, e^{jk(x\sin\theta_i\cos\phi_i + y\sin\theta_i\sin\phi_i + z\cos\theta_i)} \tag{2.78}$$

For a θ-polarized plane wave the polarization vector is given by

$$\hat{\mathbf{e}} = \hat{\mathbf{x}} \cos\theta_i \cos\phi_i + \hat{\mathbf{y}} \cos\theta_i \sin\phi_i - \hat{\mathbf{z}} \sin\theta_i \tag{2.79}$$

A ϕ-polarized wave has the polarization vector:

$$\hat{\mathbf{e}} = -\hat{\mathbf{x}} \sin\phi_i + \hat{\mathbf{y}} \cos\phi_i \tag{2.80}$$

Comparing (2.78) to the Luneberg-Kline expansion (2.10), $\mathbf{E}_0$ clearly is simply $\hat{\mathbf{e}}$ (independent of $\mathbf{r}$ in this special case), and

$$\Psi(\mathbf{r}) = -(x \sin\theta_i \cos\phi_i + y \sin\theta_i \sin\phi_i + z \cos\theta_i)$$

with higher-order terms:

$$\mathbf{E}_1 = \mathbf{E}_2 = \ldots = 0$$

Thus, as expected, the uniform plane wave is most certainly a GO field. The ray picture of such a uniform wave is shown in Figure 2.6, with all rays normal to the equiphase surfaces. It emphasizes the fact that, as far as a ray representation is concerned, such a uniform plane wave "consists of" infinitely many parallel rays propagating in direction $\hat{\mathbf{s}}$. Each of these is the central ray of a tube of the plane wave form (2.67).

A particular case that will be used constantly throughout the text is that for which $\theta_i = \pi/2$, or in other words for which $\hat{\mathbf{s}}$ lies in the $x=y$ plane as shown in

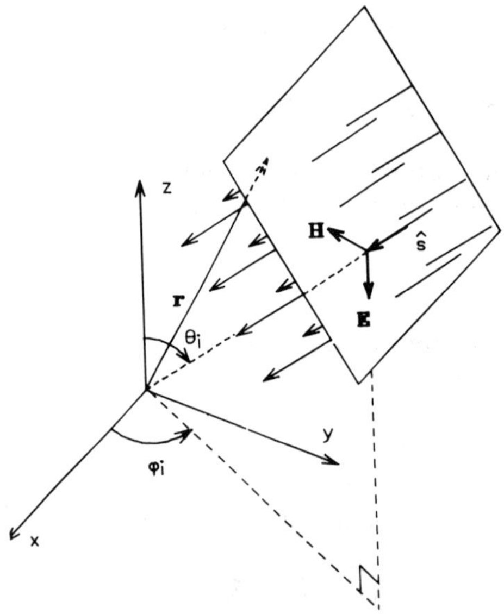

Figure 2.6 Uniform plane wave and its ray representation; we are at **r**.

Figure 2.7. Then expression (2.78) becomes

$$\mathbf{E}(\mathbf{r}) = \hat{\mathbf{e}}\, e^{jk(x\cos\phi_i + y\sin\phi_i)} \tag{2.81}$$

We have decided that in these cases we will measure directions of incidence not via the usual spherical coordinate angle, ϕ_i, but by ϕ' as illustrated in Figure 2.7. Because $\phi' = \phi_i + \pi/2$, $\sin\phi_i = -\cos\phi'$ and $\cos\phi_i = \sin\phi'$, and (2.81) becomes

$$\mathbf{E}(\mathbf{r}) = \hat{\mathbf{e}}\, e^{jk(x\sin\phi' + y\cos\phi')} \tag{2.82}$$

Suppose also that we specify the position of observation (x, y) by coordinates s and ϕ, as shown. Clearly,

$$\begin{aligned} x &= s\cos(\phi - \pi/2) \\ &= s\sin\phi \end{aligned}$$

and

$$\begin{aligned} y &= s\cos(\pi - \phi) \\ &= -s\cos\phi \end{aligned}$$

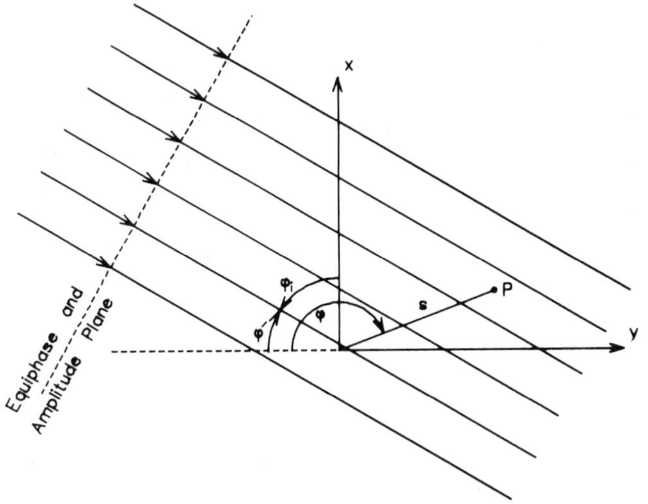

Figure 2.7 Uniform plane wave ray picture for the special case, $\theta_i = \pi/2$. Note that ϕ is not the usual spherical or cylindrical coordinate system angle. Only one ray passes through the observation point.

Thus,

$$x \sin\phi' - y \cos\phi' = s \sin\phi \sin\phi' + s \cos\phi' \cos\phi \\ = s \cos(\phi - \phi') \tag{2.83}$$

so that, finally, we have

$$\mathbf{E}(s, \phi) = \hat{\mathbf{e}}\, e^{jks\cos(\phi - \phi')} \tag{2.84}$$

When $\theta_i = \pi/2$, θ-polarization has $\hat{\mathbf{e}} = -\hat{\mathbf{z}}$, and thus corresponds to a wave with $-\hat{\mathbf{z}}$ polarization. A ϕ-polarized wave corresponds to a ϕ'-polarized one.

2.4.3 The Fields of Electric and Magnetic Line Sources

Consider first the electric line source of strength I^e, illustrated in Figure 2.8 (a). This geometry is two-dimensional, and the electric field at any distance ρ from the line source is given by [1, p. 224],

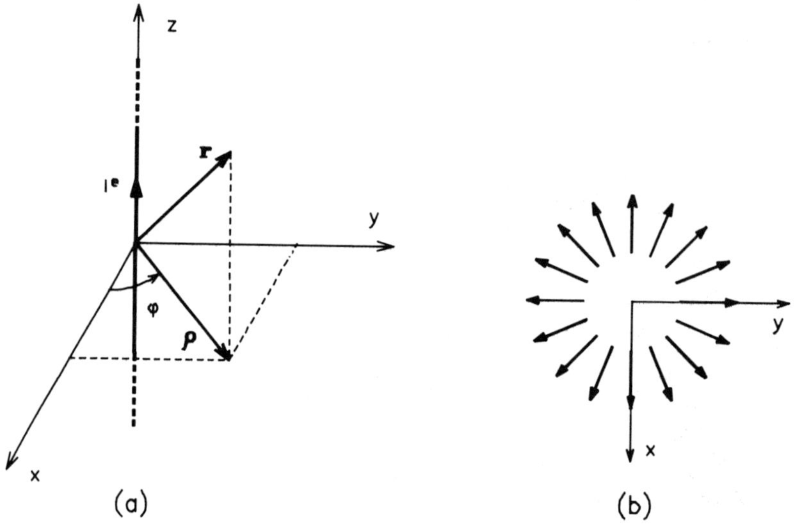

Figure 2.8 Electric or magnetic line source coordinate system (a). Ray representation of line source radiation (b). Inside the circle is a transition region within which the fields are not accurately represented as GO fields.

$$E_z(\rho) = \frac{-k^2 I^e}{4\omega\epsilon} H_o^{(2)}(k\rho) \tag{2.85}$$

At a sufficiently large electrical distance from the line source, the argument $k\rho$ of the Hankel function $H_o^{(2)}(k\rho)$ will be such that the large argument form:

$$H_o^{(2)}(k\rho) \sim \sqrt{2/\pi k}\, e^{j\pi/4} \sqrt{e^{-jk\rho}/\rho} \tag{2.86}$$

of the function can be used. Then, at sufficient distance from the line source, we have

$$E_z(\rho) = \frac{C_1 e^{-jk\rho}}{\sqrt{\rho}} \tag{2.87}$$

where

$$C_1 = ZI^e\sqrt{k/8\pi}\, e^{j\pi/4} \tag{2.88}$$

The associated magnetic field is

$$H_\phi(\rho) = \frac{C_2\, e^{-jk\rho}}{\sqrt{\rho}} \tag{2.89}$$

with

$$C_2 = C_1/Z \tag{2.90}$$

These are cylindrical waves because they have cylindrical planes of constant amplitude and phase, with a z-directed electric field and ϕ-directed magnetic field. Comparison of these expressions with the Luneberg-Kline expansion (2.10) reveals that only the lowest-order term of the latter is not zero, and that (2.87) and (2.89) are fields of the GO type. As far as the ray representation is concerned, for any fixed value of the z-coordinate there is a ray in each direction $\hat{\mathbf{s}} = \hat{\rho}$, as illustrated in Figure 2.8 (b). Each ray is surrounded by ray tubes cylindrical wave form as described by (2.72). For practical computations the large argument form of the Hankel function is valid when $k\rho$ is on the order of 0.95 or larger.

The duality principle of electromagnetic theory allows us to write dual expressions to the preceding for a magnetic line source of strength I^m lying along the z-axis:

$$H_z(\rho) = \frac{C_3\, e^{-jk\rho}}{\sqrt{\rho}} \tag{2.91}$$

$$C_3 = -\, YI^m \sqrt{\frac{k}{8\pi}}\, e^{j\pi/4} \tag{2.92}$$

$$E_\phi(\rho) = \frac{C_4\, e^{-jk\rho}}{\sqrt{\rho}} \tag{2.93}$$

$$C_4 = C_3/Y \tag{2.94}$$

2.4.4 The Fields of a Hertzian Dipole

The fields of a $\hat{\mathbf{z}}$-directed Hertzian dipole positioned at the origin of the rectangular coordinate system in Figure 2.9 are [1, p. 79]

$$E_r = A_o\, e^{-jkr}\left[\frac{2\nu}{(j\omega)r^2} + \frac{2\nu^2}{(j\omega)^2 r^3}\right]\cos\theta \tag{2.95}$$

$$E_\theta = A_o\, e^{-jkr}\left[\frac{1}{r} + \frac{\nu}{(j\omega)r^2} + \frac{\nu^2}{(j\omega)^2 r^3}\right]\sin\theta \tag{2.96}$$

$$E_\phi = 0 \tag{2.97}$$

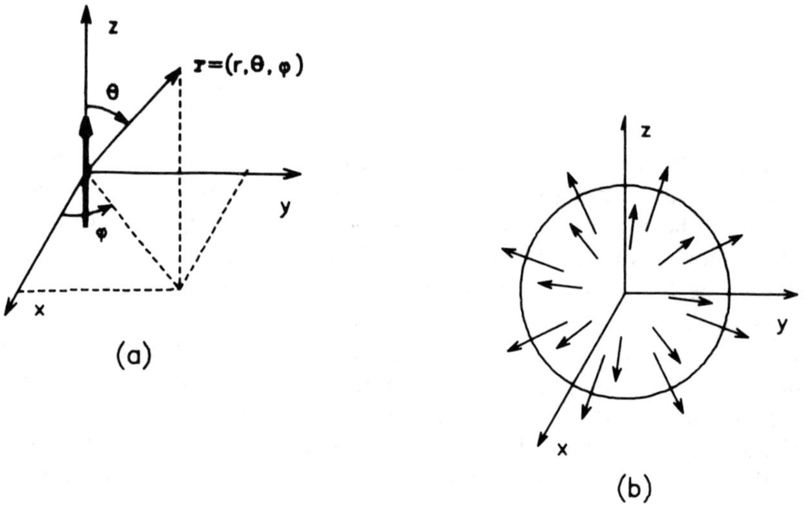

Figure 2.9 Hertzian dipole geometry (a) and ray representation (b).

Thus, by inspection, the Luneberg-Kline expansion terms for the preceding field are given by

$$\Psi(\mathbf{r}) = r \tag{2.98}$$

$$\mathbf{E}_0(\mathbf{r}) = \frac{A_o}{r} \sin\theta\hat{\theta} \tag{2.99}$$

$$\mathbf{E}_1(\mathbf{r}) = A_o\frac{2\nu \cos\theta}{r^2}\hat{\mathbf{r}} + A_o\frac{\nu \sin\theta}{r^2}\hat{\theta} \tag{2.100}$$

$$\mathbf{E}_2(\mathbf{r}) = A_o\frac{2\nu^2 \cos\theta}{r^3}\hat{\mathbf{r}} + A_o\frac{\nu^2 \sin\theta}{r^2}\hat{\theta} \tag{2.101}$$

$$\mathbf{E}_3(\mathbf{r}) = \mathbf{E}_4(\mathbf{r}) = \ldots = 0$$

The Hertzian dipole fields thus are ray optic (i.e., they can be written in the form of a Luneberg-Kline expansion), but, because there are $\mathbf{E}_1(\mathbf{r})$ and $\mathbf{E}_2(\mathbf{r})$ terms that are not zero, they are not geometrical optics type of fields. However, in the far-zone of the Hertzian dipole the term $\mathbf{E}_0(\mathbf{r})$ is dominant because of its $1/r$ dependence, which represents an amplitude decay very much less than the higher-order Luneberg-Kline terms, and therefore is the only one that needs to be retained. In the far-zone, the Hertzian dipole fields are GO fields, of the form:

$$\mathbf{E}(\mathbf{r}, \omega) = \hat{\boldsymbol{\theta}}A_o \sin\theta\frac{e^{-jkr}}{r} \tag{2.102}$$

The equiphase planes are spherical, but the planes of constant amplitude are not, as a result of the pattern factor sin θ. As far as the ray representation is concerned, however, there is a radially directed ray in each direction $\hat{\mathbf{s}} = (\theta, \phi)$, and each such ray is surrounded by a spherical wave ray tube of the form (2.74). The pattern factor essentially specifies the relative "starting values" of the amplitude and phase of each ray; these will be different on each different ray.

2.4.5 The Far-Zone Fields of Horn Antennas

Figure 2.10 shows the geometry of a smooth-walled conical horn antenna excited with a TE_{11} circular waveguide mode. The electromagnetic fields in the far-zone, in normalized form, are

$$\mathbf{E}(r, \theta, \phi) = [A(\theta) \sin\phi\hat{\theta} + B(\theta) \cos\phi\hat{\phi}]\frac{e^{-jkr}}{r} \tag{2.103}$$

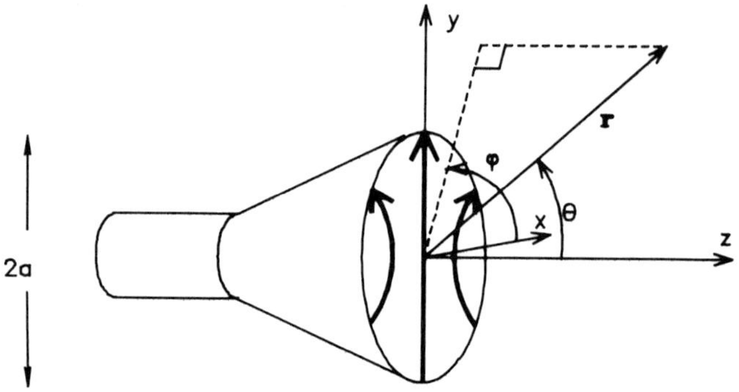

Figure 2.10 Conical horn antenna geometry.

$$\mathbf{H}(r, \theta, \phi) = Y\hat{\mathbf{r}} \times \mathbf{E}(r, \theta, \phi) \tag{2.104}$$

$$Y = \sqrt{\epsilon/\mu_0} \tag{2.105}$$

where $A(\theta)$ and $B(\theta)$ are the principal plane patterns of the horn antenna. Expressions for $A(\theta)$ and $B(\theta)$ are [21]

$$A(\theta) = \frac{J_1(ka \sin\theta)}{ka \sin\theta} \tag{2.106}$$

$$B(\theta) = \frac{\cos\theta \, J_1'(ka \sin\theta)}{1 - (ka \sin\theta/1.841)^2} \tag{2.107}$$

where $J_1(x)$ is the first-order Bessel function of the first kind, and $J_1'(x)$ its derivative. (Such expressions assume that the horn aperture radiates through an infinitely large, flat, conducting ground plane, but they usually are in excellent agreement with actual patterns of horns without ground planes, at least over the main beam and first sidelobe or two of the pattern of the horn antenna.) For this type of antenna, the terms of the Luneberg-Kline expansion are

$$\Psi(\mathbf{r}) = r \tag{2.108}$$

$$\mathbf{E}_0(\mathbf{r}) = [A(\theta) \sin\phi\hat{\phi} + B(\theta) \cos\phi\hat{\phi}]/r \tag{2.109}$$

$$\mathbf{E}_1(\mathbf{r}) = \mathbf{E}_2(\mathbf{r}) = \ldots = 0 \tag{2.110}$$

and the fields thus are of the GO type. In terms of the way in which the radiation patterns of such horn antennas will be measured in practice, the co- and cross-polarized fields are [22]

$$E_{\text{co}}(r, \theta, \phi) = F(\theta, \phi)\frac{e^{-jkr}}{r} \tag{2.111}$$

$$E_{\text{cross}}(r, \theta, \phi) = G(\theta, \phi)\frac{e^{-jkr}}{r} \tag{2.112}$$

$$F(\theta, \phi) = E_\theta \sin\phi + E_\phi \cos\phi \tag{2.113}$$

$$G(\theta, \phi) = E_\phi \cos\phi - E_\phi \sin\phi \tag{2.114}$$

These fields have spherical equiphase surfaces but not spherical equiamplitude surfaces due to the presence of the pattern factors $F(\theta, \phi)$ and $G(\theta, \phi)$. The field therefore is not a perfect spherical wave. However, there is a radially directed ray propagating in each direction $\hat{s} = (\theta, \phi)$, and each such ray is surrounded by a ray tube of the spherical wave form (2.74). The pattern factors provide the information on the "starting values" of the amplitude and phase for each ray.

For the so-called dual-mode or Potter conical horn [21], excited by a combination of the TE_{11} and TM_{11} modes of circular waveguide, the far-zone fields are of the same form as in (2.103), except that $A(\theta) = B(\theta)$, and these now have the form:

$$A(\theta) = \frac{(w + \cos\theta)J_1'(v)}{1 - (v/u)^2} \tag{2.115}$$

with $J_1'(u) = 0$, $w = [1 - (u/c)^2]$, and $v = (2a/\lambda)\sin\theta$, where λ is the wavelength in the medium into which radiation occurs.

Conical corrugated horns excited by the HE_{11} mode of circular corrugated waveguide also have $A(\theta) = B(\theta)$ and far-zone fields of the same form as (2.104). Integral expressions for the pattern factor $A(\theta)$ of such horn antennas can be found in [21].

Finally, with pyramidal horns fed by rectangular waveguide with the dominant TE_{10} mode, a form similar to (2.103) applies for the far-zone fields, with the exception that we now have forms $A(\theta, \phi)$ and $B(\theta, \phi)$; that is, the latter are functions of both θ and ϕ. Useful expressions for these factors are [21], in the geometry of Figure 2.11:

$$A(\theta, \phi) = \frac{\cos u}{u^2 - (\pi/2)^2}\frac{\sin v}{v} \tag{2.116}$$

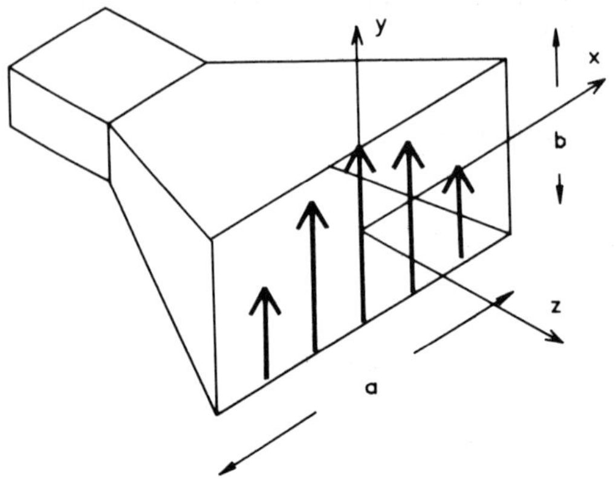

Figure 2.11 Pyramidal horn antenna geometry.

$$B(\theta, \phi) = A(\theta, \phi) \cos\theta \tag{2.117}$$

where $u = (ka/2) \sin\theta \cos\phi$ and $v = (kb/2) \sin\theta \sin\phi$, and the fields remain of the GO type with spherical wave ray tubes.

2.4.6 The Fields of a Piecewise-Sinusoidal Dipole

Consider the electric dipole of length $2d$ shown in Figure 2.12. On the segment

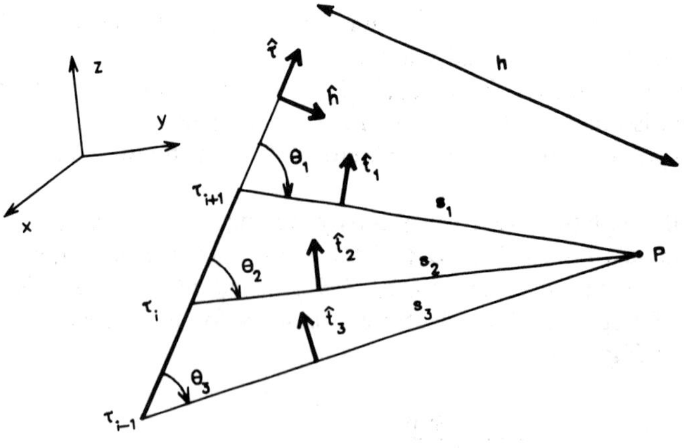

Figure 2.12 Electric dipole with piecewise-sinusoidal current distribution.

$\tau_{i-1} \leqslant \tau \leqslant \tau_{i+1}$, with τ_i at the center of the segment, the with $\hat{\tau}$-directed current distribution is given by

$$I(\tau) = \frac{I_0 \sin[k(d - |\tau - \tau_i|)]}{\sin(kd)} \tag{2.118}$$

At the observation point, the rigorous near-zone electric field of such an electric current distribution is given by [23]

$$E_h = C_o[\mathrm{e}^{-jks_1}\cos\theta_1 - 2\cos(kd)\,\mathrm{e}^{-jks_2}\cos\theta_2 + \mathrm{e}^{-jks_3}\cos\theta_3] \tag{2.119}$$

$$E_\tau = -C_o d\left[\frac{\mathrm{e}^{-jks_1}}{s_1} - \frac{2\cos(kd)\,\mathrm{e}^{-jks_2}}{s_2} + \frac{\mathrm{e}^{-jks_3}}{s_3}\right] \tag{2.120}$$

$$C_o = \frac{jZI_o}{4\pi d\sin(kd)} \tag{2.121}$$

$$Z = \sqrt{\mu_0/\epsilon} \tag{2.122}$$

Although (2.119) and (2.120) apparently do not describe GO fields, as has been shown in [24], the total field at the observation point may be written as

$$\mathbf{E} = E_h\hat{h} + E_\tau\hat{\tau} = E_a\hat{\mathbf{t}}_1 + E_b\hat{\mathbf{t}}_2 + E_c\hat{\mathbf{t}}_3 \tag{2.123}$$

where

$$E_a = \frac{-C_o}{\sin\theta_1}\frac{\mathrm{e}^{-jks_1}}{s_1} \tag{2.124}$$

$$E_b = \frac{2C_o\cos(kd)}{\sin\theta_2}\frac{\mathrm{e}^{-jks_2}}{s_2} \tag{2.125}$$

$$E_c = \frac{-C_o}{\sin\theta_3}\frac{\mathrm{e}^{-jks_3}}{s_3} \tag{2.126}$$

The sum of these three terms exactly represents the field, and can be interpreted as spherical waves with $1/\sin\theta_i$ pattern factors, emanating from the center and end points of the dipole. All three terms satisfy the property of GO ray fields in that each has no field components in the direction of the ray.

For such a source we are able to apply the GTD by individually considering reflection and diffraction of each of the component GO terms and then adding them to obtain the final answer. This process is sometimes referred to as *source decomposition.*

2.4.7 Sources with Fields That Are Not Geometrical Optics or Ray-Optic Fields

The user of the ray methods forming the subject of this text cannot be warned often enough that these techniques are valid only when an illuminating source generates a strictly GO field or, in some cases, at least a ray-optic field. For instance, the fields of a horn antenna with a radiated field that has a very rapid phase variation cannot always be written in the form of a GO field. Neither are the near-zone fields of most antennas GO fields, nor are the fields in closed waveguiding structures and hence in the apertures of waveguide-fed horn antennas, although we may be able to decompose these into the form of a sum of GO fields. Even the far-zone fields of certain complex arrays of horn antennas or wire radiators are not GO fields; in this particular case as well, we often can separately treat each element of the array, compute all the necessary GTD effects, and finally add the contributions of each element. We will encounter further examples of nonray-optic fields in later chapters of this text (e.g., edge diffracted fields in transition regions). Fortunately, there are very many instances of great practical importance where the fields from sources are indeed GO fields.

2.4.8 Further Comment

At this stage we have ascertained that the fields of a number of "practical" sources of electromagnetic fields produce, albeit under restricted circumstances, either cylindrical or spherical wave GO fields. These, together with the ubiquitous uniform plane wave, are special cases of the astigmatic ray tube expressions (2.63) and (2.65). The general astigmatic GO fields (that is, $\rho_1 \neq \rho_2$) arises in practice when plane, cylindrical, or spherical wave GO fields encounter an obstacle in their path, the resulting reflected and diffracted fields receding from the obstacle generally are astigmatic.

2.5 REDUCTION OF RESULTS TO TWO-DIMENSIONAL RAY TUBES

For two-dimensional ray tubes, the fields are independent of one of the three Cartesian coordinates (assumed here without loss of generality to be the z-coordinate), and assume a "cylindrical" character. We will refer to such GO fields as 2D with respect to the z-axis. One of the radii of curvature of any 2D ray tube must be infinite; let this be ρ_2. The spreading factor $A(s)$ in (2.61) then reduces to

$$A(s) = \sqrt{\rho_1/(\rho_1 + s)}$$

Relabeling ρ_1 simply as ρ (the radius of curvature of the wavefront in the x-y principal plane), for the general 2D astigmatic ray tube, we have

$$\mathbf{E}(s) = \mathbf{E}(0)\sqrt{\rho/(\rho + s)}\, e^{-jks} \tag{2.127}$$

which of course, is, identical to that in (2.68). Note that we need only "work in" the x-y cross-sectional plane in such instances, the geometric details of which are illustrated in Figure 2.13. A similar expression to (2.127) holds for the magnetic field $\mathbf{H}(s)$. For 2D situations expression (2.66) can be written as

$$\mathbf{H} = Y\hat{\mathbf{s}} \times \mathbf{E} \tag{2.128}$$

If the electric field has only a z-component E_z, then from the relation (2.128) for GO fields we know that the magnetic field must be $\mathbf{H} = (E_z/Z)\hat{\mathbf{s}} \times \hat{\mathbf{z}}$. This usually is called the *TM case* as the magnetic (M) field is transverse (T) to the z-axis; sometimes we see it referred to as the E-*polarization* for the 2D problem (as the electric field is z-directed).

An example of such a field is that in the far-zone of the electric line source in Section 2.4.3. Similarly, for a magnetic field that has only a z-component the electric field can be found from (2.128). The latter is known as the TE, or the H-*polarization,* case. An example of a TE field is the far-zone of the magnetic line source discussed in Section 2.4.4. Pattern factors with an angular dependence, of course, can be added to the fields from such line sources, and these will remain

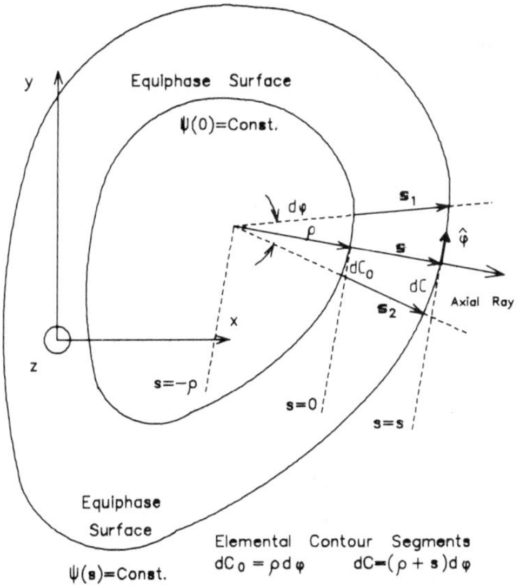

Figure 2.13 Geometrical picture of two-dimensional ray tube.

2D fields. Plane waves with appropriate polarizations also will be two-dimensional TE or TM fields.

An arbitrarily polarized electric field always can be written as a linear combination of the TE and TM states. These cases therefore can be dealt with separately, and the advantage gained is that the problem becomes scalar. If we agree therefore to work with electric field E_z for TM cases and the magnetic field H_z for TE cases, we can in effect deal with both cases simultaneously (always remembering that the two field quantities have different boundary conditions) by defining a scalar, $U(s)$, as

$$U(s) = \begin{cases} E_z, & \text{for the TM case} \\ H_z, & \text{for the TE case} \end{cases} \tag{2.129}$$

and writing the 2D ray tube field expression in the scalar form:

$$U(s) = U(0)\sqrt{\rho/(\rho + s)}\, \mathrm{e}^{-jks} \tag{2.130}$$

A still better way is to write this in the "fail-safe" form, as for the full 3D astigmatic ray tube in Section 2.2.8. For convenience, we stress again the complete set of conventions here, and then immediately study some instructive examples. Let us assume that we may calculate the 2D GO fields by using the form:

$$U(s) = U(0)\sqrt{|\rho/(\rho + s)|}\, \mathrm{e}^{-jks} \tag{2.131}$$

Then, the sign convention on the radius of curvature ρ in the cross-sectional plane is as follows: at the selected reference front $s = 0$, a *positive* (*negative*) radius of curvature implies *diverging* (*converging*) paraxial rays in the cross-sectional plane. Regarding the correct phase shift through a caustic line, with phase referenced to the position $s = 0$ on the axial ray of the ray tube:

(a) If we move along a ray through a caustic line, in the direction of propagation, we multiply the ray field expression (2.131) by $\mathrm{e}^{j\pi/2}$.
(b) If we move along a ray through a caustic line, in a direction opposite to the propagation direction, we multiply the ray field expression by $\mathrm{e}^{-j\pi/2}$.

Example 2.3

In Figure 2.14 (a) the rays converge at $s = 0$, so that $\rho = -s_1$. Given $U(0)$, we wish to determine the field at $s = s_1 + s_2$. Because in moving from $s = 0$ to $s = s_1 + s_2$ we pass through a caustic line in the direction of propagation, the multiplicative factor is $\mathrm{e}^{j\pi/2}$. Also, at $s = s_1 + s_2$ we have

$$\sqrt{|\rho/(\rho + s)|} = \sqrt{|-s_1/(-s_1 + s_1 + s_2)|} = \sqrt{s_1/s_2}$$

$$\mathrm{e}^{-jks} = \mathrm{e}^{-jk(s_1+s_2)}$$

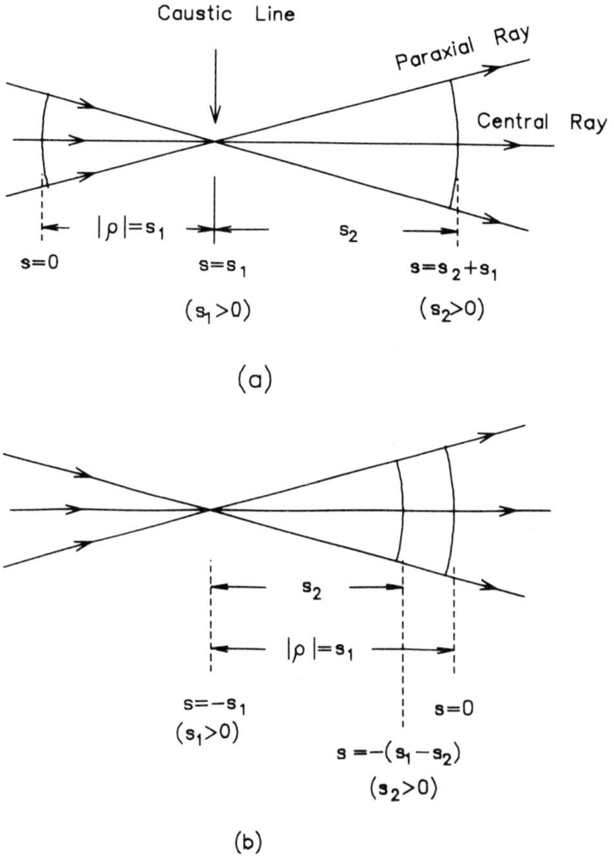

Figure 2.14 Two-dimensional ray tube with (a) $\rho < 0$, and (b) $\rho > 0$.

so that

$$U(s_1 + s_2) = U(0)\sqrt{s_1/s_2}\, e^{-jk(s_1+s_2)}\, e^{j\pi/2}$$

Example 2.4

In Figure 2.14 (b) the rays diverge at $s = 0$, so that $\rho = s_1$. No caustic lines are crossed in moving from $s = 0$ to the selected observation point at $s = -(s_1 - s_2)$. Therefore, we have

$$\sqrt{|\rho/(\rho + s)|} = \sqrt{|s_1/(s_1 - s_1 + s_2)|} = \sqrt{s_1/s_2}$$

$$e^{-jks} = e^{jk(s_1 - s_2)}$$

so that

$$U(-s_1 + s_2) = U(0)\sqrt{s_1/s_2}\, e^{jk(s_1 - s_2)}$$

The intensity law (2.59) in the 2D case becomes

$$|U(s)| = |U(0)|\sqrt{dC_0/dC} \tag{2.132}$$

which represents conservation of power per unit length along the 2D ray tube.

Clearly, the 2D GO field expressions are less complicated than their 3D counterparts, while nevertheless exhibiting the same principal features. In view of the simplifications afforded, perhaps an easier approach is to study first the 2D results, obtain an understanding of the essential principles of a phenomenon, and then proceed to the 3D forms. This fact, along with the connection shown between the 2D and 3D results, provides us with the pedagogical motivation for treating in certain sections of the text the 2D and 3D cases in the reverse order to that used in this chapter.

2.6 RAYS IN LOSSY MEDIA

In Section 2.2 we saw that in lossless media the ray direction as defined by $\nabla\Psi$ and the Poynting vector direction coincide. This fact allowed us to define ray trajectories relatively easily. If the media were lossy, however, this coincidence of directions may not occur and special steps must be taken. This leads to the subject of inhomogeneous wave tracking [25], where the term *inhomogeneous* does not refer to the properties of the media but rather to the fact that the fields behave like locally inhomogeneous plane waves. Such considerations are beyond the scope of this text. For a discussion of these and other situations where ray tracking is not straightforward, the paper by Felsen [26] is recommended.

2.7 CONCLUDING REMARKS

In the rigorous formulations of electromagnetic theory, the fields at a selected observation point can be exactly determined only if the fields over an entire surface enclosing that point are known completely. However, we saw in this chapter that, in a region that can be considered large in terms of the operating wavelength, the field in the proximity of a selected observation point depends predominantly on only a restricted portion of some initial surface. Energy is transported from this portion of the initial surface through a restricted sector of space (ray tube) that surrounds the propagation path. This path is the ray trajectory of geometrical optics, and along it the field behavior is locally plane. The process is quantified in the expressions summarized in Section 2.3. The latter results were obtained in Section 2.2 by adopting the Luneberg-Kline asymptotic expansions as anticipated solutions for high-frequency electromagnetic fields, thereby reducing Maxwell's equations to the simpler eikonal and transport equations.

More detailed and very general treatments of the foundations of geometrical optics can be found in the texts by Kline and Kay [4] and Born and Wolf [17], and the recent survey by Cornbleet [16].

2.8 A TASTE OF THINGS TO COME

This chapter has described geometrical optics electromagnetic fields that propagate in unbounded media. Of great interest in electromagnetic engineering is the behavior of such fields when conducting scatterers (both wanted and unwanted) are present. An electromagnetic field incident on some scatterer induces on it currents that are responsible for the scattered fields. If we were to use rigorous methods to solve such a scattering problem, we would (not without some effort) obtain a solution for the scattered field as a whole. What would not be immediately clear was whether certain geometrical aspects of the scatterer could be identified as responsible for definite portions of the overall scattered field. Indeed, if the scatterers were electrically small, to do this at all would often not be possible. However, for objects large in terms of the operating wavelength, use of asymptotic methods permits distinct geometrical features of a scatterer to be unambiguously associated with a portion or component of the scattered field. Thus, we are able to identify different wave processes such as reflection, edge diffraction, or curved surface diffraction. The next chapter considers the reflection of geometrical optics fields by smooth conducting surfaces. The additional scattering phenomena mentioned form the subject of later chapters of this text.

That the reader be aware of the manner in which terms such as *incident field* and *scattered field* are defined in electromagnetic theory is essential. In rigorous scattering formulations there is no consideration of rays, and the meaning of such terms is generally not the same as that used in the geometrical theory of diffraction. Nor would the terms always coincide with what we might loosely expect them to be, and so we should elaborate on the matter at this point.

The prevailing terminology in rigorous formulations is treated first. Consider a source radiating in some region as shown in Figure 2.15. It establishes a field $\mathbf{E}_i(\mathbf{r})$ in this region: $\mathbf{E}_i(\mathbf{r})$ is called the *incident field*, with $\mathbf{r}$ the position vector. Let us now introduce a conducting obstacle into the region. The incident field $\mathbf{E}_i$ is considered (in fact, it is thus defined) to be unperturbed, and continues to exist throughout the region, even inside the obstacle. However, as a result of the presence of the incident field $\mathbf{E}_i$, currents are induced on the obstacle. These induced

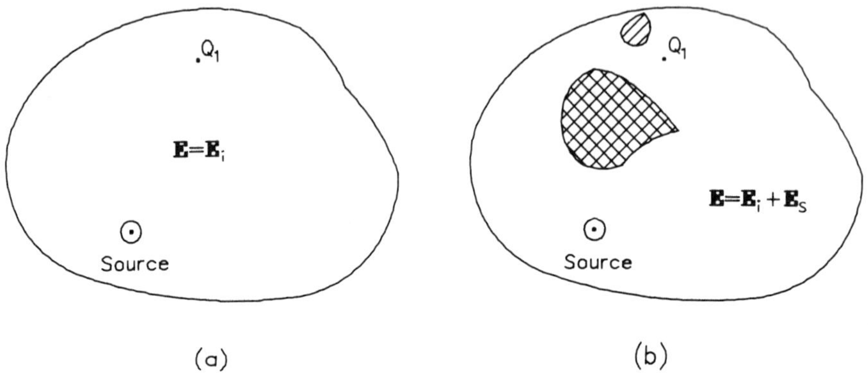

Figure 2.15 Rigorous electromagnetic theory definitions of the incident, scattered, and total field terms. In (a) no obstacle is present, whereas in (b) the fields are shown in the presence of an obstacle.

currents give rise to an additional field, called the *scattered field*, throughout the region (i.e., both outside and inside the scatterer). With the scatterer absent, the field in the region is simply $\mathbf{E}_i$. With the scatterer present, the resultant field in the region is the sum

$$\mathbf{E}(\mathbf{r}) = \mathbf{E}_i(\mathbf{r}) + \mathbf{E}_s(\mathbf{r})$$

The induced currents always will be such that inside the scatterer they produce a scattered field that is the negative of the incident field. Thus, the resultant field there is zero, as expected inside a conducting object. Especially important is to note that, even with the obstacle present, the incident field at point Q_1 (which is not visible from the source as a result of the presence of the obstacle) remains as it was before the scatterer was introduced. The obstacle has no effect whatsoever on this incident field. Concepts such as shadowing are not applicable, at least in rigorous electromagnetic scattering formulations. Neither do we speak of individual physical mechanisms such as reflection; all the effects of the scatterer on the resultant field are lumped into one phenomenon called *scattering*.

Now consider the same situation but from a GTD point of view. Here, we must remember that what we mean by an electromagnetic field at some point $\mathbf{r}$, of course, is the field proceeding along some ray (or rays, as there may be more than one) that passes through that point. We begin with the arrangement illustrated in Figure 2.16(a), in which no scatterers are present. At points Q_1 and Q_2, we denote the incident fields by $\mathbf{E}^i(Q_1)$ and $\mathbf{E}^i(Q_2)$, respectively (note that superscripts are now used), and imply the GO field carried along a ray that is able to reach these points. In the particular case shown, the incident fields at points Q_1 and Q_2

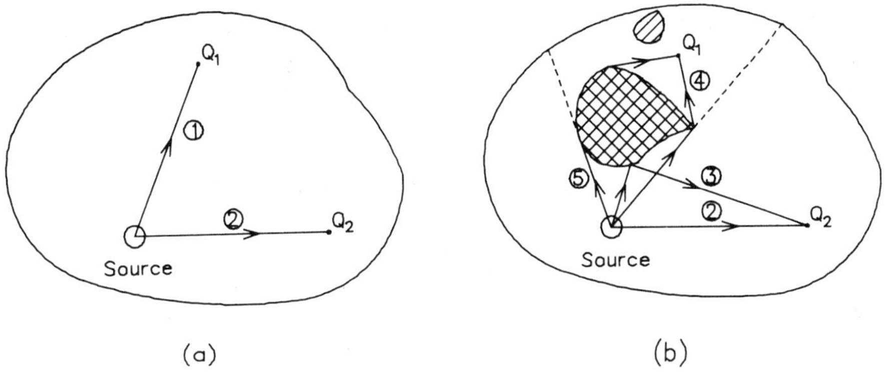

Figure 2.16 Geometrical theory of diffraction definitions of the terms incident (direct), reflected, and diffracted rays. In (a) no obstacle is present, whereas in (b) the fields are shown in the presence of an obstacle.

proceed *directly* from the impressed source of the electromagnetic field (along ray 1 and ray 2) and are therefore termed *direct rays*.

This is in contradistinction to rays incident on a selected observation point, which may arrive via some circuitous route, having been reflected or diffracted, if some scatterer has been introduced. If the scatterer indeed is now placed in the path of any ray originally incident on points Q_1 and Q_2, as shown in Figure 2.16(b), it will alter the ray trajectories and the latter rays may no longer be able to reach the particular observation points directly. In other words, unlike the "all pervading" incident fields of rigorous electromagnetic field formulations, incident ray fields (and these are always the "incident fields" we have in mind when applying the GTD) indeed are shadowed or otherwise diverted by obstacles placed in their paths. To stress this fact, some authors refer to such fields as "incident fields in the sense of geometrical optics." For the configuration of Figure 2.16(b), no direct ray is able to reach any point situated in the shaded region shown, this being called the *shadow region* for the direct rays. Its boundaries are referred to as the *shadow boundaries* (SB). Obviously, no direct ray is able to reach Q_1.

Additional rays may now reach Q_1 and Q_2 as a result of reflection and diffraction by the scatterer. As a foretaste of topics to be considered in the chapters to follow, Figure 2.16(b) shows ray 3, which reaches Q_2 through reflection at a given point on the smooth surface of the obstacle. The incident field (here and hereafter this is meant in the GTD and GO sense) thus is the sum of that brought to Q_2 by both direct ray 2 and reflected ray 3. Also indicated is ray 4, which reaches Q_1 after being diffracted at the edge on the obstacle. Finally, ray 5 creeps along the scatterer, shedding rays continually as it travels, but with one particular shed ray reaching Q_1. This process is called curved surface diffraction.

Clearly, the GTD definitions appear closer to our everyday understanding, doubtless because the only electromagnetic scattering phenomena we are able to observe without the help of sophisticated measurement instruments, those in the visible portion of the electromagnetic spectrum, are high-frequency phenomena to which the GTD is eminently suitable. However, the alternative definitions used elsewhere are most definitely advantageous for rigorous boundary value problem formulations.

As is clear from the preceding discussion, what is most important in the application of the GTD is carefully determining which rays are prevented from reaching a given point directly and which are not. This necessity leads us to some further discussion of shadowing and *shadow boundaries.*

The concept of a shadow usually is considered to be obvious. As a matter of fact, in the 1885 edition of his book on light, Tyndall [27] assures us that

> the ancients were aware of the rectilineal propagation of light. They knew that an opaque body, placed between the eye and a point of light, intercepted the light of the point."

Nevertheless, at the risk of appearing uninformed, let us formally consider what happens when we apply our existing GO knowledge to the problem of a line source illuminating a flat conducting strip. We know from Section 2.2.7 (and the ancients!) that rays travel in straight lines in a homogeneous medium. Because they cannot penetrate the conductor, the state of affairs is as shown in Figure 2.17. Note that

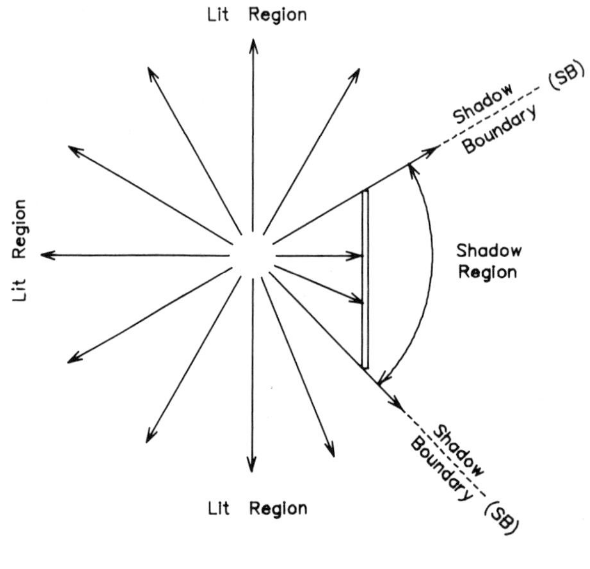

Figure 2.17 Line source illuminating a flat conducting strip. Note the transition region near the source where the fields are not ray optic.

there is a *lit region* and a *shadow region*. The GO description would have us believe that the resultant field amplitudes were zero in the shadow region, and at the shadow boundaries fall abruptly to zero from their nonzero values in the lit region. We know from observation that this is not the case and, whereas the fields in the shadow region may be small, they are not everywhere zero (i.e., it is dark there, but not pitch dark). Furthermore, the variation in the amplitude of the total field as we move from the lit region into the shadow region is more gradual. The GTD plan, rather than abandoning the GO approach, seeks to correct it by adding additional rays.

Suppose now that the conducting obstacle takes the form of the smooth cylinder shown in Figure 2.18. There are two shadow boundaries that delineate the shadow region. Some prevailing terminology is indicated. The shadow boundaries of Figures 2.17 and 2.18, controlling as they do the fate of the incident rays, are usually called *incident shadow boundaries* (ISB). In Chapter 3 shadow boundaries will be introduced that delineate the regions of space in which reflected rays do or do not exist; these obviously will be called *reflection shadow boundaries* (RSB).

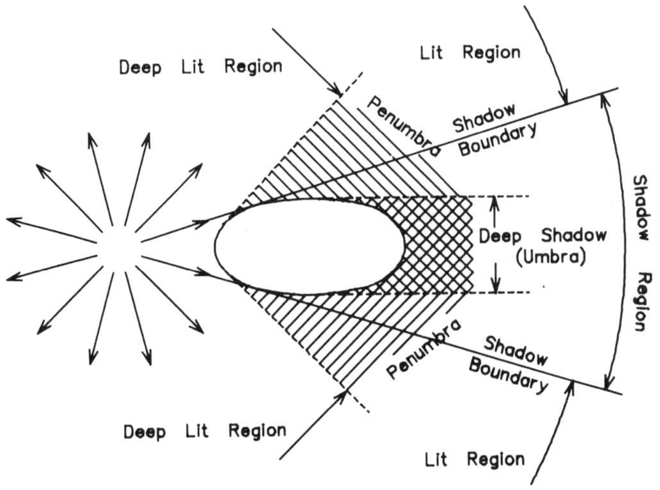

Figure 2.18 Line source illuminating a smooth cylindrical structure.

PROBLEMS

2.1 Derive in full the eikonal and transport equations of Section 2.2.4 and 2.2.5.

2.2 Use the duality principle to write expressions for the near-zone fields of a magnetic dipole with magnetic current given by equation (2.118) of Section 2.5.6. Suggest where such a magnetic dipole may arise as an analytical model for a practical situation.

2.3 Derive in full equations (2.43) and (2.44).

2.4 Repeat the derivation given in Section 2.4.6 for the eikonal equation, but use the alternative form of the Luneberg-Kline expansion to obtain $|\nabla S|^2 = n^2$, where n is the refractive index of the medium involved.

2.5 Concerning the line sources of Section 2.4.3, plot the exact Hankel function $H_0^{(2)}\ (k\rho)$, and its large argument (asymptotic) form $\sqrt{2/(\pi k\rho)}\ e^{j\pi/4}\ e^{-jk\rho}$, as a function of $k\rho$ on the same graph. Determine the value of $k\rho$ larger than which the two forms give values within one percent of each other.

2.6 For the fields of the magnetic line source given in (2.91) and (2.94), by inspection write down the terms of the Luneberg-Kline expansions (2.10) and (2.11).

2.7 Show for the piecewise-sinusoidal electric dipole of Section 2.4.6 that expressions (2.119) and (2.120) indeed can be written as the superposition of the fields in (2.124), (2.125) and (2.126). The following results from trigonometry may be useful:

$$E_i = E_t(s_i, \theta_i) \sin\theta_i - E_h(s_i, \theta_i) \cos\theta_i$$

$$h = s_i \sin\theta_i$$

$$E_t = E_1 \sin\theta_1 + E_2 \sin\theta_2 + E_3 \sin\theta_3$$

$$E_h = -(E_1 \cos\theta_1 + E_2 \cos\theta_2 + E_3 \cos\theta_3)$$

REFERENCES

[1] R.F. Harrington, *Time-Harmonic Electromagnetic Fields*, McGraw-Hill, New York, 1961.

[2] L.J. Chu, "Physical Limitations of Omni-Directional Antennas," *J. Appl. Phys.*, Vol. 19, December 1978, pp. 1163–1175.

[3] M. Kline, "An Asymptotic Solution of Maxwell's Equations," *Comm. Pure Appl. Math.*, Vol. 4, 1951, pp. 225–262.

[4] M. Kline and I. Kay, *Electromagnetic Theory and Geometrical Optics*, Interscience, New York, 1965.

[5] A. Sommerfeld and J. Runge, "Anwending der Vektorrechnung auf die Grundlagen der geometrischen Optik," *Ann. Phys.*, Vol. 35, 1911, pp. 277–298.

[6] R.M. Luneberg, *Mathematical Theory of Optics*, Brown University Press, Providence, RI, 1944.

[7] R.G. Kouyoumjian, "Asymptotic High-frequency Methods," *Proc. IEEE*, Vol. 53, August 1965, pp. 864–876.

[8] D.S. Jones, *Methods in Electromagnetic Wave Propagation*, Oxford University Press, London, 1979.

[9] S. Silver, *Microwave Antenna Theory and Design*, McGraw-Hill, New York, 1949, pp. 110–128.
[10] S.W. Lee, Y. Rahmat-Samii, and R.C. Menendez, "GTD, Ray Field, and Comments on Two Papers," *IEEE Trans. Antennas and Propagation*, Vol. AP-26, No. 2, March 1978, pp. 39–41.
[11] R.G. Kouyoumjian and P.H. Pathak, "A Uniform Geometrical Theory of Diffraction for an Edge in a Perfectly Conducting Surface," *Proc. IEEE*, November 1974, pp. 1448–1461.
[12] R.M. Lewis, "Geometrical Optics and the Polarisation Vectors," *IEEE Trans. Antennas and Propagation*, Vol. AP-14, 1966, pp. 100–101.
[13] W.V.T. Rusch, A.C. Ludwig, and W.C. Wong, "Theory of Quasi-Optical Antennas," in *The Handbook of Antenna Design*, A.W. Rudge *et al.*(eds.), Peter Peregrinus, London, 1986, pp. 65–67.
[14] S.W. Lee, "Electromagnetic Reflection from a Conducting Surface: Geometrical Optics Solution," *IEEE Trans. Antennas and Propagation*, Vol. AP-23, No. 2, March 1975, pp. 184–191.
[15] G.A. Deschamps, "Ray Techniques in Electromagnetics," *Proc. IEEE*, Vol. 6, No. 9, September 1972, pp. 1022–1035.
[16] S. Cornbleet, "Geometrical Optics Reviewed: A New Light on an Old Subject," *Proc. IEEE*, Vol. 71, No. 4, April 1983, pp. 471–502.
[17] M. Born and E. Wolf, *Principles of Optics,* 5th Edition, Pergamon, London, 1975.
[18] H. Hochstadt, *Differential Equations: A Modern Approach*, Dover, New York, 1964, pp. 37–38.
[19] I. Kay and J.B. Keller, "Asymptotic Evaluation of the Field at a Caustic," *J. Appl. Phys.*, Vol. 25, 1954, pp. 876–883.
[20] A.M. Arthurs, *Calculus of Variations*, Routledge and Kegan Paul, London, 1975.
[21] A.W. Love (ed.), *Electromagnetic Horn Antennas*, IEEE Press, New York, 1976.
[22] A.C. Ludwig, "On the Definition of Cross-Polarisation," *IEEE Trans. Antennas and Propagation*, Vol. AP-21, January 1973, pp. 116–119.
[23] J.H. Richmond and N.H. Geary, "Mutual Impedance between Coplanar-Skew Dipoles," *IEEE Trans. Antennas and Propagation*, Vol. AP-18, May 1970, pp. 414–416.
[24] E.P. Ekelman and G.A. Thiele, "A Hybrid Technique for Combining the Moment Method Treatment of Wire Antennas with the GTD for Curved Surfaces," *IEEE Trans. Antennas and Propagation*, Vol. AP-28, November 1980, pp. 831–839.
[25] S. Choudhary and L.B. Felsen, "Asymptotic Theory for Inhomogeneous Waves," *IEEE Trans. Antennas and Propagation*, Vol. AP-21, 1973, pp. 827–842.
[26] L.B. Felsen, "Novel Ways of Tracking Rays," *J. Opt. Soc. Am.*, Vol. 2, No. 6, June 1965, pp. 954–963.
[27] J. Tyndall, *Six Lectures on Light*, Longmans, London, 1885.

Chapter 3
Geometrical Optics
Reflected Fields

"Reflexibility of rays, is their disposition to be reflected or turned back into the same medium from any other medium upon whose surface they fall. And rays are more or less reflexible, which are turned back more or less easily."

Sir Isaac Newton.
Opticks, 1704

3.1 INTRODUCTION

3.1.1 Initial Remarks

The examples in the previous chapter demonstrate that GO fields are inclusive of those produced by a large number of sources of importance in electromagnetic engineering. It is only natural therefore that we consider what happens when we permit a high-frequency field to be incident on some conducting surface. Although the anticipated incident and reflected fields in general will be ray-optic fields that can be expressed in the form of Luneberg-Kline expansions, once again the *zero*th-order or GO terms will be our prime concern. Before proceeding though, we first enter into a qualitative discussion on the theme of this chapter.

3.1.2 A Stroll in the Sun

Anyone who has walked alongside a large expanse of water will have noticed that light from the sun is scattered (let us deliberately not yet use the term *reflected*) from the water surface. The sun is effectively a point source that causes a spherical electromagnetic wave to be incident on the water. Our background in electromagnetic theory informs us that a current therefore is induced over the entire surface of the water. As a consequence, the entire surface of the water radiates a scattered field that, together with the light coming directly from the sun, makes up the resultant field reaching our eyes. However, even a cursory glance at the water (if our eyes can stand the glare) reveals that there appears to be a particular restricted area of water (let us call it Q_r) that seems to be doing all the work in bringing scattered light to our eyes. If we "mark" this apparently industrious area of water, walk some distance away from that observation point, and then make the same observation, we will find that the area of water now contributing the majority of scattered light into our eyes is not the same one noticed from the first observation point. The position Q_r will have changed with the observation point.

A similar situation (a shift in Q_r) would be observed if we kept our observation position fixed and patiently recorded the location of Q_r at roughly the same time each day over a period long enough for the apparent horizontal position of the sun to change. We then would be able to conclude that the location of Q_r depended not only on the observation point but also on the position of the source.

Suppose now that we remain in one observation position, establish the location of Q_r once more, and a boat is moving over the water. While at some distance from Q_r the boat will have no effect on the scattering we observe from this area, despite the fact that it is disturbing the induced currents at other locations on the water surface. Only when it is directly over the area in the vicinity of Q_r, which is providing the scattering in the direction of our point of observation, will the scattered light that we observe be affected. We say that the scattering is a local effect, depending principally on the properties of the scatterer (the water in this case) in the vicinity of the special point (or region), Q_r.

The explanation for (actually, as in all theories of physical science, it is just a model for) this scattering mechanism is as follows. All the currents induced over the entire surface of the water indeed reradiate, but for a particular observation point their contributions tend to interfere destructively, except for those over the small area of water from which the majority of the scattered light appears to emanate. This behavior occurs only because the wavelength of the light is so small relative to the size of the scatterer. The point Q_r is called a *stationary point* (in mathematical terms) or *point of reflection* (in optical terms) because there we see the reflection of the sun in the water, and the location of Q_r certainly is a function of the position of the source of the light and the observation point.

In general, smooth objects placed in the path of an electromagnetic wave reradiate what we call *scattered fields* in all directions. When we can identify, as in the preceding illustration, some localized region (or regions) on the scatterer as furnishing the principal proportion of the scattered field at a particular observation point, only then may we really claim to have separated out a scattering mechanism that we call *reflection* (from the Latin *flectere*, "to bend"). The reflection point Q_r also sometimes is called the *specular point* (from the Latin *specere*, "to look").

3.1.3 A Strategy for This Chapter

In the context of geometrical optics the reflected field is constructed by extending the methods of the previous chapter. We can obtain an overview of how this is done by considering an astigmatic ray tube propagating in free space from some source, with a central ray that impacts a smooth surface at a point Q_r, as illustrated in Figure 3.1. This ray tube has as ray coordinate s^i measured along the central ray, and is completely described by its central ray vector $\hat{\mathbf{s}}^i$, principal radii of curvature ρ_1^i and ρ_2^i at the selected reference point Q_s, and a knowledge of the initial GO field $\mathbf{E}^i(Q_s)$ at the reference point Q_s. Then, at the point Q_r on the reflecting surface, at distance s^i from the incident ray reference point Q_s, the incident GO field is given by (with superscript i signifying the incident field),

$$\mathbf{E}^i(Q_r) = \mathbf{E}^i(Q_s)\sqrt{\rho_1^i \rho_2^i/(\rho_1^i + s^i)(\rho_2^i + s^i)}\ \mathrm{e}^{-jks^i} \tag{3.1}$$

We should always remember that reference to the principal radii of curvature of any wavefront would be meaningless unless we specified at which position in space these had been measured. As a wavefront propagates, its radii of curvature change

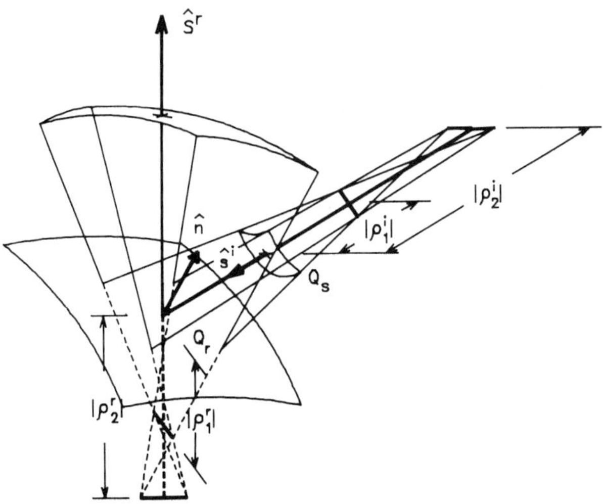

Figure 3.1 Incident and reflected GO field ray tubes. Q_s is the reference point of the incident ray tube, and $\hat{\mathbf{n}}$ is the unit normal to the surface at Q_r. Q_r is the reflection point and the reference point for the reflected ray tube.

constantly. Therefore, simply to write ρ_1^i and ρ_2^i as "the principal radii of curvature of the incident wavefront" is not sufficient; their values at Q_s will be different from those at Q_r, and so on.

In (3.1), we have used the convention that ρ_1^i and ρ_2^i, written without further elaboration as to the point at which they are measured, are the principal radii of curvature of the incident wavefront at its reference position Q_s. We have seen in Chapter 2 that this information must be quite clear if we are to predict correctly the fields along any ray tube. Later in this chapter, we often will want to know the principal radii of curvature of the incident wavefront at Q_r, and we will write these as $\rho_1^i(Q_r)$ and $\rho_2^i(Q_r)$. In later chapters, the values $\rho_1^i(Q_e)$ and $\rho_2^i(Q_e)$ at some edge diffraction point Q_e will be required, and so on. As we become more comfortable with the GO and GTD, we will find that we are able to become less fussy about such matters and we will instinctively know at which point the radii of curvature are required for application in certain formulas. Then, such points may not need to be indicated explicitly.

Next, we must determine the reflected GO field $\mathbf{E}^r(s^r)$ at some distance s^r from the point Q_r (e.g., at observation point P in Figure 3.1). The conventional approach would be to determine first the induced current distribution on the surface of the object and then to evaluate the scattered field due to these currents by some radiation integral. At sufficiently high frequencies (i.e., when the frequencies of operation are high enough so that the scatterer is relatively large in terms of the wavelength) these integrals can be evaluated asymptotically (see Appendix D and Sections 3.3.10 and 3.3.11), and it transpires that the contribution to the field at P comes to be chiefly from the currents in the neighborhood of Q_r, the *stationary phase point* that is the reflection or specular point of GO. Currents in fact are induced over the entire surface of the scattering object, but those currents in the vicinity of the specular point Q_r provide the chief contributions to the scattered field. Equivalently, we can say that the currents in the first few Fresnel zones about Q_r furnish the dominant contribution to the scattered field at the particular observation point [1].

Here, however, we wish to determine the reflected GO field at P without having to languish through the process of finding the induced currents on the entire surface of each new scatterer. When the GO reflected field leaves the reflecting surface, it again will be an astigmatic ray tube traveling through free space, as indicated in Figure 3.1. Suppose that the point Q_r is selected as the reference for the reflected ray tube. If the reflected field $\mathbf{E}^r(Q_r)$ at Q_r is known, then

$$\mathbf{E}^r(P) = \mathbf{E}^r(Q_r)\sqrt{\rho_1^r\rho_2^r/(\rho_1^r + s^r)(\rho_2^r + s^r)}\, e^{-jks^r} \tag{3.2}$$

where ρ_1^r and ρ_2^r are the principal radii of curvature of the reflected ray tube at Q_r, and s^r is the distance from Q_r to the observation point P.

Naturally, before we can apply (3.2) in a complete quantitative analysis, the quantities $\mathbf{E}^r(Q_r)$, ρ_1^r, and ρ_2^r must be known. To achieve this, we will relate $\mathbf{E}^r(Q_r)$ to the known $\mathbf{E}^i(Q_r)$ through a reflection coefficient $\mathbf{R}$ in the form

$$\mathbf{E}^r(Q_r) = \mathbf{E}^i(Q_r) \cdot \mathbf{R}$$

The coefficient $\mathbf{R}$ will have to be a dyadic quantity, in general, because $\mathbf{E}^r(Q_r)$ and $\mathbf{E}^i(Q_r)$ are both vector quantities. The form of $\mathbf{R}$ is determined through a solution based on the use of the Luneberg-Kline series and the electromagnetic boundary conditions at the surface of a conductor. Second, we need to obtain expressions for determining ρ_1^r and ρ_2^r, given the properties at Q_r of the incident wavefront and the reflecting surface. Finally, with P and Q_s specified, we must find some expedient method to locate the reflection point Q_r. We will see later in this chapter that one method positions Q_r so that the optical path length $k(Q_sQ_r + Q_rP)$ is extremized (i.e., either a maximum or a minimum, depending on the attributes of the reflecting surface). This of course is a statement of a generalized Fermat principle mentioned in Section 2.2.9. With Q_r known, so is $\hat{\mathbf{s}}^r$, and hence the distance s^r.

Note that, although ray tube diagrams such as Figure 3.1 are necessary for performing derivations of mathematical formulas for quantitative information on the GO field components, such diagrams very easily become excessively cluttered in applications. Thus, single ray representations showing only the central rays often are used. The single ray representation of the situation represented in Figure 3.1 is shown in Figure 3.2.

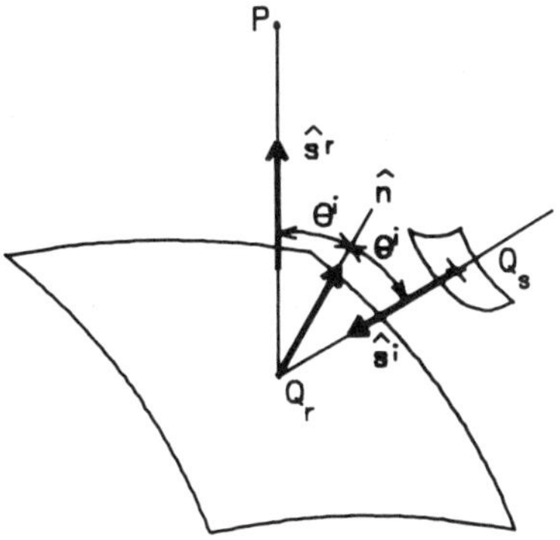

Figure 3.2 Single ray representation of the reflection mechanism in Figure 3.1.

There are several possible ways of implementing this strategy. Here we combine fundamental electromagnetic theory with the Luneberg-Kline expansion to obtain the *law of reflection*, which provides the required information on the trajectories of reflected rays, as well as the polarization and phase properties of the fields transported along reflected rays. These results will be valid for general 3D geometries, of which 2D ones are a special case. To complete the expressions for the reflected GO fields we need to quantify the amplitude continuation along reflected GO ray tubes. This is done first for problems of a 2D nature, and the results are illustrated by a number of examples that afford us increased familiarity with the essential ideas. The procedure leads logically to a discussion of the full GO reflected field expressions for 3D geometries and their application.

At the beginning of the next section we will define certain coordinate systems used for the reflection problem. In view of the utmost importance of securing an understanding of the geometry, we suggest that the reader dwell on these issues until he or she is able to reproduce the essential definitions and coordinate relations without the slightest hesitation.

3.2 THE LAW OF REFLECTION, POLARIZATION PROPERTIES, AND PHASE FUNCTIONS

3.2.1 The Definition of Certain Geometrical Terms and Coordinate Systems

Surface-Fixed Coordinate System

Consider a smooth portion of some general 3D surface such as shown in Figure 3.3, the differential geometry of which is discussed in Appendix C. At each point Q_r on the surface we can define a set of principal planes, a pair of principal directions, and a unit normal vector $\hat{\mathbf{n}}$. The line identified by the A in Figure 3.3, for example, is the line of intersection between the principal plane containing the unit principal vector $\hat{\mathbf{U}}_1$ and $\hat{\mathbf{n}}$ at Q_r, and the surface. The radii of curvature of the

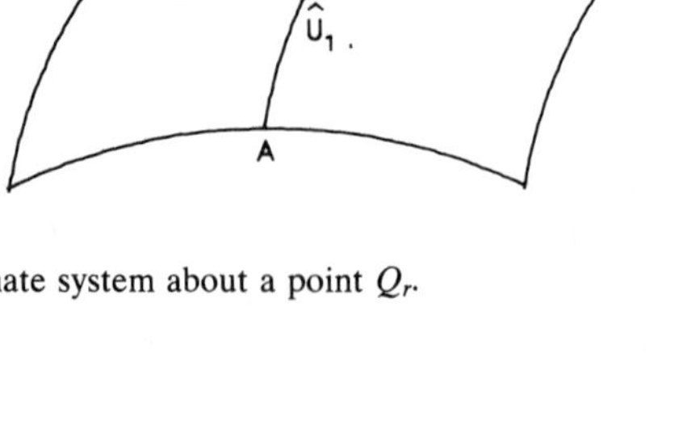

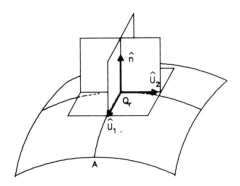

Figure 3.3 Surface-fixed coordinate system about a point Q_r.

surface in these principal planes at Q_r are referred to as the *principal radii of curvature* of the surface at Q_r, denoted a_1 and a_2, and will depend on the particular point on the surface at which they are measured. The corresponding unit principal directions or vectors of the surface at Q_r (each lying in a principal plane) are indicated in Figure 3.3 by the symbols $\hat{\mathbf{U}}_1$ and $\hat{\mathbf{U}}_2$. The principal planes at Q_r are defined unambiguously by the unit normal to the surface at Q_r, and these are the principal direction vectors of the surface at Q_r. We repeatedly tagged the qualifier "at Q_r" to emphasize that it makes no sense to speak of such surface properties without stating precisely to which point on the surface they refer.

Contemplate the central ray of some ray tube incident on the surface at Q_r, with the unit vector in its direction of propagation denoted by $\hat{\mathbf{s}}^i$. The plane containing unit vectors $\hat{\mathbf{n}}$ and $\hat{\mathbf{s}}^i$ is called the *plane of incidence* of the ray at Q_r. A ray incident at Q_r from a different direction will have a different plane of incidence. The angle of incidence θ^i is measured in the plane of incidence, and is given by

$$\theta^i = -\cos^{-1}(\hat{\mathbf{n}} \cdot \hat{\mathbf{s}}^i)$$

The notation of Kouyoumjian and Pathak [2], which by now is standard, is used in what follows. The principal directions of the incident wavefront at Q_r are denoted by $\hat{\mathbf{X}}_1^i$ and $\hat{\mathbf{X}}_2^i$, with associated principal radii of curvature ρ_1^i and ρ_2^i, respectively, at the point Q_r. In other words, ρ_1^i and ρ_2^i are the radii of curvature at Q_r of the incident wavefront in its principal planes defined by the vector pairs $(\hat{\mathbf{n}}, \hat{\mathbf{X}}_1^i)$ and $(\hat{\mathbf{n}}, \hat{\mathbf{X}}_2^i)$, respectively. Note that the plane of incidence of a ray in general may be different from the principal planes of the reflecting surface at the point of reflection. Moreover, the principal directions of the incident wavefront generally are distinct from those of the reflecting surface.

The unit tangent vector $\hat{\mathbf{t}}$ is selected to lie in the plane of incidence at Q_r, as shown in Figure 3.4. There is an angle α between the plane of incidence $(\hat{\mathbf{n}}, \hat{\mathbf{s}}^i)$ and the principal plane of the surface $(\hat{\mathbf{n}}, \hat{\mathbf{U}}_2^i)$ at Q_r, so that

$$\cos\alpha = \hat{\mathbf{U}}_2^i \cdot \hat{\mathbf{t}}$$

Hence, the other tangent vector is

$$\hat{\mathbf{b}} = \hat{\mathbf{t}} \times \hat{\mathbf{n}} = -\hat{\mathbf{n}} \times \hat{\mathbf{s}}^i$$

The reader must realize that $\hat{\mathbf{X}}_1^i$ and $\hat{\mathbf{X}}_2^i$ generally are not related to the polarization of the incident GO field. That is, they do not describe the orientations of the GO fields along the incident ray tube.

Consider Figure 3.5. The unit vector $\hat{\mathbf{s}}^r$ indicates the direction of the reflected ray at Q_r. The vector pair $(\hat{\mathbf{n}}, \hat{\mathbf{s}}^r)$ defines the *plane of reflection* at Q_r. The principal

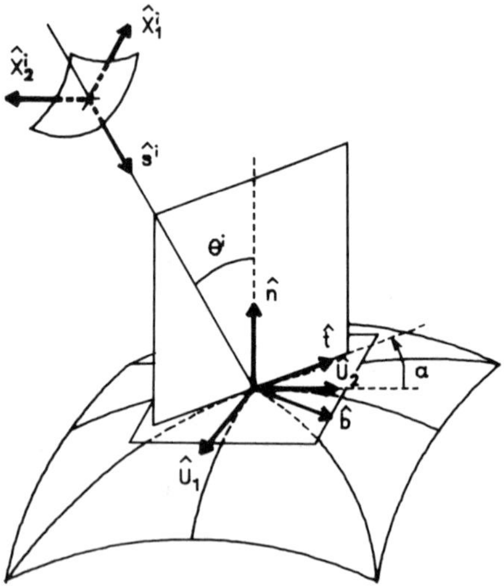

Figure 3.4 Incident ray at a point on a surface. The plane of incidence, the principal directions of the incident field, and the unit tangent vectors are defined.

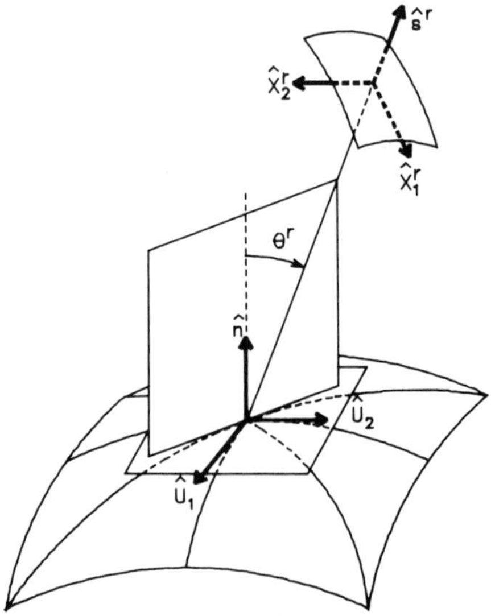

Figure 3.5 Reflected ray at a point on a surface. The plane of reflection and the principal directions of the reflected ray are defined.

directions of the reflected wavefront at Q_r are denoted by $\hat{\mathbf{X}}_1^r$ and $\hat{\mathbf{X}}_2^r$; ρ_1^r and ρ_2^r are the radii of curvature at Q_r of the reflected wavefront in its principal planes $(\hat{\mathbf{n}}, \hat{\mathbf{X}}_1^r)$ and $(\hat{\mathbf{n}}, \hat{\mathbf{X}}_2^r)$, respectively. Note that $\hat{\mathbf{X}}_1^r$ and $\hat{\mathbf{X}}_2^r$ in general are not related to the polarization of the reflected GO fields. Once more, note that the principal planes of the reflected wavefront, the principal planes of the surface, and the plane of reflection, at Q_r generally will not be coincident.

Ray-Fixed Coordinate System

We next wish to select some way of describing the polarization of the incident and reflected GO fields. Consider therefore the situation shown in Figure 3.6. The unit vector $\hat{\mathbf{s}}^i$ is the direction vector of the particular GO ray of the incident field that strikes the reflecting surface at the specific point Q_r on the surface. At that same point, Q_r, let $\hat{\mathbf{n}}$ be the unit normal to the surface, as earlier, with the unit vectors $\hat{\mathbf{s}}^i$ and $\hat{\mathbf{n}}$ thus defining the plane of incidence. The incident electric field $\mathbf{E}^i$ at Q_r always can be resolved into components parallel $(\hat{\mathbf{e}}_\parallel^i)$ and perpendicular $(\hat{\mathbf{e}}_\perp^i)$ to the plane of incidence. Note that $\hat{\mathbf{e}}_\parallel^i$ and $\hat{\mathbf{e}}_\perp^i$ in general are different from $\hat{\mathbf{X}}_1^i$ and $\hat{\mathbf{X}}_2^i$ defined earlier. The reflected electric field $\mathbf{E}^r$ at Q_r similarly can be resolved into components parallel $(\hat{\mathbf{e}}_\parallel^r)$ and perpendicular $(\hat{\mathbf{e}}_\perp^r)$ to the plane of reflection. From purely geometrical considerations, we can show that the following relations hold true [3]:

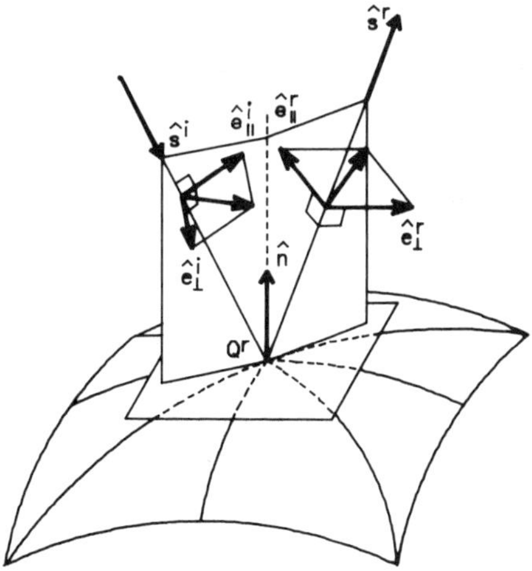

Figure 3.6 Ray-fixed coordinate system used to define the polarization of the incident field in the GO-GTD schema.

$$\hat{\mathbf{e}}^i_\perp \times \hat{\mathbf{s}}^i = \hat{\mathbf{e}}^i_\parallel \tag{3.3}$$

$$\hat{\mathbf{e}}^r_\perp \times \hat{\mathbf{s}}^r = \hat{\mathbf{e}}^r_\parallel \tag{3.4}$$

$$\hat{\mathbf{e}}^i_\parallel = \frac{\hat{\mathbf{s}}^i \times (\hat{\mathbf{n}} \times \hat{\mathbf{s}}^i)}{|\hat{\mathbf{s}}^i \times (\hat{\mathbf{n}} \times \hat{\mathbf{s}}^i)|} \tag{3.5}$$

$$\hat{\mathbf{e}}^r_\parallel = \frac{\hat{\mathbf{s}}^r \times (\hat{\mathbf{n}} \times \hat{\mathbf{s}}^r)}{|\hat{\mathbf{s}}^r \times (\hat{\mathbf{n}} \times \hat{\mathbf{s}}^r)|} \tag{3.6}$$

In addition, given $\hat{\mathbf{s}}^i$ and $\hat{\mathbf{s}}^r$, we can write

$$\hat{\mathbf{n}} \times \hat{\mathbf{e}}^i_\parallel = (\hat{\mathbf{n}} \cdot \hat{\mathbf{s}}^i)\hat{\mathbf{e}}_\perp \tag{3.7}$$

$$\hat{\mathbf{n}} \times \hat{\mathbf{e}}^r_\parallel = (\hat{\mathbf{n}} \cdot \hat{\mathbf{s}}^r)\hat{\mathbf{e}}_\perp \tag{3.8}$$

This system is a *local ray-fixed coordinate system* because it is inextricably linked to the ray directions $\hat{\mathbf{s}}^i$ and $\hat{\mathbf{s}}^r$, and the unit normal $\hat{\mathbf{n}}$ at each Q_r. These results from differential geometry will be usefully employed in the remainder of this text. In Chapter 4 an additional geometrical concept, that of the edge-fixed coordinate system, will be introduced.

Note: We have specified the principal direction vectors and polarization vectors of the incident and reflected GO fields as being measured at Q_r. Although this is indeed so, it is worth mentioning that, unlike the radii of curvature of the associated wavefronts, these quantities will be independent of position for a given ray. But this is so only because the media through which all rays propagate before and after reflection in this text are assumed homogeneous.

3.2.2 The Law of Reflection

The law of reflection provides information on the relationship between $\hat{\mathbf{n}}$, $\hat{\mathbf{s}}^i$, and $\hat{\mathbf{s}}^r$, by appealing to fundamental electromagnetic theory. Consider therefore a known field $\mathbf{E}_i(\mathbf{r})$[1] to be incident on a smooth surface *S*, as shown in Figure 3.7 (a), the object being to establish the details of the reflected field $\mathbf{E}_r(\mathbf{r})$. We know from basic electromagnetic theory that the total electric field $\mathbf{E}(\mathbf{r})$ must satisfy the boundary condition

$$\hat{\mathbf{n}} \times \mathbf{E}(\mathbf{r}) = 0$$

[1]The Luneberg-Kline expansion is used here, so once again we must distinguish the GO term by the 0 subscript. This $\mathbf{E}_i(\mathbf{r})$ denotes the conventional incident electric field. A flag will be raised at the end of the derivation to indicate when consideration of only the GO term is resumed and the zero subscript is dropped once more.

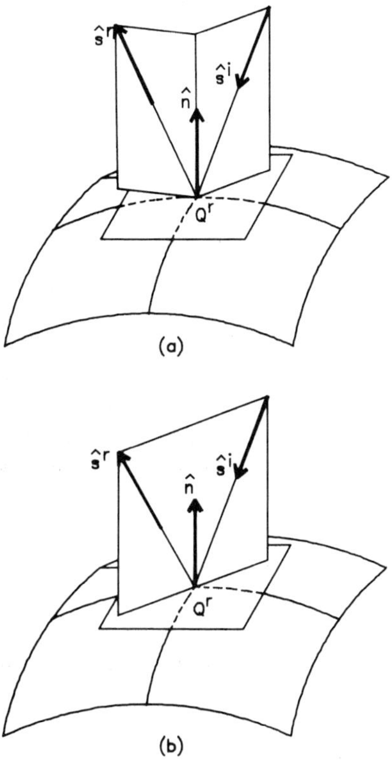

Figure 3.7 The planes of incidence and reflection are depicted in (a) to be different, but as in (b), in fact are coincident.

at all points **r** on the conducting surface. In this case, the total field at all points **r** on S (be sure to appreciate the emphasis "on S") is the sum of the incident and reflected fields

$$\mathbf{E}(\mathbf{r}) = \mathbf{E}_i(\mathbf{r}) + \mathbf{E}_r(\mathbf{r})$$

The use of *subscripts* i and r indicate that these represent not the incident and reflected GO fields but the conventional ones discussed in Section 2.8; hence, we have the requirement:

$$\hat{\mathbf{n}} \times [\mathbf{E}_i(\mathbf{r}) + \mathbf{E}_r(\mathbf{r})] = 0 \tag{3.9}$$

for all points **r** on S. Adopting the Luneberg-Kline asymptotic expansion (as an ansatz) for the impinging and reflected ray-optic (but not purely GO) fields, we have for any point **r** (not necessarily on S):

$$\mathbf{E}_i(\mathbf{r}, \omega) \sim e^{-jk\Psi_i(\mathbf{r})} \sum_{n=0}^{\infty} \frac{\mathbf{E}_n^i(\mathbf{r})}{(j\omega)^n} \tag{3.10}$$

$$\mathbf{E}_r(\mathbf{r}, \omega) \sim e^{-jk\Psi_r(\mathbf{r})} \sum_{n=0}^{\infty} \frac{\mathbf{E}_n^r(\mathbf{r})}{(j\omega)^n} \tag{3.11}$$

Recall that the $n = 0$ terms are the GO fields. Note that the phase functions (eikonals) in the two expansions are not the same functions at all points in space. Restricting **r** to points on the surface once more, if (3.10) and (3.11) are substituted into boundary condition (3.9), and like terms in $(j\omega)$ equated in a manner similar to that done in Section 2.2.4, we obtain the equation (which applies only at points **r** on S, of course):

$$\hat{\mathbf{n}} \times \mathbf{E}_n^i(\mathbf{r})\, e^{-jk\Psi_i(\mathbf{r})} = -\hat{\mathbf{n}} \times \mathbf{E}_n^r(\mathbf{r})\, e^{-jk\Psi_r(\mathbf{r})}, \quad \text{for each } n = 0, 1, 2, \ldots \tag{3.12}$$

Because (3.12) must be satisfied for all n, we can conclude that (i.e., we may phase match):

$$\Psi_i(\mathbf{r}) = \Psi_r(\mathbf{r}), \quad \text{for all } \mathbf{r} \text{ on } S \tag{3.13}$$

and therefore also that

$$\hat{\mathbf{n}} \times \mathbf{E}_n^i(\mathbf{r}) = -\hat{\mathbf{n}} \times \mathbf{E}_n^r(\mathbf{r}), \quad \text{for all } \mathbf{r} \text{ on } S \tag{3.14}$$

Furthermore, the fact that (3.13) holds true at *all* points *on the surface* guarantees that the derivatives of $\Psi_i(\mathbf{r})$ and $\Psi_r(\mathbf{r})$ in any direction *along* the smooth surface (but *only* along the surface, not normal to it) also must be equal. In particular, this must be the case at the point Q_r. So if $\hat{\mathbf{t}}$ is a unit vector tangent to the surface at Q_r we may write

$$\nabla\Psi_i(Q_r) \cdot \hat{\mathbf{t}}(Q_r) = \nabla\Psi_r(Q_r) \cdot \hat{\mathbf{t}}(Q_r) \tag{3.15}$$

From Section 2.2.6 we recall that the gradient of the phase function of a ray-optic field is a unit vector that coincides with the direction of propagation of its GO part. Consequently we may write $\nabla\Psi_i(Q_r) = \hat{\mathbf{s}}^i(Q_r)$ and $\nabla\Psi_r(Q_r) = \hat{\mathbf{s}}^r(Q_r)$, so that (3.15) can be presented in the more revealing form:

$$\hat{\mathbf{s}}^i \cdot \hat{\mathbf{t}}(Q_r) = \hat{\mathbf{s}}^r \cdot \hat{\mathbf{t}}(Q_r) \tag{3.16}$$

This implies that $\hat{\mathbf{s}}^i$ and $\hat{\mathbf{s}}^r$ have the same projection on the plane tangent to the surface S at the reflection point Q_r. Recall that, although $\hat{\mathbf{t}}$ can be any unit tangent at Q_r, we will select it to be the one that lies in the plane of incidence at Q_r. By inspection of Figure 3.7(a) we can deduce from geometrical considerations that (3.16) at once implies that

$$\hat{\mathbf{n}} \cdot \hat{\mathbf{s}}^i = -\hat{\mathbf{n}} \cdot \hat{\mathbf{s}}^r \tag{3.17}$$

This means that, in terms of Figure 3.7:

$$-\hat{\mathbf{n}}(Q_r) \cdot \hat{\mathbf{s}}^i(Q_r) = \cos\theta^i = \cos\theta^r$$

and accordingly that

$$\theta^i = \theta^r \tag{3.18}$$

which is recognized as the well-known rule from elementary physics that the angle of incidence equals the angle of reflection. This also is sometimes referred to as Snell's law for reflection. Moreover, because the cross product of the surface unit normal $\hat{\mathbf{n}}$ with any vector will be tangent to the surface, (3.12) in fact also concedes the result:

$$\hat{\mathbf{n}} \times \hat{\mathbf{s}}^i = \hat{\mathbf{n}} \times \hat{\mathbf{s}}^r \tag{3.19}$$

Expression (3.19), together with relation (3.17) is a statement that $\hat{\mathbf{n}}$, $\hat{\mathbf{s}}^i$, and $\hat{\mathbf{s}}^r$ are coplanar. Thus, the plane of incidence and plane of reflection are the same at any point of reflection Q_r. Therefore, Figure 3.7(a) *must* be altered to the situation depicted in Figure 3.7(b); the same applies to Figure 3.6, making $\hat{\mathbf{e}}^i_\perp = \hat{\mathbf{e}}^r_\perp$, as will be seen in Section 3.2.4.

Expression (3.17), or its equivalent (3.18), along with the coplanarity condition on the planes of incidence and reflection that follows from (3.19) will be referred to as the *law of reflection*, obtained by satisfying the boundary conditions exactly and Maxwell's equations approximately [3].

Note: In the remainder of this chapter it is once more only the GO fields with which we will be concerned. For notational convenience the subscript 0 is abandoned for the second time, and we simply write $\mathbf{E}^i$ and $\mathbf{E}^r$ for the GO (*zero*th-order terms of (3.10) and (3.11) only) incident and reflected fields.

3.2.3 Trajectories of Reflected Rays

The vector diagram in Figure 3.8 shows unit vectors $\hat{\mathbf{s}}^i$ and $\hat{\mathbf{s}}^r$ resolved into components normal to the surface at Q_r (namely, $\hat{\mathbf{n}} \cdot \hat{\mathbf{s}}^i$ and $\hat{\mathbf{n}} \cdot \hat{\mathbf{s}}^r$) and tangential to the surface (namely $\hat{\mathbf{t}} \cdot \hat{\mathbf{s}}^i$ and $\hat{\mathbf{t}} \cdot \hat{\mathbf{s}}^r$), where a vector $\hat{\mathbf{t}}$ tangential to the surface at Q_r has been selected to lie in the plane of incidence. In symbolic form:

$$\hat{\mathbf{s}}^i = (\hat{\mathbf{n}} \cdot \hat{\mathbf{s}}^i)\hat{\mathbf{n}} + (\hat{\mathbf{t}} \cdot \hat{\mathbf{s}}^i)\hat{\mathbf{t}} \tag{3.20}$$

$$\hat{\mathbf{s}}^r = (\hat{\mathbf{n}} \cdot \hat{\mathbf{s}}^r)\hat{\mathbf{n}} + (\hat{\mathbf{t}} \cdot \hat{\mathbf{s}}^r)\hat{\mathbf{t}} \tag{3.21}$$

From (3.17) we know that $\hat{\mathbf{n}} \cdot \hat{\mathbf{s}}^i = -\hat{\mathbf{n}} \cdot \hat{\mathbf{s}}^r$, and from inspection of the geometry in Figure 3.8, bearing in mind that all the vectors are unit vectors, the following is obvious:

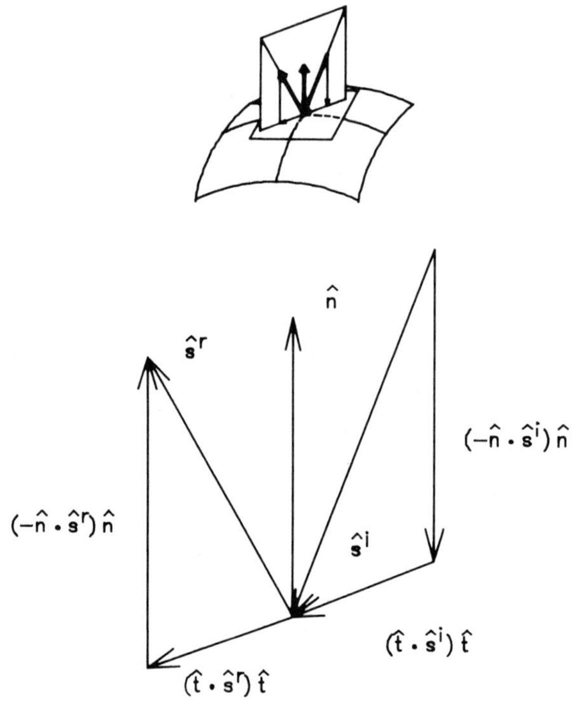

Figure 3.8 Tangential and normal projections of $\hat{\mathbf{s}}^i$ and $\hat{\mathbf{s}}^r$.

$$(\hat{\mathbf{t}} \cdot \hat{\mathbf{s}}^i)\hat{\mathbf{t}} = (\hat{\mathbf{t}} \cdot \hat{\mathbf{s}}^r)\hat{\mathbf{t}}$$

Substitution of these two equalities into (3.21) yields the relation:

$$\hat{\mathbf{s}}^r = -(\hat{\mathbf{n}} \cdot \hat{\mathbf{s}}^i)\hat{\mathbf{n}} + (\hat{\mathbf{t}} \cdot \hat{\mathbf{s}}^i)\hat{\mathbf{t}} \tag{3.22}$$

However, rearrangement of (3.20) yields

$$(\hat{\mathbf{t}} \cdot \hat{\mathbf{s}}^i)\hat{\mathbf{t}} = \hat{\mathbf{s}}^i - (\hat{\mathbf{n}} \cdot \hat{\mathbf{s}}^i)\hat{\mathbf{n}}$$

which, when substituted into (3.22) once more, provides us with the final result:

$$\hat{\mathbf{s}}^r = \hat{\mathbf{s}}^i - 2(\hat{\mathbf{n}} \cdot \hat{\mathbf{s}}^i)\hat{\mathbf{n}} \tag{3.23}$$

Given the direction of propagation of the incident GO field and the unit normal to the surface at the point of reflection Q_r (for the time being assumed known, but dealt with in Section 3.3.6), relation (3.23) allows us to find the direction in which the reflected GO ray will travel. Once this reflected ray has left the surface it once more is in a homogeneous medium and must obey the requirement of Section 2.2.7 that its trajectory be a straight line.

3.2.4 Polarization of Reflected Rays

In this section, the reflected GO field $\mathbf{E}^r(Q_r)$ at the point of reflection is determined in the form $\mathbf{E}^r(Q_r) = \mathbf{E}^i(Q_r) \cdot \mathbf{R}$, where $\mathbf{R}$ is the dyadic reflection coefficient. A certain amount of detail is included, because we use this opportunity to become quite familiar with the dyadic notation used liberally in later chapters on diffraction.

The ray-fixed coordinate system introduced in Section 3.2.1 is convenient and conventional to use, the reason for which will soon become clear. With the knowledge that we now have regarding the coincidence of the planes of incidence and reflection, we can immediately deduce that

$$\hat{\mathbf{e}}^i_\perp = \hat{\mathbf{e}}^r_\perp \tag{3.24}$$

and henceforth simply write them both as $\hat{\mathbf{e}}_\perp$. However, although unit vectors $\hat{\mathbf{e}}^i_\parallel$ and $\hat{\mathbf{e}}^r_\parallel$ now also lie in the same plane, note in general they are not equal. To stress this fact (which is quite easily forgotten by users of the GTD) let us give this inequality a number as well; namely,

$$\hat{\mathbf{e}}^i_\parallel \neq \hat{\mathbf{e}}^r_\parallel \tag{3.25}$$

We may resolve the incident GO field into components:

$$\mathbf{E}^i = (\mathbf{E}^i \cdot \hat{\mathbf{e}}_{\parallel}^i)\hat{\mathbf{e}}_{\parallel}^i + (\mathbf{E}^i \cdot \hat{\mathbf{e}}_{\perp})\hat{\mathbf{e}}_{\perp} = \mathbf{E}_{\parallel}^i\hat{\mathbf{e}}_{\parallel}^i + \mathbf{E}_{\perp}^i\hat{\mathbf{e}}_{\perp} \tag{3.26}$$

at the reflection point Q_r on the surface. Similarly, for the reflected field *at* Q_r we have

$$\mathbf{E}^r = (\mathbf{E}^r \cdot \hat{\mathbf{e}}_{\parallel}^r)\hat{\mathbf{e}}_{\parallel}^r + (\mathbf{E}^r \cdot \hat{\mathbf{e}}_{\perp})\hat{\mathbf{e}}_{\perp} = \mathbf{E}_{\parallel}^r\hat{\mathbf{e}}_{\parallel}^r + \mathbf{E}_{\perp}^r\hat{\mathbf{e}}_{\perp} \tag{3.27}$$

In familiar matrix notation:

$$E^i = \begin{bmatrix} E_{\parallel}^i \\ E_{\perp}^i \end{bmatrix} \tag{3.28}$$

$$E^r = \begin{bmatrix} E_{\parallel}^r \\ E_{\perp}^r \end{bmatrix} \tag{3.29}$$

$$[R] = \begin{bmatrix} R_{11} & R_{12} \\ R_{21} & R_{22} \end{bmatrix} \tag{3.30}$$

In this matrix format, the dyadic relation:

$$\mathbf{E}^r(Q_r) = \mathbf{E}^i(Q_r) \cdot \mathbf{R}$$

implies that

$$\mathbf{E}_{\parallel}^r = R_{11}E_{\parallel}^i + R_{21}E_{\perp}^i \tag{3.31}$$

$$\mathbf{E}_{\perp}^r = R_{21}E_{\parallel}^i + R_{22}E_{\perp}^i \tag{3.32}$$

The result (3.14), which was obtained through application of the boundary conditions at the surface, requires of the GO field that

$$\hat{\mathbf{n}} \times \mathbf{E}^i(Q_r) = -\hat{\mathbf{n}} \times \mathbf{E}^r(Q_r) \tag{3.33}$$

Substitution of (3.26) and (3.27) into (3.33) yields, at any reflection point Q_r,

$$(\mathbf{E}^i \cdot \hat{\mathbf{e}}_{\parallel}^i)\hat{\mathbf{n}} \times \hat{\mathbf{e}}_{\parallel}^i + (\mathbf{E}^i \cdot \hat{\mathbf{e}}_{\perp})\hat{\mathbf{n}} \times \hat{\mathbf{e}}_{\perp}$$

$$= -(\mathbf{E}^r \cdot \hat{\mathbf{e}}_{\parallel}^r)\hat{\mathbf{n}} \times \hat{\mathbf{e}}_{\parallel}^r - (\mathbf{E}^r \cdot \hat{\mathbf{e}}_{\perp})\hat{\mathbf{n}} \times \hat{\mathbf{e}}_{\perp} \tag{3.34}$$

and from (3.7), (3.8), and (3.17):

$$(E_\parallel^i - E_\parallel^r)\hat{\mathbf{e}}_\perp + (E_\perp^i + E_\perp^r)(\hat{\mathbf{n}} \times \hat{\mathbf{e}}_\perp) = 0 \tag{3.35}$$

Because $\hat{\mathbf{e}}_\perp$ and $\hat{\mathbf{n}} \times \hat{\mathbf{e}}_\perp$ are orthogonal, we may equate the individual terms to zero independently, from which follows:

$$E_\parallel^r = E_\parallel^i \tag{3.36}$$

$$E_\perp^r = -E_\perp^i \tag{3.37}$$

Comparison of (3.31) with (3.36), and (3.32) with (3.37), admits without difficulty that $R_{11} = 1$, $R_{22} = -1$, and $R_{12} = R_{21} = 0$, so that finally we are able to write

$$[R] = \begin{bmatrix} R_h & 0 \\ 0 & R_s \end{bmatrix} \tag{3.38}$$

with $R_h = 1$ and $R_s = -1$. The notations s and h stand for "soft" and "hard," respectively. This is a carry-over from the field of acoustics, which also uses many of the results demonstrated for electromagnetics problems in this text. In dyadic notation, (3.38) is

$$\mathbf{R} = \hat{\mathbf{e}}_\parallel^i\hat{\mathbf{e}}_\parallel^r - \hat{\mathbf{e}}_\perp\hat{\mathbf{e}}_\perp = R_h\hat{\mathbf{e}}_\parallel^i\hat{\mathbf{e}}_\parallel^r + R_s\hat{\mathbf{e}}_\perp\hat{\mathbf{e}}_\perp \tag{3.39}$$

The dyadic reflection coefficient $\mathbf{R}$ in general would be a 3×3 matrix. The reduction to a 2×2 matrix results from the selection of the local ray-fixed coordinate system. The adoption of an edge-fixed coordinate system in later chapters on edge diffraction similarly will reduce the dyadic diffraction coefficient to a 2×2 matrix.

Note that $R_s = -1$ applies to $E_\perp$, which always is tangential to the reflecting surface; $R_h = 1$ is associated with $E_\parallel$, which is parallel, not to the surface, but to the plane of incidence.

3.2.5 Phase Continuation along Reflected Rays

The point Q_r serves as the reference point for the ray tube reflected at this point. The reflected electric field at Q_r has been determined as $\mathbf{E}_r(Q_r)$. Once the reflected ray tube leaves the surface, it travels through a homogeneous medium once more, and the results from Section 2.2.7 can immediately be used to fix the phase function as

$$e^{-jks^r}$$

where s^r is the distance from the particular specular point Q_r to the observation point.

3.2.6 Invocation of the Locality Principle

At the very beginning of this section we adopted the assumption that the entire surface with which we are dealing is a smooth one. At this stage we invoke the *locality principle* to which we alluded in Section 3.1; namely, that the earlier results apply as long as the surface is sufficiently smooth in the neighborhood of the particular reflection point at which the geometrical arrangement requires the results to be applied.

3.2.7 More about Shadowing

Elementary ideas on shadowing and shadow boundaries were introduced at the end of Chapter 2. There, as in most of classical optics, our concern was with the manner in which an obstacle prevents the rays from some source from directly reaching areas of space called *shadow regions*. These shadow regions were bordered by *incident shadow boundaries* (ISBs). In modern geometrical optics (and hence the GTD framework), where a quantitative capability exists, these concepts are expanded to include a second type of shadow boundary, the *reflection shadow boundary* (RSB).

For purposes of illustration consider the cross-sectional geometry shown in Figure 3.9, which shows the ray representation of a plane wave obliquely incident on a long conducting strip that lies normal to the paper. The ISBs have been located as indicated; they are different from those in Figure 2.17 because of the change in the form of the incident field from a cylindrical to a plane wavefront, even though the obstacle is the same. Here, we wish to determine which regions of space will contain rays *reflected* from the conducting strip. We begin at the bottom of the figure and work our way up through the numbered rays singled out for consideration. (of course, there are infinitely many of them). Ray 1 is not reflected from the strip because it does not strike it at all. Ray 2 is the "first" to impact the conducting strip and is reflected in accordance with the law of reflection. Rays 3 to 5 behave in the same way, but any ray "after" 5 (e.g., ray 6) will not be reflected, as it travels straight past the strip. Thus, only in the region of space *between* the reflected rays labeled *RSB* will there exist a reflected GO field, the latter rays forming the reflection shadow boundaries for this particular obstacle and specific incident wavefront, incident at the given angle with respect to the obstacle.

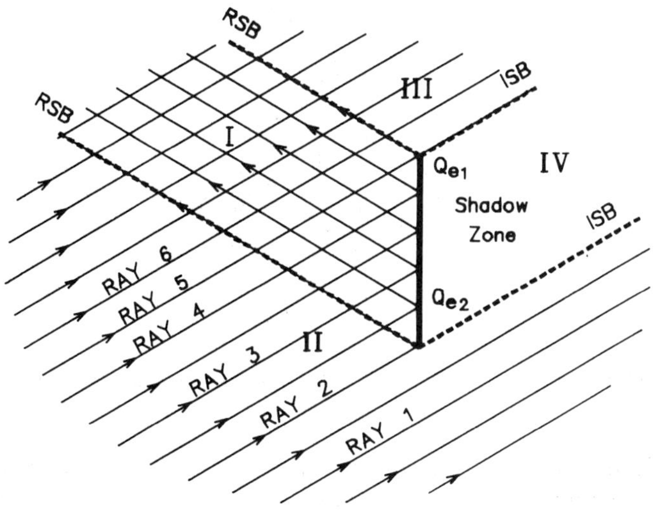

Figure 3.9 Incident and reflection shadow boundaries associated with a strip illuminated by a plane wave.

In summary, there is a reflected GO field in region I; a direct (incident) GO field in regions I, II, and III; and neither direct nor reflected rays in region IV. Just as the incident GO field falls abruptly (stepwise) to zero as the observer crosses in ISB into the shadow region, so, too, is the reflected GO field discontinuous at a RSB. If the angle of incidence of the plane wavefront in Figure 3.9 is altered, of course, there will be an accompanying shift of the SBs.

In Figure 3.10, a conducting strip illuminated by a line source is considered, as has been done in Figure 2.17. The ISBs, of course, are identical to those indicated in Figure 2.17 but the RSBs now have been added. There are direct rays in regions I, II, and III, and reflected rays in region I, but neither direct nor reflected rays in region IV. In other words, given the source location and obstacle position, if the observation point is outside region I there will be no point Q_r on the obstacle from which a ray is reflected to the observer. If we keep this source but change its position relative to the obstacle, the SBs once more will be altered.

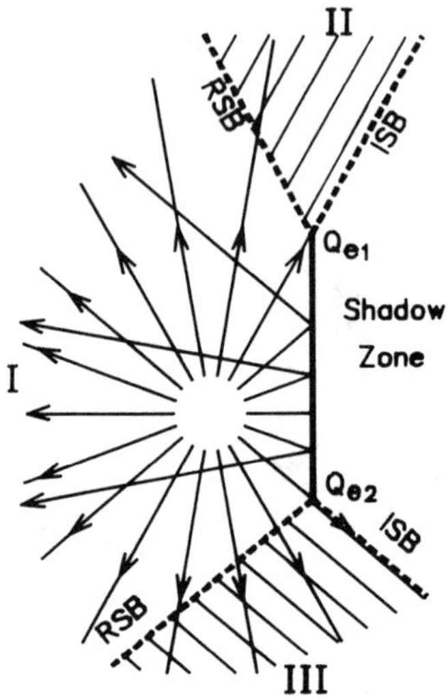

Figure 3.10 Incident and reflection shadow boundaries associated with a strip illuminated by a line source.

As a final illustration in this section (because the correct location of shadow boundaries is important) take the case of a point source near an aperture in an otherwise semi-infinite conducting half-plane, as depicted in Figure 3.11(a). The ISB and RSBs are shown in Figure 3.11(b). Regions I and III contain reflected rays. There are direct rays in regions I, II, III, IV, and VI. Regions V and VII have neither direct nor reflected rays.

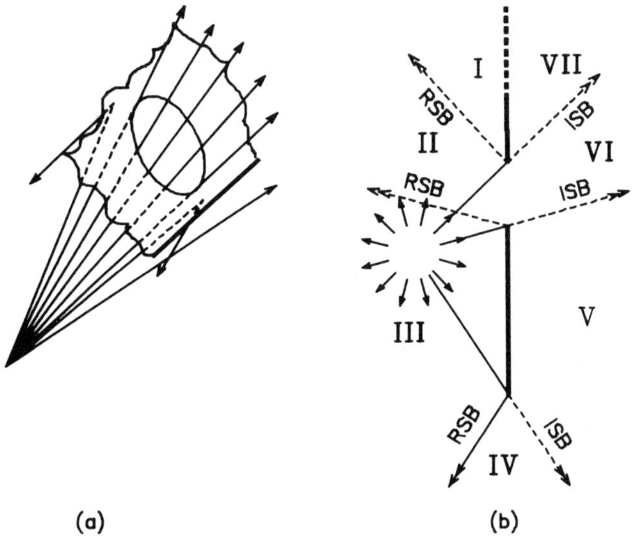

Figure 3.11 (a) Aperture in half-plane, illuminated by a plane wave. (b) Cross section through the geometry of (a), showing the associated incident and reflection shadow boundaries.

3.2.8 Geometrical Optics Surface Currents

The surface current density $\mathbf{J}_s$ at any point Q_r on a surface is given by

$$\mathbf{J}_s(Q_r) = \hat{\mathbf{n}} \times \mathbf{H}(Q_r) \tag{3.40}$$

where $\mathbf{H}(Q_r)$ is the total magnetic field at Q_r. In this instance we are interested in the surface current $\mathbf{J}_{go}$ due solely to the GO field terms. We know that the total geometrical optics magnetic field at points Q_r may be written as

$$\mathbf{H}(Q_r) = \mathbf{H}^i(Q_r) + \mathbf{H}^r(Q_r) \tag{3.41}$$

and therefore we have

$$\mathbf{J}_{go}(Q_r) = \hat{\mathbf{n}} \times [\mathbf{H}^i(Q_r) + \mathbf{H}^r(Q_r)] \tag{3.42}$$

Recall from Section 2.2.4 that GO electric and magnetic fields are related as

$$\mathbf{E}^i = -Z\,\hat{\mathbf{s}}^i \times \mathbf{H}^i \tag{3.43}$$

$$\mathbf{E}^r = -Z\,\hat{\mathbf{s}}^r \times \mathbf{H}^r \tag{3.44}$$

Substitution of (3.43) and (3.44) into (3.33) gives

$$\hat{\mathbf{n}} \times (\hat{\mathbf{s}}^i \times \mathbf{H}^i) = -\hat{\mathbf{n}} \times (\hat{\mathbf{s}}^r \times \mathbf{H}^r) \tag{3.45}$$

Furthermore, use of Luneberg-Kline expansions similar to (3.10) and (3.11), but for the magnetic field, in the boundary condition $\hat{\mathbf{n}} \cdot \mathbf{H} = 0$, which applies at all points on the reflecting surface, along with arguments along the same lines as those used to obtain (3.14), gives the result

$$\hat{\mathbf{n}} \cdot \mathbf{H}^i(\mathbf{r}) = -\hat{\mathbf{n}} \cdot \mathbf{H}^r(\mathbf{r}), \quad \text{for all } \mathbf{r} \text{ on } S \tag{3.46}$$

Use of the general vector identity:

$$\mathbf{A} \times (\mathbf{B} \times \mathbf{C}) = \mathbf{B}(\mathbf{A} \cdot \mathbf{C}) - \mathbf{C}(\mathbf{A} \cdot \mathbf{B}) \tag{3.47}$$

on both the left- and right-hand sides of (3.45) and the subsequent application of (3.46), (3.17), and (3.19) gives the result:

$$\hat{\mathbf{n}} \times \mathbf{H}^i = \hat{\mathbf{n}} \times \mathbf{H}^r, \quad \text{for all } \mathbf{r} \text{ on } S \tag{3.48}$$

Note that (3.48) does not hold in general, but only for fields of the GO type. Upon substituting (3.48) into (3.42), we obtain

$$\mathbf{J}_{go}(Q_r) = 2\hat{\mathbf{n}} \times \mathbf{H}(Q_r) \tag{3.49}$$

$\mathbf{H}^i$ is a GO field, and in the directly illuminated region of the surface, there is a current given by (3.49). On the section of the surface facing into the shadow region, this current density is zero, the field $\mathbf{H}^i$ being zero there, as illustrated in Figure 3.12.

3.2.9 An Alternative Interpretation of the Form of R and the Law of Reflection

Equation (3.49) looks suspiciously like the induced current on an infinitely large perfectly conducting flat plate, when a true (globally) plane wave is incident on it. If we were to solve such a problem, for a true plane wave incident in direction $\hat{\mathbf{s}}^i$, then we would discover that the reflected field also is a true plane wave, propagating in direction $\hat{\mathbf{s}}^r$ given by (3.23). The law of reflection is satisfied, with the reflection coefficient identical to (3.39), from which the initial form (polarization and phase) of the reflected true plane wave at the surface could be determined. We could then argue as follows:

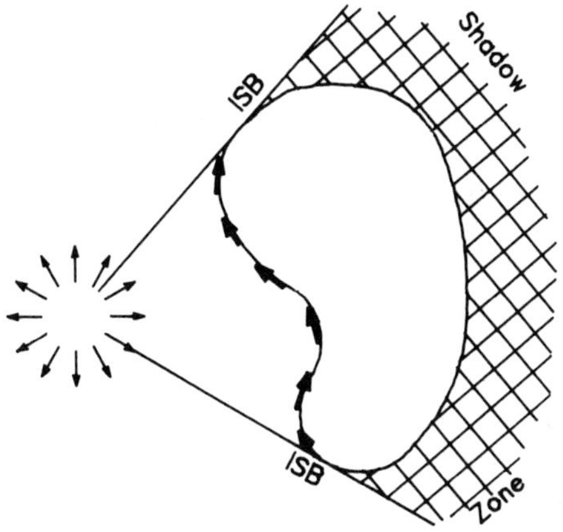

Figure 3.12 Geometrical optics surface currents.

> Because a general GO field is locally plane (is a TEM field), if it impinges on some curved surface that has a sufficiently large curvature at the point of reflection that it can be locally approximated by an infinite plane surface tangent to the curved surface there, then the results derived for **R** and $\hat{\mathbf{s}}^r$ in the preceding *canonical problem* can be assumed to hold for a GO field incident on the general curved surface.

Indeed, this indirect approach would be an equally valid approach to that which in Section 3.2.2 used the Luneberg-Kline expansion.

In summary, the initial field value is determined through solution of some canonical problem, in which the incident locally plane GO field is replaced by a true plane wave field and the scattering object by a simpler geometry that retains the correct local scattering properties. In the case discussed in this section, the simpler geometry was a flat plane; in Section 8.2.2, the results for **R** obtained via a slightly more sophisticated canonical geometry will be discussed. This will enable us, among other things, to quantify what is "sufficiently large curvature."

3.2.10 What More Do We Need?

Though we have inferred much about the nature of the reflected GO field our knowledge still is incomplete. Given the properties of the incident ray tube and those of the surface at Q_r, we are not yet able to determine the caustic distances, and hence the spreading factor, of the reflected ray tube once it leaves the surface. This information is not supplied (at least not without some extra work that solicits the assistance of differential geometry) by either the method using the Luneberg-Kline expansion or by the canonical problem solution. Expressions enabling us to do this however are provided in the following sections. These are derived in full in Section 3.3 for the 2D case, and reveal the physics of the problem. The 3D results may be obtained in a manner based on identical principles but are geometrically intricate, resulting in an interesting but rather cluttered derivation. Therefore, they simply will be quoted in a later section and immediately applied to various examples. References where the interested reader may find a complete derivation however will be given.

3.3 THE EXPRESSIONS FOR THE GEOMETRICAL OPTICS FIELD REFLECTED FROM SMOOTH CONDUCTING SURFACES: TWO-DIMENSIONAL PROBLEMS

3.3.1 When Is a Problem of a Two-Dimensional Nature?

The concept of a ray tube (GO field) that is 2D with respect to some axis was described in Section 2.5. Here, we explain what is meant by a 2D *problem*. For a problem to be of a 2D nature two conditions, one on the source fields and one on the scatterer geometry, are essential:

(a) The scatterer geometry must be uniform along some axis (i.e., 2D with respect to some axis). Here, the said axis will be assumed to be the z-axis.
(b) The incident GO field must have the form of a ray tube that is 2D with respect to the same axis as the scatterer in (a). The most general 2D source is an electric or magnetic line source with a pattern factor.

Note that the conditions do not impose any restriction on the polarization of the incident GO field. It is worth considering a number of possible 2D geometries, only some of which will be seen to qualify as 2D problems. The examples in Figure 3.13(a) and 3.13(b) are easily seen to satisfy the conditions just given and therefore are 2D problems. On the other hand, the problem in Figure 3.13(c) is not, because although the source fulfills condition (b), the scatterer is of a 3D form. The same can be said for the situation in Figure 3.12(d). In Figure 3.12(e) the geometry satisfies condition (a) but the point source generates a spherical wavefront, which violates condition (b). Although one might hope that the cases in Figures 3.13(f) and 3.13(g) are 2D, this is not true; whereas the source fields and scatterer geometries are independently 2D with respect to some axis, this axis is not the same for the source and the scatterer. Observation of the incident field along the z-axis in Figure 3.13(f) will reveal a phase change with z; and for Figure 3.13(g) there will be both a phase and amplitude variation with z. This nonuniformity of the incident fields along the scatterer geometry violates condition (b).

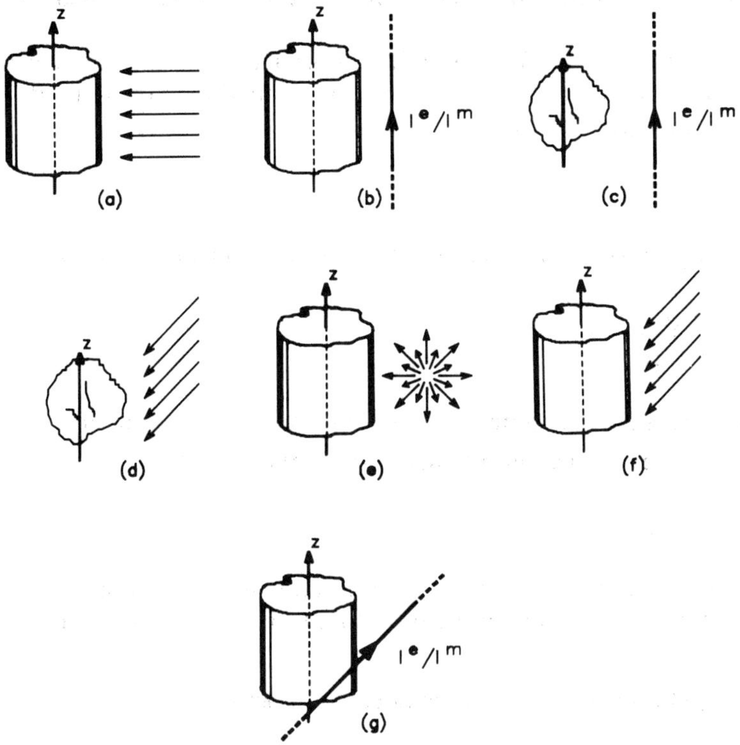

Figure 3.13 (a) and (b) are examples of two-dimensional problems; (c) through (g) are examples of problems that are not two-dimensional.

If (a) and (b) indeed are satisfied, the problem takes on a 2D character, which allows us to consider the entire question in the *x-y* plane only. In view of these simplifications we first will complete the theory for, and study reflection of, GO fields for 2D problems. Only then will we proceed to the more intricate 3D forms. A similar approach (and for the same reasons) will be adopted in the chapters on diffraction.

A trivial remark is that all real scatterers are of a 3D nature. Nevertheless, to assume that the 2D results should be forsaken once the 3D ones are later established would be wrong. Many problems of great practical importance can be studied through the use of 2D models that emphasize particular aspects of the authentic 3D reality.

3.3.2 Description of the Two-Dimensional Reflecting Surface Geometry

The cross section of the reflecting surface associated with a 2D problem is a planar curve. Thus, using the concepts discussed in Appendix C, it can be described in terms of a single parameter as

$$\mathbf{r}(\gamma) = x(\gamma)\,\hat{\mathbf{x}} + y(\gamma)\,\hat{\mathbf{y}} \tag{3.50}$$

The unit tangent to the curve (and hence the 2D surface) is then found as

$$\hat{\mathbf{t}} = \frac{\mathbf{r}'(\gamma)}{|\mathbf{r}'(\gamma)|} = (t_1,\, t_2) \tag{3.51}$$

where the prime signifies differentiation with respect to the parameter γ. Because $\hat{\mathbf{n}} \cdot \hat{\mathbf{t}} = 0$, where $\hat{\mathbf{n}}$ is the unit normal to the curve, we have

$$\hat{\mathbf{n}} = (\pm t_2,\, \mp t_1) \tag{3.52}$$

The $\pm$ and $\mp$ selection depends on which side of the curve is illuminated. To determine this by inspection of the problem geometry usually is quite easy.

3.3.3 Simplifications for Two-Dimensional Problems

From the definition of a 2D source field, $\hat{\mathbf{s}}^i$ will always lie in the cross-sectional x-y plane, and the x-y plane will be one of the principal planes of this incident field. Futhermore, the 2D geometry requirements imply that unit normal $\hat{\mathbf{n}}$ always is in the x-y plane (which is one of the principal planes of the reflecting surface), and so the x-y plane is the plane of incidence for all 2D problems as well. The coplanarity of the planes of incidence and reflection therefore implies that $\hat{\mathbf{s}}^r$ also lies in the x-y plane. The radius of curvature of the incident wavefront in the plane of incidence (which without ambiguity simply can be referred to as *the* radius of curvature of the incident wavefront) simply will be denoted ρ^i, and that for the reflected wavefront ρ^r. The radius of curvature of the 2D reflecting surface in the x-y principal plane will be denoted by a_0; that in the other principal plane is infinite.

3.3.4 Simplification of the Polarization Description of Reflected GO Fields for Two-Dimensional Problems

In the 2D case the cross-sectional x-y plane always is the plane of incidence, a principal plane of the incident wavefront, a principal plane of the reflecting surface, and a principal plane of the reflected wavefront. Any incident field of course can be resolved into components perpendicular and parallel to the plane of incidence (i.e., the x-y plane), and these components can be treated separately. Because the x-y plane is the plane of incidence, $\hat{\mathbf{e}}_\perp = \hat{\mathbf{z}}$.

If the incident wavefront has an electric field that everywhere is perpendicular to the plane of incidence (i.e., $\mathbf{E}^i = E^i_\perp \hat{\mathbf{e}}_\perp = E^i_z \hat{\mathbf{z}}$), from Section 3.2.4 we know that the reflected field at the point of reflection also will be z-polarized, with a reflection coefficient of -1, and so $E^r_z(Q_r) = -E^i_z(Q_r)$. From expression (2.127) the reflected field at a point P at a distance s^r from Q_r in the direction $\hat{\mathbf{s}}^r$ can be written as the 2D ray tube given by

$$E^r_z(P) = -E^i_z(Q_r)\sqrt{\rho^r/(\rho^r + s^r)}\, e^{-jks^r} \tag{3.53}$$

This is called the *TM case* because the magnetic field is transverse to the z-axis (e.g., excitation by an electric line source).

Alternatively, an incident field with polarization $\mathbf{E}^i = E^i_\| \hat{\mathbf{e}}^i_\|$ will have an associated reflected field amplitude at the point of reflection Q_r given by $E^r_\|(Q_r) = E^i_\|(Q_r)$. But polarizations of the fields $E^r_\|(Q_r)$ and $E^i_\|(Q_r)$ (namely, $\hat{\mathbf{e}}^i_\|$ and $\hat{\mathbf{e}}^r_\|$ respectively) will not be the same. However, from the local plane wave property (2.128), we know that their associated magnetic fields both will be z-directed. Therefore, $E^r_\|(Q_r) = E^i_\|(Q_r)$ is equivalent to the alternative scalar form, $H^r_z(Q_r) = H^i_z(Q_r)$, and so the reflected magnetic field at an observation point P a distance s^r from Q_r, in direction $\hat{\mathbf{s}}^r$, can be written as

$$H^r_z(s^r) = H^i_z(Q_r)\sqrt{\rho^r/(\rho^r + s^r)}\, e^{-jks^r} \tag{3.54}$$

This is the TE case because the electric field is transverse to the z-axis (e.g., excitation by a magnetic line source), and for such cases, to work with the magnetic field will be more convenient because its vector orientation does not change on reflection. The associated reflected electric field can then be obtained by using

$$\mathbf{E}^r(s^r) = -Z\, H^r_z\, \hat{\mathbf{s}}^r \times \hat{\mathbf{z}} \tag{3.55}$$

By using the scalarized notation of Section 2.5, we can combine (3.53) and (3.54) into the form:

$$U^r_z(s^r) = R_{s,h}\, U^i_z(Q_r)\sqrt{\rho^r/(\rho^r + s^r)}\, e^{-jks^r} \tag{3.56}$$

where $R_s = -1$ is used for the TM case and $R_h = +1$ for the TE case.

Now we need to find an expression that will enable us to determine ρ^r, given the properties of the incident GO field, and the surface, at the point of reflection (specular point) Q_r.

3.3.5 Amplitude Continuation along Two-Dimensional Reflected Ray Tubes

Derivation Using Differential Geometry

In high-frequency electromagnetic scattering problems from smooth surfaces, we can take advantage of the infinitesimal nature of the ray tube and confine our attention to the local properties of the surface in the neighborhood of the point of reflection Q_r of the central ray. Then, instead of using the exact representation of the surface, we may approximate it in the vicinity of Q_r by a circular arc of radius equal to the radius of curvature of the actual surface. In the limit of an infinitesimally narrow ray tube the expressions obtained from such an approximate representation of the surface will be identical to that obtained from the exact representation, as the curvature information at Q_r obtained from both representations will be the same.

Let $a_0(Q_r)$ denote the radius of curvature of the surface at Q_r and $\rho^i(Q_r)$ the radius of curvature at Q_r of the central ray of the incident ray tube. The point Q_r is selected as the reference point for the reflected ray tube. The radius of curvature of the central ray of the reflected ray tube at its reference point Q_r, (i.e., the caustic distance of the reflected ray tube) is denoted by ρ^r. In the remainder of this section a_0 and ρ^i always should be assumed to be measured at Q_r, but for notational convenience will not be denoted specifically as such. To construct Figure 3.14(a), the law of reflection simply is applied to the central ray, and the two paraxial rays shown. The normal congruence of rays incident on the surface then remains a normal congruence after reflection (theorem of Malus). Addition of a host of geometric details to Figure 3.14(a) provides Figure 3.14(b), and this is enlarged around the point of reflection to become Figure 3.14(c). From Figure 3.14(b), we can write

$$dC = a_0 d\phi \tag{3.57}$$

Because we are dealing with infinitesimal quantities, we may easily see from Figure 3.14(c) that

$$dC \cos\theta^i = \rho^i \, d\xi$$

or, in other words, that

$$dC = \rho^i d\xi / \cos\theta^i \tag{3.58}$$

and, similarly, from

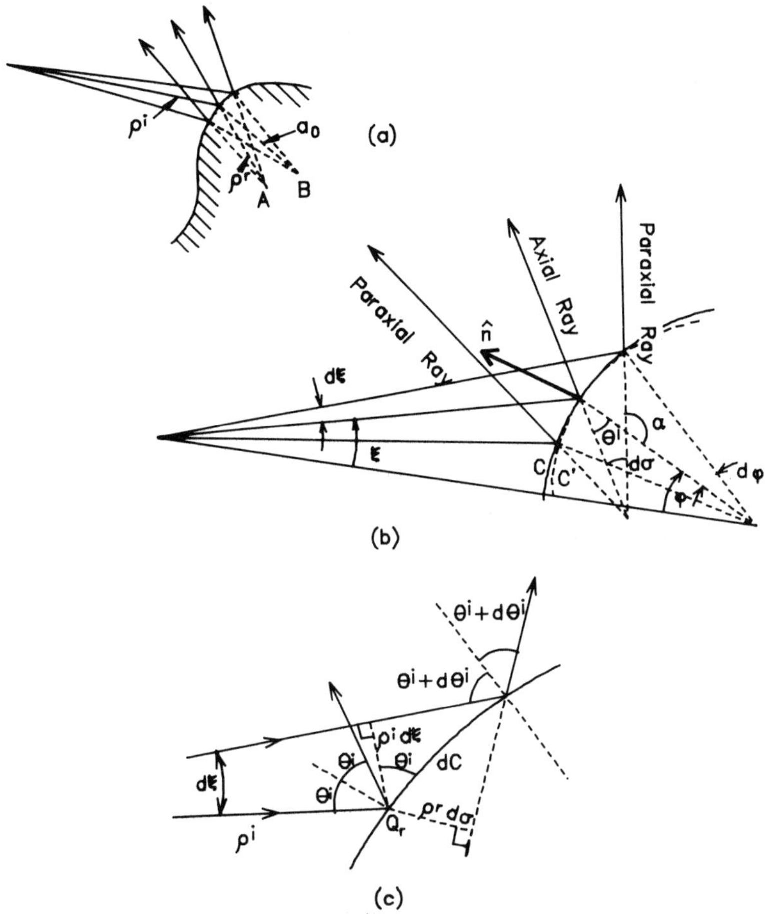

Figure 3.14 Reflection at a two-dimensional surface: (a) Construction of incident and reflected ray tubes according to the law of reflection. The caustic of the reflected ray tube lies at A, and the center of curvature of the cylindrical approximation to the surface at Q_r lies at B. (b) The situation in (a) is shown with full geometrical annotation. The surface C is approximated quadratically by C′ at Q_r. (c) A portion of (b) is enlarged.

$$dC \cos\theta^i = \rho^r d\sigma$$

that

$$dC = \rho^r d\sigma/\cos\theta^i \tag{3.59}$$

Furthermore, from simple geometry, we have from Figure 3.14(a) that $\theta^i = \xi + \phi$, so that $d\xi = d\theta^i - d\phi$, from which it follows that

$$d\xi = d\theta^i - d\phi \tag{3.60}$$

From (3.58) and (3.60), we have the relation:

$$dC = \rho^i(d\theta^i - d\phi)/\cos\theta^i \tag{3.61}$$

which combined with (3.57) yields

$$a_0 d\phi = \rho^i(d\theta^i - d\phi)/\cos\theta^i \tag{3.62}$$

If $d\phi$ is made the subject of expression (3.62), it reads

$$d\phi = \rho^i d\theta^i/(a_0 \cos\theta^i + \rho^i) \tag{3.63}$$

The next step is to combine (3.59) with (3.57) to obtain

$$a_0 d\phi = \rho^r d\sigma/\cos\theta^i \tag{3.64}$$

Once more, from the geometry of Figure 3.14(b) and 3.14(c) we have the relations:

$$\alpha + d\sigma + \theta^i = \alpha + d\phi + (\theta^i + d\theta^i)$$

and hence that

$$d\sigma = d\phi + d\theta^i \tag{3.65}$$

Substituting (3.65) into (3.64), and making $d\phi$ the subject of the expression, we obtain

$$d\phi = \rho^r d\theta^i/(a_0 \cos\theta^i - \rho^r) \tag{3.66}$$

Finally, equating (3.66) and (3.63), we have

$$\frac{\rho^i d\theta^i}{(a_0 \cos\theta^i + \rho^i)} = \frac{\rho^r d\theta^i}{(a_0 \cos\theta^i - \rho^r)} \tag{3.67}$$

Straightforward algebraic manipulation then yields the final expression for the caustic distance of the reflected field:

$$\frac{1}{\rho^r} = \frac{1}{\rho^i(Q_r)} + \frac{2}{a_0(Q_r)\cos\theta^i} \tag{3.68}$$

with Q_r the reference point on the reflected ray tube. The preceding result is given by, among others, Kouyoumjian, Peters, and Thomas [4], and in Felsen and Marcuvitz [5].

Before proceeding, a brief word of caution in is order regarding the use of (3.68). In the derivation there is an inherent assumption that $\theta^i = \pi/2$ (the case of grazing incidence when the incident ray is tangential to the surface at Q_r) is forbidden; otherwise, $\cos\theta^i = 0$ and expressions such as (3.58) and those that follow it are meaningless. More will be said about this grazing incidence case in Chapter 4 and Chapter 8.

Special Case of Plane Wave Incidence

If a plane wave ray tube is incident on the surface, then $\rho^i \rightarrow \infty$, and (3.68) gives the radius of curvature of the reflected ray tube as

$$\rho^r = \frac{1}{2} a_0(Q_r)\cos\theta^i \tag{3.69}$$

Sign Convention for Surfaces

By convention, if a scatterer presents a concave (convex) surface to an incident ray the radius of curvature $a_0(Q_r)$ of the scatterer surface is taken to be a negative (positive) quantity. Additional discussion on this is given in Section 4.4.2.

3.3.6 The Classical Geometrical Optics Interpretation

The results derived in Chapter 2 conveniently provide us with information on the properties of GO fields, and it is a good idea at this time to recall two of the facts we learned there. We learned that for GO fields there is power conservation along any ray tube and that the ray trajectories minimize the optical path length over which the ray travels. Classical geometrical optics proceeds from these as basic assumptions. The classical geometrical optics interpretation of reflection therefore involves application of the Fermat principle generalized to include the reflection point Q_r on the ray trajectory, as we mentioned briefly in Section 2.2.9. Let us here see how application of this Fermat principle (albeit only for the 2D situation) yields the law of reflection result (3.17).

Consider therefore the geometry shown in Figure 3.15. A line source is positioned at point (x_s, y_s), and the observation point is at (x_o, y_o). The desired reflection point is at $Q_r(x_r, y_r)$ on the cross-sectional curve C of the 2D scatterer, described by the parameterized form:

$$\mathbf{r}(\gamma) = x(\gamma)\,\hat{\mathbf{x}} + y(\gamma)\,\hat{\mathbf{y}}$$

If $\gamma = \gamma_r$ at Q_r, then the electrical path length along the reflected ray becomes

$$\begin{aligned} L &= k(s^i + s^r) \\ &= k\sqrt{[x(\gamma_r) - x_s]^2 + [y(\gamma_r) - y_s]^2} \\ &\quad + k\sqrt{[x_o - x(\gamma_r)]^2 + [y_o - y(\gamma_r)]^2} \end{aligned} \tag{3.70}$$

The Fermat Principle requires γ_r to be such that L has a stationary value, and we thus require that $\partial L/\partial\gamma_r = 0$. Straightforward differentiation makes this requirement:

$$\begin{aligned} \partial L/\partial\gamma_r &= \frac{k[x(\gamma_r) - x_s]x'(\gamma_r) + k[y(\gamma_r) - y_s]y'(\gamma_r)}{s^i} \\ &\quad - \frac{k[x_o - x(\gamma_r)]x'(\gamma_r) + k[y_o - y(\gamma_r)]y'(\gamma_r)}{s^r} = 0 \end{aligned} \tag{3.71}$$

where the prime indicates differentiation with respect to γ_r. If we next observe that

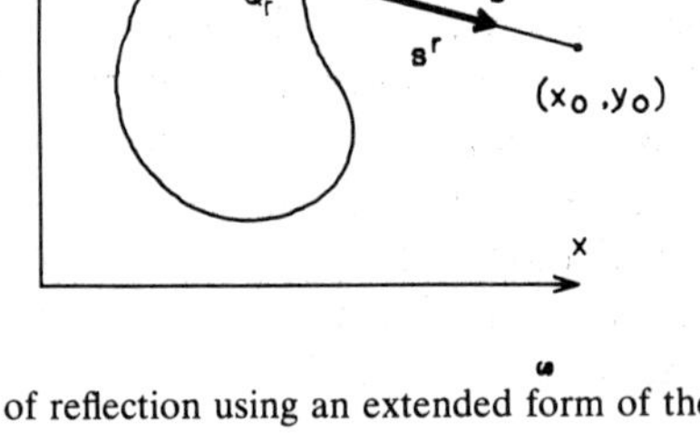

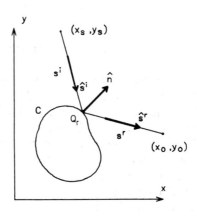

Figure 3.15 Derivation of the law of reflection using an extended form of the Fermat principle.

$$\hat{\mathbf{s}}^i = \frac{[x(\gamma_r) - x_s]\hat{\mathbf{x}} + [y(\gamma_r) - y_s]\hat{\mathbf{x}}}{s^i} \tag{3.72}$$

$$\hat{\mathbf{s}}^r = \frac{[x_o - x(\gamma_r)]\hat{\mathbf{x}} + [y_o - y(\gamma_r)]\hat{\mathbf{y}}}{s^r} \tag{3.73}$$

$$\hat{\mathbf{n}} = \frac{x'(\gamma_r)\hat{\mathbf{x}} + y'(\gamma_r)\hat{\mathbf{y}}}{|x'(\gamma_r)\hat{\mathbf{x}} + y'(\gamma_r)\hat{\mathbf{y}}|} \tag{3.74}$$

then we easily see (3.71) to be equivalent to the requirement $\hat{\mathbf{s}}^i \cdot \hat{\mathbf{n}} = -\hat{\mathbf{s}}^r \cdot \hat{\mathbf{n}}$, which is the law of reflection result desired.

Equation (3.68) also can be obtained through application of the intensity law by requiring the location of the reflected field caustic to be such that the power contained in the reflected ray tube equals that of the incident ray tube. This, of course, is subject to the assumption (which always is taken for granted in this text) that the scatterer is a perfect conductor so that there is no ohmic loss of power on the reflecting surface. This derivation is instructive and included as an exercise in the problem set at the end of this chapter.

Finally, note that although Fermat's principle tells us the trajectory followed by the reflected ray, it says nothing of what happens to the vector fields on reflection. If it were the only approach we had followed from the beginning, the latter information would have *had* to have been extracted from a supplementary canonical problem.

3.3.7 Summary of Reflected Field Expressions for Two-Dimensional Problems

(a) The z-axis is taken to be the axis with respect to which the scatterer geometry and the source fields are uniform.

(b) Let the GO field incident on the reflection point Q_r be specified by $U_z^i(Q_r)$, with the phase of this incident field at Q_r referred to some chosen point (e.g., Q_p).

(c) The unit vector in the direction of the reflected ray ($\hat{\mathbf{s}}^r$) is related to the incident ray direction ($\hat{\mathbf{s}}^i$) and the surface unit normal $\hat{\mathbf{n}}$ at Q_r through the expression:

$$\hat{\mathbf{s}}^r = \hat{\mathbf{s}}^i - 2(\hat{\mathbf{s}}^i \cdot \hat{\mathbf{n}}) \tag{3.75}$$

(d) The reflected GO field at an observation point P a distance s^r (in the direction $\hat{\mathbf{s}}^r$) from the point of reflection Q_r is described by the expression:

$$U_z^r(P) = R_{s,h}\, U_z^i(Q_r)\, \sqrt{\rho^r/(\rho^r + s^r)}\, e^{-jks^r} \tag{3.76}$$

its phase being referred to the same point Q_p as that of $U_z^i(Q_r)$.

(e) $R_s = -1$, the soft reflection coefficient, is used for the TM case, where $U_z = E_z$.

(f) $R_h = +1$, the hard reflection coefficient, is used for the TE case, where $U_z = H_z$.

(g) The reflected ray tube caustic distance ρ^r (i.e., the radius of curvature of the reflected ray at its reference point Q_r) is obtained from the relation:

$$\frac{1}{\rho^r} = \frac{1}{\rho^i(Q_r)} + \frac{2}{a_0(Q_r)\cos\theta^i} \tag{3.77}$$

where $\rho^i\,(Q_r)$ is the radius of curvature of the incident wavefront at the point of reflection Q_r, $a_0(Q_r)$ is the radius of curvature of the reflecting surface at Q_r, and $\cos\theta^i = -\hat{\mathbf{n}} \cdot \hat{\mathbf{s}}^i$ at Q_r.

Remark 1: If the caustic distance ρ^r of the reflected ray tube is finite, and we observe this ray tube field at a sufficiently large distance from the reflection point Q_r, that $s^r \gg \rho^r$, then an acceptable approximation for the spreading factor is

$$\sqrt{\rho^r/(\rho^r + s^r)} \approx \sqrt{\rho^r/s^r} \tag{3.78}$$

We will refer to this as "the usual far-zone approximation for the spreading factor."

Remark 2: Similarly, a usual far-zone approximation for the phase factor is associated with the observation point at a very large distance. Consider the conditions in Figure 3.16 (a), where Q_1 and Q_2 either locate actual sources or are points of

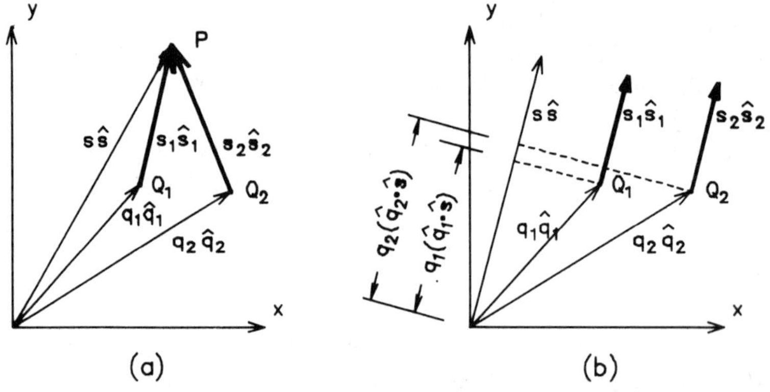

Figure 3.16 Geometry for the far-zone approximation.

reflection on some surface. These points are located, relative to the origin of the coordinate system, by vectors $q_1\hat{\mathbf{q}}_1$ and $q_2\hat{\mathbf{q}}_2$ respectively. Vectors $s_1\hat{\mathbf{s}}_1$ and $s_2\hat{\mathbf{s}}_2$ are measured *from* Q_1 and Q_2, respectively, to some observation point P, for example. Observation point P is located by vector $s\hat{\mathbf{s}}$. Now suppose that the observation point is moved to a very large distance from the sources and scatterers, so that s, s_1, and s_2 are much larger than q_1 and q_2. Then, for all practical purposes, vectors $s_1\hat{\mathbf{s}}_1$, $s_2\hat{\mathbf{s}}_2$, and $s\hat{\mathbf{s}}$ can be assumed to be parallel (in which case their associated *unit* vectors are equal, $\hat{\mathbf{s}}_1 = \hat{\mathbf{s}}_2 = \hat{\mathbf{s}}$), as indicated in Figure 3.16 (b). The geometry then assures us that

$$s_1 \approx s - q_1(\hat{\mathbf{q}}_1 \cdot \hat{\mathbf{s}}) \tag{3.79}$$

$$s_2 \approx s - q_2(\hat{\mathbf{q}}_2 \cdot \hat{\mathbf{s}}) \tag{3.80}$$

If this is so, then the propagation phase terms for rays from Q_1 and Q_2 can be written:

$$e^{-jks_1} = e^{-jk[s - q_1(\hat{\mathbf{q}}_1 \cdot \hat{\mathbf{s}})]} \tag{3.81}$$

$$e^{-jks_2} = e^{-jk[s - q_2(\hat{\mathbf{q}}_2 \cdot \hat{\mathbf{s}})]} \tag{3.82}$$

This is the usual far-zone approximation for the phase factor.

Remark 3: Regarding Figure 3.16(b), with the observation point in the far-zone, the following assumptions usually are made with respect to terms of the form $1/\sqrt{s_1}$ as far as the *amplitude* of the field is concerned:

$$\frac{1}{\sqrt{s_1}} = \frac{1}{\sqrt{s - q_1(\hat{\mathbf{q}}_1 \cdot \hat{\mathbf{s}})}} \approx \frac{1}{\sqrt{s}} \tag{3.83}$$

We will call this the usual far-zone approximation for the amplitude factor. If we are observing the reflected field in the far-zone, then the preceding approximation most likely will be combined with the usual far-zone approximation for the spreading factor of expression (3.78).

Remark 4: In the summary of the 2D field expressions listed immediately before these remarks, we stressed that the phase of the fields were referred to *some chosen point* Q_p. In applications, when we combine direct, reflected, and (later on) diffracted rays, the phase of each ray must be referenced to the same point. In this way, we can shift the phase reference of a ray from one point to another at will.

Figure 3.17 shows a source (it could be a reflection point reference on some reflected ray, of course) located at Q_s, which position is described with respect to the point Q_p via the vector $s_1\hat{\mathbf{s}}_1$. The field from the source is to be observed at P, which is oriented with respect to Q_p by the vector $s_2\hat{\mathbf{s}}_2$. The phase factor of the source field at P, referenced to the source position Q_s, then is

$$e^{-jks^i}$$

Suppose now that we want to reference the phase factor of this same source field, again observed at P, to the point Q_p. Then the phase factor will read, in terms of the vectors previously defined:

$$e^{-jk\hat{\mathbf{s}}_1\cdot(s_1\hat{\mathbf{s}}_1+s_i\hat{\mathbf{s}}^i)} \tag{3.84}$$

In many instances, we wish to reference all phases to the origin of the coordinate system, in which case Q_p simply is the point (0, 0).

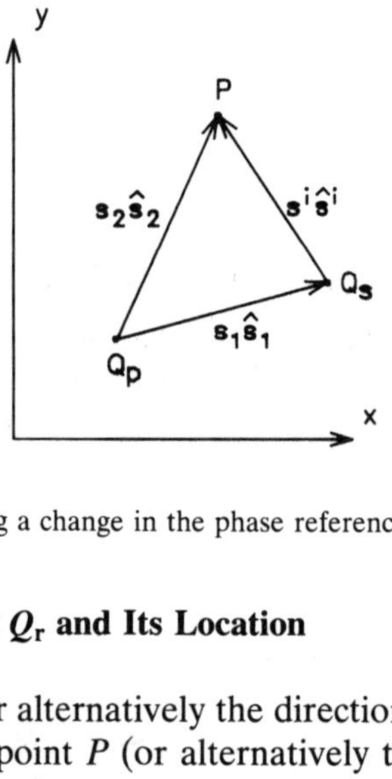

Figure 3.17 Geometry for calculating a change in the phase reference point.

3.3.8 On the Specular Point Q_r and Its Location

Given the source point Q_s (or alternatively the direction of incidence of some ray-optic field), the observation point P (or alternatively the observation direction if the field point is in the far-zone) and the properties of the surface at the reflection point Q_r, we can use the expressions just summarized to find the reflected field at P. This assumes we know the location of Q_r for the given Q_s and P.

To emphasize that, for a given surface, a different Q_r generally is required for each different source point and observation point combination, let us examine Figure 3.18. Here we have a line source at Q_s launching a cylindrical wavefront,

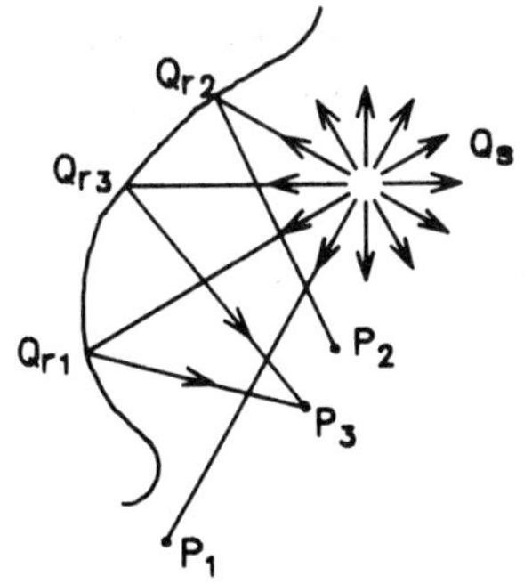

Figure 3.18 Demonstration that there are different specular points for different observation points.

with rays in all directions. Distinct specular points Q_r are shown for a number of observation points P_m. Observation point P_3 receives reflected rays from two points, Q_{r1} and Q_{r3}. Field point P_1 lies on the trajectory of a direct ray but not on any reflected ray path. In other words, for a specific observation point, there may be more than one pertinent reflection point, or there may be none at all.

Clearly, the search for Q_r must be based on the known law of reflection, or the Fermat principle can be applied directly if necessary. Initially we will consider this aspect on a problem-to-problem basis as the need arises. We should realize that the determination of reflection points often is one of the dominant consumers of time in the computer implementation of electromagnetic analysis techniques that utilize geometrical optics in some way. Further remarks on this topic will be made in Section 3.6.

3.3.9 Initial Two-Dimensional Problem Examples

The descriptions of the first few examples purposely will be lengthy and detailed. This is important when we apply the GO for the first time. Later, problems will be described in a more terse form.

Example 3.1: Far-zone radiation pattern of a line source in the presence of a conducting circular cylinder

Rough Sketch

A good idea before starting is to apply GO to make a rough sketch of the physical arrangement and locate, even if at first only in an approximate manner, the position of any shadow boundaries. Accordingly, the sketch for this example is given in Figure 3.19(a). We are able to identify the shadow zone immediately and realize that, in the lit region, there will be both direct and reflected rays in all observation directions. Also, recognize that, because of symmetry, only the pattern in the upper half of the *x*-*y* plane need be considered. After gaining such an initial understanding of the problem, the next step is to construct a diagram of the geometry that contains only one specimen of each ray type being considered.

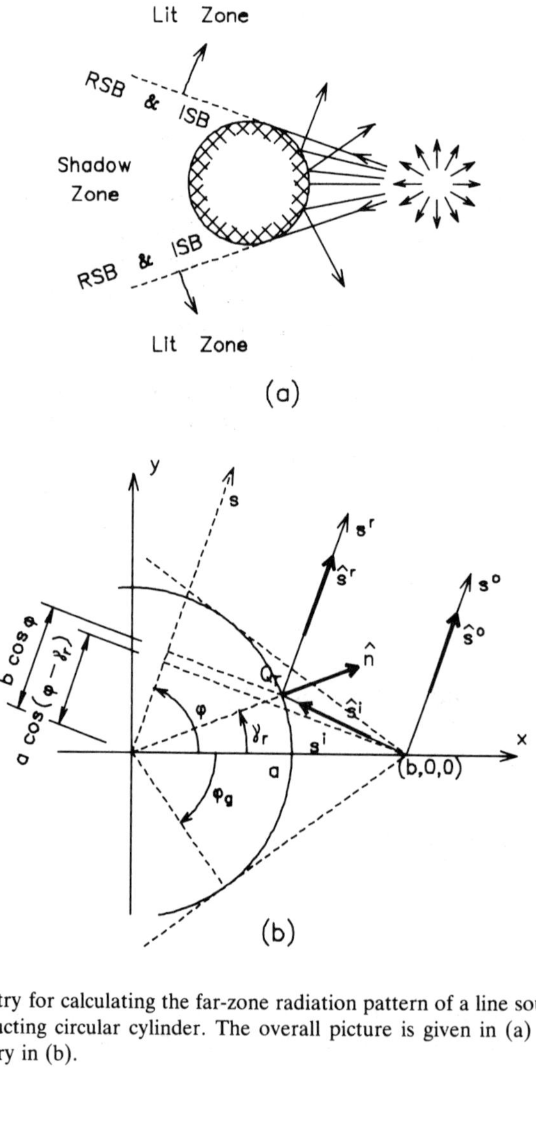

Figure 3.19 Geometry for calculating the far-zone radiation pattern of a line source in the presence of a conducting circular cylinder. The overall picture is given in (a) and the details of the geometry in (b).

Geometrical Relationships

Figure 3.19(b) now is appropriate. Because we are interested in the far-zone fields, the direct and reflected rays in any given observation direction ϕ may be assumed parallel. The unit vector in the observation direction is

$$\hat{\mathbf{s}} = \cos\phi\,\hat{\mathbf{x}} + \sin\phi\hat{\mathbf{y}} \tag{3.85}$$

Let $\hat{\mathbf{s}}^o$ denote the unit vector from the line source in the observation direction. Similarly we denote by $\hat{\mathbf{s}}^r$ the unit vector from a reflection point Q_r in the direction of the observer. The far-zone assumption then implies that

$$\hat{\mathbf{s}} = \hat{\mathbf{s}}^o = \hat{\mathbf{s}}^r = \cos\phi\,\hat{\mathbf{x}} + \sin\phi\,\hat{\mathbf{y}} \tag{3.86}$$

The surface is described by the parameterized form (with single parameter γ),

$$\mathbf{r}(\gamma) = a\cos\gamma\,\hat{\mathbf{x}} + a\sin\gamma\,\hat{\mathbf{y}} \tag{3.87}$$

where a is the radius of the cylinder. In this particular instance, it is possible and convenient to select parameter γ to be the angular variable measured from the x-axis, as indicated in Figure 3.19(b). Clearly, γ is the same as ϕ, but in this problem we will use γ when locating points on the reflecting surface, and ϕ for the observation angle. Taking the derivative of $\mathbf{r}$ with respect to γ:

$$\mathbf{r}'(\gamma) = -a\sin\gamma\,\hat{\mathbf{x}} + a\cos\gamma\,\hat{\mathbf{y}} \tag{3.88}$$

from whence it follows that the unit tangent $\hat{\mathbf{t}}$ is given by

$$\hat{\mathbf{t}} = \mathbf{r}'(\gamma)/|\mathbf{r}'(\gamma)| = -\sin\gamma\,\hat{\mathbf{x}} + \cos\gamma\,\hat{\mathbf{y}} \tag{3.89}$$

and thus the unit normal to the surface by

$$\hat{\mathbf{n}} = \cos\gamma\,\hat{\mathbf{x}} + \sin\gamma\,\hat{\mathbf{y}} \tag{3.90}$$

Location of the Specular Point

Given observation direction $\hat{\mathbf{s}}$, and hence $\hat{\mathbf{s}}^r$, plus the source position $(b, 0, 0)$, we next find the associated reflection point Q_r. Let Q_r be that point on the surface at which the parameter $\gamma = \gamma_r$. The unit vector in the direction of the ray from the source point $(b, 0, 0)$ to reflection point $(a\cos\gamma_r, a\sin\gamma_r, 0)$ is obtained easily by referring to Figure 3.19(b) as

$$\hat{\mathbf{s}}^i = \frac{(a\cos\gamma_r - b)\,\hat{\mathbf{x}} + a\sin\gamma_r\,\hat{\mathbf{y}}}{\sqrt{(a\cos\gamma_r - b)^2 + (a\sin\gamma_r)^2}} \tag{3.91}$$

From (3.91) and (3.90) we have at Q_r

$$\hat{\mathbf{n}} \cdot \hat{\mathbf{s}}^i = \frac{a - b\cos\gamma_r}{\sqrt{a^2 + b^2 - 2ab\cos\gamma_r}} \tag{3.92}$$

Similar use of (3.86) and (3.90) provides

$$\hat{\mathbf{n}} \cdot \hat{\mathbf{s}}^r = \cos(\phi - \gamma_r) \tag{3.93}$$

Enforcement of the law of reflection $\hat{\mathbf{n}} \cdot \hat{\mathbf{s}}^i = -\hat{\mathbf{n}} \cdot \hat{\mathbf{s}}^r$ then yields the following transcendental equation for γ_r:

$$\frac{a - b\cos\gamma_r}{\sqrt{a^2 + b^2 - 2ab\cos\gamma_r}} = -\cos(\phi - \gamma_r) \tag{3.94}$$

For each observation angle ϕ, this equation has to be solved for γ_r to locate the correct point of reflection on the surface of the cylinder. Methods of doing this are available in a number of texts on numerical analysis, of which [6] is particularly useful and readable. Most solution methods for this type of problem require some initial guess at the result; the better the guess the more rapidly do we arrive at such a solution. When we compute radiation patterns the increment in observation angle ϕ usually is relatively small, so the value of γ_r for the previous ϕ can be used as a good initial guess for the current value of ϕ.

Determination of the Reflection Caustic Distance

At Q_r we know that $\cos\theta^i = \hat{\mathbf{n}} \cdot \hat{\mathbf{s}}^r$, which from (3.85) means that

$$\cos\theta^i = \cos(\phi - \gamma_r) \tag{3.95}$$

In finding the caustic distance ρ^r for the reflected field we note that the radius of curvature of the incident wavefront at Q_r is

$$\rho^i(Q_r) = s^i = \sqrt{a^2 + b^2 - 2ab\cos\gamma_r} \tag{3.96}$$

and that the radius of curvature of the surface at Q_r always is simply $a_0(Q_r) = a$. Therefore,

$$\frac{1}{\rho^r} = \frac{1}{\sqrt{a^2 + b^2 - 2ab\cos\gamma_r}} + \frac{2}{a\cos(\phi - \gamma_r)} \tag{3.97}$$

from which ρ^r obviously can be obtained.

Direct Field at Observation Point

We choose to reference the phases of all fields to the line source position. From the geometry shown in the sketch (Figure 3.19(a)), the distances s^r and s^o clearly are related to s as

$$s^r = s - a\cos(\phi - \gamma_r) \tag{3.98}$$

$$s^o = s - b\cos\phi \tag{3.99}$$

The field propagating directly from the source to the far-zone is given by

$$U_z^i(s, \phi) = U_o \frac{e^{-jks^o}}{\sqrt{s^o}} \tag{3.100}$$

The usual far-zone approximation for the amplitude factor is made in the denominator in (3.100); namely,

$$\frac{1}{\sqrt{s^o}} \approx \frac{1}{\sqrt{s}} \tag{3.101}$$

Use of this result along with (3.99) converts (3.100) to

$$U_z^i(s, \phi) = U_o\, e^{jkb\cos\phi} \frac{e^{-jks}}{\sqrt{s}} \tag{3.102}$$

Incident Field at Reflection Point

The incident field at Q_r, with its phase referenced to the source position, is

$$U_z^i(Q_r) = U_o \frac{e^{-jks^i}}{\sqrt{s^i}} \tag{3.103}$$

with s^i given by (3.96).

Reflected Field at Observation Point

Applying this value for the incident field at Q_r, the reflected field, with its phase therefore also referenced to the source position, is given by

$$U_z^r(s, \phi) = R_{s,h}\, U_o \frac{e^{-jks^i}}{\sqrt{s^i}}\, e^{-jks^r} \sqrt{\rho^r/(\rho^r + s^r)} \tag{3.104}$$

If the usual far-zone approximation (3.78) for the spreading factor is applied to that in (3.104), it becomes

$$U_z^r(s, \phi) = R_{s,h}\, U_o \frac{e^{-jks^i}}{\sqrt{s^i}}\, e^{-jks^r} \sqrt{\frac{\rho^r}{s^r}} \tag{3.105}$$

Next, make the usual far-zone approximations for the phase factor e^{-jks^r} and amplitude factor $1/\sqrt{s^r}$; these reduce (3.105) to the form:

$$U_z^r(s, \phi) = R_{s,h}\, U_o \sqrt{\frac{\rho^r}{s^i}}\, e^{-jks^i}\, e^{jka\cos(\phi - \gamma_r)} \frac{e^{-jks}}{\sqrt{s}} \tag{3.106}$$

Position of Shadow Boundaries

The angle γ_g is that value of γ for which $\hat{\mathbf{n}} \cdot \hat{\mathbf{s}}^i = 0$. From (3.92) this yields $\gamma_g = \cos^{-1}(a/b)$. This result enables us to determine that the shadow boundary is at

$$\phi_g = (\pi/2) + \cos^{-1}(a/b)$$

Total Field at Observation Point

The total field in any direction ϕ then is given by the sum of (3.102) and (3.106), for $|\phi| < \phi_g$. The computational procedure will be as follows. For a given ϕ value, compute $U_z^i(\phi)$ from (3.102). Then find γ_r by solving (3.94), and hence obtain ρ^r from (3.97). $U_z^r(\phi)$ is found from (3.106) and added to $U_z^i(\phi)$ to obtain the total field in direction ϕ.

Numerical Results

Figures 3.20 and 3.21 show computed results for both the case of magnetic and electric line source excitation. Note that these are normalized patterns, with the

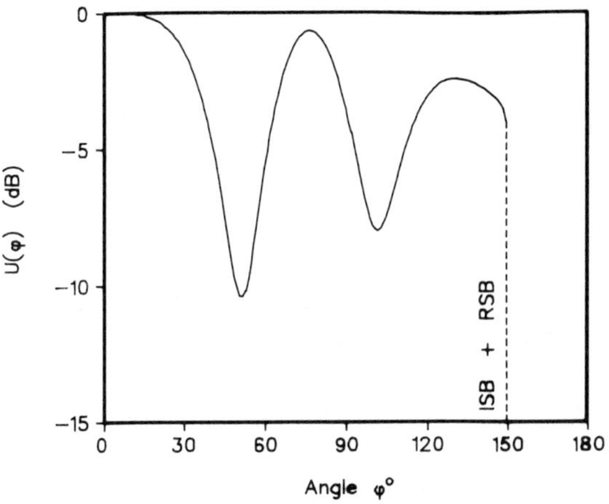

Figure 3.20 Computed pattern of the cylinder in Figure 3.19 illuminated by a magnetic line source, with $a = \lambda$ and $b = 2\lambda$.

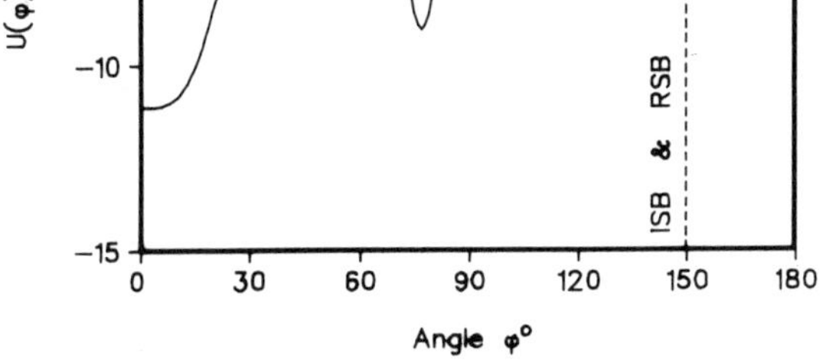

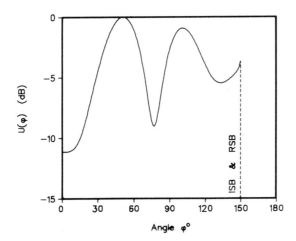

Figure 3.21 Computed pattern of the cylinder in Figure 3.19 illuminated by an electric line source, with $a = \lambda$ and $b = 2\lambda$.

common factor $e^{-jks}/\sqrt{s}$ suppressed and numerical values $a = \lambda$ and $b = 2\lambda$. These results clearly show the abrupt manner in which the GO fields fall to zero at the SB. Examination of the exact eigenfunction solution for the same problem [7] reveals that this GO solution is sufficient except close to the SB, and that the abrupt discontinuity at the SB is not physical, as we would expect. In Chapter 8, we will reconsider this example, where the addition of curved surface diffraction terms will correct the deficiency.

Example 3.2: Bistatic far-zone scattering of a plane wave incident on a conducting cylinder of circular cross section

Rough Sketch

Figure 3.22(a) shows the initial sketch of the problem. In the far-zone there will be a shadow boundary (coincidence of an ISB and an RSB) in the $\phi = \pi$ direction. For all other observation angles there will be reflected rays, but only one in each direction ϕ. Now we could solve this problem by letting the line source in the previous example recede to infinity ($b \rightarrow \infty$). However, this is not always the best or easiest route to follow in more complex situations, although if closed form expressions are not required the present example could be "simulated" numerically by making distance b in Example 3.1 large enough. We nevertheless will redo the problem, making use of as many results from Example 3.1 as we can.

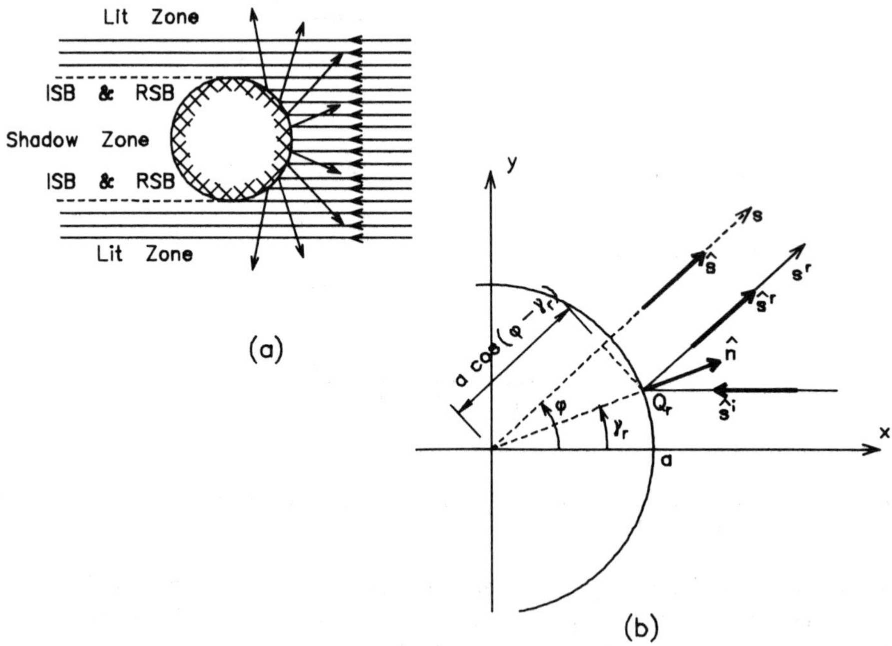

Figure 3.22 Calculation of the bistatic far-zone scattered field of a circular cylinder illuminated by a plane wave. The geometry is shown in (a), with the detail in (b).

Geometrical Relationships

Figure 3.22(b) shows the essentials of the geometry and ray arrangement. Because the observer is in the far-zone, the reflected ray unit vector $\hat{\mathbf{s}}^r$ coincides with the unit vector in the observation direction $\hat{\mathbf{s}}$, as explained in Remark 2 of the previous section. Hence,

$$\hat{\mathbf{s}}^r = \hat{\mathbf{s}} = \cos\phi\ \hat{\mathbf{x}} + \sin\phi\ \hat{\mathbf{y}} \qquad (3.107)$$

The surface equations, unit normal, and radius of curvature remain unchanged from the previous example.

Locating the Reflection Point Q_r

Given the observation direction ϕ, we must find the associated reflection point Q_r. For plane wave incidence, at any Q_r the incident ray unit vector simply is

$$\hat{\mathbf{s}}^i = -\hat{\mathbf{x}} \qquad (3.108)$$

Let the point Q_r be that point on the surface at which the parameter $\gamma = \gamma_r$. From (3.90) and (3.108), we therefore have at Q_r:

$$\hat{\mathbf{n}} \cdot \hat{\mathbf{s}}^i = -\cos\gamma_r \qquad (3.109)$$

The quantity $\hat{\mathbf{n}} \cdot \hat{\mathbf{s}}^r$ at Q_r is unchanged from that in (3.93), so

$$\hat{\mathbf{n}} \cdot \hat{\mathbf{s}}^r = \cos(\phi - \gamma_r) \qquad (3.110)$$

The law of reflection condition $\hat{\mathbf{n}} \cdot \hat{\mathbf{s}}^r = -\hat{\mathbf{n}} \cdot \hat{\mathbf{s}}^i$ then supplies the condition:

$$\cos(\phi - \gamma_r) = \cos\gamma_r$$

for which the solution is

$$\gamma_r = \phi/2 \qquad (3.111)$$

Caustic Distance for Reflection

Because $\cos\theta^i = -\hat{\mathbf{n}} \cdot \hat{\mathbf{s}}^i$ it follows from (3.110) and (3.111) that

$$\cos\theta^i = \cos(\phi/2) \qquad (3.112)$$

Because for plane wave incidence $\rho^i \to \infty$, we have $1/\rho^i(Q_r) \to 0$, with the result that $\rho^r = (a/2)\cos\theta^i$. Use of (3.112) gives the final form:

$$\rho^r = (a/2)\cos(\phi/2) \tag{3.113}$$

Incident Field at the Reflection Point

From Section 2.4.2, the expression for the field of the incident plane wave, with its phase referred to the origin of the coordinate system, is simply e^{jkx}. At all points on the cylinder $x = a\cos\gamma$. Therefore, at Q_r we have $x = a\cos\gamma_r$, and so the incident field at Q_r is

$$e^{jka\cos\gamma_r} \tag{3.114}$$

Reflected Field at the Observation Point

From Figure 3.22 (b), we have the geometrical result:

$$s^r = s - a\cos(\phi - \gamma_r) \tag{3.115}$$

Incorporation of (3.111) reduces this to

$$s^r = s - a\cos(\phi/2) \tag{3.116}$$

This means that the usual far-zone approximation for the phase factor of the reflected field becomes

$$e^{-jks^r} = e^{-jks}\, e^{jka\cos(\phi/2)} \tag{3.117}$$

Application of the usual far-zone approximation for the spreading factor of the reflected field means that

$$\sqrt{\rho^r/(\rho^r + s^r)} \approx \sqrt{\rho^r/s^r}$$

Subsequent use of the usual far-zone approximation for the amplitude factor $1/\sqrt{s^r} \approx 1/\sqrt{s}$ then gives the reflected field in direction ϕ as

$$U_z^r(\phi) = R_{s,h}\sqrt{\frac{a\cos(\phi/2)}{2}}\, e^{j2ka\cos(\phi/2)}\, \frac{e^{-jks}}{\sqrt{s}}$$

This result is for bistatic scattering from a cylinder for the case of a plane wave incident from the $\phi = 0$ direction. Note that at $\phi = \pi$, $\cos(\phi/2) = 0$, and the reflected field is zero as expected according to the GO. The backscattered field simply is the expression in (3.117) evaluated at $\phi = 0$, which is

$$U_z^r(0) = R_{s,h}\sqrt{a/2}\,e^{j2ka}\frac{e^{-jks}}{\sqrt{s}} \qquad (3.118)$$

Surface Current Density

We derive here for later use an expression for the GO surface current density on the cylinder in the present problem. In particular, we are interested in this current density for the TM case for which

$$\mathbf{E}^i = \hat{\mathbf{z}}\,e^{jka\cos\gamma}$$

at all points on the illuminated portion of the cylinder surface for $|\gamma| < |\pi/2|$. $\mathbf{H}^i$ at all points on the cylinder can be found from

$$\mathbf{H}^i = Y(\hat{\mathbf{s}}^i \times \mathbf{E}^i)$$

and hence from (3.42) the GO surface current density by

$$\mathbf{J}_{go} = 2\hat{\mathbf{n}} \times \mathbf{H}^i = 2Y\hat{\mathbf{n}} \times (s^i \times \mathbf{E}^i)$$

With $\hat{\mathbf{s}}^i = -\hat{\mathbf{x}}$, and

$$\hat{\mathbf{n}} = \cos\gamma\hat{\mathbf{x}} + \sin\gamma\hat{\mathbf{y}}$$

from (3.90), we thus have

$$\mathbf{J}_{go} = \hat{\mathbf{z}}2Y\cos\gamma\,e^{jka\cos\gamma} \qquad (3.119)$$

at all points $|\gamma| < |\pi/2|$ on the cylinder.

Example 3.3: Near-zone fields of a line source in the presence of a conducting strip

GO reflection also can be used to find reflected fields in the near-zone of an object, so let us consider the problem of a line source at a distance d above a flat conducting strip of width w. The initial sketch is shown in Figure 3.23 (a). The details of surface and ray geometry are given in Figure 3.23 (b).

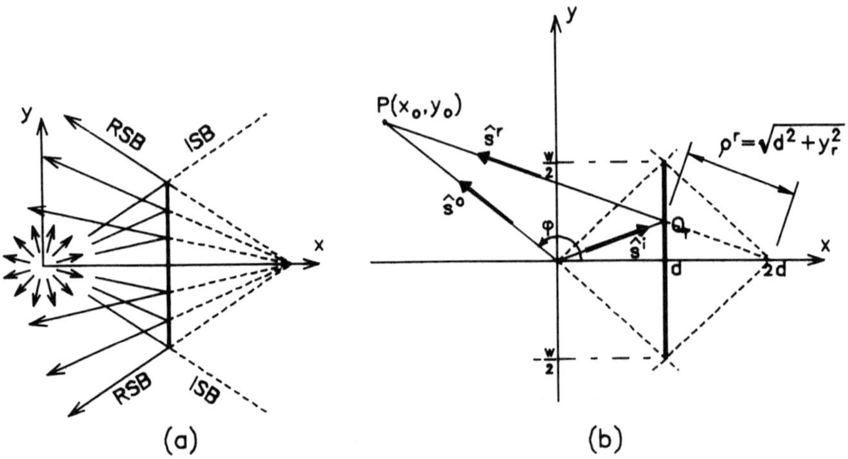

Figure 3.23 Calculation of the near-zone fields of a line source in the presence of a conducting strip; the geometry is shown in (a), and the details in (b).

Geometrical Relationships

$P(x_o, y_o)$ is the point of observation. The distance from the origin to P is denoted s^o, with ϕ the observation angle as indicated. At all points on the surface of the conducting strip the unit normal is

$$\hat{\mathbf{n}} = -\hat{\mathbf{x}} \tag{3.120}$$

and the radius of curvature is infinite ($a_0 \rightarrow \infty$). For a given observation point P the associated reflection point Q_r has rectangular coordinates (x_r, y_r).

Location of Shadow Boundaries

The incident shadow boundaries are along the constant $\phi = \pm\phi_{sb}$ lines, where ϕ_{sb} is found from the geometry to be

$$\phi_{sb} = \tan^{-1}(w/2d) \tag{3.121}$$

A shadow zone thus exists for those values of ϕ for which $|\phi| < |\phi_{sb}|$. The reflection shadow boundary is located at

$$\phi = \pm (\pi - \phi_{sb})$$

Direct Field at P

The direct field from the source, with its phase referred to the origin of the coordinate system (which also happens to be the source position), is given at P by

$$U_z^i(P) = U_o \frac{e^{-jks^o}}{\sqrt{s^o}} \tag{3.122}$$

for $|\phi| < |\phi_{sb}|$, with s^o given by

$$s^o = \sqrt{x_o^2 + y_o^2} \tag{3.123}$$

Incident Field at Q_r

The distance from the source to Q_r is $s^i = \sqrt{x_r^2 + y_r^2}$, and because $x_r = d$ always, this becomes

$$s^i = \sqrt{d^2 + y_r^2} \tag{3.124}$$

and the incident field at Q_r, its phase referenced to the origin, is

$$U_z^i(Q_r) = U_o \frac{e^{-jks^i}}{\sqrt{s^i}} \tag{3.125}$$

Location of the Specular Point Q_r

Given the observation point P, we need to find the appropriate reflection point Q_r. To this end we note that the unit vector in the direction of the incident ray at Q_r is

$$\hat{\mathbf{s}}^i = \frac{d\hat{\mathbf{x}} + y_r\hat{\mathbf{y}}}{\sqrt{d^2 + y_r^2}} \tag{3.126}$$

Thus, using (3.120) and (3.126), we have

$$\hat{\mathbf{n}} \cdot \hat{\mathbf{s}}^i = \frac{-d}{\sqrt{d^2 + y_r^2}} \tag{3.127}$$

The distance s^r from Q^r to P is

$$s^r = \sqrt{(x_0 - d)^2 + (y_o - y_r)^2} \tag{3.128}$$

The unit vector $\hat{\mathbf{s}}^r$ in the direction of the reflected ray at Q_r therefore is

$$\hat{\mathbf{s}}^r = \frac{(x_o - d)\hat{\mathbf{x}} + (y_o - y_r)\hat{\mathbf{y}}}{\sqrt{(x_o - d)^2 + (y_o - y_r)^2}} \tag{3.129}$$

and so we have the result:

$$\hat{\mathbf{n}} \cdot \hat{\mathbf{s}}^r = \frac{-(x_o - d)}{\sqrt{(x_o - d)^2 + (y_o - y_r)^2}} \tag{3.130}$$

Hence the law of reflection requirement $\hat{\mathbf{n}} \cdot \hat{\mathbf{s}}^i = -\hat{\mathbf{n}} \cdot \hat{\mathbf{s}}^r$, after some simple algebraic manipulation, becomes

$$(x_o^2 - 2dx_o)y_r^2 + (2d^2y_o)y_r - d^2y_o^2 = 0 \tag{3.131}$$

For a specified observation point P, in other words a given (x_o, y_o), equation (3.131) is a simple quadratic equation for the point y_r, and hence the reflection point (d, y_r).

Caustic Distance of the Reflected Field

The general expression (3.77) for the caustic distance of the reflected ray simplifies here because $a_0(Q_r) \to \infty$, so that we simply have $\rho^r = \rho^i(Q_r)$, with $\rho^i(Q_r)$ equal to s^i in (3.124). Hence,

$$\sqrt{\rho^r/(\rho^r + s^r)} = \sqrt{s^i/(s^i + s^r)}$$

Reflected Field at P

Collecting the preceding information, the reflected field at P is found to be described by

$$U_z^r(P) = R_{s,h}\, U_o \frac{\mathrm{e}^{-jk(s^i + s^r)}}{\sqrt{s^i + s^r}} \tag{3.132}$$

Total Field at P

The computational procedure would be as follows. Given an observation point (x_o, y_o), find s^o from (3.120) and hence $U_z^i(P)$ from (3.123). The y_r is found by solving (3.130). Expression (3.128) then enables us to determine s^r and thus $U_z^r(P)$ can be computed from (3.132). The total field at P is the sum:

$$U_z^t(P) = U_z^i(P) + U_z^r(P)$$

Additional Comment

We saw earlier that for this problem $\rho^r = s^i = \sqrt{d^2 + y_r^2}$. Because this is the caustic distance of the reflected field, or the radius of curvature of the reflected field at Q_r, we can adopt the interpretation that the reflected field appears to come from the image of the line source, the image being positioned on the x-axis at $x = 2d$. This is supported by the form of the reflected field in (3.132). Furthermore, the reflection shadow boundaries are seen to be the incident shadow boundaries of the fields of this image source.

Numerical Results

The total field (for the case of illumination by an electric line source), normalized to the maximum value, is plotted as a function of ϕ in Figure 3.24 for $d = \lambda$, w

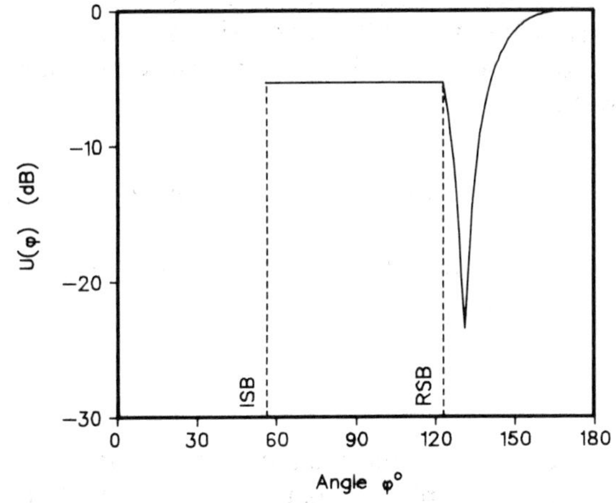

Figure 3.24 Computed pattern of the conducting strip in Figure 3.23 illuminated by an electric line source, with $d = \lambda$, $w = 3\lambda$, and $s = 5\lambda$.

= 3λ and s^o = 5λ. The disturbing jump in the computed fields at both the incident and reflection GO shadow boundaries certainly is not what we would observe with practical antennas. The advantage of GO-GTD techniques is that, as in the present case, such nonphysical behavior is obvious from a visual inspection of the results. We immediately can see if some ray contribution is missing (in the present case the diffracted rays from the edges of the strip, which is discussed in Chapter 4).

Remark 5: Definition of Radar Width. Consider a uniform plane wave to be incident on some object. We select some point (e.g., point on the scattering object or the origin of the coordinate system) as the phase reference for both the incident (U^i) and *backscattered* (U^r) fields. If s is the distance from the phase reference to some far-zone observation point, then the radar width is defined as

$$\sigma^w = \lim_{s \to \infty} 2\pi s \left| \frac{U^r}{U^i} \right|^2 \tag{3.133}$$

Because the determination of radar width is concerned with backscatter, the direction of a reflected ray at some point Q_r on the scattering object is the negative of the ray incident in the object at that point. Thus $\theta^i = 0$ for such cases, and so $\cos\theta^i = 1$. Furthermore, the law of reflection can be stated conveniently in terms of the unit tangent vector to the surface at Q_r as

$$\hat{\mathbf{s}}^i \cdot \hat{\mathbf{t}}(Q_r) = 0 \tag{3.134}$$

Note that reference to radar width always also implies plane wave incidence. The quantity σ^w in (3.133) often is referred to as the *monostatic* radar width, to distinguish it from the *bistatic* radar width that considers scattered fields in directions other than that of the incident field. In this text the term *radar width* will imply monostatic radar width.

Example 3.4: The radar width of a conducting cylinder of elliptical cross section

In this example we determine the radar width of an elliptical cylinder whose geometry is shown in Figure 3.25.

Incident Field

We will reference the phase of all fields to the origin. From Section 2.4.2 we know that at position (r, ϕ) a unit amplitude plane wave incident at angle ϕ_i is specified by

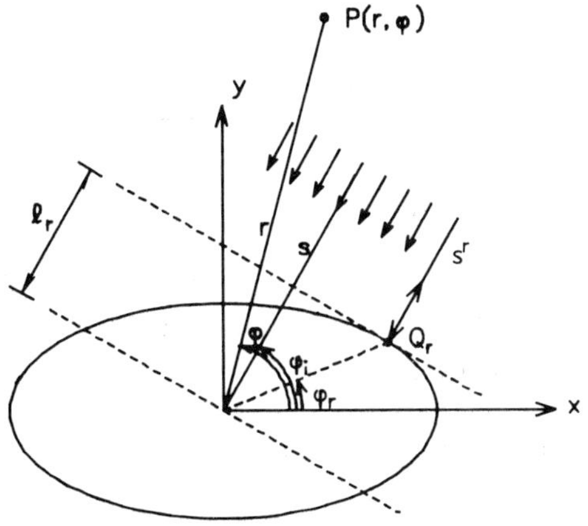

Figure 3.25 Plane wave reflection from a conducting elliptical cylinder.

$$U^i(r, \phi) = e^{jkr\cos(\phi - \phi_i)} \tag{3.135}$$

Geometry

The geometrical description of the elliptical cylinder that we will use here is discussed in some detail in Appendix C. We simply quote a number of final expressions here. The elliptical curve is defined by

$$\mathbf{r}(\gamma) = a \cos\gamma \, \hat{\mathbf{x}} + b \sin\gamma \, \hat{\mathbf{y}} \tag{3.136}$$

where γ is a parameter ($0 \leq \gamma \leq 2\pi$) related to angle ϕ as

$$\tan\phi = \frac{b}{a} \tan\gamma \tag{3.137}$$

Furthermore, the radius of curvature at any point γ on the elliptical surface is

$$a_0(\gamma) = \frac{(a^2 \sin^2 \gamma + b^2 \cos^2 \gamma)^{3/2}}{ab} \tag{3.138}$$

and the unit tangent to the surface is given by

$$\hat{\mathbf{t}} = -a \sin\gamma \, \hat{\mathbf{x}} + b \cos\gamma \, \hat{\mathbf{y}} \tag{3.139}$$

Reflected Field

Because the plane wave is incident on the elliptical cylinder from a direction ϕ_i, it follows that the direction of the incident ray at the point of reflection Q_r is

$$\hat{\mathbf{s}}^i = -\cos\phi_i \, \hat{\mathbf{x}} - \sin\phi_i \, \hat{\mathbf{y}} \tag{3.140}$$

For backscatter the direction of reflected rays is $-\hat{\mathbf{s}}^i$. Therefore, if for a specified ϕ_i we locate the point of reflection Q_r by $\gamma = \gamma_r$, the law of reflection requires that $\hat{\mathbf{s}}^i \cdot \hat{\mathbf{t}}(\gamma_r) = 0$. This means that

$$a \sin\gamma_r \cos\phi_i - b \cos\gamma_r \sin\phi_i = 0 \tag{3.141}$$

which implies that Q_r is given by

$$\gamma_r = \tan^{-1}\left[\frac{b}{a} \tan\phi_i\right] \tag{3.142}$$

or, in other words, by the coordinates:

$$x_r = a \cos\gamma_r \tag{3.143}$$

$$y_r = b \sin\gamma_r \tag{3.144}$$

Note that (3.142) is not the same as (3.137). The incident field at Q_r, with its phase referenced to the origin, according to (3.135), is given by

$$U^i(Q_r) = e^{jkl_r} \tag{3.145}$$

where

$$l_r = \sqrt{x_r^2 + y_r^2} \cos(\phi_r - \phi_i) \tag{3.146}$$

with ϕ_r related to γ_r by (3.137) as

$$\phi_r = \tan^{-1}\left[\frac{b}{a} \tan\gamma_r\right] \tag{3.147}$$

Caustic Distance of the Reflected Field

The caustic distance of the reflected field ρ^r is given by (3.68). For plane wave incidence $\rho^i \to \infty$, and the fact that we are considering the case of backscatter means that $\theta^i = 0$ and hence $\cos\theta^i = 1$. Thus, (3.68) becomes

$$\rho^r = \frac{a_0(\gamma_r)}{2} \tag{3.148}$$

Spreading and Phase Factors of the Reflected Field

The distance from the specular point Q_r to a far-zone observation point is

$$s^r = s - l_r \tag{3.149}$$

and we thus have

$$e^{-jks^r} = e^{-jks}\, e^{jkl_r} \tag{3.150}$$

For far-zone observation of the reflected field $s^r \gg 1$, and hence

$$\sqrt{\rho^r/(\rho^r + s^r)} \approx \sqrt{\rho^r/s^r} \tag{3.151}$$

However, because $s \gg l_r$, we may write

$$1/\sqrt{s^r} \approx 1/\sqrt{s} \tag{3.152}$$

Expression for the Far-Zone Reflected Field

The far-zone reflected field is

$$U^r(s, \phi_i) = U^i(Q_r)\, R_{s,h}\sqrt{\rho^r}\, \frac{e^{-jks^r}}{\sqrt{s^r}} \tag{3.153}$$

Using (3.145), (3.148), (3.150), and (3.152), this becomes

$$U^r(s, \phi_i) = R_{s,h}\, e^{j2kl_r}\, \sqrt{a_0(\gamma_r)/2}\, \frac{e^{-jks}}{\sqrt{s}} \tag{3.154}$$

Radar Width

Inserting (3.154) and (3.145) into (3.133), we finally obtain the radar width of the elliptic cylinder as

$$\sigma^{w}(\phi_i) = \pi\, a_0(\gamma_r) \tag{3.155}$$

with $a_0(\gamma_r)$ as given in equation (3.138). The expression (3.155) holds for both TM and TE polarizations.

3.3.10 Interpretation in Terms of Fundamental Electromagnetic Theory

Let us consider a specific scattering problem from the viewpoint of classical electromagnetics. Before doing so we note that reference will be made to certain mathematical concepts that can be classed under the heading of asymptotic techniques. A brief discussion on, and references to, such methods is given in Appendix D. The inclusion here of manipulations of a purely mathematical nature (and these easily become tedious) simply would obscure an understanding of the scattering physics we are discussing.

We examine again what happens when a true plane wave is incident in the $-\hat{\mathbf{x}}$ direction upon an infinitely long conducting cylinder, with the incident electric field $\mathbf{E}_i$ parallel to the axis of the cylinder, as was the case in Figure 3.22. Note that i is now a subscript, as this field is the incident field of rigorous electromagnetic theory that does not undergo shadowing, as explained in Section 2.8. An expression for the incident electric field, therefore, is

$$\mathbf{E}_i = \hat{\mathbf{z}} E_o\, e^{jkx} = \hat{\mathbf{z}} E_o\, e^{jk\rho\cos\phi} \tag{3.156}$$

The eigenfunction expansion technique [8–10] supplies the solution for the *total* electric field at any angle ϕ as

$$\mathbf{E}(\rho, \phi) = \hat{\mathbf{z}} E_o \sum_{n=-\infty}^{\infty} \left[J_n(k\rho) - \frac{J_n(ka)\, H_n^{(2)}(k\rho)}{H_n^{(2)}(ka)} \right] (j)^n\, e^{jn\phi} \tag{3.157}$$

where $J_n(x)$ is the nth-order Bessel function of the first kind and $H_n^{(2)}(x)$ the nth-order Hankel function of the second kind. The associated induced surface current density on the cylinder is given by

$$\mathbf{J}(a, \phi) = \hat{\mathbf{z}} \frac{2E_o}{Z\pi ka} \sum_{n=-\infty}^{\infty} \frac{(j)^n\, e^{jn\phi}}{H_n^{(2)}(ka)} \tag{3.158}$$

Application of the Watson transformation [9] to transform these summations to contour integrals (for observation angles in the range $-\pi/2 < \phi < \pi/2$), and the retention of certain significant terms, yields [9] the following asymptotic expression for (3.156), valid for large ka:

$$\mathbf{E}(\rho, \phi) \sim \hat{\mathbf{z}}E_o\left[e^{jk\rho\cos\phi} - \frac{e^{-jk\rho}}{\sqrt{\rho}}\sqrt{\frac{a\cos(\phi/2)}{2}}\, e^{j2ka\cos(\phi/2)}\right] \tag{3.159}$$

The first term is just the incident field, and the second clearly is identical to the reflected GO field result obtained in Example 3.2. An examination of the eigenfunction solution (3.158) for the induced current density on the cylinder reveals [11, pp. 378–379] that for increasing $ka = 2\pi a/\lambda$ the major portion of the induced current density indeed appears to become concentrated over the illuminated portion of the surface, as found by GO. The values obtained from the exact expression for $\mathbf{J}(a, \phi)$ in (3.158) approaches that of the GO current expression

$$\mathbf{J}_{go} = \hat{\mathbf{z}}\, 2Y\cos\phi\, e^{jka\cos\phi}$$

obtained from Example 3.2. There is excellent agreement between the exact and GO currents over the illuminated portion of the cylinder, except at those points where the incident GO ray is at grazing incidence. Indeed, if asymptotic techniques similar to those used to obtain (3.159) are applied [8] to the eigenfunction solution (3.158), the GO expression is obtained for the surface current density over illuminated portions of the surface away from the points of grazing incidence.

Although the current over the back surface of the cylinder that is not illuminated in the GO sense certainly is small compared to that over the illuminated front surface facing the incident plane wave, an examination [11] of (3.158) shows that it is small but not zero and will be a significant contributor to the fields in the GO shadow zone. These effects are accounted for by curved surface diffraction concepts discussed in Chapter 8.

These considerations are useful in that they indicate how the GO and GTD fit into the context of electromagnetic theory as a whole. The geometrical optics, geometrical theory of diffraction, and rigorous electromagnetic theory clearly are not disparate treatments of the same physical phenomena.

3.3.11 Relationship to Physical Optics

A technique that has found widespread application in the computation of electromagnetic wave scattering, as well as in the calculation of radiation patterns for reflector antennas, is called *physical optics* (PO). The PO method assumes that the induced surface current density on some scatterer or antenna surface is given by the geometrical optics current density $\mathbf{J}_{go} = 2\hat{\mathbf{n}} \times \mathbf{H}^i$ over those portions of the surface directly illuminated (in the GO sense) by the incident field, and zero over shadowed sections of the surface. This current then is used in a radiation integral [5] to find the scattered or radiated fields.

Consider once more the problem of the plane wave incident on the circular cylinder in Figure 3.22. The GO surface current density was found in Example 3.2 to be

$$\mathbf{J}_{go} = \begin{cases} \hat{\mathbf{z}}\, 2Y \cos\gamma\, e^{jka\cos\gamma}, & |\gamma| < |\pi/2| \\ 0, & \text{otherwise} \end{cases} \tag{3.160}$$

where γ is the angular position of a point on the cylinder surface, measured counterclockwise from the x-axis. The TM incidence case, for which the incident electric field is z-directed, is considered. The 2D radiation integral [10, p. 229] for the scattered field in any direction ϕ is then given by the following integral of the GO current over the illuminated portion ($-\pi/2 < \gamma < \pi/2$) of the cylindrical surface,

$$E_z^s(\phi) = C_0 \int_{-\pi/2}^{\pi/2} \cos\gamma\, e^{jka\cos\gamma}\, e^{jka\cos(\phi-\gamma)} d\gamma \tag{3.161}$$

where

$$C_0 = aZ\sqrt{k/2\pi}\, e^{j\pi/4} \frac{e^{-jks}}{\sqrt{s}}$$

This integral is evaluated in Appendix D by using the method of stationary phase to yield

$$E_z^s(\phi) \sim \sqrt{\frac{a\cos(\phi/2)}{2}}\, e^{j2ka\cos(\phi/2)} - \frac{C_0\sqrt{2}}{ka\,(\cos\phi + 1)}\, e^{j(ka\sin\phi - \pi/2)} - \frac{C_0\sqrt{2}}{ka\,(\cos\phi + 1)}\, e^{-j(ka\sin\phi + \pi/2)} \tag{3.162}$$

Equation (3.162) shows three separate terms for the scattered field in the direction ϕ. The first term is precisely the GO result found in Example 3.3. The second and third terms (the end-point contributions) arise because of the abrupt discontinuity in the GO surface current density at the edges of the illuminated region [12], and are not physical. The GO surface currents are particularly bad approximations to the true surface currents at such boundaries, as well as near sharp edges; more will be said about the latter deficiency in the next chapter. Finally, with its reliance on the GO currents, the PO does not provide us with the true fields in the GO shadow zone.

The question, of course, arises as to the validity of our use of the method of stationary phase for the integration in (3.161). We therefore write (3.161) in the form:

$$E_z^s(\phi) = C_0 \int_{-\pi/2}^{\pi/2} \cos\gamma \cos\Psi(\gamma)\, d\gamma + jC_0 \int_{\pi/2}^{\pi/2} \cos\gamma \sin\Psi(\gamma)\, d\gamma$$

where

$$\Psi(\gamma) = ka[\cos\gamma - \cos(\phi - \gamma)]$$

and examine the second term. Because $\cos\gamma$ is slowly varying over the integration range, the value of the integral will be determined largely by the $\sin\Psi(\gamma)$ factor. Plots of this function are shown in Figure 3.26 for the cases $ka = 10$ and $ka =$

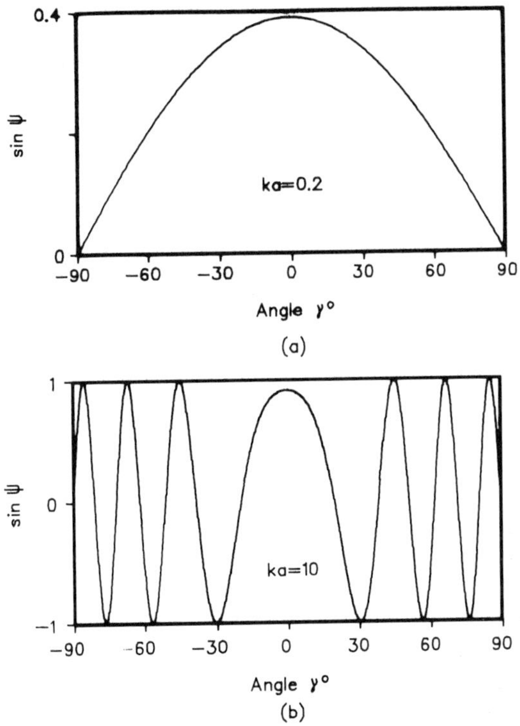

Figure 3.26 Graphic illustration of the self-cancelling effect exploited by the method of stationary phase.

0.2. To generate numerical values for plotting, a value for the observation direction was chosen without loss of generality to be $\phi = 0$. For this particular observation angle, the stationary phase points (at which the phase Ψ has zero a derivative) are centered about the GO specular point Q_r located at $\gamma = 0$. Consider first the $ka = 10$ case. Observe that apart from the stationary phase regions about Q_r the phase term varies rapidly. The effect on the integral is that these rapid phase excursions of opposite sign tend to interfere destructively, with only that about the stationary phase point remaining to provide the scattered field in the given direction. A similar conclusion results upon examination of that portion of the integral containing the cos $\Psi(\gamma)$ factor. This is a quantitative account of the same phenomenon discussed in Section 3.1 regarding the scattering of the sun's rays from an expanse of water.

For the case of the electrically small cylinder, $ka = 0.2$, the appropriate curve in Figure 3.26 clearly shows that a stationary phase evaluation of the PO integral would not be valid, as there are no terms to cause phase cancellation of the type that takes place in the $ka = 10$ (or larger) case. Indeed, the cylinder is electrically too small for the GO current to be a suitable approximation to the true current, and the use of PO is questionable in the first place.

3.3.12 Comments on GO Reflected Fields about Shadow Boundaries

Consider again the results in Figures 3.20 and 3.21 for the total far-zone fields of the line source and cylinder combination geometry of Figure 3.19. By reciprocity, this also is the form of the angular distribution of the total field about the cylinder at a distance b from the origin, when a plane wave is incident on the cylinder. We can see [7] that the agreement between the GO and exact solution is worst in that portion of the lit zone just before the shadow boundary, and of course in the shadow zone itself. One cause of the latter discrepancy is clear; there simply is no GO field in the shadow zone. The reason for the lack of agreement in the lit region about the shadow boundary is not as obvious.

To understand this problem we consider the equation (3.77) for the caustic distance ρ^r of the reflected ray tube

$$\frac{1}{\rho^r} = \frac{1}{\rho^i(Q_r)} + \frac{2}{a_0(Q_r)\cos\theta^i} \tag{3.163}$$

Recall that the point of reflection Q_r is the reference point on the reflected ray tube at which the initial field is computed for continuation along the ray, and that $|\rho^r|$ is the distance of the caustic of this reflected ray from the reference point Q_r. When θ^i is close to grazing incidence (at which θ^i is precisely $\pi/2$), $\cos\theta^i \approx 0$, the second term in (3.163) becomes very large, and hence $\rho^r \approx 0$. This means that the caustic of the reflected ray tube moves uncomfortably close to the reference point Q_r. Now, we know that GO does not reliably predict the fields in the vicinity of a caustic. Thus, our GO estimate $\mathbf{E}^r(Q_r)$ in (3.33) for the reflected field at Q_r, and thus also the reflection coefficient $\mathbf{R}$ in (3.39), which we obtained by using it, will be inaccurate near those points Q_r at which $\cos\theta^i \approx 0$.

We observed in Section (3.3.6) that equations (3.33) and hence (3.39), because $\mathbf{E}^i$ is locally plane, may receive the equally valid interpretation of being derived by studying the canonical problem of a true plane wave incident on an infinite plane surface tangent to the curved surface at the point of reflection. Note that this viewpoint does not rely on a GO approximation to the field at the point Q_r at all,but that the curved surface have a sufficiently large curvature at the point of reflection that it can be approximated locally with sufficient accuracy by the infinite plane surface there. We might surmise that we could obtain an improved reflection coefficient by studying asymptotic solutions to the more sophisticated canonical problem of a true plane wave incident on a circularly cylindrical surface that approximates some actual surface at the point of reflection. This is precisely the canonical problem discusssed in Section 3.3.10, and for such a problem Pathak [13] obtained a more accurate asymptotic solution for the reflected fields than the one given there. This more appropriately will be discussed in Chapter 8.

3.4 FURTHER EXAMPLES OF TWO-DIMENSIONAL REFLECTED FIELD PROBLEMS

Example 3.5: The near-zone fields of a plane wave incident on a semiinfinite planar conducting wedge

The geometry of a plane wave incident on a planar conducting wedge is shown in Figure 3.27. We will express our results in the coordinate system (s, ϕ) that is

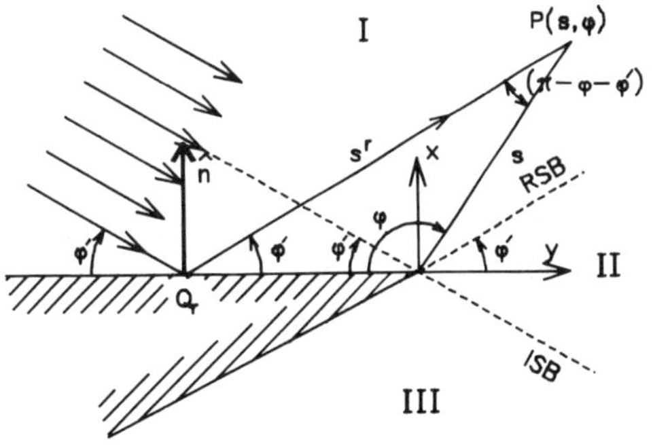

Figure 3.27 Plane wave reflection at a semiinfinite wedge.

fixed to the edge of the wedge. In preparation for a standard notation used in edge diffraction problems, the wedge interior angle is defined as $(2 - n)\pi$, although n need not be an integer. We will reference the phases of all fields to the edge.

A plane wave is incident at an angle ϕ' with respect to the face of the wedge; and from Section 2.4.2 this incident field at any observation point $P(s, \phi)$ can be expressed in the form $U_z^i(s, \phi) = e^{jks\cos(\phi - \phi')}$. With the ISB situated at $\phi = \phi' + \pi$, there is clearly no direct GO field in region III. Thus, we write for the direct GO field at any point P

$$U_z^i(s, \phi) = \begin{cases} e^{jks\cos(\phi - \phi')}, & 0 \leq \phi < \phi' + \pi \\ 0, & n\pi > \phi > \phi + \pi \end{cases} \tag{3.164}$$

At observation point $P(s, \phi)$ there will be two rays: the direct ray expressed by (3.164), and a ray reflected from some point Q_r on the surface of the wedge. For each $P(s, \phi)$ there will be a distinct Q_r. Because at all points Q_r the unit normal $\hat{\mathbf{n}} = \hat{\mathbf{x}}$, the reflected ray picture shown in Figure 3.27 will be identical for all reflected rays; only the position of Q_r will change as $P(s, \phi)$ is altered. Let us identify the location of Q_r by coordinates $s = q$ and $\phi = 0$. Then the incident field at Q_r is given by

$$U_z^i(Q_r) = e^{jkq\cos\phi'} \tag{3.165}$$

But from the law of sines of elementary trigonometry we know that

$$\frac{s}{\sin\phi'} = \frac{s^r}{\sin\phi} = \frac{q}{\sin(\pi - \phi - \phi')} \tag{3.166}$$

$\text{Sin}(\pi - \phi - \phi') = \sin(\phi + \phi')$; hence, from (3.166), (3.165) can be written as

$$U_z^i(Q_r) = e^{jks[\sin(\phi + \phi')\cos\phi'/\sin\phi']} \tag{3.167}$$

Because the face of the wedge is planar, the radius of curvature at all points $a_0(Q_r) \to \infty$. Thus, $2/a_0 \cos\theta^i \to 0$, and the caustic distance of the reflected wavefront simply is $\rho^r = \rho^i(Q_r)$. However, for plane wave incidence $\rho^i(Q_r) \to \infty$, and so $\rho^r \to \infty$. Therefore, the spreading factor:

$$\sqrt{\frac{\rho^r}{\rho^r + s^r}} \to 1$$

The reflected field at $P(s, \phi)$ is therefore given by

$$U_z^r(P) = R_{s,h}\, U_z^i(Q_r)\, \mathrm{e}^{-jks^r} \tag{3.168}$$

From (3.166) we have the intermediate result:

$$s^r = s\, \sin\phi/\sin(\phi')$$

If this is substituted, along with the $U_z^i(Q_r)$ of (3.167), into (3.168), it follows that

$$U_z^r(P) = R_{s,h}\, \mathrm{e}^{jks[\sin(\phi+\phi')\cos\phi'/\sin\phi' + \sin\phi/\sin\phi']} \tag{3.169}$$

Straightforward trigonometric manipulation reduces the argument of the exponential to $jks\,\cos(\phi + \phi')$, and so (3.169) becomes

$$U_z^r(s,\, \phi) = \begin{cases} R_{s,h}\, \mathrm{e}^{jks\cos(\phi+\phi')}, & 0 < \phi < \pi - \phi' \\ 0, & \pi - \phi' < \phi < n\pi \end{cases} \tag{3.170}$$

There is no reflected field outside region I; hence, the form of (3.170).

The total field at any observation point $P(s,\, \phi)$ is given by the sum:

$$U_z^t(s,\, \phi) = U_z^i(s,\, \phi) + U_z^r(s,\, \phi)$$

There are abrupt changes in the fields at the RSB and ISB, due to the omission from the calculation of the contribution of those rays arising from the mechanism of edge diffraction. This will be corrected by the edge diffracted fields introduced in Chapter 4; and the same problem geometry is considered there in Examples 4.1 and 4.2.

Example 3.6: Near-zone fields of a section of a parabolic cylinder illuminated by a line source

We wish to find the total field, as a function of distance y_a, at any point x_a on a vertical line near the parabolic cylinder shown in Figure 3.28. The equations for the parabolic cylinder can be obtained from those for the paraboloid given in Appendix C. In addition to the defining equations written in terms of rectangular coordinates, a form in terms of a coordinate system $(\tau,\, \phi)$, centered at the focal line of the parabolic cylinder, will be used at different stages in the solution of the problem.

It is important to remember that unit vectors $\hat{\tau}$ and $\hat{\phi}$ do not have fixed directions, which of course is the case with many of the coordinate systems we will encounter in this text and which are usually used simply because they are especially

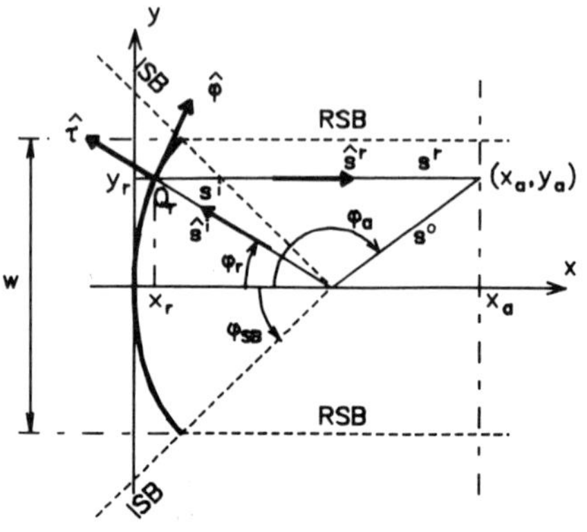

Figure 3.28 Parabolic cylinder illuminated by a line source.

convenient for some particular problem. In most instances we utilize such "convenient" coordinate systems along with the immovable rectangular system, transforming back and forth to simplify the problem. Note also that $\hat{\tau}$ is not normal to the parabolic cylinder.

Note that the line source in the problem is located at the focal line $(f, 0)$ of the parabolic cylinder, which has a width w. From the equation

$$x = y^2/4f$$

of the parabolic curve we easily can deduce that, at $y = \pm w/2$,

$$x = w^2/16f$$

and so the angle ϕ_{sb} subtended by the parabolic curve at its focal line is

$$\phi_{sb} = 2 \cot^{-1}(4f/w)$$

This also enables us to locate the incident shadow boundaries. The direction of the reflected rays is derived as follows.

Let $Q_r(x_r, y_r)$ be some point of reflection on the curve, specified by (τ_r, ϕ_r). Then, in terms of the coordinate directions $\hat{\tau}$ and $\hat{\phi}$, we have the unit normal at Q_r

$$\hat{\mathbf{n}}(Q_r) = -\cos(\phi_r/2)\,\hat{\tau} + \sin(\phi_r/2)\hat{\phi} \tag{3.171}$$

and the radius of curvature is given by

$$a_o(Q_r) = \frac{-4\sqrt{2}\, f}{(1 + \cos\phi_r)^{3/2}} \tag{3.172}$$

Because the line source is positioned at the focal line of the parabolic cylinder, from which the coordinate τ is measured, any ray from the line source incident on the surface has a ray vector:

$$\hat{\mathbf{s}}^i = \hat{\tau}(Q_r) \tag{3.173}$$

From (3.171) and (3.173), therefore, the dot product at Q_r is

$$\hat{\mathbf{n}} \cdot \hat{\mathbf{s}}^i = -\cos(\phi_r/2) \tag{3.174}$$

We next use equation (3.75) for the direction vector of the reflected ray at Q_r:

$$\hat{\mathbf{s}}^r = \hat{\mathbf{s}}^i - 2(\hat{\mathbf{s}}^i \cdot \hat{\mathbf{n}})$$

Upon substitution of (3.149) and (3.152) this becomes

$$\hat{\mathbf{s}}^r = -\cos\phi_r\, \hat{\tau} + \sin\phi_r\, \phi \tag{3.175}$$

which easily is shown from simple geometry to be equivalent to

$$\hat{\mathbf{s}}^r = \hat{\mathbf{x}} \tag{3.176}$$

This implies that, as long as the source is at the focal line of the parabolic cylinder, *any* ray reflected from the parabolic cylinder has a trajectory parallel to the x-axis, and the reflection shadow boundaries therefore may be located as shown. The reflected ray coordinate s^r, measured from Q_r in the direction $\hat{\mathbf{s}}^r$ of the reflected ray, can never be negative for this problem. So s^r always will be equal to the distance from Q^r to the observation point P:

$$s^r = \sqrt{(x_a - x_r)^2 + (y_a - y_r)^2} \tag{3.177}$$

An alternative expression for the reflected ray direction vector (the unit vector from Q_r to point $P(x_a, y_a)$) is

$$\hat{\mathbf{s}}^r = [(x_a - x_r)\hat{\mathbf{x}} + (y_a - y_r)\hat{\mathbf{y}}]/s^r \tag{3.178}$$

Comparison of (3.176) and (3.178) immediately shows that

$$y_r = y_a \tag{3.179}$$

and the equation of the paraboloid then demands that

$$x_r = y_a^2/4f \tag{3.180}$$

Thus, given (x_a, y_a), we may use (3.179) and (3.180) to locate Q_r. Because, from the defining equation of the parabolic curve:

$$\tau_r = \sqrt{x_r^2 + y_r^2} = f \sec^2(\phi_r/2)$$

we also have

$$\phi_r = 2 \sec^{-1} \left[\frac{\sqrt{(x_r^2 + y_r^2)}}{f} \right]^{1/2} \tag{3.181}$$

In some instances (3.181) may be more convenient than just (3.180). For any (x_a, y_a) expressions (3.179), (3.180), and (3.181) allow us to find the associated ϕ_r. The incident field at Q_r is

$$U_z^i(Q_r) = U_0 \frac{e^{-jks^i}}{\sqrt{s^i}} \tag{3.182}$$

and, because $s^i = \tau_r = f \sec^2(\phi_r/2)$, (3.182) becomes

$$U_z^i(Q_r) = U_0 \frac{e^{-jkf \sec^2(\phi_r/2)}}{\sqrt{f} \sec(\phi_r/2)} \tag{3.183}$$

We still need an expression for the caustic distance ρ^r of the reflected field at Q_r. The radius of curvature of the incident wavefront at Q_r is simply

$$\rho^i(Q_r) = \tau_r = f \sec^2(\phi_r/2)$$

so that

$$1/\rho^i(Q_r) = \cos^2(\phi_r/2)/f \tag{3.184}$$

Also, from (3.174), we have

$$\cos\theta^i = \cos(\phi_r/2) \tag{3.185}$$

The radius of curvature of the parabolic cylinder at Q_r is known from (3.172). If we combine this with (3.185) and use the trigonometric identity:

$$\cos^2(\phi/2) = (1 + \cos\phi)/2$$

we find that

$$2/[a_0(Q_r)\cos\theta^i] = -\cos^2(\phi_r/2)/f \tag{3.186}$$

and so (3.77) predicts that

$$\rho^r(Q_r) \to \infty \tag{3.187}$$

In other words, the reflected wavefronts are plane waves. The spreading factor therefore reduces to unity, as in the previous example. The reflected field at (x_a, y_a) accordingly is given by

$$U_z^r(x_a, y_a) = R_{s,h} U_z^i(Q_r)\, e^{-jks_r} \tag{3.188}$$

Substitution of equations (3.177) and (3.183) into (3.188) yields the reflected field at any observation point P:

$$U_z^r(x_a, y_a) = R_{s,h}\, U_0 \cdot \frac{e^{-jkf\sec^2(\phi_r/2)}\, e^{-jk\sqrt{(x_a-x_r)^2+(y_a-y_r)^2}}}{\sqrt{f}\sec(\phi_r/2)} \tag{3.189}$$

but only for the range $|y_a| < w/2$, because of the reflection shadow boundaries. For y_a values outside of this frame the reflected fields are zero.

Because the distance from the source to the point P is

$$s^o = \sqrt{(x_a - f)^2 + y_a^2} \tag{3.190}$$

the direct field reaching P from the line source is described by

$$U_z^i(x_a, y_a) = U_0 \frac{e^{-jks^o}}{\sqrt{s^o}} \tag{3.191}$$

The computation of the total field $U_z^t(P)$ therefore proceeds as follows. Given (x_a, y_a), we locate y_r from (3.179), and thus x_r and ϕ_r from (3.180) and (3.181), respectively. Then $U_z^r(x_a, y_a)$ is found from (3.189). With s^i determined from (3.190), equation (3.191) provides $U_z^i(x_a, y_a)$.

If the line source has a pattern factor $g(\phi)$ then these results remain essentially the same, except that the reflected field in (3.189) is multiplied by the factor $g(\phi_r)$ and the direct field in (3.191) by $g(\phi_a)$, where the angle:

$$\phi_a = \pi - \tan^{-1}[y_a/(x_a - f)]$$

is shown in Figure 3.28. In antenna applications the pattern factor usually will be such that most of the energy from the line source is directed toward the reflector, so that the direct field along the line $x = x_a$ will be small in relation to the reflected field. If we therefore examine the reflected field $U_z^r(x_a, y_a)$ in the aperture, we note that the expression (3.189) can be written:

$$U_z^r(x_a, y_a) = R_{s,h} U_0 \, g(\phi_r) \frac{e^{-jk(s^i + s^r)}}{\sqrt{f} \sec(\phi_r/2)} \tag{3.192}$$

From the definition of the paraboloid, for any observation point on the line $x = x_a$ the length $(s^i + s^r)$ will be a constant equal to $f + x_a$. Thus we deduce that the phase of the reflected field is uniform along this line, also known as the *aperture* of the parabolic reflector. Omitting for the moment the line source pattern factor $g(\phi_r)$, the reflected field along the $x = x_a$ line (aperture field distribution) is tapered toward the ends of the aperture. The ratio

$$U_z^r(y_a = w)/U_z^r(y_a = 0)$$

called the *space taper* of the parabolic reflector, can be shown from the earlier results to be inversely proportional to the quantity f/w. Notice that if we include a pattern factor $g(\phi_r)$ that decreases in magnitude toward the ends of the reflector this will add an additional taper to the reflected fields in the aperture.

Example 3.7: Near-zone fields of a plane wave incident on a section of a parabolic cylinder

We hope that Example 3.6 has familiarized the reader with some of the properties of the parabolic reflector. By reciprocity we know that, because a line source at the focal line of the parabolic cylinder generates a plane wave in the x direction, a plane wave incident from the x direction will cause all the reflected rays to be focused through a caustic at the focal line. Indeed, this is the origin of the term *focal line*. However, even with this hindsight, let us faithfully apply the GO to the case of a plane wave incident in the $-\hat{\mathbf{x}}$ direction on the parabolic cylinder, and examine the formation of the caustic.

Consider a typical point of reflection (x_r, y_r), or equivalently (τ_r, ϕ_r), on the reflector, as shown in Figure 3.29. Then at this reflection point Q_r the incident ray vector simply is

$$\hat{\mathbf{s}}^i = -\hat{\mathbf{x}}$$

which in terms of the (τ, ϕ) coordinate system is

$$\hat{\mathbf{s}}^i = \cos\phi_r \hat{\tau} - \sin\phi_r \hat{\phi} \tag{3.193}$$

From (3.171), the unit normal to the parabolic cylinder at Q_r is

$$\hat{\mathbf{n}}(Q_r) = -\cos(\phi_r/2)\hat{\tau} + \sin(\phi_r/2)\hat{\phi}$$

and its radius of curvature there is

$$a_0(Q_r) = \frac{-4\sqrt{2}f}{(1 + \cos\phi_r)^{3/2}}$$

Therefore at Q_r we have the dot product $\hat{\mathbf{n}} \cdot \hat{\mathbf{s}}^i = -\cos(\phi_r/2)$. If this is inserted into the expression $\hat{\mathbf{s}}^r = \hat{\mathbf{s}}^i - 2\,(\hat{\mathbf{s}}^i \cdot \hat{\mathbf{n}})$ we obtain the reflected ray direction vector as

$$\hat{\mathbf{s}}^r = -\hat{\tau} \tag{3.194}$$

As surmised, the reflected ray from Q_r is in the direction that takes its trajectory through the focal line.

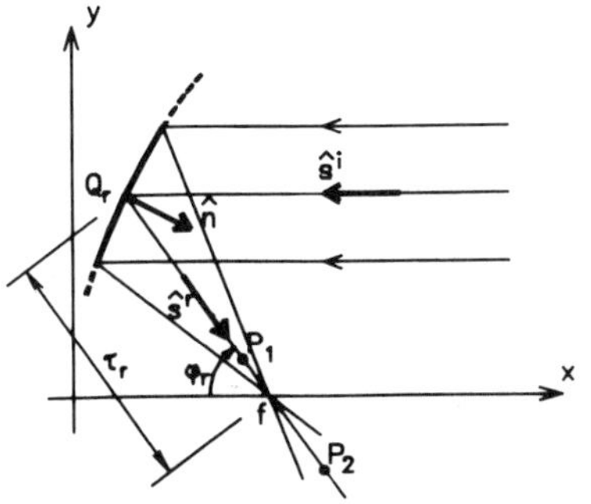

Figure 3.29 Parabolic cylinder illuminated by a plane wave.

Next consider the caustic distance of the reflected ray at Q_r. Because the incident field is a plane wave, $\rho^i\,(Q_r) \to \infty$, so that $1/\rho^i(Q_r) = 0$. The expression for ρ^r simplifies then to $\rho^r = a_0(Q_r)\cos\theta^i/2$. Because $\cos\theta^i = -\hat{\mathbf{n}} \cdot \hat{\mathbf{s}}^i$, we have $\cos\theta^i = \cos(\phi_r/2)$. Use of this fact, plus the form for $a_0(Q_r)$ given earlier, yields

$$\rho^r = -2f/[1 + \cos\phi_r] \tag{3.195}$$

In effect, the radius of curvature of the reflected ray at its reference point Q_r is negative, meaning that the ray tube is convergent at this point. If s^r is the reflected ray coordinate measured from Q_r in the direction of the reflected ray $\hat{\mathbf{s}}^r = \hat{\boldsymbol{\tau}}$, the spreading factor of the reflected wavefront is

$$\sqrt{\left|\frac{\rho^r}{\rho^r + s^r}\right|} = \sqrt{\left|\frac{2f}{s^r(1 + \cos\phi_r) - 2f}\right|} \tag{3.196}$$

There will be a caustic at that value of s^r for which the denominator:

$$s^r(1 + \cos\phi_r) - 2f = 0$$

or, in other words, when

$$s^r = 2f/(1 + \cos\phi_r)$$

The definition of the parabolic cylinder tells us that this is simply the position of the focal line.

The incident field at Q_r, with its phase referenced to the origin, is

$$U_z^i(Q_r) = e^{jkx_r} \tag{3.197}$$

The reflected field at an observation point P is

$$U_z^r(P) = R_{s,h}\, e^{jkx_r} \sqrt{\left|\frac{2f}{s^r(1 + \cos\phi^r) - 2f}\right|}\; e^{-jks^r}\, e^{j\delta\pi/2} \tag{3.198}$$

where $\delta = 0$ if the observation point P is before the focal line (e.g., at P_1), and $\delta = 1$ if P is beyond the focal line (e.g., at P_2).

Remark 6: Useful Caustics. The convex reflecting surfaces in Example 3.1 to 3.4 were all "well-behaved," in the sense that they did not generate real caustics. The only caustics are virtual ones ("images") located inside the conducting scatterer where we do not bother to compute the fields, which are known to be zero there anyway. On the other hand, for plane wave incidence, the concave parabolic cylinder of Example 3.7 gave rise to a real caustic at its focal line. It is this very fact that makes it so useful in the design of reflector antennas where some sort of focusing action is exploited. Other geometric shapes particularly useful in the design of reflector antennas are the hyperbolic cylinder, elliptic cylinder, and concave circular cylinder, and their 3D counterparts. References to such applications are given in Appendix E. Although reflector antennas have reached an advanced state of development, their surface shapes still are *designed* on the basis of GO reflection [14]. Essential to the final determination of their radiation patterns, however, is that the diffracted rays we discuss in later chapters be incorporated in the GTD formulation. Alternatively, we may use the PO approach, which, as we have seen, utilizes the GO currents in a radiation integral. Use may also be made of the aperture field integration approach, which relies on a knowledge of the GO reflected field in the aperture of the reflector antenna. Overall, if we analyze or design reflector antennas, GO is indispensable.

3.5 GENERAL EXPRESSIONS FOR THE REFLECTED FIELDS FROM THREE-DIMENSIONAL SMOOTH CONDUCTING SURFACES

3.5.1 Introduction

In (3.2) the GO reflected field from a smooth surface at some observation point P was written in the form:

$$\mathbf{E}^r(P) = \mathbf{E}^r(Q_r)\sqrt{\frac{\rho_1^r\rho_2^r}{(\rho_1^r + s^r)(\rho_2^r + s^r)}}\,e^{-jks^r} \tag{3.199}$$

We then presented a means of obtaining $\mathbf{E}^r(Q_r)$ in terms of the incident GO field at Q_r as

$$\mathbf{E}^r(Q_r) = \mathbf{E}^i(Q_r)\cdot\mathbf{R}$$

From the properties of the reflecting surface and the direction $\hat{\mathbf{s}}^i$ of an incident ray, the law of reflection provided all the information necessary to determine associated reflected ray directions or locations of Q_r. At the conclusion of Section 3.2, all that we needed to complete the picture and allow us actually to use (3.199) in computations was an ability to find ρ_1^r and ρ_2^r, the principal radii of curvature at Q_r of the reflected wavefront. We postponed a discussion of the full 3D expressions for these radii of curvature in favor of first gaining some experience with the simpler 2D forms. We now proceed to the general case.

The first usable results enabling us to perform amplitude continuation along GO rays reflected from a 3D surface were reported by Deschamps [15]. Utilizing these results, Kouyoumjian and Pathak [2] derived concise expressions for the direct determination of the principal radii of curvature and principal directions of the reflected wavefronts necessary for problem solving. This was done by diagonalizing the so-called curvature matrices obtained by Deschamps [15]. Alternative compact forms for the final expressions have been presented by Pathak, Burnside, and Marhefka [7], which provide increased physical insight into the 3D reflection process. In this section, the results of [2] and [7] will be stated simply as fact (which they are) and the constituent terms discussed, their application being our primary goal. The interested reader is referred to [15] and [2] for an outline of their derivation.

Often it is erroneously thought that the final results of Deschamps [15] are of an approximate nature. This assumption is made because quadratic approximations to the wavefronts and reflecting surface in the neighborhood of the point of reflection are used in their derivation [2, 15]. But this means that along the central rays of the incident and reflected ray tubes the second derivatives of the exact and quadratic representations of the surface are equal. Because these very second derivatives provide the curvature information for the wavefronts and reflecting surface, in the limit of an infinitesimally narrow ray tube, the exact and approximate surface and wavefront representations will provide identical results. The results in [15] thus are exact within the framework of GO, and hence also those of [2]and[7].

The relevant geometry remains that given in Figures 3.1 to 3.8, for which the appropriate terminology already has been defined. It will be worthwhile to revise some of the jargon of Section 3.2.1 at this point. Additionally, in the expressions that follow certain factors should be noted:

(a) Reference to any principal radii of curvature, principal planes, principal directions, or other unit vector associated with the wavefronts or the reflecting surface implies that these are measured *at* Q_r. (The diagrams become illegibly cluttered if we attempt to draw all the unit vectors at Q_r.)

(b) Remember that the plane of incidence ($\hat{\mathbf{n}}$, $\hat{\mathbf{s}}^i$) need not coincide with either of the principal planes of the surface ($\hat{\mathbf{n}}$, $\hat{\mathbf{U}}_1$) or ($\hat{\mathbf{n}}$, $\hat{\mathbf{U}}_2$). We denote the angle between $\hat{\mathbf{s}}^i$ and $\hat{\mathbf{U}}_1$ by θ_1, and that between $\hat{\mathbf{s}}^i$ and $\hat{\mathbf{U}}_2$ by θ_2. Thus,

$$\theta_1 = \cos^{-1}(-\hat{\mathbf{n}} \cdot \hat{\mathbf{U}}_1) \tag{3.200}$$

$$\theta_2 = \cos^{-1}(-\hat{\mathbf{n}} \cdot \hat{\mathbf{U}}_2) \tag{3.201}$$

(c) Without loss of generality, we may consider the angle α limited to the range $0 \leq \alpha \leq \pi/2$. Therefore,

$$\alpha = \begin{cases} \cos^{-1}(\hat{\mathbf{t}} \cdot \hat{\mathbf{U}}_2) = \cos^{-1}(\hat{\mathbf{b}} \cdot \hat{\mathbf{U}}_1), & \text{if } \hat{\mathbf{t}} \cdot \hat{\mathbf{U}}_2 > 0 \\ \pi - \cos^{-1}(\hat{\mathbf{t}} \cdot \hat{\mathbf{U}}_2) = \pi - \cos^{-1}(\hat{\mathbf{b}} \cdot \hat{\mathbf{U}}_1), & \text{otherwise} \end{cases} \tag{3.202}$$

(d) The radius of curvature of the surface in the plane of incidence will be denoted a_t, and given in terms of the principal radii a_1 and a_2 by (Euler's theorem),

$$\frac{1}{a_t} = \frac{\sin^2\alpha}{a_1} + \frac{\cos^2\alpha}{a_2} \tag{3.203}$$

The radius of curvature of the surface in the plane $(\hat{\mathbf{n}}, \hat{\mathbf{b}})$ will be denoted a_b and given by

$$\frac{1}{a_b} = \frac{\cos^2\alpha}{a_1} + \frac{\sin^2\alpha}{a_2} \tag{3.204}$$

Recall that, as $\hat{\mathbf{t}}$ (in the plane of incidence) and $\hat{\mathbf{b}}$ are orthogonal, the plane $(\hat{\mathbf{n}}, \hat{\mathbf{b}})$ will be orthogonal to the plane of incidence.

(e) The principal directions $\hat{\mathbf{X}}_1^i$ and $\hat{\mathbf{X}}_2^i$ of the incident wavefront are orthogonal to $\hat{\mathbf{s}}^i$ and to each other. Their relationship to any of the other vectors shown depends on the properties of the incident GO field considered in a specific problem. Thus, in general they have an arbitrary orientation with respect to the plane of incidence. We should not think that $\hat{\mathbf{X}}_1^i$ must lie in the plane of incidence, for example, although it indeed might do so. A similar remark may be made for $\hat{\mathbf{X}}_2^i$. Remember that $\hat{\mathbf{X}}_1^i$ and $\hat{\mathbf{X}}_2^i$ do not describe the polarization of the incident wavefront. Furthermore, $\hat{\mathbf{X}}_1^i$ and $\hat{\mathbf{X}}_2^i$ in general do not coincide with the principal directions of the surface $\hat{\mathbf{U}}_1$ and $\hat{\mathbf{U}}_2$. For later use, we adopt the following notation:

$$[\Gamma] = \begin{bmatrix} \Gamma_{11} & \Gamma_{12} \\ \Gamma_{21} & \Gamma_{22} \end{bmatrix} = \begin{bmatrix} \hat{\mathbf{X}}_1^i \cdot \hat{\mathbf{U}}_1 & \hat{\mathbf{X}}_1^i \cdot \hat{\mathbf{U}}_2 \\ \hat{\mathbf{X}}_2^i \cdot \hat{\mathbf{U}}_1 & \hat{\mathbf{X}}_2^i \cdot \hat{\mathbf{U}}_2 \end{bmatrix} \tag{3.205}$$

(f) The polarization of the incident field is described in terms of the hard and soft directions $\hat{\mathbf{e}}_\parallel^i$ and $\hat{\mathbf{e}}_\perp$, respectively. These vectors are defined with respect to the plane of incidence. Because they also are orthogonal to the incident ray direction $\hat{\mathbf{s}}^i$, if we stand at Q_r and look in the direction of the incoming ray, then $\hat{\mathbf{e}}_\parallel^i$ and $\hat{\mathbf{e}}_\perp$ in general will be oriented with respect to principal directions $\hat{\mathbf{X}}_1^i$ and $\hat{\mathbf{X}}_2^i$ as shown in Figure 3.30. The angle δ will appear in the formulas for the caustic distances of the reflected ray in Section 3.5.3.

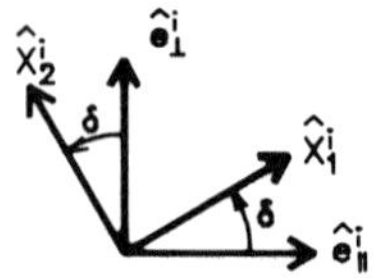

Figure 3.30 Angle δ between the polarization vectors and principal directions of a ray tube.

(g) Statements similar to those in (e) and (f) also apply in connection with the properties of the reflected wavefront.

3.5.2 Principal Radii of Curvature of Reflected Ray Tube at Q_r—First Format

The expressions for the principal radii of curvature at Q_r of the reflected wavefront were shown by Kouyoumjian and Pathak [2] to be of the form:

$$\frac{1}{\rho_1^r} = \frac{1}{2}\left[\frac{1}{\rho_1^i} + \frac{1}{\rho_2^i}\right] + \frac{1}{f_1} \tag{3.206}$$

$$\frac{1}{\rho_2^r} = \frac{1}{2}\left[\frac{1}{\rho_1^i} + \frac{1}{\rho_2^i}\right] + \frac{1}{f_2} \tag{3.207}$$

where

$$\begin{aligned}\frac{1}{f_{1,2}} = {}& \frac{\cos\theta^i}{(\det[\Gamma])^2}\left[\frac{(\Gamma_{22})^2 + (\Gamma_{12})^2}{a_1}\right.\\ &\left. + \frac{(\Gamma_{21})^2 + (\Gamma_{11})^2}{a_2}\right] \pm \frac{1}{2}\left[\left(\frac{1}{\rho_1^i} - \frac{1}{\rho_2^i}\right)^2\right.\\ &+\left(\frac{1}{\rho_1^i} - \frac{1}{\rho_2^i}\right)\frac{4\cos\theta^i}{(\det[\Gamma])^2} \times \left[\frac{(\Gamma_{22})^2 - (\Gamma_{12})^2}{a_1} + \frac{(\Gamma_{21})^2 - (\Gamma_{11})^2}{a_2}\right]\\ &+ \frac{4\cos^2\theta^i}{(\det[\Gamma])^4}\left\{\left[\frac{(\Gamma_{22})^2 + (\Gamma_{12})^2}{a_1}\right.\right.\\ &\left.\left.\left. + \frac{(\Gamma_{21})^2 + (\Gamma_{11})^2}{a_2}\right]^2 - \frac{4(\det[\Gamma])^2}{a_1 a_2}\right]\right\}^{1/2}\end{aligned} \tag{3.208}$$

in which the plus sign is associated with f_1 and the minus sign with f_2. The Γ_{ij} are defined in (3.183).

3.5.3 Principal Radii of Curvature of Reflected Ray Tube at Q_r—Second Format

An alternative (though entirely consistent) form of the expressions for ρ_1^r and ρ_2^r have been provided by Pathak, Burnside, and Marhefka [7]:

$$\frac{1}{\rho_{1,2}^r} = \frac{1}{\rho_m^i} + \frac{f}{a_t \cos\theta^i}\left\{1 \pm \left[\frac{a_t^2 \cos^2\theta^i}{4f^2}\left(\frac{1}{\rho_1^i} - \frac{1}{\rho_2^i}\right)^2 \right.\right.$$
$$+ \frac{a_t^2 \cos\theta^i}{f^2}\left(\frac{1}{\rho_1^i} - \frac{1}{\rho_2^i}\right)\left\{\frac{g \cos^2\delta}{a_t}\right.$$
$$\left. - \sin 2\delta \sin 2\alpha \cos\theta^i\left(\frac{1}{a_1} - \frac{1}{a_2}\right)\right\} \tag{3.209}$$
$$\left.\left. + 1 - \frac{4a_t^2 \cos^2\theta^i}{f^2 a_1 a_2}\right]^{1/2}\right\}$$

in which the plus sign is associated with ρ_1^r and the minus sign with ρ_2^r. The term ρ_m^i is

$$\frac{1}{\rho_m^i} = \frac{1}{2}\left(\frac{1}{a_1} + \frac{1}{a_2}\right) \tag{3.210}$$

where the subscript m indicates the mean radius of curvature of the incident wavefront at Q_r. Other terms are

$$f = 1 + \frac{a_t}{a_b}\cos^2\theta^i \tag{3.211}$$

$$g = 1 - \frac{a_t}{a_b}\cos^2\theta^i \tag{3.212}$$

and a_t and a_b are defined in (3.203) and (3.204), respectively. The advantage of this second format is that the role of the radii of curvature a_t and a_b, as well as the angle δ, are shown explicitly.

The results in this and the previous section are of a very general nature. For many problems in practice the situation simplifies. Useful special cases will be presented in the next section.

3.5.4 Important Special Cases

Plane of Incidence Coincident with a Principal Plane of the Surface at Point Q_r

Consider the case for which the plane of incidence defined by $(\hat{\mathbf{n}}, \hat{\mathbf{s}}^i)$, and the principal plane of the surface at Q_r defined by the vector pair $(\hat{\mathbf{n}}, \hat{\mathbf{U}}_2)$, coincide. Then $\alpha = 0$, and expressions (3.206) to (3.209) reduce to

$$\frac{1}{\rho_1^r} = \frac{1}{\rho_1^i} + \frac{2\cos\theta^i}{a_1} \tag{3.213}$$

$$\frac{1}{\rho_2^r} = \frac{1}{\rho_2^i} + \frac{2}{a_2\cos\theta^i} \tag{3.214}$$

Always remember that in (3.213) and (3.214) the principal plane $(\hat{\mathbf{n}}, \hat{\mathbf{U}}_2)$ of the surface at Q_r has been assumed coincident with the plane of incidence. Thus a_2 is the radius of curvature of the surface in its principal plane that coincides with the plane of incidence, and a_1 is the principal radius of curvature of the surface at Q_r in the companion principal plane defined by the vector pair $(\hat{\mathbf{n}}, \hat{\mathbf{b}})$.

Expressions (3.213) and (3.214) are valid for *any* incident astigmatic ray tube *provided the conditions on the coincident planes are satisfied* as stated. If, *in addition to* the requirement regarding the coincidence of planes $(\hat{\mathbf{n}}, \hat{\mathbf{s}}^i)$ and $(\hat{\mathbf{n}}, \hat{\mathbf{U}}_2)$, the incident wavefronts are spherical (that is, e.g., $\rho_1^i = \rho_2^i = s^i$), then (3.213) and (3.214) become

$$\frac{1}{\rho_1^r} = \frac{1}{s^i} + \frac{2\cos\theta^i}{a_1} \tag{3.215}$$

$$\frac{1}{\rho_2^r} = \frac{1}{s^i} + \frac{2}{a_2\cos\theta^i} \tag{3.216}$$

Finally, if this incident wavefront in fact is a plane wave ($s^i \to \infty$), (3.215) and (3.216) reduce to

$$\frac{1}{\rho_1^r} = \frac{2\cos\theta^i}{a_1} \tag{3.217}$$

$$\frac{1}{\rho_2^r} = \frac{2}{a_2 \cos\theta^i} \tag{3.218}$$

If we select the case for which the plane of incidence, and the principal plane of the surface at Q_r defined by the vector pair ($\hat{\mathbf{n}}$, $\hat{\mathbf{U}}_1$), coincide, then $\alpha = \pi/2$, and instead of (3.213) and (3.214) we have

$$\frac{1}{\rho_1^r} = \frac{1}{\rho_1^i} + \frac{2}{a_1 \cos\theta^i} \tag{3.219}$$

$$\frac{1}{\rho_2^r} = \frac{1}{\rho_2^i} + \frac{2 \cos\theta^i}{a_2} \tag{3.220}$$

with the associated simplifications for spherical and plane wave incidence.

Incident Spherical Wavefront

Let us return to the general expressions (3.206) and (3.209); this time we will *not* make any assumptions regarding the plane of incidence coinciding with one of the principal planes of the reflecting surface. However, we are interested in the important case of spherical wave incidence ($\rho_1^i = \rho_2^i = s^i$). Then, the orientation of the principal planes of the incident wavefront can be selected just as we wish; that is, the principal directions $\hat{\mathbf{X}}_1^i$ and $\hat{\mathbf{X}}_2^i$ of the incident wavefront can be chosen in any convenient way. Following Kouyoumjian and Pathak [2] we select $\hat{\mathbf{X}}_1^i$ to lie in the plane of incidence; then $\hat{\mathbf{X}}_2^i$ is parallel to the plane tangent to the surface at the reflection point Q_r. If we then recognize from the geometry that [2]

$$\hat{\mathbf{X}}_1^i = -\cos\theta^i \sin\alpha\hat{\mathbf{U}}_1 + \cos\theta^i \cos\alpha\hat{\mathbf{U}}_2 + \sin\theta^i\hat{\mathbf{n}} \tag{3.221}$$

$$\hat{\mathbf{X}}_2^i = \cos\alpha\hat{\mathbf{U}}_1 + \sin\alpha\hat{\mathbf{U}}_2 \tag{3.222}$$

We may easily determine the dot products Γ_{ij} in (3.205). Hence, for spherical wave incidence (all other conditions being quite general), we have

$$\frac{1}{\rho_1^r} = \frac{1}{s^i} + \frac{1}{f_1} \tag{3.223}$$

$$\frac{1}{\rho_2^r} = \frac{1}{s^i} + \frac{1}{f_2} \tag{3.224}$$

where

$$f_{1,2} = \frac{1}{\cos\theta^i}\left[\frac{\sin^2\theta_2}{a_1} + \frac{\sin^2\theta_1}{a_2}\right] \pm \sqrt{\frac{1}{\cos^2\theta^i}\left[\frac{\sin^2\theta_2}{a_1} + \frac{\sin^2\theta_1}{a_2}\right]^2 - \frac{4}{a_1 a_2}} \tag{3.225}$$

in which the plus sign is associated with ρ_1^r and the minus sign with ρ_2^r, and where [2]

$$\sin^2\theta_1 = \cos^2\alpha + \sin^2\alpha \cos^2\theta^i \tag{3.226}$$

$$\sin^2\theta_2 = \sin^2\alpha + \cos^2\alpha \cos^2\theta^i \tag{3.227}$$

angles θ_1 and θ_2 being the same as those whose general definition is given by (3.200) and (3.201).

If we allow s^i to become infinite, which implies *plane wave incidence*, (3.226) and (3.227) transform to

$$\rho_1^r = f_1 \tag{3.228}$$

$$\rho_2^r = f_2 \tag{3.229}$$

The special cases just considered are summarized in the form of a flow chart in Figure 3.31.

Remark 7: If we next let the plane of incidence coincide with principal plane $(\hat{\mathbf{n}}, \hat{\mathbf{U}}_2^i)$ of the surface, $\theta_2 = 0$ and (3.223) and (3.224) simplify to (3.215) and (3.216). Under the same conditions equations (3.228) and (3.229) reduce to (3.217) and (3.218).

3.5.5 Principal Directions of the Reflected Wavefront

For the 2D problems previously considered the principal planes of both the incident and reflected wavefronts were restricted to the x-y plane and the plane orthogonal to it. In the more general situations presently under consideration, this is not the case. At Q_r, the principal directions of the incident wavefront and those of the reflecting surface will be known, as this information is needed to unambiguously specify the problem in the first place. However, the principal directions of the reflected wavefront must be found as part of solving the problem.

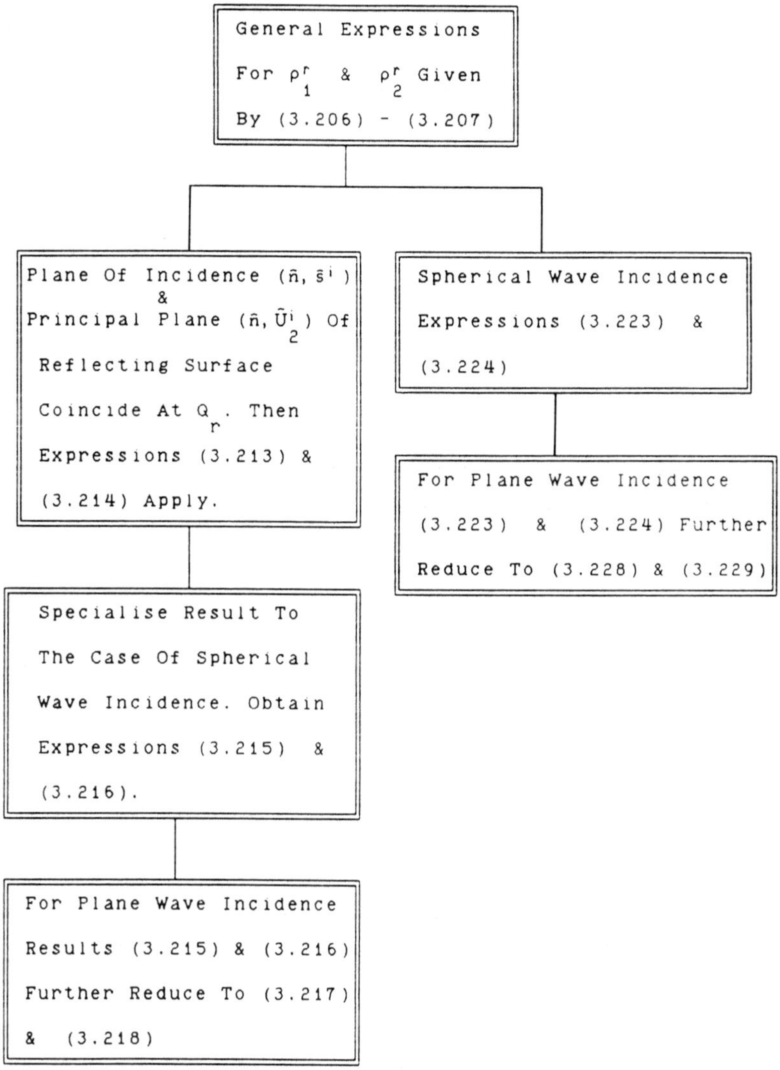

Figure 3.31 Guide to special cases of the formulas for the caustic distances of a reflected wavefront.

If the vectors $\hat{\mathbf{X}}_1^i$ and $\hat{\mathbf{X}}_2^i$ are reflected in the plane tangent to the surface at the reflection point Q_r, we obtain the intermediate vectors:

$$\hat{\sigma}_1 = \hat{\mathbf{X}}_1^i - 2(\hat{\mathbf{n}} \cdot \hat{\mathbf{X}}_1^i)\hat{\mathbf{n}} \tag{3.230}$$

$$\hat{\sigma}_2 = \hat{\mathbf{X}}_2^i - 2(\hat{\mathbf{n}} \cdot \hat{\mathbf{X}}_2^i)\hat{\mathbf{n}} \tag{3.231}$$

However, although these indeed are orthogonal to the reflected ray, they are not the principal directions of the reflected wavefront. The principal directions $\hat{\mathbf{X}}_1^r$ and $\hat{\mathbf{X}}_2^r$ are obtained by transforming $\hat{\sigma}_1$ and $\hat{\sigma}_2$ using the following results [2]:

$$\hat{\mathbf{X}}_1^r = \frac{\left[Q_{22}^r - 1/\rho_1^r\right]\hat{\sigma}_1 - Q_{12}^r\hat{\sigma}_2}{\sqrt{(Q_{22}^r - 1/\rho_1^r)^2 + (Q_{22}^r)^2}} \tag{3.232}$$

and so we must have

$$\hat{\mathbf{X}}_2^r = -\hat{\mathbf{s}}^r \times \hat{\mathbf{X}}_1^r \tag{3.233}$$

as $\hat{\mathbf{X}}_1^r$ and $\hat{\mathbf{X}}_2^r$ are both orthogonal to $\hat{\mathbf{s}}^r$. Quantities $\hat{\mathbf{X}}_1^r$ and $\hat{\mathbf{X}}_2^r$ are clearly linear combinations of $\hat{\sigma}_1$ and $\hat{\sigma}_2$. The elements Q_{ij}^r are those of the curvature matrix [2]. We write these in the compact form:

$$Q_{11}^r = \frac{1}{\rho_1^i} + \frac{2}{a_t \cos\theta^i} \tag{3.234}$$

$$Q_{12}^r = Q_{21}^r = -\sin 2\alpha\left[\frac{1}{a_1} - \frac{1}{a_2}\right] \tag{3.235}$$

$$Q_{22}^r = \frac{1}{\rho_2^i} + \frac{2\cos\theta^i}{a_b} \tag{3.236}$$

Note that when $\alpha = 0$ or $\alpha = \pi/2$, which implies a coincidence of the plane of incidence and one of the principle planes of the reflecting surface, $Q_{12}^r = Q_{21}^r = 0$, and the principal directions of the reflected wavefront are $\hat{\sigma}_1$ and $\hat{\sigma}_2$. If the reflecting surface is planar, so that a_1 and a_2 are infinite, then, too, we have $Q_{12}^r = Q_{21}^r = 0$. We say that Q_{12}^r and Q_{21}^r account for the torsional effect that the reflecting surface has on the incident ray tube. When the wavefront travels through free space en route to Q_r, and when it travels away from Q_r after reflecting there, it does not experience any torsion.

3.5.6 Alternative Form for the Reflected GO Field at the Specular Point Q_r

To work in terms of the ray-fixed coordinate system $(\hat{\mathbf{e}}_\parallel^i, \hat{\mathbf{e}}_\perp)$ may not always be convenient as far as the polarization of the incident and reflected fields are concerned. Therefore, we next derive an alternative expression to those given for $\mathbf{E}^r(Q_r)$ in Section 3.2.4. Once more, we will exploit the local plane wave (TEM) properties of GO fields.

If at the point Q_r on the reflecting surface we take the dot product of both sides of equation (3.43) with $\hat{\mathbf{n}}$, we obtain

$$\hat{\mathbf{n}} \cdot \mathbf{E}^i = -Z\hat{\mathbf{n}} \cdot (\hat{\mathbf{s}}^i \times \mathbf{H}^i)$$

which, upon application of the vector identity:

$$\mathbf{A} \cdot (\mathbf{B} \times \mathbf{C}) = \mathbf{B} \cdot (\mathbf{C} \times \mathbf{A})$$

yields

$$\hat{\mathbf{n}} \cdot \mathbf{E}^i = -Z\hat{\mathbf{s}}^i \cdot (\hat{\mathbf{n}} \times \mathbf{H}^i) \tag{3.237}$$

Doing the same to (3.44) gives us

$$\hat{\mathbf{n}} \cdot \mathbf{E}^r = -Z\hat{\mathbf{s}}^r \cdot (\hat{\mathbf{n}} \times \mathbf{H}^r)$$

which from (3.48) becomes

$$\hat{\mathbf{n}} \cdot \mathbf{E}^r = -Z\hat{\mathbf{s}}^r \cdot (\hat{\mathbf{n}} \times \mathbf{H}^i) \tag{3.238}$$

Replacing $\hat{\mathbf{s}}^r$ in the right-hand side of (3.238) by its form in expression (3.23), and recognizing that $\hat{\mathbf{n}} \cdot (\hat{\mathbf{n}} \times \mathbf{H}^r)$ always is zero, hence (3.238) is equivalent to

$$\hat{\mathbf{n}} \cdot \mathbf{E}^r = -Z\hat{\mathbf{n}} \cdot (\hat{\mathbf{s}}^i \times \mathbf{H}^i) \tag{3.239}$$

From a comparison of (3.239) and (3.237), we draw the obvious conclusion that, at Q_r,

$$\hat{\mathbf{n}} \cdot \mathbf{E}^i = \hat{\mathbf{n}} \cdot \mathbf{E}^r \tag{3.240}$$

Remark 8: Remember that this is not the boundary condition that applies to the normal component of electric fields in general. It pertains here only because of the locally plane wave property of Go fields.

As the next step we take the vector product of both sides of condition (3.33) with $\hat{\mathbf{n}}$, that is,

$$\hat{\mathbf{n}} \times (\hat{\mathbf{n}} \times \mathbf{E}^i) = -\hat{\mathbf{n}} \times (\hat{\mathbf{n}} \times \mathbf{E}^r) \tag{3.241}$$

Application of the vector identity:

$$\mathbf{A} \times (\mathbf{B} \times \mathbf{C}) = \mathbf{B}(\mathbf{A} \cdot \mathbf{C}) - \mathbf{C}(\mathbf{A} \cdot \mathbf{B})$$

to both sides of (3.219) then yields the result:

$$\mathbf{E}^r = -\mathbf{E}^i + (\hat{\mathbf{n}} \cdot \mathbf{E}^i + \hat{\mathbf{n}} \cdot \mathbf{E}^r)\hat{\mathbf{n}} \tag{3.242}$$

Incorporation of the intermediate result (3.240) then gives the desired result written in its final form as

$$\mathbf{E}^r(Q_r) = -\mathbf{E}^i(Q_r) + 2[\hat{\mathbf{n}} \cdot \mathbf{E}^i(Q_r)]\hat{\mathbf{n}} \tag{3.243}$$

A similar derivation may be done for the GO magnetic field, yielding

$$\mathbf{H}^r(Q_r) = \mathbf{H}^i(Q_r) - 2[\hat{\mathbf{n}} \cdot \mathbf{H}^i(Q_r)]\hat{\mathbf{n}} \tag{3.244}$$

3.5.7 Comments on the Expressions for the Reflected GO Field

We have now assembled all the tools necessary to find the GO reflected fields from smooth conducting objects by using the expression:

$$\mathbf{E}^r(P) = \mathbf{E}^r(Q_r)\sqrt{\frac{\rho_1^r \rho_2^r}{(\rho_1^r + s^r)(\rho_2^r + s^r)}}\, e^{-jks^r} \tag{3.245}$$

Let us examine the conduct of (3.245) for certain distinctive cases that frequently arise in typical engineering applications of GO. We divide our discussion according to the observation conditions. First, we deal with those instances in which the observation point is in the far-zone; hence, s^r is extremely large. Then, we consider the case where we wish to find the reflected GO fields in the near-zone of the scatterer.

Observation Point in the Far-Zone

(a) Suppose that the reflected GO field is to be observed in the far-zone of the scatterer, so that $s^r \gg 1$. Assume, furthermore, that at all specular points Q_r of interest the *radii of curvature of the reflected ray tubes are finite*. Here, we have $s^r \gg \rho_1^r$ and $s^r \gg \rho_2^r$, and the usual far-zone approximation of the spreading factor, similar to that in (3.78) for the 2D case, is

$$\sqrt{\frac{\rho_1^r \rho_2^r}{(\rho_1^r + s^r)(\rho_2^r + s^r)}} \approx \frac{\sqrt{\rho_1^r \rho_2^r}}{s^r}$$

Expression (3.245) reduces to the widely used form:

$$\mathbf{E}^r(P) = \mathbf{E}^r(Q_r)\sqrt{\rho_1^r \rho_2^r}\,\frac{e^{-jks^r}}{s^r} \tag{3.246}$$

If, *in addition*, the incident field at Q_r is a plane wave, (3.228) and (3.229) give us, in terms of the principal radii of curvature of the surface at Q_r:

$$\sqrt{\rho_1^r \rho_2^r} = \frac{\sqrt{a_1 a_2}}{2}$$

and the reflected field in (3.246) becomes

$$\mathbf{E}^r(P) = \mathbf{E}^r(Q_r)\frac{\sqrt{a_1 a_2}}{2}\,\frac{e^{-jks^r}}{s^r} \tag{3.247}$$

(b) Whether the radii of curvature ρ_1^r or ρ_2^r of the reflected wavefront are finite depends of course on the properties of both the incident wavefront and the surface at Q_r. If either or both of ρ_1^r and ρ_2^r become infinite, then (3.246) predicts infinite fields in the given observation direction, and therefore an invalid description of the fields scattered in this direction. In other words, there is a caustic in that direction under the particular set of conditions. In such circumstances, the equivalent current concepts of Chapter 7 must be utilized.

Observation Point in the Near-Zone

If s^r is finite (that is, the reflected fields are observed in the near-zone) GO may be used to find the reflected field even if ρ_1^r, ρ_2^r, or both are infinite. If only $\rho_1^r \to \infty$ then the spreading factor becomes that of a cylindrical ray tube.

$$\sqrt{\frac{\rho_2^r}{\rho_2^r + s^r}} \tag{3.248}$$

If in addition $\rho_2^r \to \infty$, then the spreading factor further reduces to unity, which is that of a plane wave ray tube.

3.5.8 Alternative Determination of Principal Radii of Curvature of the Reflected Wavefront

An alternative and often more convenient (especially in reflector antenna applications) way of determining the caustic distances ρ_1^r and ρ_2^r has been given by Rusch [16]. The procedure is as follows. Let $\mathbf{r} = x(u, v)\hat{\mathbf{x}} + y(u, v)\hat{\mathbf{y}} + z(u, v)\hat{\mathbf{z}}$ be the parameterized coordinate vector describing the reflecting surface, with u and v any curvilinear parameters. The first step is to determine the reflected field propagation vector $\hat{\mathbf{s}}^r(u, v)$ from (3.23). Then, the following quantities must be found:

$$M = (\partial\hat{\mathbf{s}}^r/\partial u) \cdot (\partial\hat{\mathbf{s}}^r/\partial u) \tag{3.249}$$

$$N = (\partial\hat{\mathbf{s}}^r/\partial u) \cdot (\partial\hat{\mathbf{s}}^r/\partial v) \tag{3.250}$$

$$Q = (\partial\hat{\mathbf{s}}^r/\partial v) \cdot (\partial\hat{\mathbf{s}}^r/\partial v) \tag{3.251}$$

$$m = (\partial\mathbf{r}/\partial u) \cdot (\partial\hat{\mathbf{s}}^r/\partial u) \tag{3.252}$$

$$n = (\partial\mathbf{r}/\partial v) \cdot (\partial\hat{\mathbf{s}}^r/\partial u) \tag{3.253}$$

$$q = (\partial\mathbf{r}/\partial v) \cdot (\partial\hat{\mathbf{s}}^r/\partial v) \tag{3.254}$$

The two roots, ξ_1 and ξ_2, of the quadratic equation:

$$(MQ - N^2)\xi^2 + (Mq - 2Nn + Qm)\xi + (mq - n^2) = 0 \tag{3.255}$$

then supply the caustic distances of the reflected wavefront as

$$\rho_1^r = -\xi_1 \tag{3.256}$$

$$\rho_2^r = -\xi_2 \tag{3.257}$$

For a problem with axial symmetry this simplifies [16] to the expressions:

$$\rho_1^r = m/M \tag{3.258}$$

$$\rho_2^r = q/Q \tag{3.259}$$

Remark 9: Definition of Radar Cross Section. The *radar cross section* (RCS) σ of a 3D object is defined in a manner similar to the radar width σ^w of a 2D one. We assume uniform plane wave incidence on the scattering object and observation of the field backscattered in the direction from whence this plane wave was incident. The phases of the incident (E^i) and *backscattered* (E^r) fields are referred to the same reference point. If r is the distance from the phase reference to some far-zone observation point, then the RCS is defined as

$$\sigma = \lim_{r\to\infty} 4\pi r^2 \left|\frac{E^r}{E^i}\right|^2 \tag{3.260}$$

As with the determination of radar width, $\theta^i = 0$ in RCS calculations, and hence $\cos\theta^i = 1$. Also, the law of reflection can be used conveniently in terms of the unit tangent vector to the surface at Q_r as $\hat{\mathbf{s}}^i \cdot \hat{\mathbf{t}}(Q_r) = 0$. We in fact have defined monostatic RCS. Although it indeed is possible to find the bistatic RCS of an object, when we refer to RCS in this text we refer to the monostatic case.

3.6 EXAMPLES OF THREE-DIMENSIONAL REFLECTED FIELD PROBLEMS

Example 3.8: On the radar cross section of bodies with finite radii of curvature

Consider a plane wave incident on a 3D conducting object from direction (θ_i, ϕ_i). The unit vector in the direction of any incident ray associated with this wave thus is given by

$$\hat{\mathbf{s}}^i = \sin\theta_i \cos\phi_i \hat{\mathbf{x}} + \sin\theta_i \sin\phi_i \hat{\mathbf{y}} + \cos\theta_i \hat{\mathbf{z}} \tag{3.261}$$

The scatterer surface can be defined in parametric form by

$$\mathbf{r}(u, v) = x(u, v)\,\hat{\mathbf{x}} + y(u, v)\,\hat{\mathbf{y}} + z(u, v)\hat{\mathbf{z}} \tag{3.262}$$

with $u_{\min} \leqslant u \leqslant u_{\max}$ and $v_{\min} \leqslant v \leqslant v_{\max}$. The vectors $\mathbf{r}_u$ and $\mathbf{r}_v$, as defined by expressions (C.34) and (C.35) of Appendix C, are tangent to the surface along the u-curve and v-curve, as will be shown in Figure C.3. They thus are orthogonal to the normal $\hat{\mathbf{n}}$, and hence to the incident vector $\hat{\mathbf{s}}^i$. Therefore, if the point of monostatic backscatter on the surface, for a given (θ_i, ϕ_i) is specified by the parameters (u_s, v_s), then we must have

$$\mathbf{r}_u(u_s, v_s) \cdot \hat{\mathbf{s}}^i = 0 \tag{3.263}$$

$$\mathbf{r}_v(u_s, v_s) \cdot \hat{\mathbf{s}}^i = 0 \tag{3.264}$$

Simultaneous solution of (3.263) and (3.264) locates the specular point for any incidence direction.

As a specific example consider the ellipsoid shown in Figure 3.32. It is described in parametric form by

$$\mathbf{r}(u, v) = a \sin u \cos v \hat{\mathbf{x}} + b \sin u \sin v \hat{\mathbf{y}} + c \cos u \hat{\mathbf{z}} \tag{3.265}$$

with the limits to the parameters given by

$$0 \leqslant u \leqslant \pi \tag{3.266}$$

$$0 \leqslant v \leqslant 2\pi \tag{3.267}$$

The parameters u and v are not the spherical coordinate angles θ and ϕ, respectively. We have

$$\mathbf{r}_u(u, v) = -a \sin u \sin v \hat{\mathbf{x}} + b \sin u \cos v \hat{\mathbf{y}} \tag{3.268}$$

$$\mathbf{r}_v(u, v) = a \cos u \cos v \hat{\mathbf{x}} + b \cos u \sin v \hat{\mathbf{y}} - c \sin u \hat{\mathbf{z}} \tag{3.269}$$

From (3.261), (3.263), and (3.269), it follows that

$$\sin u_s \sin\theta_i (a \cos\phi_i \sin v_s - b \sin\phi_i \cos v_s) = 0$$

and hence that

$$\tan v_s = (b/a) \tan\phi_i \tag{3.270}$$

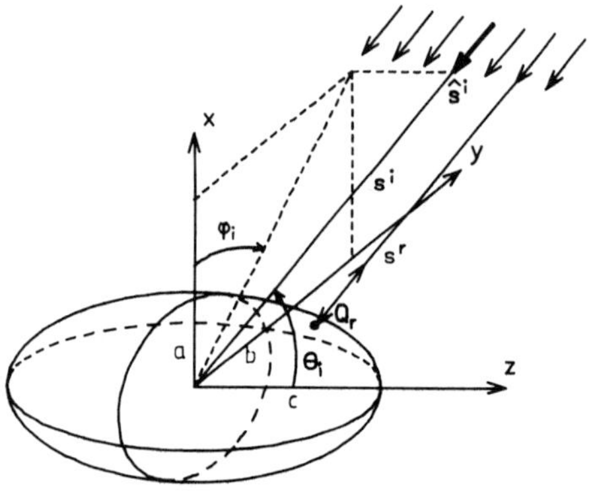

Figure 3.32 Ellipsoid illuminated by a plane wave from direction (θ_i,ϕ_i).

so that

$$v_s(\theta_i, \phi_i) = \tan[(b/a)\tan\phi_i] \tag{3.271}$$

$$\sin v_s = \frac{(b/a)\tan\phi_i}{\sqrt{1 + (b/a)^2 \tan^2\phi_i}} \tag{3.272}$$

$$\cos v_s = \frac{1}{\sqrt{1 + (b/a)^2 \tan^2\phi_i}} \tag{3.273}$$

Similarly, it follows from (3.261), (3.263), and (3.268) that

$$\cos u_s \sin\theta_i (a \cos v_s \cos\phi_i + b \sin v_s \cos\phi_i) = c \sin u_s \cos\theta_i \tag{3.274}$$

and thus that

$$\tan u_s = \frac{\tan\theta_i [a \cos\phi_i + (b^2/a) \sin\phi_i \tan\phi_i]}{c\sqrt{1 + (b/a)^2 \tan^2\phi_i}} \tag{3.275}$$

$$u_s = \tan^{-1}\left[\frac{\tan\theta_i [a \cos\phi_i + (b^2/a) \sin\phi_i \tan\phi_i]}{c\sqrt{1 + (b/a)^2 \tan^2\phi_i}}\right] \tag{3.276}$$

from which we can easily find

$$\sin u_s = \frac{\tan u_s}{\sqrt{1 + \tan^2 u_s}} \tag{3.277}$$

$$\cos u_s = \frac{1}{\sqrt{1 + \tan^2 u_s}} \tag{3.278}$$

The specular point thus is given by

$$\mathbf{r}_s(\theta_i, \phi_i) = a \sin u_s \cos v_s \hat{\mathbf{x}} + b \sin u_s \sin v_s \hat{\mathbf{y}} + c \cos u_s \hat{\mathbf{z}} \tag{3.279}$$

with $\sin(u_s)$, $\cos(u_s)$, $\cos(v_s)$, and $\sin(v_s)$ given by (3.277), (3.278), (3.272), and (3.273), respectively. The methods of Section C.2.2 of Appendix C can be used to find the principal radii of curvature of the ellipsoid at any point (u_s, v_s); the principal radii of curvature of the reflected wavefront determined from (3.228) and (3.229); expression (3.247) used to find $\mathbf{E}^r$; and the definition (3.260) used to find the RCS.

Example 3.9: Field reflected from a hyperboloid

The hyperboloid forms the subreflector of the classical Cassegrain reflector antenna geometry [17]. In this example we assume that the hyperboloid, which is rotationally symmetric, is illuminated by a source that has an axially symmetric pattern. We then may concentrate consideration on any cross section of the hyperboloid, as shown in Figure 3.33.

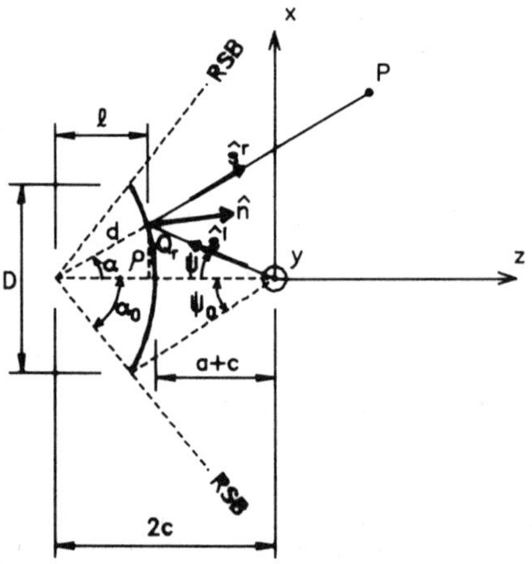

Figure 3.33 Hyperboloid illuminated by a spherical wave source placed at its real focus.

Geometrical Relationships

The defining equation of the hyperboloid is given by

$$\mathbf{r} = \rho \cos\phi \hat{\mathbf{x}} + \rho \sin\phi \hat{\mathbf{y}} - [c + (a/b)\sqrt{\rho^2 + b^2}]\hat{\mathbf{z}} \tag{3.280}$$

where ρ and ϕ are the usual cylindrical coordinates. The unit normal at any point (ρ, ϕ) on the surface can be found using the techniques of Appendix C to be

$$\hat{\mathbf{n}} = n_x\hat{\mathbf{x}} + n_y\hat{\mathbf{y}} + n_z\hat{\mathbf{z}} \tag{3.281}$$

with

$$n_x = \frac{a\rho \cos\phi}{\sqrt{b^2(\rho^2 + b^2) + a^2\rho^2}} \tag{3.282}$$

$$n_y = \frac{a\rho \sin\phi}{\sqrt{b^2(\rho^2 + b^2) + a^2\rho^2}} \tag{3.283}$$

$$n_z = \frac{b\sqrt{\rho^2 + b^2}}{\sqrt{b^2(\rho^2 + b^2) + a^2\rho^2}} \tag{3.284}$$

Incident Field

The hyperboloid is assumed to be illuminated by a corrugated conical horn whose axially symmetric electric fields are of the GO type, as given in Section 2.4.5. So that the horn antenna's main beam illuminate the hyperboloid, we write the expression for the horn antenna's field as

$$\mathbf{E}^i(s^i, \Psi, \phi) = \frac{e^{-jks^i}}{s^i} A(\Psi)[\sin\phi\hat{\Psi} + \cos\phi\hat{\phi}] \tag{3.285}$$

where s^i is the distance from the horn antenna's phase center, the horn assumed to be positioned at the origin of the coordinate system (the real focus of the hyperboloid [17]), with $A(\Psi)$ having its maximum value at $\Psi = 0$. From (3.280) we have

$$s^i = |\mathbf{r}| = \sqrt{\rho^2 + [c + (a/b)\sqrt{\rho^2 + b^2}]^2} \tag{3.286}$$

The incident ray vector therefore is

$$\hat{\mathbf{s}}^i = \frac{\mathbf{r}}{|\mathbf{r}|} \tag{3.287}$$

Caustic Distances of the Reflected Field

We apply the method for finding the caustic distances of the reflected field described in Section 3.6.8. Applying (3.23) we find that the unit vector in the direction of a reflected ray is

$$\hat{\mathbf{s}}^r = \sin\alpha \cos\phi\hat{\mathbf{x}} + \sin\alpha \sin\phi\hat{\mathbf{y}} + \cos\alpha\hat{\mathbf{z}} \tag{3.288}$$

where

$$\sin\alpha = \frac{\rho}{\sqrt{l^2 + \rho^2}} \tag{3.289}$$

$$\cos\alpha = \frac{l}{\sqrt{l^2 + \rho^2}} \tag{3.290}$$

$$l = c - (a/b)\sqrt{\rho^2 + b^2} \tag{3.291}$$

Because of the rotational symmetry of the geometry, and the fact that we have described the hyperboloid in terms of cylindrical coordinates (ρ, ϕ), we need compute only quantities M and m from (3.249) and (3.252), respectively, to find the caustic distances of the reflected field as

$$\rho_1^r = m/M = (c/b)\sqrt{\rho^2 + b^2} - a \tag{3.292}$$

For this particular geometry, we have

$$\rho_2^r = q/Q = m/M = \rho_1^r \tag{3.293}$$

From the geometry of Figure 3.33, this is equivalent to

$$\rho_1^r = \rho_2^r = \sqrt{l^2 + \rho^2} = d \tag{3.294}$$

The reflected wavefront thus is spherical and appears to emanate from the virtual focus [17] of the hyperboloid.

Reflected Field

From (3.294) the spreading factor of the reflected field becomes

$$A(s^r) = \frac{d}{d + s^r} \tag{3.295}$$

To find $\mathbf{E}^r(Q_r)$, equation (3.243) is possibly the most convenient form to be used. The incident field $\mathbf{E}^i(Q_r)$ is given by (3.285). We easily can see that angle Ψ and the spherical coordinate angle θ are related as

$$\Psi = \pi - \theta \tag{3.296}$$

Furthermore, the unit vector

$$\hat{\Psi} = -\hat{\theta} \tag{3.297}$$

Thus, in the usual spherical coordinates, we have

$$\mathbf{E}^i(Q_r) = \frac{e^{-jks^i}}{s^i} A(\pi - \theta)[-\sin\phi\hat{\theta} + \cos\phi\hat{\phi}] \tag{3.298}$$

Converting $\hat{\mathbf{n}}$ in (3.281) to spherical coordinate form and using it with (3.298), the reflected field at Q_r, $\mathbf{E}^r(Q_r)$ can be found from (3.243). Given θ, it of course will be necessary to find the associated α (only in the far-zone is $\alpha = \theta$) and hence the appropriate value of ρ (which locates Q_r) for the specific θ, using (3.289) and (3.290), for example, to compute the various parameters at Q_r.

Example 3.10: Near-zone scattering of a spherical wavefront from a parabolic cylinder

This geometry has been considered previously in the 2D context in Example 3.6. For the case of line source illumination considered there the reflected wavefront was seen to be plane. The difference here is that, although the reflecting surface is an infinite cylinder, it is illuminated by a point source, and the problem therefore is of a 3D nature. The relevant geometry is shown in Figure 3.34. We here simply wish to determine the properties of the reflected wavefront.

We study the problem of reflected rays in the x-y plane only. Further discussion of the use of parabolic reflectors may be found in [18]. The point Q_r (x_r, y_r) is a point of reflection, whereas Q_s is a point on the directrix of the parabolic surface. From Appendix C we can find the normal vector at Q_r to be

$$\hat{\mathbf{n}}(Q_r) = \frac{-(1/2f)x_r\hat{\mathbf{x}} + \hat{\mathbf{y}}}{\sqrt{1 + (1/2f)^2x_r^2}} \tag{3.299}$$

and the principal radii of curvature at Q_r are given by

$$a_1 = -2f[1 + (1/2f)^2x_r^2]^{3/2} \tag{3.300}$$

$$a_2 \rightarrow \infty \tag{3.301}$$

The unit vectors in the principal directions of the surface can be found by using the methods of Appendix C to be given by

$$\hat{\mathbf{U}}_1 = \frac{\hat{\mathbf{x}} + (1/2f)x_r\hat{\mathbf{y}}}{\sqrt{1 + (1/2f)^2x_r^2}} \tag{3.302}$$

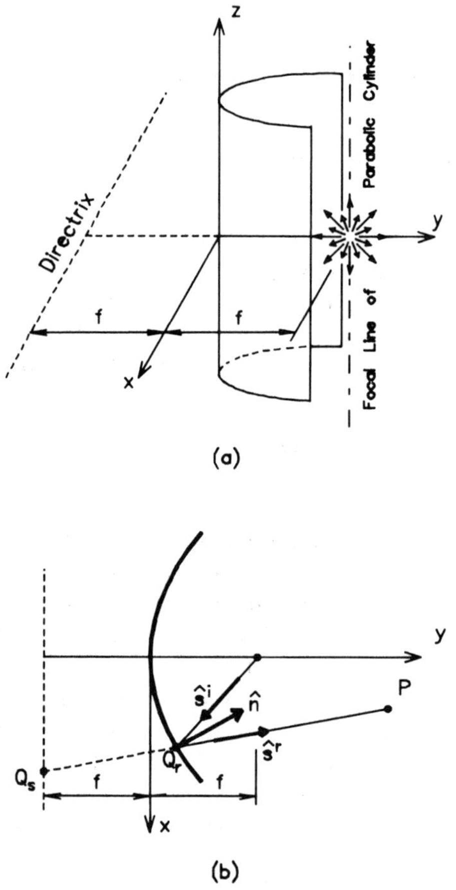

Figure 3.34 (a) Parabolic cylinder illuminated by a spherical wave source placed on its focal line and in the $x - y$ plane. (b) Cross section, in the $x - y$ plane, of the geometry in (a).

$$\hat{\mathbf{U}}_2 = \hat{\mathbf{z}} \tag{3.303}$$

Because we are confining our attention to incident and reflected rays in the x-y plane, the principal planes of the incident rays and those of the surface coincide, and we may use expressions (3.215) and (3.216) to find the radii of curvature of the reflected ray tubes at Q_r. The fact that $a_2 \rightarrow \infty$ immediately reduces (3.215) to

$$\rho_2^r = s^i \tag{3.304}$$

The distance s^i easily is seen to be given by

$$s^i = \sqrt{x_r^2 + (f - y_r)^2} \tag{3.305}$$

However, as the definition of the parabolic surface implies that

$$x_r = \sqrt{4fy_r} \tag{3.306}$$

we have

$$s^i = (f + y_r) \tag{3.307}$$

The unit vector in the direction of the ray incident at Q_r thus is

$$\hat{\mathbf{s}}^i = \frac{x_r\hat{\mathbf{x}} - (f - y_r)\hat{\mathbf{y}}}{(f + y_r)} \tag{3.308}$$

Therefore,

$$\cos\theta^i = -\hat{\mathbf{n}} \cdot \hat{\mathbf{s}}^i = \frac{-(1/2f)x_r^2 - (f - y_r)}{(f + y_r)\sqrt{1 + (1/2f)^2 x_r^2}} \tag{3.309}$$

Thus, (3.215) becomes

$$\begin{aligned} \frac{1}{\rho_1^r} &= \frac{1}{s^i} + \frac{2\cos\theta^i}{a_1} \\ &= \frac{1}{(f + y_r)} - \frac{2[(1/2f)x_r^2 + f - y_r]}{2f(f + y_r)[1 + (1/2f)^2 x_r^2]} \end{aligned} \tag{3.310}$$

Use of the defining equation (3.306) of the parabolic surface reduces the right-hand side of (3.310) to zero, which, in turn, implies that

$$\rho_1^r \to \infty \tag{3.311}$$

The reflected ray tube thus is cylindrical (with its axis orthogonal to that of the parabolic surface) with radius of curvature at Q_r given by s^i. Furthermore, we can see [18] that $s^i = |Q_s Q_r|$, which means that the caustic of the reflected ray tube appears to be situated on the directrix of the parabolic cylinder.

Remark 10: Although from our experience with problems of a 2D nature the location of specular points may have seemed an incidental task, for a complex 3D surface this process might be a tedious and computer-time–consuming one. To make use of special case results, if possible, always is advisable. One such case is that of a surface of revolution, for which an efficient Q_r search technique has been given by Lee and Cramer [19]. If not possible in closed form, the brute force determination of the specular point locations is not recommended. Efficient schemes applicable to very general (e.g., even numerically described surfaces for which closed form expressions are not available) reflecting surfaces are available [20]. The difficulty of securing the positions of specular points for general complex 3D surfaces is evidenced by the fact that methods have been devised for first determining, for a given source-observation point combination, whether a specular point exists at all, and only then attempting to find the point itself [21].

3.7 CONCLUDING REMARKS

Both in this and the previous chapter the reader may have been able to view the classical geometrical optics approach with a certain amount of chronological snobbery. To be fair, we must admit that classical geometrical optics indeed appears to arrive at most of the expressions we have been able to obtain through our more sophisticated Luneberg-Kline asymptotic expansion methods. Yet, surely we can manage without classical geometrical optics and dismiss it as an interesting but nonessential interpretation. Moreover, use of the Luneberg-Kline ansatz in Maxwell's equations provided us with all the necessary equations needed to fully describe the field on a ray tube as it propagates through a homogeneous medium. Furthermore, use of this same expansion, plus the well known boundary conditions at a conducting surface and some differential geometry, supplies the law of reflection for such a ray tube. There seems to be no need to appeal to such concepts as the Fermat principle. Such an unsympathetic attitude toward classical GO nonetheless would be decidedly unwise. The reasons are perhaps best summarized as follows [8]:

> If the curved surface has a shadow boundary where the rays of the incident field are tangent to the surface, geometrical optics fails to account for the diffraction of energy into the shadow region. Also, if the surface has an edge or vertex, geometrical optics again fails to predict the resulting diffraction. An examination of available asymptotic solutions for fields diffracted from edges and shadow boundaries reveals that [unlike the terms in the Luneberg-Kline expansion] they contain fractional powers of ω. Furthermore, one notes that caustics of the diffracted field are located at the boundary surface. From both these considerations, it is evident that the Luneberg-Kline series cannot be used to treat diffraction. At the present time it appears that we must use heuristic methods to improve the geometrical optics approximation.

The pioneers of the GTD have combined a number of possible viewpoints with much ingenuity to arrive at satisfactory and usable results. For instance, we will see in later chapters that the interpretation of the law of reflection in terms of Fermat's principle, and the subsequent generalization of this principle by Keller [22], is a key postulate of the GTD; and is the way we arrive at the GTD laws of diffraction. The additional information required to complete the edge diffracted or curved surface diffracted field expressions cannot be obtained through imposition of a Luneberg-Kline series, but rather via the canonical problem approach mentioned in Section 3.2. But it involves more than just rigorous mathematical techniques. As part of the solution process itself "some asymptotic approximations are introduced heuristically" [13]. The result of this approach has been the development of a *uniform theory of diffraction* (UTD), which provides a means for solving electromagnetic problems of engineering interest in a most reliable fashion.

We conclude this chapter by repeating what was said in Section 3.1. The incident ray-optic field from a source point Q_s propagates in the unbounded medium (in this text, free space) to the reflecting surface. This field is represented by a system of rays (ray tube), with each of the rays striking the surface being scattered. The nature of each scattered ray depends on the properties of the surface at the point at which the ray impacts it. Consider just the central ray of some ray tube that impacts the surface at the specific point Q_r. We always know what the incident GO field on this ray is at Q_r, namely, $\mathbf{E}^i(Q_r)$. Using the dyadic reflection coefficient $\mathbf{R}$, the reflected GO field *at the point of* Q_r *on the surface* then is given as $\mathbf{E}^r(Q_r) = \mathbf{E}^i(Q_r) \cdot \mathbf{R}$. From the properties of the surface at Q_r and the properties of the incident wavefront there, the divergence factor $A(s^r)$ of the reflected wavefront can be found. The reflected GO field, which once more propagates in free space after it has been detached from the surface, can be found at any observation point P *off the surface* as

$$\mathbf{E}^r(P) = \mathbf{E}^i(Q_r) \cdot \mathbf{R}A(s^r)\frac{e^{-jks^r}}{s^r}$$

with s^r measured from Q_r in the direction $\hat{\mathbf{s}}^r$ of the reflected ray. If these remarks now make more sense to the reader than the similar ones in the first section, then this chapter will have achieved its aim.

We have come a long way as regards our ability to quantitatively describe GO fields reflected from obstacles. This alone permits us to deal with a large number of problems regularly requiring solution in electromagnetic engineering practice. But let us not be too pleased with ourselves as yet. Remember that the direct and reflected GO fields not only predict zero fields in the shadow regions, but are discontinuous at boundaries between the various lit and shadow regions, a behavior that generally is not physical. The GTD fortunately is designed to overcome such awkward situations by introducing the concept of diffracted rays. The treatment of edge diffraction is developed in the following four chapters, with curved surface diffraction introduced in Chapter 8.

PROBLEMS

3.1 Exercise your mastery of geometry by deriving the results expressed by equations (3.3) to (3.8).

3.2 Derive the relation (3.68) for the caustic distance of the reflected field for 2D problems by using the method of power conservation suggested in Section 3.3.6. (*Hint:* Because this is a 2D problem we of course must work not with power but with power per unit length. Perform the derivation for the TM case, working with the z-directed electric field E_z. The power per unit length at any point then is $|E_z|^2$ multiplied by the length of the arc across the ray tube at that point.)

3.3 Consider the problem of the far-zone fields of an elliptical cylinder plus magnetic line source combination. Check your solutions by setting the major and minor axes of the elliptical cylinder equal to each other, so that it reduces to a circular cylinder, and comparing the results with those of Example 3.1.

3.4 Repeat Example 3.7 for convex hyperbolic cylinders and concave elliptical cylinders [17].

3.5 Explain why the expression (3.214) for the caustic distance of the reflected ray tube in the 3D case is identical to that given in (3.68) for problems of a 2D nature.

3.6 Apply the Fermat principle to determine the law of reflection for the full 3D case. Do not forget to draw a sketch to assist in your understanding and interpretation of the geometry. [*Hint:* Work in rectangular coordinates. Specify the source, reflection, and observation points as triplets (x_s, y_s, z_s), (x_r, y_r, z_r), and (x_o, y_o, z_o), respectively. Express the total distance along the ray path in terms of these coordinates using analytic geometry and minimize this distance with respect to x_r, y_r, and z_r.]

3.7 Repeat Example 3.1 for the case where the observation point is in the near-zone of the cylinder.

3.8 By specializing the results for the ellipsoid in Example 3.8 to the case of a spheroid, and using the results derived for the spheroid in Example C.3 of Appendix C, derive an expression for the RCS of a spheroid in terms of the direction of incidence.

REFERENCES

[1] J.S. Hollis, T.J. Lyon, and L. Clayton (eds.), *Microwave Antenna Measurements,* Scientific Atlanta, Inc., Atlanta, GA.

[2] R.G. Kouyoumjian and P.H. Pathak, "A Uniform Geometrical Theory of Diffraction for an Edge in a Perfectly Conducting Surface," *Proc. IEEE,* Vol. 62, November 1984, pp. 1448–1481.

[3] H. Bremmer and S.W. Lee, "Geometrical Optics Solution of Reflection from an Arbitrarily Curved Surface," *Radio Science,* Vol. 17, September–October 1982, pp. 1117–1131.

[4] R.G. Kouyoumjian, L. Peters, and D.T. Thomas, "A Modified Geometrical Optics Method for Scattering by Dielectric Bodies," *IEEE Trans. Antennas and Propagation,* November 1963, pp. 690–703.

[5] L.B. Felsen and N. Marcuvitz, *Radiation and Scattering of Waves,* Prentice-Hall, Englewood Cliffs, NJ, 1973.

[6] W.H. Press, B.P. Flemmin, S.A. Teukolsky and W.T. Vetterling, *Numerical Recipes,* Cambridge University Press, London, 1986.

[7] P.B. Pathak, W.D. Burnside, and R.J. Marhefka, "A Uniform GTD Analysis of the Diffraction of Electromagnetic Waves by a Smooth Convex Surface," *IEEE Trans. Antennas and Propagation,* Vol. AP-28, No. 5, September 1980, pp. 631–642.

[8] R.G. Kouyoumjian, "Asymptotic High-frequency Methods," *Proc. IEEE,* Vol. 53, August 1965, pp. 864–876.

[9] J.A. Kong, *Electromagnetic Wave Theory,* John Wiley and Sons, New York, 1986.

[10] R.F. Harrington, *Time-Harmonic Electromagnetic Fields,* McGraw-Hill, New York, 1961.

[11] J. van Bladel, *Electromagnetic Fields,* Hemisphere Publishing Corp., New York, 1985.

[12] I.J. Gupta, C.W.I. Pistorius, and W.D. Burnside, "An Efficient Method to Compute Spurious End Point Contributions in PO Solutions," *IEEE Trans. Antennas and Propagation,* Vol. AP-35, No. 12, December 1987, pp. 1426–1435.

[13] P.H. Pathak, "An Asymptotic Analysis of the Scattering of Plane Waves by a Smooth Convex Cylinder," *Radio Science,* Vol. 14, No. 3, May–June 1979, pp. 419–435.

[14] R. Mittra and V. Galindo-Israel, "Shaped Dual Reflector Synthesis," *IEEE Antennas and Propagation Society Newsletter,* August 1980, pp. 5–9.

[15] G.A. Deschamps, "Ray Techniques in Electromagnetics," *Proc. IEEE,* Vol. 60, No. 9, September 1972, pp. 1022–1035.

[16] W.V.T. Rusch, "Notes for the Short Course on Reflector Antenna Theory and Design," University of Southern California, Los Angeles, 1982.

[17] P.W. Hannan, "Microwave Antennas Derived from the Cassegrain Telescope," *IRE Trans. Antennas and Propagation,* Vol. AP-9, March 1961, pp. 140–153.

[18] M.S.A. Sanad and L. Shafai, "Performance and Design Procedure of Dual Parabolic Cylindrical Antennas," *IEEE Trans. Antennas and Propagation,* Vol. AP-36, No. 3, March 1988, pp. 331–338.

[19] S.W. Lee and P.W. Cramer, "Determination of Specular Points on a Surface of Revolution," *IEEE Trans. Antennas and Propagation,* Vol. AP-29, No. 4, July 1981, pp. 662–664.

[20] R. Mittra and A. Rushdi, "An Efficient Approach for Computing the Geometrical Optics Field Reflected from a Numerically Specified Surface," *IEEE Trans. Antennas and Propagation,* Vol. AP-27, No. 6, November 1979, pp. 871–877.

[21] W.V.T. Rusch and O. Sørenson, "On Determining if a Specular Point Exists," *IEEE Trans. Antennas and Propagation,* Vol. AP-27, No. 1, January 1979, pp. 99–101.

[22] J.B. Keller, "Geometrical Theory of Diffraction," *J. Opt. Soc. Amer.,* Vol. 52, 1962, pp. 116–130.

[23] R.G. Kouyoumjian, P.H. Pathak and W.D. Burnside, "A uniform GTD for the diffraction by edges, vertices and convex surfaces" in J.K. Skwirzynski (ed.), *Theoretical Methods For Determining The Interaction Of Electromagnetic Waves With Structures* Sijthoff and Noordhoff, Amsterdam, 1981, pp. 497–561.

Chapter 4
Two-Dimensional Wedge Diffraction

"Diffraction is the process whereby light propagation differs from the predictions of geometrical optics."

Joseph B. Keller,
"One Hundred Years of Diffraction Theory,"
IEEE Transactions on Antennas and Propagation, Vol. AP-33, No. 2,
February 1985

4.1 INTRODUCTION

In Chapters 2 and 3, we indicated that geometrical optics is incapable of predicting the field in the shadow regions, and is inaccurate in the vicinity of the shadow boundaries. For a long time we have known that the fields in the shadow regions are caused by diffraction from the edges that create shadow boundaries. In fact, many exact and approximate solutions for the diffracted fields have been found. In 1801, Thomas Young showed from experiment that light consists of waves. He attempted to explain diffraction by means of rays and postulated that diffraction was an edge effect. Young's experiments, in which the interference patterns due to slits in a screen caused alternating bright and dark bands, are well known. In 1896, Sommerfeld derived an exact solution for the diffraction of a plane wave by a perfectly conducting half-plane [1]. His solution, however, was not a ray optical solution. Since then, the exact solution also had been derived by other methods, notably by Clemmow [2]. More recent discussions can be found in [3–5].

The failure of GO to predict the correct fields in the shadow regions is a serious shortcoming. Furthermore, because electromagnetic fields have to be smooth and continuous everywhere, the discontinuities across the shadow boundaries cannot occur in nature and constitute another reason why GO fails to describe the total electromagnetic field in the presence of structures. However, because of the ray tracing on which the GO is based, it has a unique simplicity and lends itself to illuminating physical insight. It indeed is a powerful technique for analyzing electromagnetic problems, and consequently we would much rather find a cure for its failures than reject the method altogether.

In this chapter we introduce the uniform theory of diffraction in an attempt to compensate for the shortcomings in the GO. The geometrical optics fields can be classified as direct, reflected, or refracted. A field point is considered to be illuminated by a direct ray if there is an unobstructed path from the source to that field point. Reflected rays, on the other hand, are reflected from structures before illuminating a field point. These fields satisfy the *Law of Reflection*; that is, the angle of incidence is equal to the angle of reflection. In scalar form the reflected field is given by

$$U_z^r(s) = U_z^i(Q_r)RA(s)\,\mathrm{e}^{-jks} \tag{4.1}$$

where $U_z^i(Q_r)$ is the incident field at the reflection point (Q_r), R is a reflection coefficient, $A(s)$ is a spreading factor, and s is the distance from Q_r to the field point. It is important to remember that GO reflection is a localized effect, in the sense that the geometry in the immediate area of the reflection point has a dominant effect on the value of the reflected field.

Refracted fields have propagated across the boundary between two different dielectric media, and the direction of propagation of the transmitted ray is determined by Snell's law. Although dielectric structures, both lossless and lossy, are important, we shall consider only perfectly conducting structures, with the exception of a few applications involving EM absorbing material. Perfectly conducting structures, of course, are impenetrable. Only direct and reflected GO fields (as well as diffracted fields, of course) therefore will be considered, as refracted fields do not exist for such structures. Because the structures are assumed to be in free space, which is a lossless homogeneous medium, all rays travel in straight lines.

Rather than consider a general three-dimensional body of arbitrary constitutive parameters, we shall start our investigation of the UTD solution for diffracted fields by simplifying the diffraction problem. Therefore, to illustrate the concepts of diffracted fields and rays, we shall restrict ourselves initially to two-dimensional geometries (except for a small digression in Section 4.3). It is much easier to develop an understanding of the UTD and associated diffraction coefficients by first considering the two-dimensional case. An understanding of the physical phenomena of diffraction and the two-dimensional UTD ideas allows us to grasp the additional complications of three dimensions more easily. It is too cumbersome to keep track of all the additional parameters associated with the three-dimensional case while still in the process of learning the essential workings of the UTD.

In the two-dimensional case, electromagnetic fields have either transverse electric or transverse magnetic components. As earlier we consider the two-dimensional plane we are working in to be the x-y plane, so that electric fields that are perpendicular to this plane are E_z fields; that is,

$$\mathbf{E} = E_z\hat{\mathbf{z}} \tag{4.2}$$

Similarly, magnetic fields that are perpendicular to this plane are H_z fields. As earlier where an expression is applicable to either E_z or H_z, we use the symbol U_z; later on the subscript z will be dropped as it becomes cumbersome. Although there are tremendous advantages in describing fields as vector quantities, we will simplify matters (without loss of generality) by working with the scalar fields in the two-dimensional case.

The total GO field illuminating a field point is given by

$$U_z^t = U_z^i + U_z^r \tag{4.3}$$

where U_z^i is the incident field and U_z^r is the reflected field. Consider now a semiinfinite conducting wedge illuminated by a line source as shown in Figure 4.1. In this example we shall consider both faces of the wedge to be flat, that is, they have infinite radii of curvature. The line source radiates a field:

$$U_z(\rho) = C\frac{e^{-jk\rho}}{\sqrt{\rho}} \tag{4.4}$$

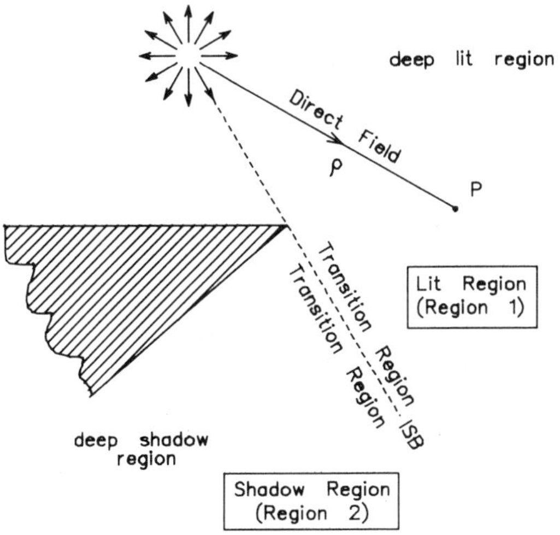

Figure 4.1 Creation of an incident shadow boundary.

where ρ is the distance from the line source to the field point (P), C is a constant complex amplitude, and k is the free space propagation constant:

$$k = \frac{2\pi}{\lambda} \tag{4.5}$$

with λ the wavelength in free space.

The region in space illuminated by a given field is referred to as the *lit region,* whereas the region not illuminated is known as *shadow region.* The direct lit region (region 1) and shadow region (region 2) are clearly indicated in Figure 4.1. On the lit side of the ISB the GO field is given by (4.4), whereas the GO field is zero on the shadow side of the ISB. The region close to the shadow boundary is known as the *transition region,* whereas the regions removed from the shadow boundary are known as the *deep lit* and *deep shadow regions,* respectively. We can thus express the direct GO field in the presence of the wedge as

$$U_z^i(\rho) = \begin{cases} C\dfrac{e^{-jk\rho}}{\sqrt{\rho}}, & \text{in Region 1} \\ 0, & \text{in Region 2} \end{cases} \tag{4.6}$$

The reflected lit region (region 3) and reflected shadow region (region 4) are separated by the *reflection shadow boundary* (RSB), shown in Figure 4.2. The GO reflected field in the presence of the wedge can be expressed as

$$U_z^r(r) = \begin{cases} R_{s,h}C\dfrac{e^{-jk(r+r')}}{\sqrt{r+r'}}, & \text{in Region 3} \\ 0, & \text{in Region 4} \end{cases} \tag{4.7}$$

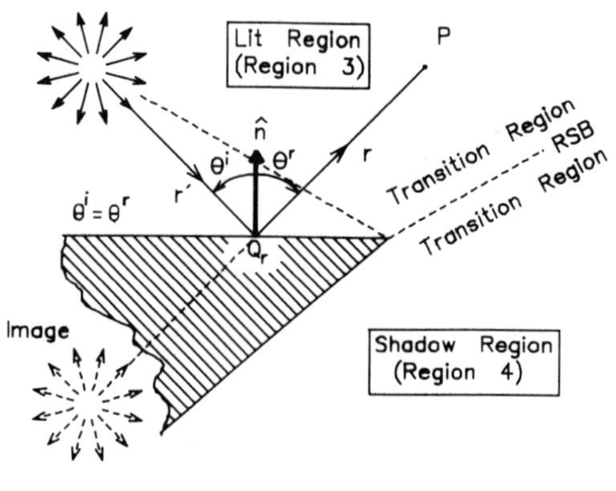

Figure 4.2 Creation of a reflection shadow boundary.

where r' is the distance from the line source to the reflection point (Q_r), and r is the distance from the reflection point to the field point. The reflection coefficient is given by

$$R_{s,h} = \mp 1 \tag{4.8}$$

Equation (4.8) is an approximation that is valid when the radius of curvature of the surface at the point of reflection is not too small in terms of wavelength. A more exact reflection coefficient will be discussed in Chapter 8.

4.2 DIFFRACTION BY HUYGENS' PRINCIPLE

To illustrate the effect of diffracted fields, consider a perfectly conducting half-plane illuminated by a plane wave at normal incidence as shown in Figure 4.3. The incident plane wave in the region $x > 0$ can be expressed as

$$E^i(x, y) = \begin{cases} E_0 \, e^{-jkx}, & \text{in the region } y > 0 \\ 0, & \text{in the region } y < 0 \end{cases} \tag{4.9}$$

where the phase reference has been chosen in the $x = 0$ plane. By using Huygens' principle, we can show [6] that the total field in the region $x > 0$ is given by

$$E^t(x, y) = \frac{E_0 \, e^{-j(kx - \pi/4)}}{\sqrt{2}} \int_{-\gamma y}^{\infty} e^{-j(\pi/2)u^2} \, du \tag{4.10}$$

where

$$\gamma = 2/\sqrt{\lambda x} \tag{4.11}$$

Note that (4.10) is an integral formulation of the total field and not a ray solution.

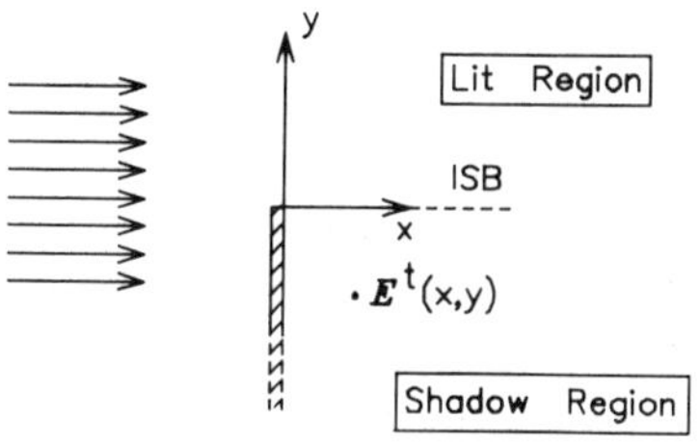

Figure 4.3 Diffraction from a perfectly conducting half-plane illuminated by a plane wave at normal incidence.

The integral in (4.10) can be expressed in terms of Fresnel integrals, where the Fresnel cosine integral and Fresnel sine integral, respectively are given by

$$C(x) = \int_0^x \cos\frac{\pi}{2}u^2 \, du \tag{4.12}$$

$$S(x) = \int_0^x \sin\frac{\pi}{2}u^2 \, du \tag{4.13}$$

so that

$$C(x) - jS(x) = \int_0^x e^{-j(\pi/2)u^2} \, du \tag{4.14}$$

The Fresnel integrals are discussed in Appendix B. Their values are readily calculated by computer; a subroutine is given in Appendix F. We find that

$$\lim_{x \to \infty} [C(x) - jS(x)] = \frac{1}{2} - j\frac{1}{2} \tag{4.15}$$

so that (4.10) can be formulated as

$$E^t(x, y) = \frac{E_0 \, e^{-j(kx - \pi/4)}}{\sqrt{2}} \left\{ \left[\frac{1}{2} - C(-\gamma y) \right] - j\left[\frac{1}{2} - S(-\gamma y) \right] \right\} \tag{4.16}$$

Figure 4.4 shows a plot of $E^t(x, y)$ with $x = 100$ m and a frequency of 10 GHz. From this figure it is clear that a field exists in the shadow region ($x > 0$, $y < 0$). However, in (4.9) we postulated that there is no incident field in this region; therefore, we conclude that the field in the region ($x > 0$, $y < 0$) shown in Figure 4.4 must be a diffracted field. The total field in the region $x > 0$ thus can be expressed as

$$E^t(x, y) = \begin{cases} E^i + E^d, & \text{in the region } y > 0 \\ E^d, & \text{in the region } y < 0 \end{cases} \tag{4.17}$$

where the total field is given in (4.16) and the incident field in (4.9).

The diffracted field can be found by subtracting the incident field from the total field, also shown in Figure 4.4. Notice that the amplitude of the total field on the shadow boundary ($y = 0$) is one-half the amplitude of the incident field and reduces as we move away from the shadow boundary into the shadow region ($y < 0$). Furthermore, the amplitude of the diffracted field is a maximum in the direction along the shadow boundary. In the lit region ($y > 0$) there is a ripple in the total field that is caused by the interference of the diffracted field with the

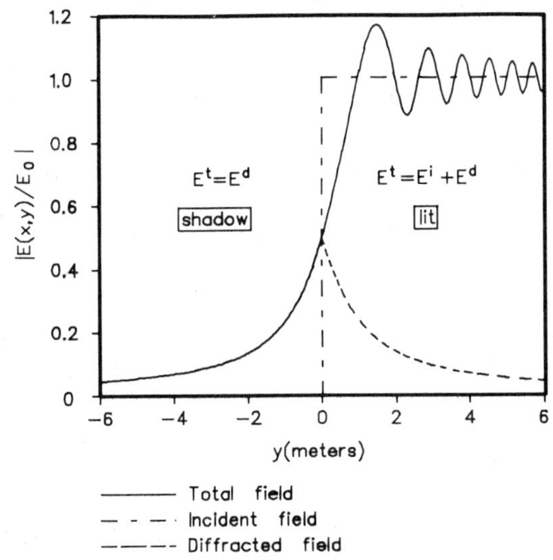

Figure 4.4 Diffraction from a half-plane illuminated by a plane wave at normal incidence; the field is evaluated at $x = 100$ m at a frequency of 10 GHz.

incident field. As we move away from the shadow boundary into the deep lit region ($y \gg 0$), the amplitude of the diffracted field decreases. Consequently the ripple in the total field also decreases so that the amplitude of the total field in the region $y \gg 0$ approaches that of the incident plane wave.

Let us now return to the GO fields. We could surmise that if somehow it were possible to supplement the geometrical optics by adding a ray solution to account for the diffracted fields, the GO would be enhanced considerably. This is exactly what the UTD does: it adds diffracted rays to geometrical optics, so that the fields in the shadow regions are accounted for. The UTD therefore is a ray method in which there are basically three kinds of rays: the direct and reflected GO rays as well as diffracted rays. In a later chapter we shall also discuss so-called creeping waves that arise as a result of the diffraction by smooth convex surfaces.

4.3 KELLER'S ORIGINAL GTD

The geometrical theory of diffraction was developed by Keller in the 1950s as an extension to geometrical optics. His earlier work is described in [7–11] but the classic paper on GTD was published in 1962 [12]. By adding diffracted rays Keller succeeded in correcting the deficiency in the GO that predicts zero fields in the shadow regions. Keller's GTD, however, still had shortcomings. It was not uniform in the sense that the diffracted fields become singular in certain volumetric regions, notably the transition regions surrounding the shadow boundaries. Furthermore, the GTD, like GO, cannot predict field values at caustics. Bear in mind, though, that caustics are isolated points and lines in space, rather than volumetric regions like shadow regions. Also, being a ray method, field values can be calculated only in the directions of the ray paths.

Keller postulated that diffracted rays exist and that they are produced when geometrical optics rays illuminate edges, corners, and vertices of boundary surfaces or when GO rays graze such surfaces, in fact, whenever a structure causes a discontinuity in a GO field by creating shadow regions. Based on a generalized Fermat's principle, Keller formulated a law of diffraction that determines the location of a diffraction point and the direction of propagation of a diffracted ray in a manner analogous to the way the law of reflection determines the location of a reflection point and the direction of propagation of a GO reflected ray. The generalized Fermat's principle states that an edge diffracted ray between a point S and another point P is a curve that has a stationary optical length among all curves between S and P with one point on the edge. Keller noted that, in the case where the incident field is oblique on an edge, Sommerfeld's solution indicated that the diffracted wave is conical; that is, the diffracted rays propagate along parallel cones with the edge as their common axis. He therefore postulated the *law of diffraction* as follows [12]:

> A diffracted ray and the corresponding incident ray make equal angles with the edge at the point of diffraction, provided they are in the same medium. They lie on opposite sides of the plane normal to the edge at the point of diffraction.

To illustrate the law of diffraction, we digress briefly from the two-dimensional case into three dimensions. Figure 4.5 illustrates the concept of the diffraction cone. A ray is incident at an angle β_0' relative to an edge and consequently the

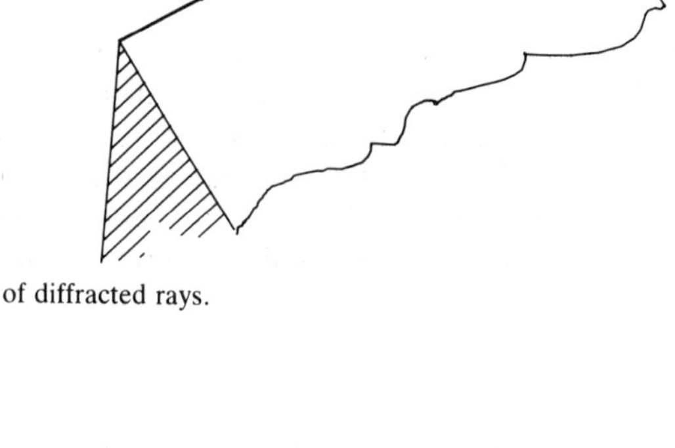

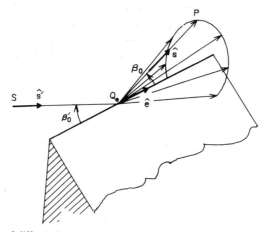

Figure 4.5 Cone of diffracted rays.

diffracted rays form a cone with half-angle β_0 as shown. From the law of diffraction we know that $\beta_0 = \beta_0'$. One incident ray thus will result in an infinite number of diffracted rays lying on the diffraction cone. Note that, although the edge has been shown straight in Figure 4.5, it in fact can be curved. At the point of diffraction (Q_e), we define a unit vector $\hat{\mathbf{e}}$ that is tangential to the edge. If the incident ray propagates in the direction defined by the unit vector $\hat{\mathbf{s}}'$ and the diffracted ray propagates along a direction defined by a unit vector $\hat{\mathbf{s}}$, we find that

$$\sin\beta_0 = |\hat{\mathbf{s}}' \times \hat{\mathbf{e}}| = |\hat{\mathbf{s}} \times \hat{\mathbf{e}}| \tag{4.18}$$

As $\sin\beta_0$ in (4.18) is a double-valued function, we do better to use

$$\cos\beta_0 = \hat{\mathbf{s}}' \cdot \hat{\mathbf{e}} \tag{4.19}$$

for computational purposes. Note that

$$0° \leqslant \beta_0 \leqslant 180° \tag{4.20a}$$

so that

$$0 \leqslant \sin\beta_0 \leqslant 1 \tag{4.20b}$$

Given the direction of propagation of the incident ray and the position of the field point, the diffraction point thus will be the point on the edge that satisfies (4.18). In the case where $\beta_0 = 90°$, the incident ray is perpendicular to the edge and the diffraction cone degenerates to a disk as shown in Figure 4.6. In the two-dimensional case, all rays are by definition perpendicular to the edge; that is, $\sin\beta_0 = 1$. The existence of the diffraction cone has been determined experimentally [13].

Keller postulated that the high-frequency diffracted rays would depend strongly on the geometry in the immediate vicinity of the point of diffraction. In GO reflection, we assume that the surface is locally plane and not affected by the geometry some distance away from the point of reflection. In the diffraction case, we make a similar assumption; that is, that diffraction is a *local* phenomenon. Keller's diffracted rays satisfy all the fundamental principles of ordinary geometrical optics, and as such, they are characterized by phase, amplitude, and polarization. The phase is proportional to the optical path length (which, in a homogeneous medium, is proportional to the physical distance), the amplitude varies in accordance with the conservation of power in a ray tube, and the field is polarized perpendicular to the direction of propagation.

As with all GO rays, however, the problem is to determine the initial values of the amplitude, phase and polarization at the point where the ray is launched;

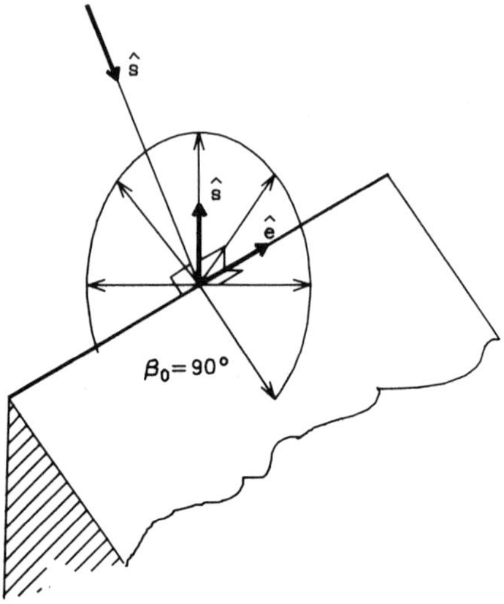

Figure 4.6 Disk of diffracted rays: normal incidence.

that is, the point of diffraction. Following the analogy of GO reflected rays (given in (4.1)), Keller's diffracted rays are determined by the field incident on the diffraction point multiplied by a diffraction coefficient, a spreading factor, and a phase term. The diffracted field will, in general, have the following form:

$$\mathbf{E}^d(s) = \mathbf{E}^i(Q_e) \cdot \mathbf{D}A(s)\, e^{-jks} \tag{4.21}$$

where $\mathbf{E}^i(Q_e)$ is the field incident on the point of diffraction (Q_e) on an edge, $\mathbf{D}$ is a dyadic diffraction coefficient, $A(s)$ is a spreading factor, and s is the distance from Q_e to the field point. We see that the form of (4.21) is very similar to that of the GO reflected field studied in Chapter 3.

To derive the diffraction coefficient, Keller started with the case of a two-dimensional half-plane illuminated by an incident plane wave as shown in Figure 4.7. As the entire edge acts as a virtual line source for the diffracted rays, the

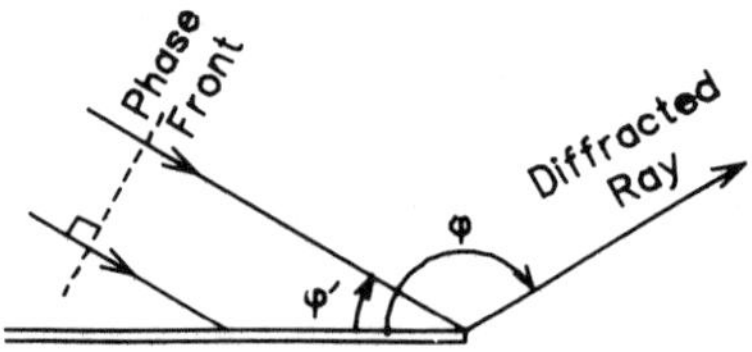

Figure 4.7 Plane wave incident on a half-plane.

wave fronts of the diffracted fields are cylindrical and hence the spreading factor becomes

$$A(s) = \frac{1}{\sqrt{s}} \tag{4.22}$$

Keller compared his solution for the diffracted field as given in (4.21) to an asymptotically expanded form of Sommerfeld's exact solution and found that for large values of the parameter *ks* the scalar diffraction coefficients are given by

$$D^k_{s,h}(\phi, \phi') = -\frac{e^{-j\pi/4}}{2\sqrt{2\pi k}}\left[\frac{1}{\cos\frac{(\phi - \phi')}{2}} \mp \frac{1}{\cos\frac{(\phi + \phi')}{2}}\right] \tag{4.23}$$

where ϕ', and ϕ are the angles of the incident and diffracted rays, respectively, as indicated in Figure 4.7, and measured in a plane perpendicular to the edge at Q_e. The superscript k is used in this text to indicate Keller's original diffraction coefficients. The minus or plus sign actually is the reflection coefficient (see (4.8)) and corresponds to the soft and hard boundary conditions mentioned in Chapter 3, that is, E_z and H_z, respectively. Sommerfeld's solution, however, also solved the case of a field incident at an oblique angle ($\beta_0 \neq \pi/2$) on an infinitely long wedge with interior angle α, as shown in Figure 4.5. By comparing his solution to an asymptotic expansion of Sommerfeld's solution, Keller found that the scalar diffraction coefficients for this case were given by

$$D^k_{s,h}(\phi, \phi', \beta, n) = \frac{-e^{-j\pi/4}\sin(\pi/n)}{2n\sqrt{2\pi k}\sin\beta_0}\left[\left(\frac{1}{\cos\frac{\pi}{n} - \cos\frac{\phi - \phi'}{n}}\right) \mp \left(\frac{1}{\cos\frac{\pi}{n} - \cos\frac{\phi + \phi'}{n}}\right)\right] \tag{4.24}$$

where the interior wedge angle (α) would be related to the parameter n by

$$\alpha = (2 - n)\pi \tag{4.25}$$

or

$$n = \frac{2\pi - \alpha}{\pi} \tag{4.26}$$

with α measured in radians; the exterior wedge angle is equal to $n\pi$. Note that n need not be an integer. We now return to the two-dimensional case, where $\beta_0 = \pi/2$.

Several commonly encountered wedges are shown in Figure 4.8. The half-plane has a wedge angle of $\alpha = 0°$, and is infinitely thin and straight; that is, it has an infinite radius of curvature. The curved screen, on the other hand, is similar to the half-plane in the sense that $\alpha = 0°$, and it also is infinitely thin, but has a finite radius of curvature (a_0) as shown in Figure 4.8(c). In the case of a half-plane, (4.24) reduces to (4.23). Keller's diffraction coefficients do not distinguish between a half-plane and a curved screen as his diffraction coefficients are not functions of the radii of curvature of the wedge faces.

In the case of the full plane, we find that substitution of $n = 1$ into (4.24) yields

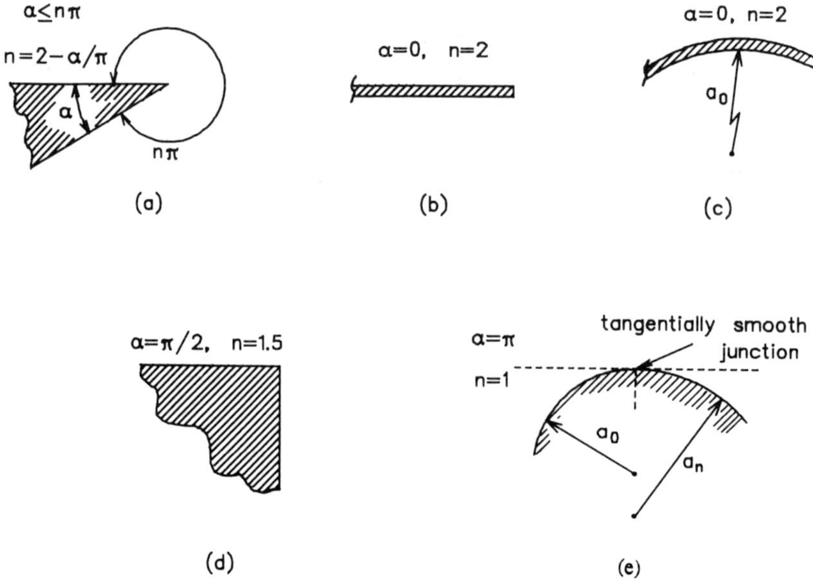

Figure 4.8 Several commonly encountered wedges: (a) arbitrary wedge with flat faces; (b) half-plane; (c) curved screen; (d) 90° wedge; (e) full plane.

$$D^k_{s,h} \equiv 0 \tag{4.27}$$

Several accounts of the GTD have been published since Keller's original papers, notably the book by James [3] and a chapter by Jones [14]. Keller wrote a review paper published in 1985 in which he gives a concise history of how the GTD was developed [15].

Example 4.1: Scattering from a wedge illuminated by a plane wave using Keller's diffraction coefficients

As an example, let us consider a semi-infinite wedge with interior angle α that is illuminated by a plane wave as shown in Figure 4.9. The phase reference is chosen at the edge (Q_e). The plane wave is incident at an angle ϕ' with $\phi' < \pi - \alpha$ and the electric field perpendicular to the plane (TM case). We assume that $\phi' \neq 0$ because the incident electric field then will be tangential to the perfectly conducting wedge and vanish. We shall now use the GTD to determine the total field at a field point (P) located at a distance s and angle ϕ from the wedge, as shown.

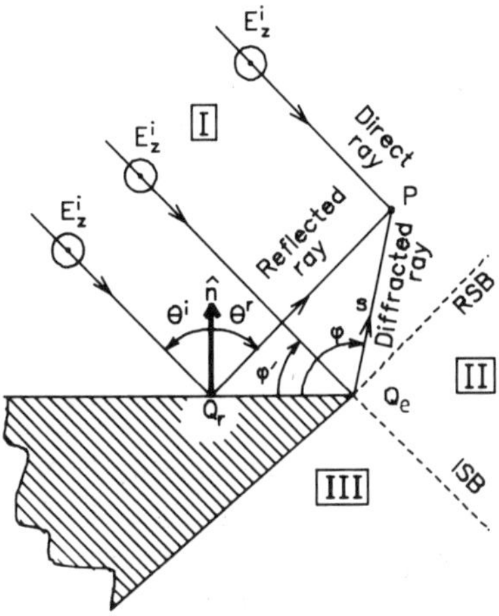

Figure 4.9 Geometry for scattering from a wedge with plane wave incidence (the regions in (4.32) are indicated by roman numerals).

The first step is to determine the location of the shadow boundaries so that the lit and shadow regions for the various field components can be established. It is clear that the ISB is located at

$$\phi_{ISB} = \pi + \phi' \tag{4.28}$$

or

$$\phi_{ISB} - \phi' = \pi \tag{4.29}$$

The RSB is located at

$$\phi_{RSB} = \pi - \phi' \tag{4.30}$$

or

$$\phi_{RSB} + \phi' = \pi \tag{4.31}$$

The term in the diffraction coefficient of (4.24) containing the expression $(\phi - \phi')$ is associated with the incident shadow boundary, and the term containing the expression $(\phi + \phi')$ is associated with the reflection shadow boundary. Note that the latter term is multiplied by the reflection coefficient of the surface adjacent to the edge.

Keller's GTD solution for the total field then is given by

$$E_z^t = \begin{cases} E_z^i + E_z^r + E_z^d, & 0 \leq \phi < \pi - \phi' \quad \text{(Region I)} \\ E_z^i + E_z^d, & \pi - \phi' < \phi < \pi + \phi' \quad \text{(Region II)} \\ E_z^d, & \pi + \phi' < \phi \leq 2\pi - \alpha \quad \text{(Region III)} \end{cases} \tag{4.32}$$

where, as has been shown in (3.164) and (3.170), respectively,

$$E_z^i(s, \phi) = C\, e^{jks\cos(\phi - \phi')} \tag{4.33}$$

$$E_z^r(s, \phi) = -C\, e^{jks\cos(\phi + \phi')} \tag{4.34}$$

Recall that C is the constant complex amplitude of the plane wave. The minus sign in (4.34) is caused by the reflection coefficient for the soft polarization case. Note that the total field is not specified on the shadow boundaries, as the diffraction coefficients given in (4.24) become singular there. Combining (4.21) and (4.22), we find that the diffracted field is given by

$$E_z^d(s) = E_z^i(Q_e) D_s^k(\phi, \phi', \pi/2, n) \frac{e^{-jks}}{\sqrt{s}} \tag{4.35}$$

Furthermore, we find that, in this case,

$$E_z^i(Q_e) = C \tag{4.36}$$

as the phase reference of the plane wave is chosen to be at the edge (Q_e).

We can make certain interesting observations concerning the diffracted field and the diffraction coefficients in particular:

(a) The diffraction coefficient D_s^k approaches zero as the observation angle (ϕ) tends to 0 and $n\pi$. This is to be expected because the tangential electric fields have to be zero on the surfaces of the perfectly conducting wedge. In the case of a magnetic field transverse to the plane of incidence (TE case), we find that the diffraction coefficient D_h^k is finite and not zero at $\phi = 0$ and $\phi = n\pi$.

(b) The diffraction coefficients ($D_{s,h}^k$) become singular along the shadow boundaries as defined in (4.29) and (4.31). Figure 4.10 shows the magnitude of the diffraction coefficients for the case where $\alpha = 40°$ ($n = 1.78$) and $\phi' = 55°$ at a frequency of 10 GHz. We see that D_s^k approaches zero ($-\infty$ dB) as ϕ tends to 0° and 320° (grazing incidence), but that D_h^k is finite at these angles. Because Keller's diffraction coefficients tend to infinity along the shadow boundaries, the GTD is *not* valid in the transition region surrounding the shadow boundaries and especially not in the immediate vicinity of the shadow boundaries. Keller's original GTD thus is not a uniform solution because it is not valid in all regions. This is a serious deficiency, as the fields on the shadow boundary and in the transition regions often are exactly those fields we are interested in.

Let us now consider the results shown in Figure 4.11, where the total and diffracted fields as given by (4.32) to (4.36) are plotted for the case where $C = 1$, $\alpha = 40°$, $\phi' = 55°$, and $s = 1$ m. The frequency is 3 GHz so that $s = 10\lambda$. Note that the total field tends to zero at $\phi = 0°$ and $\phi = 320°$. This is to be

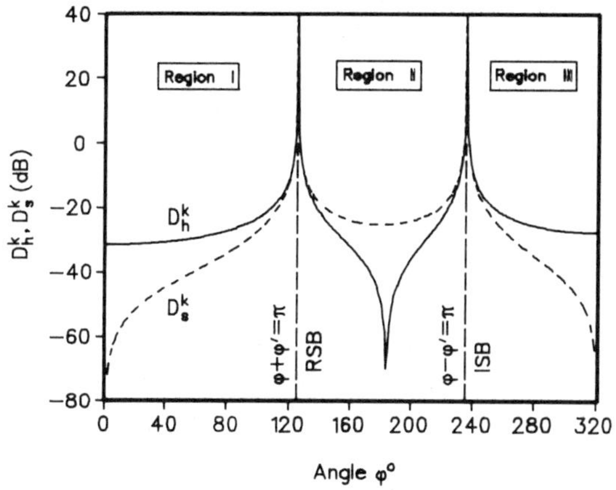

Figure 4.10 Keller's diffraction coefficients evaluated at 10 GHz for a wedge with plane wave incidence ($\phi' = 55°$ and $\alpha = 40°$).

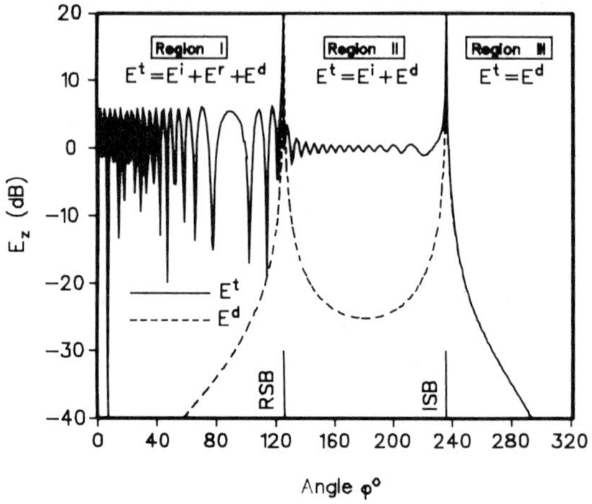

Figure 4.11 Scattered field from a wedge with plane wave incidence using Keller's diffraction coefficients. The geometry is as in Figure 4.9, with $\alpha = 40°$, $\phi' = 55°$, $f = 3$ GHz, $s = 1$ m, and soft polarization. The level of the incident field is 0 dB. Note that the diffracted fields are unbounded at the shadow boundaries.

expected because the electric field is perpendicular to the plane of incidence, and hence tangential to the surface, should vanish on the surface of the wedge. It is important to keep in mind that the solution presented in Figure 4.11 is suspect in the transition regions around the shadow boundaries because the diffraction coefficients, and hence both the diffracted and total fields, become singular there; we conclude that these solutions are invalid close to the shadow boundaries.

Apart from becoming singular in the transition regions adjacent to the shadow boundaries and at caustics, Keller's original GTD also fails when the incident field has a rapid spatial variation, when the incident field is not a ray optical field, or when the circumstances are such that the diffraction and reflection can no longer be considered to be local phenomena.

4.4 THE UNIFORM THEORY OF DIFFRACTION

We have seen that Keller's GTD was a significant improvement to GO, and that he paved the way for a very useful high frequency analysis technique. The GTD still had some serious shortcomings, though. It could predict the diffracted fields in regions away from the shadow boundaries, but became singular in the transition regions surrounding such boundaries. The same dilemma arose again—we had to either discard the GTD altogether or develop a solution whereby the deficiencies could be cured.

In 1974 Kouyoumjian and Pathak at Ohio State University wrote a landmark paper [16] in which they set out the *uniform theory of diffraction* (UTD). They had performed an asymptotic analysis and found that, by multiplying the diffraction coefficients by a transition function, the diffracted fields remain bounded across the shadow boundaries. The form of the transition function is such that it approaches zero at the same rate as that at which the diffraction coefficients become singular at the shadow boundaries, so that the resultant diffracted fields remain bounded at the shadow boundaries. Kouyoumjian and Pathak had thus succeeded in developing a ray-based uniform diffraction theory; that is, one that is valid everywhere in space. Even so, the UTD still suffers from some of the shortcomings of the GTD; namely, the theory fails when the incident field is not a ray optical field and cannot be applied when reflection and diffraction no longer are local phenomena.

Some of the shortcomings of the UTD have been compensated for. For instance, equivalent currents, which will be treated in Chapter 7, can be used in some cases to calculate fields at caustics, whereas slope diffraction is used in cases where the incident field has a rapid spatial variation. We shall study slope diffraction in Section 4.5.

Since the publication of the paper of Kouyoumjian and Pathak in 1974, a multitude of publications dealing with both the theoretical and applied aspects of the UTD have appeared in the literature. We will reference papers dealing with specific aspects of the UTD as well as applications in this text when applicable; general accounts of the UTD can be found in [17–27]. References [27 and 28] were extensively used in this text. Discussions on the numerical evaluation of diffraction coefficients are given in [29 and 30]. Many useful references to the UTD and its applications can be found in [31 and 32].

4.4.1 Shadow Boundaries

Before we investigate the detailed workings of the UTD, we need to take a closer look at shadow boundaries and in particular where they are located. It is customary to label the two faces of the wedge the *o-face* and *n-face,* respectively. Although the designation of which face is called the *o*-face and which is the *n*-face is in fact arbitrary, as a matter of convention, we will consider the angles ϕ' and ϕ to be measured from the *o*-face. The wedge is assumed to have an interior angle α. The angle α is related to the parameter n by (4.25). The *o*-face thus is located at $\phi = 0$, whereas the *n*-face is located at $\phi = n\pi$. In this text we shall restrict ourselves to wedges that have interior angles smaller or equal to 180°, that is,

$$1 \leqslant n \leqslant 2 \tag{4.37}$$

The UTD can also be extended to cases where $\alpha > 180°$ [16]. In general,

$$0 \leqslant \phi' \leqslant n\pi \tag{4.38}$$

$$0 \leqslant \phi \leqslant n\pi \tag{4.39}$$

Although this may seem an obvious remark, sometimes in complicated problems, the angles ϕ' and ϕ are calculated (perhaps by a computer) to fall outside the boundaries in (4.38) and (4.39). In such cases it is necessary to add or subtract 2π from the angle. The cases where $\phi' = 0$ and $\phi' = n\pi$ are known as *grazing incidence* and will be treated as a special case.

A closer examination of the wedges in Figure 4.12 reveals that both the *o*-face and *n*-face can create incident as well as reflection shadow boundaries. We therefore need to consider four cases.

Case 1: n-*Face Shadowed*

Figure 4.12(a) shows the case where the *n*-face is shadowed. We find that

$$0 \leqslant \phi' \leqslant (n - 1)\pi \tag{4.40}$$

so that the ISB is located at

$$\phi_{ISB} - \phi' = \pi \tag{4.41}$$

$$\pi \leqslant \phi_{ISB} \leqslant n\pi \tag{4.42}$$

Case 2: o-*Face Shadowed*

The case where the *o*-face is shadowed is shown in Figure 4.12(b). We find that

$$\pi \leqslant \phi' \leqslant n\pi \tag{4.43}$$

so that the ISB is located at

$$\phi_{ISB} - \phi' = -\pi \tag{4.44}$$

$$0 \leqslant \phi_{ISB} \leqslant (n - 1)\pi \tag{4.45}$$

In general,

$$|\phi_{ISB} - \phi'| = \pi \tag{4.46}$$

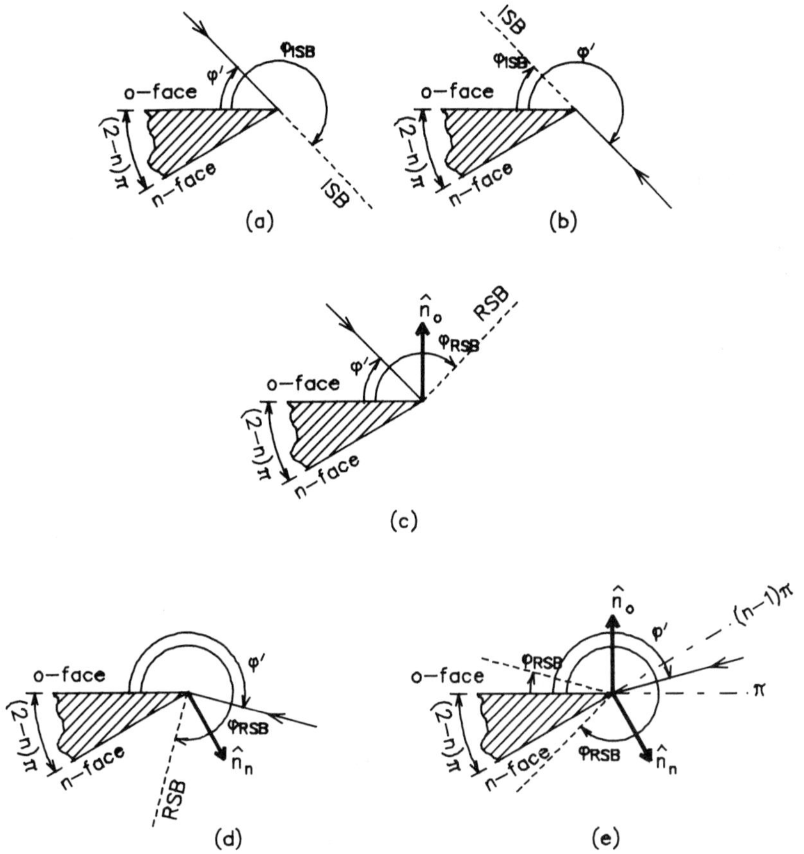

Figure 4.12 Location of shadow boundaries: (a) n-face shadowed; (b) o-face shadowed; (c) reflection from the o-face; (d) reflection from the n-face; (e) creation of two simultaneous reflection shadow boundaries ($\pi \leqslant \phi' \leqslant [n-1]\,\pi$).

where ϕ_{ISB} refers to the ISB created by either the o-face or the n-face. Figure 4.13(a) shows the location of the ISB as a function of ϕ' and n.

Case 3: Reflection from the o*-Face*

Figure 4.12(c) shows the case where there is reflection from the o-face. We find that

$$0 \leqslant \phi' \leqslant \pi \tag{4.47}$$

so that the RSB is located at

$$\phi_{RSB} + \phi' = \pi \tag{4.48}$$

$$0 \leqslant \phi_{RSB} \leqslant \pi \tag{4.49}$$

Case 4: Reflection from the n*-Face*

The case where there is reflection from the n-face is shown in Figure 4.12(d). We find that

$$(n - 1)\pi \leqslant \phi' \leqslant n\pi \tag{4.50}$$

so that the RSB is located at

$$\phi_{RSB} + \phi' = (2n - 1)\pi \tag{4.51}$$

$$(n - 1)\pi \leqslant \phi_{RSB} \leqslant n\pi \tag{4.52}$$

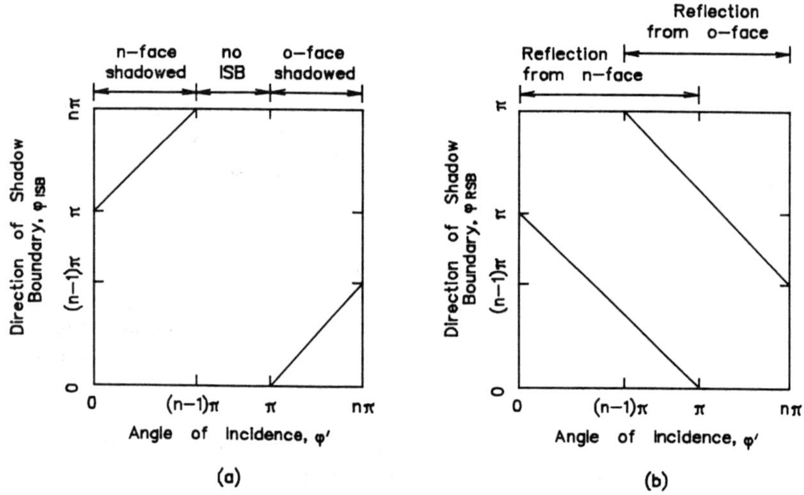

Figure 4.13 Location of the (a) incident and (b) reflection shadow boundaries.

Figure 4.13(b) shows the location of the RSB as a function of ϕ' and n. Note that when ϕ' is in the region $(n - 1)\pi \leqslant \phi' \leqslant \pi$, there is no incident shadow boundary. In this region, however, fields are reflected from both the o-face and n-face so that two simultaneous reflection shadow boundaries are created, as shown in Figure 4.12(e).

4.4.2 Two-Dimensional UTD Diffraction Coefficients

The diffraction coefficients developed by Kouyoumjian and Pathak are more general than Keller's coefficients in the sense that not only do they remain bounded along the shadow boundaries, but they also can be used to calculate the diffracted fields from a wedge with curved surfaces. Consider the two-dimensional wedge in Figure 4.14. The *o*-face of the wedge has a radius of curvature a_o at the edge, whereas the *n*-face has a radius of curvature a_n at the edge. To measure the angles ϕ' and ϕ, we construct a reference plane tangential to the *o*-face at the edge. The angles ϕ and ϕ' are then measured from the reference plane. To measure the wedge angle (α) a reference plane is constructed tangential to the *n*-face at Q_e, and α is then measured as the angle between the two reference planes.

Because diffracted rays propagate in straight lines, there will be shadow regions between the reference planes and the actual surface that the edge-diffracted rays will not illuminate. This problem will be addressed when we discuss creeping waves in Chapter 8.

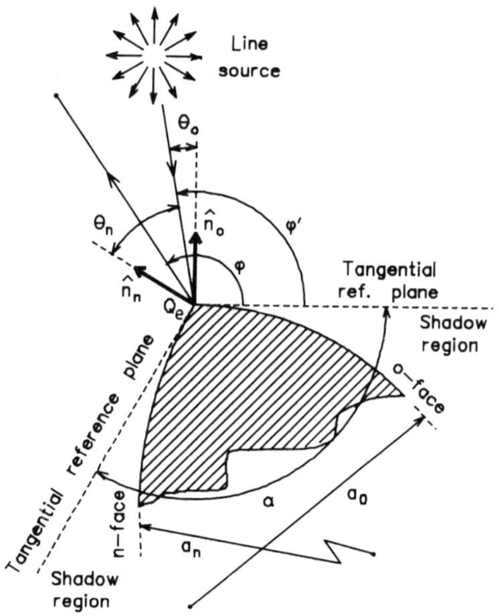

Figure 4.14 Wedge with curved faces.

The two-dimensional edge diffracted fields are given by

$$E_z^d(s) = E_z^i(Q_e)\, D_s \frac{e^{-jks}}{\sqrt{s}} \tag{4.53}$$

$$H_z^d(s) = H_z^i(Q_e)\, D_h \frac{e^{-jks}}{\sqrt{s}} \tag{4.54}$$

where Q_e is the diffraction point on the edge. As mentioned earlier, $D_{s,h}$ are referred to as the soft and hard diffraction coefficients, respectively. The $1/\sqrt{s}$ amplitude dependence indicates that the diffracted fields have cylindrical wavefronts originating at the edge. The edge thus acts as a virtual line source (and caustic) for the diffracted fields. We shall see later that the $1/\sqrt{s}$ term is a special case of a more general spreading factor that holds when the edge is straight and a cylindrical wave is incident upon the edge.

The two-dimensional UTD edge diffraction coefficients are given by

$$D_{s,h}(L^i, L^{ro}, L^{rn}, \phi, \phi', n) = D_1 + D_2 + R_{s,h}(D_3 + D_4) \tag{4.55}$$

where $R_{s,h}$ are the soft and hard reflection coefficients of the surfaces of the wedge at the edge, respectively, and L^i, L^{ro}, and L^{rn} are the so-called distance parameters; we shall return to them shortly. For a conducting wedge we find that $R_{s,h}$ is given by (4.8); that is, $R_{s,h} = \mp 1$. The components of the diffraction coefficients in (4.55) are given by

$$D_1 = \frac{-e^{-j\pi/4}}{2n\sqrt{2\pi k}} \cot\left[\frac{\pi + (\phi - \phi')}{2n}\right] F[kL^i a^+(\phi - \phi')] \tag{4.56}$$

$$D_2 = \frac{-e^{-j\pi/4}}{2n\sqrt{2\pi k}} \cot\left[\frac{\pi - (\phi - \phi')}{2n}\right] F[kL^i a^-(\phi - \phi')] \tag{4.57}$$

$$D_3 = \frac{-e^{-j\pi/4}}{2n\sqrt{2\pi k}} \cot\left[\frac{\pi + (\phi + \phi')}{2n}\right] F[kL^{rn} a^+(\phi + \phi')] \tag{4.58}$$

$$D_4 = \frac{-e^{-j\pi/4}}{2n\sqrt{2\pi k}} \cot\left[\frac{\pi - (\phi + \phi')}{2n}\right] F[kL^{ro} a^-(\phi + \phi')] \tag{4.59}$$

In the two-dimensional case it is assumed that the incident fields have cylindrical wavefronts, so that the distance parameters in the diffraction coefficients are given by

$$L^i = \frac{s's}{s + s'} \tag{4.60}$$

$$L^{ro} = \frac{\rho^{ro}s}{\rho^{ro} + s} \tag{4.61}$$

$$L^{rn} = \frac{\rho^{rn}s}{\rho^{rn} + s} \tag{4.62}$$

The reasons that the distance parameters assume the forms that they do will become apparent in the next section. Two distance parameters are required for each face of the wedge. One parameter of every face is associated with the incident shadow boundary (L^i), and the other with the reflection shadow boundary ($L^{ro,n}$). The same parameter (L^i) is used for the terms associated with the incident shadow boundaries of the two faces, but different parameters are used for the terms associated with the reflection shadow boundaries of the *n*-face (L^{rn}) and *o*-face (L^{ro}).

The distances s' and s are the distances from the line source to the diffraction point and from the diffraction point to the field point, respectively. We see that L^{ro} and L^{rn} are functions of ρ^{ro} and ρ^{rn}, respectively; the caustic distances ρ^{ro} and ρ^{rn} are for their part functions of the radii of curvature (a_o and a_n) of the *o*-face and *n*-face at Q_e, respectively, and given by

$$\frac{1}{\rho^{ro,n}} = \frac{1}{s'} + \frac{2}{a_{o,n}\cos\theta_{o,n}} \tag{4.63}$$

These caustic values also can be interpreted as the principal radii of curvature of the reflected wavefronts at the point of diffraction. Because the two faces can have different radii of curvature, a different distance parameter is required for the term associated with the RSB of each face.

The definitions of the other parameters are as follows:

$a_{o,n}$ = radii of curvature of the *o*-face and *n*-face, respectively, at the point of diffraction. Recall from Chapter 3 that $a_{o,n} > 0$ when the *o*- or *n*-face presents a convex surface to the incident ray and that $a_{o,n} < 0$ when the *o*- or *n*-face presents a concave surface to the incident ray.

ϕ' = the incident angle with respect to the *o*-face, with $0 \leq \phi' \leq n\pi$.

ϕ = the diffraction angle with respect to the *o*-face, with $0 \leq \phi \leq n\pi$.

$\hat{\mathbf{n}}_{o,n}$ = unit vectors normal to the *o*-face and *n*-face, respectively, at the point of diffraction.

θ_o = angle defined by $\hat{\mathbf{n}}_o \cdot \hat{\mathbf{s}}' = -\cos\theta_o$, with $\cos\theta_o \geq 0$.

θ_n = angle defined by $\hat{\mathbf{n}}_n \cdot \hat{\mathbf{s}}' = -\cos\theta_n$, with $\cos\theta_n \geq 0$.

Finally, we consider the definitions of the functions a^+, a^-, and F that occur in (4.56)–(4.59). Because we are dealing with an asymptotic method, we require a so-called largeness parameter κ to be bigger than a minimum value. In the case of the UTD,

$$\kappa = kL \sin^2\beta_0 \tag{4.64}$$

where L represents any of the distance parameters in (4.56)–(4.59). In the two-dimensional case, $\beta_0 = \pi/2$, so that

$$\kappa = kL \tag{4.65}$$

The limitations on κ are discussed in Section 4.4.4. To ensure the validity of the asymptotic solution, the diffraction coefficients can be used only when $\kappa > 1$, unless the parameter n is close to 1 when $\kappa > 3$ [16].

The functions $a^\pm$ are defined as

$$a^\pm(\beta^\pm) = 2\cos^2\left(\frac{2n\pi N^\pm - \beta^\pm}{2}\right) \tag{4.66}$$

where

$$\beta^\pm = \phi \pm \phi' \tag{4.67}$$

and $N^\pm$ are integers that most nearly satisfy the equations.

$$2\pi n N^+ - (\phi \pm \phi') = \pi \tag{4.68}$$

$$2\pi n N^- - (\phi \pm \phi') = -\pi \tag{4.69}$$

Note that a^+ and N^+ are associated with the n-face and that a^- and N^- are associated with the o-face. Figure 4.15 shows N^+ and N^- as functions of $\phi \pm \phi'$ and n. The trapezoidal regions bounded by the solid straight lines represent permissible values of $\phi \pm \phi'$ for $0 \leqslant \phi \leqslant n\pi$ and $0 \leqslant \phi' \leqslant n\pi$ with $1 \leqslant n \leqslant 2$. N^+ can take on the values 0 and 1, whereas N^- can take on the values -1, 0, and 1.

From Figure 4.15 we see that $N^\pm$ are stable in the regions around the shadow boundaries indicated by the dotted lines, in the sense that there are no abrupt changes in $N^\pm$ across any shadow boundary. Table 4.1 shows the values of $N^\pm$ at the four possible shadow boundaries. From Figure 4.15 we see that, for the first three cases in Table 4.1, N^+ and N^- are different from zero only at angular distances greater than π from the shadow boundaries. In such cases the field point usually is outside the transition region. Assuming that κ is not too small, as indicated, we

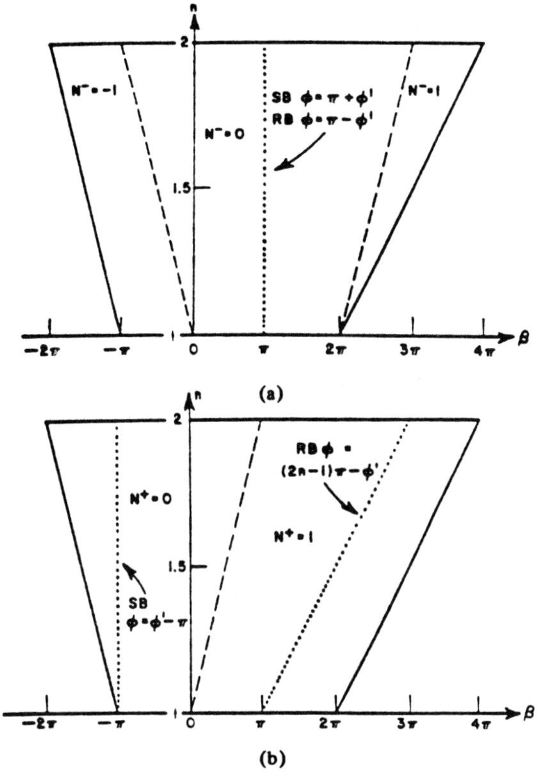

Figure 4.15 N^+ and N^- as functions of $\beta = \phi \pm \phi'$ and n for the (a) o-face and (b) n-face. The trapezoidal region bounded by the solid lines represents the permissible values of $(\phi \pm \phi')$ $1 \leq n \leq 2$ (from [16] © 1974 IEEE, reprinted with permission).

Table 4.1
Shadow Boundaries

Shadow Boundary		*Location*	*Value of N at Boundary*
n-face shadowed	ISB	$\phi_{ISB} - \phi' = \pi$	$N^- = 0$
o-face shadowed	ISB	$\phi_{ISB} - \phi' = -\pi$	$N^+ = 0$
Reflection from o-face	RSB	$\phi_{RSB} + \phi' = \pi$	$N^- = 0$
Reflection from n-face	RSB	$\phi_{RSB} + \phi' = (2n - 1)\pi$	$N^+ = 1$

are justified [16] in assuming the values of $N^{\pm}$ in Table 4.1 and replacing $a^{\pm}$ in (4.56), (4.57), and (4.59) with

$$a(\phi \pm \phi') = 2\cos^2\left(\frac{\phi \pm \phi'}{2}\right) \tag{4.70}$$

and a^+ in (4.58) with

$$a^+(\phi + \phi') = 2\cos^2\left[\frac{2\pi n - (\phi + \phi')}{2}\right] \tag{4.71}$$

The function F is the transition function given by

$$F(x) = 2j\sqrt{x}\, e^{jx} \int_{\sqrt{x}}^{\infty} e^{-ju^2}\, du \tag{4.72}$$

in which we take the principal (positive) branch of the square root. Note that the integral part of the transition function resembles a Fresnel integral (see (4.14)). The transition function is discussed in more detail in Appendix B. A computer routine is given in Appendix F, together with routines for calculating the diffraction coefficients. For values of $x < 0.3$, the small argument form of the transition function is given by

$$F(x) \approx \left(\sqrt{\pi x} - 2x\, e^{j\pi/4} - \frac{2x^2\, e^{-j\pi/4}}{3}\right) e^{j(x + \pi/4)} \tag{4.73}$$

For values of $x > 5.5$, the large argument form of the transition function is given by

$$F(x) \approx 1 + j\frac{1}{2x} - \frac{3}{4x^2} - j\frac{15}{8x^3} + \frac{75}{16x^4} \tag{4.74}$$

For values of $0.3 \leqslant x \leqslant 5.5$, an interpolation scheme can be used to determine the values of $F(x)$ [27]. When $x < 0$, the transition function is evaluated as

$$F(x) = F^*(|x|) \tag{4.75}$$

where the asterisk indicates complex conjugate [27]. The amplitude and phase of the transition function are plotted in Figure 4.16.

There are several distinct differences between the diffraction coefficients in (4.55) to (4.59) and Keller's original diffraction coefficients, given in (4.24). The most obvious differences are the transition functions ($F[x]$) and the fact that now there are four terms instead of Keller's original two. Four terms are needed, as

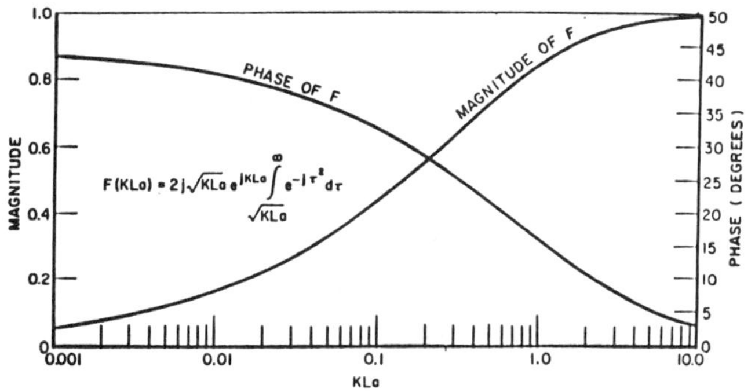

Figure 4.16 Transition function (from [16] © 1974 IEEE, reprinted with permission).

both faces can create incident shadow boundaries and reflection shadow boundaries. We see that D_1 in (4.55) is associated with the diffracted field that compensates for the discontinuity in the GO field when the o-face is shadowed; D_2, when the n-face is shadowed; D_3, when there is reflection from the n-face; and D_4, when there is reflection from the o-face. The contribution of a single face to the wedge-diffracted field is discussed in [33].

The interpretation of $\theta_{o,n}$ when one of the faces is not illuminated requires careful consideration. Although the caustic distance of reflection given in (4.63) can be readily evaluated when calculating the GO reflected field from a curved surface, evaluation for use of this term in the UTD diffraction coefficients requires greater care in some instances.

Consider for example the case where the o-face is illuminated and the n-face is not, both faces being convex, as shown in Figure 4.17(a). The unit vectors $\hat{\mathbf{n}}_o$ and $\hat{\mathbf{n}}_n$ are normal to the o- and n-faces at Q_e, respectively, with $\hat{\mathbf{n}}_n' = -\hat{\mathbf{n}}_n$. Note that the o-face is illuminated only when $\hat{\mathbf{n}}_o \cdot \hat{\mathbf{s}}' < 0$, and similarly for the n-face. The location of the reflection shadow boundary linked to the o-face can be determined from Figure 4.13. Because the n-face is not illuminated in this case, no RSB is connected with it. However, in their reply to Cashman's enquiries [34], Kouyoumjian and Pathak indicate that a *virtual reflection shadow boundary* (VRSB) exists for the n-face and that, in this case, this VRSB should be accounted for when evaluating ρ^{rn}.

Whereas the RSB for the n-face normally is found by considering the reflection of the incident ray with respect to $\hat{\mathbf{n}}_n$, the VRSB is found by considering the virtual reflection of the incident ray with respect to $\hat{\mathbf{n}}_n'$, shown in Figure 4.17(a). In essence,

$$\hat{\mathbf{n}} \cdot \hat{\mathbf{s}}' = -\cos\theta_{o,n} \tag{4.76}$$

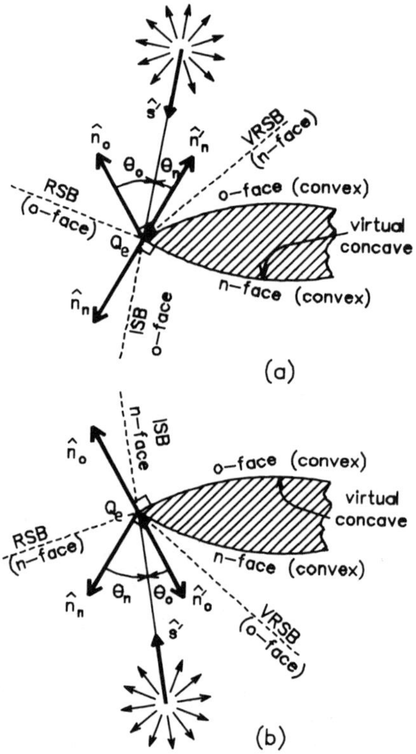

Figure 4.17 Definition of virtual reflection shadow boundaries: (a) o-face illuminated; (b) n-face illuminated.

or

$$\hat{\mathbf{n}} \cdot (-\hat{\mathbf{s}}') = \cos\theta_{o,n} \tag{4.77}$$

with

$$\cos\theta_{o,n} \geqslant 0 \tag{4.78}$$

is required.

From a geometrical vantage point, we can interpret these requirements as follows. For the n-face, the unit vector $\hat{\mathbf{n}}$ in (4.77) thus will be $\hat{\mathbf{n}}_n$ in the case of a real RSB and $\hat{\mathbf{n}}_n'$ in the case of VRSB for (4.77) to satisfy (4.78). Similar equations hold for the o-face. Thus, the VRSB, and hence $\hat{\mathbf{n}}_n'$ rather than $\hat{\mathbf{n}}_n$, is to be taken into account with regard to the n-face in the case illustrated in Figure 4.17(a). The n-face then presents a virtual concave surface to the incident ray, so that $-|a_n|$

should be substituted for the radius of curvature in (4.63) even though the n-face is physically a convex surface. On the other hand, when the n-face is illuminated and the o-face is not, as in Figure 4.17(b), the o-face presents a virtual concave surface to the incident ray so that a_o and $-|a_o|$ should be substituted for the radii of curvature in (4.63). In the region where both faces are illuminated, $a_{o,n} > 0$, and both are substituted as such for the radii of curvature in (4.63). However, we find that in the case where the VRSB of the n-face, for example, has to be taken into account (as in Figure 4.17(a)), L^{rn} is large so that the associated transition function approaches unity [16]. For all practical purposes, ρ^{rn} and L^{rn} thus need not be calculated.

Although we need not calculate $L^{ro,n}$ in certain regions, a great convenience would nevertheless be for the UTD user to have a uniform expression for $\rho^{ro,n}$ available that was valid in all regions (especially from a computer programming point of view), as this expression would obviate the need for a decision as to whether $L^{ro,n}$ was relevant. Such a uniform expression for the caustic distances would take into account the effect of the virtual shadow boundaries and accordingly adjust the signs of $a_{o,n}$. This we achieve in the derivation that follows.

Substituting (4.77) and (4.78) into (4.63), we find that

$$\frac{1}{\rho^{ro,n}} = \frac{1}{s'} + \frac{2}{b_{o,n}|\hat{\mathbf{s}}' \cdot \hat{\mathbf{n}}_{o,n}|} \tag{4.79}$$

where $b_{o,n}$ are given by

$$b_{o,n} = |a_{o,n}| \begin{cases} \hat{\mathbf{s}}' \cdot \hat{\mathbf{n}}_{o,n} < 0, & \text{and } o,\, n\text{-face is convex} \\ \hat{\mathbf{s}}' \cdot \hat{\mathbf{n}}_{o,n} > 0, & \text{and } o,\, n\text{-face is concave} \end{cases} \tag{4.80}$$

$$b_{o,n} = -|a_{o,n}| \begin{cases} \hat{\mathbf{s}}' \cdot \hat{\mathbf{n}}_{o,n} < 0, & \text{and } o,\, n\text{-face is concave} \\ \hat{\mathbf{s}}' \cdot \hat{\mathbf{n}}_{o,n} > 0, & \text{and } o,\, n\text{-face is convex} \end{cases} \tag{4.81}$$

We can consolidate equations (4.79)–(4.81) into the uniform expression:

$$\frac{1}{\rho^{ro,n}} = \frac{1}{s'} - \frac{2}{a_{o,n}(\hat{\mathbf{s}}' \cdot \hat{\mathbf{n}}_{o,n})} \tag{4.82}$$

where $a_{o,n} > 0$ if the o, n-face is convex and $a_{o,n} < 0$ if it is concave. This formulation of $\rho^{ro,n}$ yields the correct value of the reflection caustic distance to be used in the UTD diffraction coefficients in all regions and eliminates the need to account explicitly for a VRSB.

Although (4.82) is very convenient for computational and programming purposes, we stress that the equation follows from a geometrical interpretation of the discussion in [35], rather than being based on a formal analysis of the diffraction coefficients *per se*. The impetus behind the derivation of (4.82) was to find a uniformly valid expression for $\rho^{ro,n}$, specifically for use in the distance parameters of the diffraction coefficients.

Example 4.2: Calculation of reflection caustic distances for use in the distance parameters of the diffraction coefficients

We now illustrate the use of (4.82) by considering the case of the wedge with one convex face and one concave face shown in Figure 4.18. For purposes of comparison, the expressions for $\rho^{ro,n}$ are all cast in the form of (4.63), keeping in mind that $\cos\theta_{o,n} \geqslant 0$. In this case, the o-face is convex ($a_o > 0$) whereas the n-face is concave ($a_n < 0$). In Figure 4.18(a), $\hat{\mathbf{s}}' \cdot \hat{\mathbf{n}}_o < 0$ and $\hat{\mathbf{s}}' \cdot \hat{\mathbf{n}}_n > 0$; that is, the o-face is illuminated whereas the n-face is not. Hence, we find from (4.82) that

$$\frac{1}{\rho^{ro}} = \frac{1}{s'} + \frac{2}{a_o|\hat{\mathbf{s}}' \cdot \hat{\mathbf{n}}_o|} \tag{4.83}$$

$$\frac{1}{\rho^{rn}} = \frac{1}{s'} + \frac{2}{|a_n|\,|\hat{\mathbf{s}}' \cdot \hat{\mathbf{n}}_n|} \tag{4.84}$$

Because $a_o > 0$, the o-face in (4.81) presents a convex surface to the incident ray and, as $|a_n| > 0$, the n-face in (4.84) also seems to present a convex surface.

In Figure 4.18(b), $\hat{\mathbf{s}}' \cdot \hat{\mathbf{n}}_o < 0$ and $\hat{\mathbf{s}}' \cdot \hat{\mathbf{n}}_n < 0$, so that both faces are illuminated; and two simultaneous reflection shadow boundaries are created. We find from (4.82) that

$$\frac{1}{\rho^{ro}} = \frac{1}{s'} + \frac{2}{a_o|\hat{\mathbf{s}}' \cdot \hat{\mathbf{n}}_o|} \tag{4.85}$$

$$\frac{1}{\rho^{rn}} = \frac{1}{s'} + \frac{2}{(-|a_n|)|\hat{\mathbf{s}}' \cdot \hat{\mathbf{n}}_n|} \tag{4.86}$$

Because $a_o > 0$, the o-face in (4.85) presents a convex surface to the incident ray, whereas, because $-|a_n| < 0$, the n-face in (4.86) presents a concave surface.

In Figure 4.18(c), $\hat{\mathbf{s}}' \cdot \hat{\mathbf{n}}_o > 0$ and $\hat{\mathbf{s}}' \cdot \hat{\mathbf{n}}_n < 0$, that is, the n-face is illuminated whereas the o-face is not, we find that

$$\frac{1}{\rho^{ro}} = \frac{1}{s'} + \frac{2}{(-a_o)|\hat{\mathbf{s}}' \cdot \hat{\mathbf{n}}_o|} \tag{4.87}$$

$$\frac{1}{\rho^{rn}} = \frac{1}{s'} + \frac{2}{(-|a_n|)|\hat{\mathbf{s}}' \cdot \hat{\mathbf{n}}_n|} \tag{4.88}$$

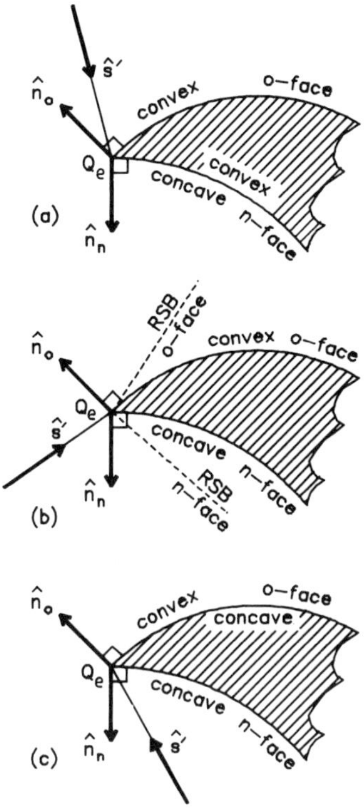

Figure 4.18 Illumination of a wedge with a convex *o*-face and concave *n*-face: (a) *o*-face illuminated; (b) *o*- and *n*-faces illuminated resulting in no incident shadow boundary but two reflection shadow boundaries; (c) *n*-face illuminated.

Because $-a_o < 0$, the *o*-face in (4.87) seems to present a concave surface to the incident ray and, because $-|a_n| < 0$, the *n*-face in (4.88) also presents a concave surface.

A few special cases of the distance parameters in (4.60) to (4.62) now will be considered.

Case 1: Plane Wave Incidence

If the incident field is a plane wave, then

$$s' \to \infty \tag{4.89}$$

so that

$$\lim_{s' \to \infty} L^i = \lim_{s' \to \infty} \frac{s'}{s'}\left(\frac{s}{1 + s/s'}\right) = s \tag{4.90}$$

We therefore have

$$L^i = s \text{ for plane wave incidence} \tag{4.91}$$

By using (4.82) and (4.90), the same argument yields

$$\rho^{ro,n} = \frac{-a_{o,n}(\hat{\mathbf{s}}' \cdot \hat{\mathbf{n}}_{o,n})}{2} \tag{4.92}$$

for plane wave incidence, which can be substituted into (4.61) and (4.62) to yield $L^{ro,n}$.

Case 2: Wedge with Infinite Radius of Curvature

Consider the case where the *o*-face of a wedge has an infinite radius of curvature. An infinite radius of curvature implies that the face of the wedge is flat; that is,

$$a_o \to \infty \tag{4.93}$$

Substituting (4.93) into (4.82), we find that

$$\rho^{ro} = s' \tag{4.94}$$

so that

$$L^{ro} = \frac{ss'}{s + s'} \tag{4.95}$$

Note that $L^{ro} = L^i$ in this case. In the case of plane wave incidence on a wedge with an infinite radius of curvature, (4.90) is applied to (4.95) and we find that

$$L^{ro} = s \tag{4.96}$$

The reflected field seems to emanate from an image source located the same distance below the surface as the source is above the surface. Hence, we find that $\rho^{ro,n} = s'$. A similar argument holds for the *n*-face.

Case 3: Cylindrical Wave Incidence with Field Point in the Far-Zone

If the incident field has a cylindrical wavefront, s' is finite. For a field point in the far-zone, $s \to \infty$, so that

$$L^i = s' \tag{4.97}$$

4.4.3 Enforcing Continuity across the Shadow Boundaries

To illustrate how the transition function enforces continuity of the total fields across the shadow boundaries, consider the wedge shown in Figure 4.19. We shall first treat continuity across an incident shadow boundary. Let the wedge be illuminated by a line source as shown in Figure 4.19(a) so that, in this case, the n-face is

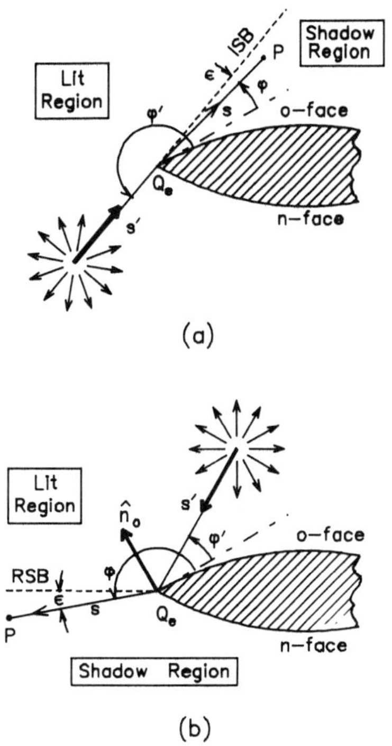

Figure 4.19 Continuity across the shadow boundaries: (a) geometry for the incident shadow boundary with the o-face shadowed; (b) reflection shadow boundary for reflection from the o-face.

shadowing the incident field. We shall evaluate the field at a small angle ϵ from the ISB and then investigate the continuity in the total field as $\epsilon \to 0$. Note that $\epsilon > 0$ places the diffracted ray in the shadow region, whereas $\epsilon < 0$ places the diffracted ray in the lit region. From (4.44) we thus find that

$$\phi - \phi' = -\pi - \epsilon \tag{4.98}$$

at an angle ϵ from the ISB. The diffracted field is given by

$$U^d(s, \phi) = U^i(Q_e)\, D_{s,h}(L^i, \phi - \phi') \frac{e^{-jks}}{\sqrt{s}} \tag{4.99}$$

where

$$U^i(Q_e) = C \frac{e^{-jks'}}{\sqrt{s'}} \tag{4.100}$$

and C is the amplitude of the line source. The diffraction coefficient is given by (4.56) as

$$D_{s,h}(L^i, \phi - \phi') = \frac{-e^{-j\pi/4}}{2n\sqrt{2\pi k}} \cot\left[\frac{\pi + (\phi - \phi')}{2n}\right] F[kL^i a^+(\phi - \phi')] + D^t_{s,h} \tag{4.101}$$

where

$$D^t_{s,h} = D_2 + R_{s,h}(D_3 + D_4) \tag{4.102}$$

We have isolated the component D_1 of the diffraction coefficient in (4.55), because this component enforces continuity across the ISB when the n-face is shadowing the incident field; that is, the o-face is shadowed. The other three components (D_2, D_3, and D_4) remain finite and continuous across the ISB under consideration and therefore have been lumped together as $D^t_{s,h}$. Because D_1 does not depend on the parameters $\phi + \phi'$, L^{ro}, and L^{rn}, they need not be specified in (4.99).

We can now evaluate the term D_1 at $\phi - \phi' = -(\pi + \epsilon)$, and we find that

$$\cot\left(\frac{\pi + (\phi - \phi')}{2n}\right) = \cot\left(\frac{-\epsilon}{2n}\right) \approx \frac{-2n}{\epsilon} \text{ as } \epsilon \to 0 \tag{4.103}$$

From Table 4.1, $N^+ = 0$ follows at this ISB. Evaluation of the function a^+ close to the ISB yields

$$a^+(\pi + \epsilon) = 2\cos^2\left(\frac{\pi + \epsilon}{2}\right) \approx \frac{\epsilon^2}{2} \tag{4.104}$$

By virtue of (4.70) we could have used the function a here instead of a^+, but for the sake of completeness a^+ was used. If we substitute (4.104) into the transition function and use the first term of the small argument form of the transition function given in (4.73), we find that

$$F(kL^i a^+) = F\left(kL^i \frac{\epsilon^2}{2}\right) \approx \sqrt{\frac{\pi k}{2}\left(\frac{ss'}{s+s'}\right)}\,|\epsilon|\, e^{j\pi/4} \tag{4.105}$$

as $\epsilon \to 0$. The diffraction coefficient then reduces to

$$\underset{\lim \epsilon \to 0}{D_1} = \frac{|\epsilon|}{2\epsilon}\sqrt{\frac{ss'}{s+s'}} \tag{4.106}$$

so that the diffracted field can be expressed as

$$U^d = C\frac{e^{-jks'}}{\sqrt{s'}}\left(\frac{|\epsilon|}{2\epsilon}\sqrt{\frac{ss'}{s+s'}}\right)\frac{e^{-jks}}{\sqrt{s}} + U^{cd} \tag{4.107}$$

If we simplify (4.107), we find that

$$U^d = \frac{|\epsilon|}{2\epsilon} C \frac{e^{-jk(s+s')}}{\sqrt{s+s'}} + U^{cd} \tag{4.108}$$

where U^{cd} is finite and continuous across the ISB under consideration and given by

$$U^{cd} = C\frac{e^{-jks'}}{\sqrt{s'}}\left[D_2 + R_{s,h}(D_3 + D_4)\right]\frac{e^{-jks}}{\sqrt{s}} \tag{4.109}$$

Because diffracted fields decay rapidly away from the shadow boundaries, it is to be expected that U^{cd} will be small if the other shadow boundaries are not too near the ISB under consideration.

The distance parameter L^i has served to convert the term $1/(\sqrt{s'}\,\sqrt{s})$ caused by the multiplication of the cylindrical spreading factors of the incident and diffracted rays into the term $1/\sqrt{s'+s}$. This term corresponds to the cylindrical spreading factor of the incident field at a distance $s + s'$ from the line source.

In the direction of the ISB, the incident field is given by

$$\underset{\lim \epsilon \to 0}{U^i} = \begin{cases} C\dfrac{e^{-jk(s+s')}}{\sqrt{s+s'}}, & \text{in the lit region } (\epsilon < 0) \quad (4.110a) \\[2ex] 0, & \text{in the shadow region } (\epsilon > 0) \quad (4.110b) \end{cases}$$

From (4.108) and (4.110), we therefore find that, in the immediate vicinity of the ISB,

$$U^d = \begin{cases} -\dfrac{U^i}{2} + U^{cd}, & \text{in the lit region } (\epsilon < 0) \\ \dfrac{U^i}{2} + U^{cd}, & \text{in the shadow region } (\epsilon > 0) \end{cases} \tag{4.111}$$

so that the total field on either side of the shadow boundary, that is, as $\epsilon \to 0$, is given by

$$U^t = \begin{cases} U^i - \dfrac{U^i}{2} + U^{cd} = \dfrac{U^i}{2} + U^{cd}, & (\epsilon < 0) \\ \dfrac{U^i}{2} + U^{cd}, & (\epsilon > 0) \end{cases} \tag{4.112}$$

with U^i given in (4.110a).

From (4.112), the total field clearly is continuous across the ISB and equal to one-half the incident field (6 dB below) plus a small but continuous component of diffracted field. Recall that in the calculation of the diffracted field using the Fresnel integrals that resulted in Figure 4.4, we also found that the total field was equal to one-half the incident field at the shadow boundary. We must recognize that, although the incident and diffracted fields individually are discontinuous across the shadow boundary, the total field—that is, the sum of incident and diffracted fields—is continuous across the shadow boundary. Bear in mind that the total field is represented numerically as a complex number and that both the amplitude and phase are continuous across the shadow boundary.

In the case where the *o*-face is shadowing the incident fields, a similar analysis can be performed to illustrate that D_2 enforces continuity across the ISB. To illustrate how the transition function enforces continuity across a reflection shadow boundary, consider the RSB created by the *o*-face of the wedge in Figure 4.19(b). The *o*-face of the wedge has a radius of curvature (a_o) at the point of diffraction (Q_e).

Let the wedge be illuminated by a line source as shown. As in the previous case, we shall evaluate the field at a small angle ϵ from the RSB and then investigate the continuity in the total field as $\epsilon \to 0$. For values of $\epsilon > 0$ the diffracted field is placed in the shadow region, whereas $\epsilon < 0$ places the diffracted field in the lit region. From (4.48), we find that

$$\phi + \phi' = \pi + \epsilon \tag{4.113}$$

at an angle ϵ from the RSB. The diffracted field is given by

$$U^d(s, \phi) = U^i(Q_e)D_{s,h}(L^{ro}, \phi + \phi')\frac{e^{-jks}}{\sqrt{s}} \tag{4.114}$$

where

$$U^i(Q_e) = C\frac{e^{-jks'}}{\sqrt{s'}} \tag{4.115}$$

and C is the amplitude of the line source. The diffraction coefficient is given as

$$\begin{aligned} D_{s,h}(L^{ro}, \phi + \phi') = \mp \frac{-e^{-j\pi/4}}{2n\sqrt{2\pi k}} \cot\left(\frac{\pi - (\phi + \phi')}{2n}\right) \\ \cdot F[KL^{ro}a^{-}(\phi + \phi')] + D^t_{s,h} \end{aligned} \tag{4.116}$$

where

$$D^t_{s,h} = D_1 + D_2 + R_{s,h}D_3 \tag{4.117}$$

The minus or plus sign in (4.116) refers to the soft and hard polarizations.

We now have isolated the component D_4 of the diffraction coefficient in (4.55), because this component enforces continuity across the RSB when there is reflection from the o-face. The other three components (D_1, D_2 and D_3) remain finite across the RSB under consideration and therefore have been lumped together as $D^t_{s,h}$.

The term D_4 can now be evaluated at $\phi + \phi' = \pi + \epsilon$. We find that

$$\cot\left(\frac{\pi - (\phi + \phi')}{2n}\right) = \cot\left(\frac{-\epsilon}{2n}\right) \approx \frac{-2n}{\epsilon} \text{ as } \epsilon \to 0 \tag{4.118}$$

From Table 4.1, $N^- = 0$ follows at this RSB. Evaluation of the function a^- yields

$$a^-(\pi + \epsilon) = 2\cos^2\left(\frac{\pi + \epsilon}{2}\right) \approx \frac{\epsilon^2}{2} \tag{4.119}$$

Due to (4.70), the function a could have been used here instead of the function a^-, but, as earlier, for the sake of completeness a^- is used. Substituting (4.119) into the transition function and using the first term of the small argument form of the transition function given in (4.73), we find that

$$F(kL^{ro}a^{-}) = F\left(kL^{ro}\frac{\epsilon^2}{2}\right) \approx \sqrt{\frac{\pi k}{2}\left(\frac{s\rho^{ro}}{s + \rho^{ro}}\right)}\,|\epsilon|\, e^{j\pi/4} \tag{4.120}$$

as $\epsilon \to 0$. Substituting (4.118), (4.119), and (4.120) into (4.59) we find that the diffraction coefficient then reduces to

$$\underset{\lim \epsilon \to 0}{D_4} = R_{s,h}\frac{|\epsilon|}{2\epsilon}\sqrt{\frac{s\rho^{ro}}{s + \rho^{ro}}} \tag{4.121}$$

so that the diffracted field is given by

$$U^d = CR_{s,h}\frac{e^{-jks'}}{\sqrt{s'}}\left(\frac{|\epsilon|}{2\epsilon}\sqrt{\frac{s\rho^{ro}}{s + \rho^{ro}}}\right)\frac{e^{-jks}}{\sqrt{s}} + U^{cd} \tag{4.122}$$

Simplification of (4.122) yields

$$U^d = CR_{s,h}\frac{|\epsilon|}{2\epsilon}\sqrt{\frac{\rho^{ro}}{s'(s + \rho^{ro})}}\, e^{-jk(s+s')} + U^{cd} \tag{4.123}$$

where U^{cd} is finite and continuous across the RSB under consideration and given by

$$U^{cd} = C\frac{e^{-jks'}}{\sqrt{s'}}\left(D_1 + D_2 + R_{s,h}D_3\right)\frac{e^{-jks}}{\sqrt{s}} \tag{4.124}$$

We already have seen that U^{cd} will be small if the RSB under consideration is not too near the other shadow boundaries.

In the direction of the RSB, the reflected field is given by

$$\underset{\lim \epsilon \to 0}{U^r} = \begin{cases} CR_{s,h}\dfrac{e^{-jks'}}{\sqrt{s'}}\sqrt{\dfrac{\rho^{ro}}{s + \rho^{ro}}}\, e^{-jks}, & \text{in the lit region } (\epsilon < 0) \quad (4.125a) \\[2ex] 0, & \text{in the shadow region } (\epsilon > 0) \quad (4.125b) \end{cases}$$

From (4.123) and (4.124), we therefore find that, in the immediate vicinity of the RSB, that is, as $\epsilon \to 0$,

$$U^d = \begin{cases} -\dfrac{U^r}{2} + U^{cd}, & \text{in the lit region } (\epsilon < 0) \\ \dfrac{U^r}{2} + U^{cd}, & \text{in the shadow region } (\epsilon > 0) \end{cases} \tag{4.126}$$

so that the total field on either side of the shadow boundary is given by

$$U^t = \begin{cases} U^r - \dfrac{U^r}{2} + U^{cd} = \dfrac{U^r}{2} + U^{cd}, & (\epsilon < 0) \\ \dfrac{U^r}{2} + U^{cd}, & (\epsilon > 0) \end{cases} \tag{4.127}$$

From (4.127), the total field clearly is continuous across the RSB and equal to one-half the reflected field plus a small but continuous component of diffracted field.

In the case where the n-face creates an RSB, a similar analysis can be performed to illustrate that D_3 enforces continuity across the RSB. Keep in mind that $N^+ = 1$ at the resulting RSB, so that we use a^+ in (4.71) rather than a in (4.70)

4.4.4 Transition Regions

In the previous section, we saw that the transition function in the diffraction coefficients enforced continuity across both incident and reflection shadow boundaries. Because the transition function approaches zero at the same rate as that at which the cotangent terms in the diffraction coefficients approach infinity at the shadow boundaries, the diffraction coefficients and hence the diffracted fields remain bounded in the immediate vicinity of the shadow boundaries. This feature makes the UTD uniform.

We need to investigate the behavior of the diffracted fields in the transition regions. Figure 4.16 and (4.74) reveal that

$$\underset{\lim x \to \infty}{F(x)} = 1 \tag{4.128}$$

The transition function is a complex function, so that (4.128) implies that the amplitude approaches unity and the angle approaches zero as $x \to \infty$. For practical purposes, however, it is reasonable to assume that [16]

$$F(kLa^{\pm}) \approx 1 \text{ if } kLa^{\pm} \geqslant 2\pi \tag{4.129}$$

As shown in (4.104) and (4.119), $a^{\pm}$ approaches zero close to the shadow boundaries. The transition region thus would be the region where

$$F(kLa^{\pm}) < 1 \tag{4.130}$$

From (4.130), this would happen when

$$La^{\pm} < \lambda \tag{4.131}$$

Outside the transition region $F(kLa^{\pm}) \approx 1$, so that the diffracted fields are ray optical. Keeping in mind that $\sin\beta = 1$ when $\beta = 90°$, we find that the diffraction coefficients in (4.55) to (4.59) then reduce to Keller's diffraction coefficient as given in (4.24). Inside the transition regions, the diffracted fields are not ray optical as they will be functions of the parameters upon which the transition functions depend.

To determine the location of the transition regions, we need to examine the region for which the equality:

$$kLa^{\pm} = 2\pi \tag{4.132}$$

is satisfied. Let us consider the transition region around the ISB when the n-face of a wedge is shadowed. Solving for (4.132) and using (4.70), we find that the transition region around the ISB is demarcated by the polar equation:

$$L = \frac{\lambda}{2\cos^2\left(\frac{\phi - \phi'}{2}\right)} \tag{4.133}$$

In a similar fashion we find that the transition region around the RSB created by reflection from the o-face is demarcated by the polar equation:

$$L = \frac{\lambda}{2\cos^2\left(\frac{\phi + \phi'}{2}\right)} \tag{4.134}$$

The shape of the transition region is illustrated by considering the specific example where a wedge is illuminated by a plane wave incident at an angle of $\phi' = 40°$ as shown in Figure 4.20. Using (4.91) and (4.96), we find that (4.133) and (4.134) reduce to

$$s = \frac{\lambda}{2\cos^2\left(\frac{\phi \mp \phi'}{2}\right)} \tag{4.135}$$

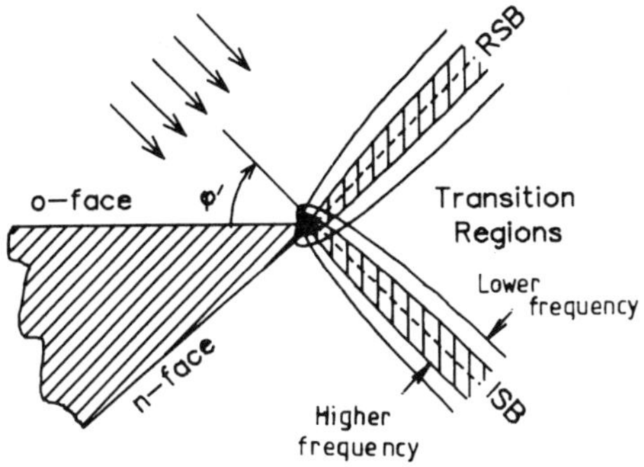

Figure 4.20 Location of the transition regions for a wedge illuminated by a plane wave. The lines demarcating the transition regions are parabolas with their foci at the edge. Outside the transition region $F \approx 1$.

Figure 4.20 shows the shape of the transition regions around the ISB and RSB at 1 GHz and 10 GHz. We see that the transition regions tend to shrink closer to the shadow boundaries as the frequency increases. Recall that when ϕ is not close to the shadow boundaries, we can make the assumption that

$$F(kLa^{\pm}) \approx 1 \tag{4.136}$$

in the diffraction coefficients in (4.55) to (4.59). We leave as an exercise for the reader to show that the lines demarcating the transition regions as shown in Figure 4.20 are parabolas with the edge at the foci.

Keep in mind that the expressions for the diffracted fields will remain valid, provided that $\kappa > 1$, or

$$\frac{L}{\lambda} > \frac{1}{2\pi} \tag{4.137}$$

This condition is imposed by the requirements of the asymptotic approximation [16]. Some authors [36] consider the asymptotic approximation to be valid when $\kappa > 1.5$. Referring to (4.60)–(4.61), we find that several factors can contribute to κ being small. For instance, the frequency can be too low, resulting in a small value of k; in addition, the distances s or s' may be too small. Before using the diffraction coefficients, we ought to ensure that (4.137) is satisfied.

Example 4.3: Scattering from a wedge illuminated by a plane wave using the UTD diffraction coefficients

As an example, we now shall use the diffraction coefficients in (4.55) to (4.59) to calculate the scattering from a wedge illuminated by a plane wave with electric field perpendicular to the plane, using the geometry of Figure 4.9 and the parameters of Example 4.1. Recall from Example 4.1 in Section 4.3 that we used Keller's diffraction coefficients to calculate the scattering from a wedge.

Because $\alpha = 40°$, we calculate

$$n = 1.778 \tag{4.138}$$

Equations (4.32) to (4.36) still apply, except that we replace D_s^k in (4.35) by D_s in (4.55). Because $a_{o,n} \to \infty$ and $s' \to \infty$, we find from (4.90) and (4.96) that

$$L = L^i = L^{ro} = L^{rn} = s \tag{4.139}$$

The transition functions in (4.56) and (4.57) have the form:

$$F(kLa) = F\left[2ks \cos^2\left(\frac{\phi - \phi'}{2}\right)\right] \tag{4.140}$$

where the transition functions in (4.58) and (4.59), respectively, have the forms:

$$F(kLa^+) = F\left\{2ks \cos^2\left[\frac{2\pi n - (\phi + \phi')}{2}\right]\right\} \tag{4.141}$$

$$F(kLa) = F\left[2ks \cos^2\left(\frac{\phi + \phi'}{2}\right)\right] \tag{4.142}$$

The total and diffracted fields are shown in Figure 4.21 (a) for the case where $C = 1$, $\alpha = 40°$, and $s = 1$ m; the frequency is 3 GHz so that $s = 10\lambda$. Because the electric field is tangential to the surface (i.e., perpendicular to the plane), the reflection coefficient is given by $R_s = -1$. The resultant field must be compared with that in Figure 4.11, in which Keller's diffraction coefficients were used. The first obvious difference between the plots in Figure 4.11 and Figure 4.21 (a) is that in the latter case the total field is continuous across the shadow boundaries (as it should be), whereas in Figure 4.11 singularities existed in the total field at the shadow boundaries, for reasons now known to us.

Figure 4.21 (b) shows the calculated field for the same situation except that the magnetic field is now perpendicular to the plane and consequently the reflection coefficient is given by $R_h = +1$. Figures 4.22 (a) and 4.22 (b) show the total and diffracted fields at 10 GHz for soft and hard polarized cases, respectively.

Examination of Figures 4.21 and 4.22 reveals a few interesting observations:

(a) The z-polarized electric fields shown in Figures 4.21 (a) and 4.22 (a) approach zero as the observation angle (ϕ) tends to 0 and $n\pi$. This is to be expected,

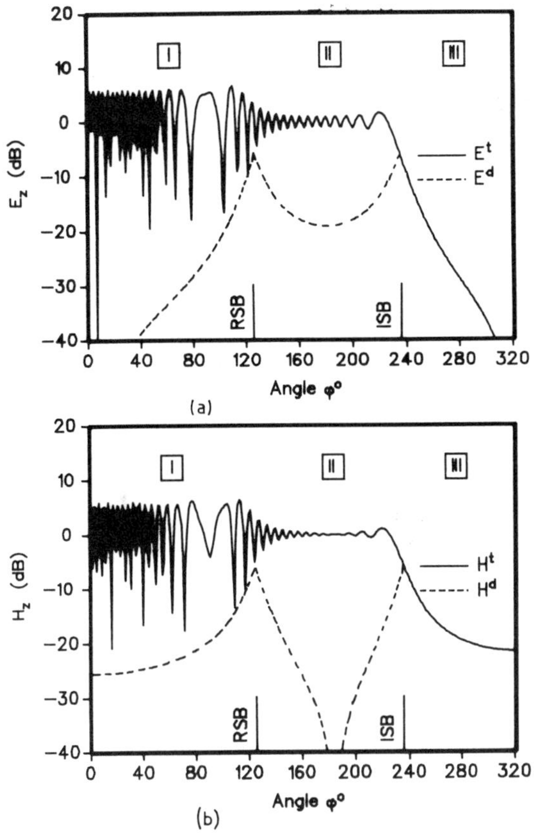

Figure 4.21 Scattered field from a wedge with plane wave incidence using the UTD diffraction coefficients. The geometry is as in Figure 4.9, with $\alpha = 40°$, $\phi' = 55°$, $f = 3$ GHz,and $s = 1$ m: (a) soft case; (b) hard case. The level of the incident field is 0 dB.

as these fields have to be zero on the surfaces of the perfectly conducting wedge. The z-polarized magnetic fields in Figures 4.21 (b) and 4.22 (b) are finite and nonzero at $\phi = 0$ and $\phi = n\pi$.

(b) In region I, we find that the total field consists mainly of the incident and reflected fields, as the diffracted field is very weak (except in the transition region). The incident and reflected fields both have amplitudes of 1 (0 dB). The maximum value of the total field in region I thus is approximately 6 dB (i.e., 20 log2), and it occurs at points where the incident and reflected fields interfere constructively. Note that the positions of the peaks in region I of

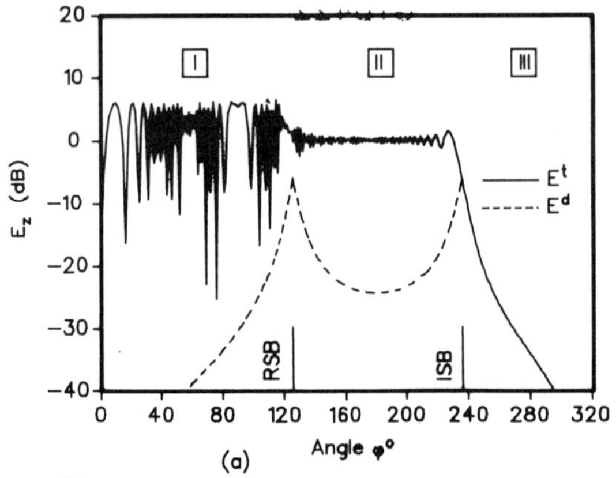

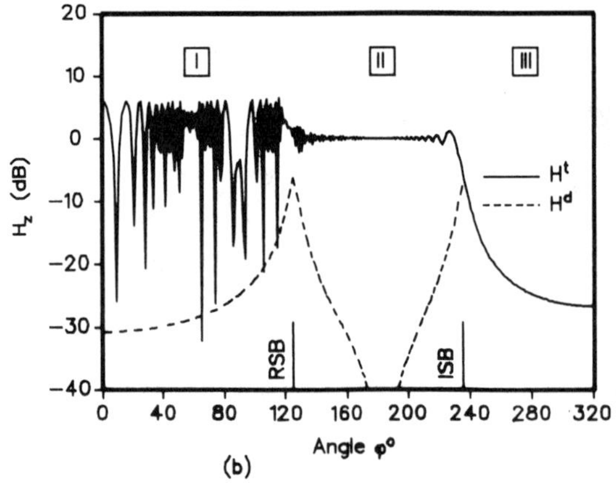

Figure 4.22 Scattered field from a wedge with plane wave incidence using the UTD diffraction coefficients. The geometry is as in Figure 4.9 with $\alpha = 40°$, $\phi' = 55°$, $f = 10$ GHz, and $s = 1$ m: (a) soft case; (b) hard case. The level of the incident field is 0 dB.

Figures 4.21 (a) and 4.22 (a) correspond to the positions of the nulls in Figures 4.21 (b) and 4.22 (b), respectively. This is due to the fact that the phase of the reflected field changes by 180° as the reflection coefficient changes from $R_s = -1$ to $R_h = +1$. In region II, only the incident and diffracted fields are present and, as the diffracted field is much weaker than the incident field, the total field exhibits only a small ripple. In region III, only the diffracted field is present and hence there is no ripple.

(c) At the shadow boundaries the amplitude of the diffracted field is approximately 6 dB below the level of the incident field, as was shown in (4.112) and (4.126).

(d) The diffracted field attains a maximum amplitude at the shadow boundaries and decays rapidly away from the shadow boundaries.

(e) Equations (4.55) to (4.59) reveal that

$$U_z^d \alpha \sqrt{\lambda} \tag{4.143}$$

Because the wavelength is inversely proportional to the frequency, we may tend to think that the amplitude of the diffracted field also will be inversely proportional to the square root of the frequency. Bear in mind, however, that the transition function also is a function of frequency. Close to the shadow boundaries the $\sqrt{k}$ term in the small argument form of the transition function cancels the $1/\sqrt{k}$ term in the diffraction coefficient, so that the amplitude of the diffracted field is independent of the frequency at the shadow boundary. We saw in (4.112) and (4.126) that it was approximately one-half that of the incident field (assuming that U^{cd} is small). We find that as the frequency increases, the transition regions tend to shrink toward the shadow boundaries, as seen in Figure 4.20. However, as the diffracted ray moves out of the transition region, the transition function approaches unity and becomes ray optical. The diffracted fields decay away from the shadow boundaries faster with increasing frequency as prescribed by (4.143). Comparison between Figures 4.21 and 4.22 verifies this fact.

In the limit as the frequency tends to infinity (and the wavelength to zero), the diffracted field will be equal to zero and the total field will contain only GO terms.

The reader is referred to [35, 37], which deal specifically with 90° wedges.

4.4.5 Grazing Incidence

The cases where $\phi' = 0$ and $\phi' = n\pi$ are known as *grazing incidence* because the incident ray grazes the surface of the wedge. Grazing incidence is a special case as far as UTD is concerned, and we have to treat it separately.

Consider a line source at a height δ above a wedge as shown in Figure 4.23. The total GO field at a point in the far-zone is given by

$$U_z^t(\theta) = U_z^i(\theta) + U_z^r(\theta) \tag{4.144}$$

The incident field is given by

$$U_z^i(\theta) = C\frac{e^{-jkr}}{\sqrt{r}} \tag{4.145}$$

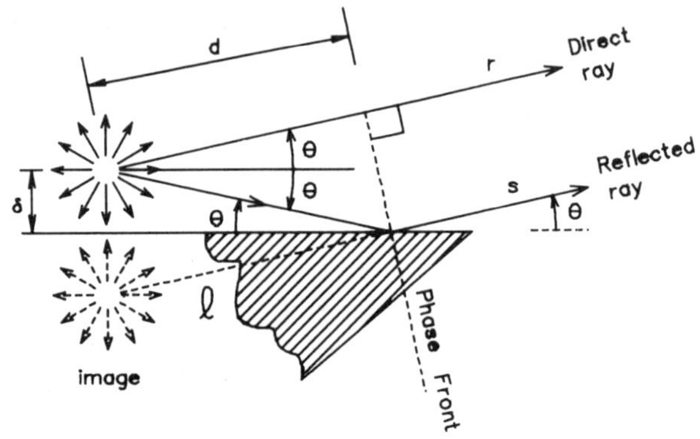

Figure 4.23 Geometry for the derivation of grazing incidence.

where r is the distance from the line source to a field point in the far-zone. The reflected field is given by

$$U_z^r(\theta) = C\frac{e^{-jkl}}{\sqrt{l}}R_{s,h}\sqrt{\frac{l}{l+s}}e^{-jks} \tag{4.146}$$

where

$$l = \frac{\delta}{\sin\theta} \tag{4.147}$$

$$s = r - d \tag{4.148}$$

$$d = \frac{\delta \cos 2\theta}{\sin\theta} \tag{4.149}$$

and $R_{s,h}$ is the reflection coefficient. It follows that

$$U_z^r(\theta) = R_{s,h}C\frac{e^{-jk(r+2\delta\sin\theta)}}{\sqrt{r + 2\delta \sin\theta}} \tag{4.150}$$

If the line source is mounted on the half-plane, as in the case of grazing incidence, $\delta \to 0$ and we find that

$$\lim_{\delta\to 0} U_z^r(\theta) = R_{s,h}C\frac{e^{-jkr}}{\sqrt{r}} = R_{s,h}U_z^i(\theta) \tag{4.151}$$

so that

$$U_z^t(\theta) = U_z^i(\theta)(1 + R_{s,h}) \tag{4.152}$$

Because $R_s = -1$ and $R_h = +1$,

$$E_z^t(\theta) = 0 \tag{4.153}$$

$$H_z^t(\theta) = 2H_z^i(\theta) \tag{4.154}$$

We expect the electric field to be zero, as the z-polarized electric field must vanish on the surface of the perfectly conducting wedge.

We can view the reflected field as produced by the image of the line source, shown in Figure 4.23. If the line source has an amplitude C, the image of an electric line source has an amplitude of $-C$, whereas the image of a magnetic line source has an amplitude of $+C$. As the line source approaches the surface, the line source and its image will merge. In the case of an electric line source, the line source and its image will cancel, whereas in the case of a magnetic line source, the line source and its image will create a resultant line source of twice the strength, that is, of amplitude $2C$.

We would consider a line source of strength C on the surface as the limiting case of a line source of strength $C/2$ approaching the surface. In this case there would be a reduction of the diffracted field from each of the diffraction terms of one-half. Consequently, it has become accepted practice to assume an amplitude of C rather than $2C$ for the line source, and divide all diffraction coefficients by two.

The incident and reflected fields have merged, and so have their shadow boundaries; the diffraction terms associated with the incident and reflection shadow boundaries also are equal. In the cases where two edges are close to another, one of the edges can be in the transition region of the other. This in turn can have implications on the validity of the assumption that the edge is illuminated by a ray optical field. The case of two nearby edges illuminated at grazing incidence is discussed in [38–41].

4.4.6 Half-Plane and Curved Screen

Not only are the special cases of a half-plane and curved screen important in themselves, but they also will be used to illustrate some concepts pertaining to grazing incidence in the next section. In the case of a half-plane, $n = 2$ (see Figure 4.8 (b)), and the radii of curvature of both the o-face and n-face are equal to infinity. As shown in (4.95),

$$L^{ro,n} = L^i = \frac{ss'}{s + s'} \tag{4.155}$$

follows and (4.71) reduces to (4.70). Furthermore, we can show that

$$\cot\frac{\pi + \beta}{2n} + \cot\frac{\pi - \beta}{2n} = \frac{-2\sin(\pi/n)}{\cos(\pi/n) - \cos(\beta/n)} \tag{4.156}$$

Substituting $n = 2$ into (4.56) to (4.59), we find that, in the case of a half-plane, the diffraction coefficients are given by

$$D_{s,h}(L^i, \phi, \phi') = \frac{-e^{-j\pi/4}}{2\sqrt{2\pi k}}\left\{\frac{F[kL^i a(\phi - \phi')]}{\cos[(\phi - \phi')/2]} \mp \frac{F[kL^i a(\phi + \phi')]}{\cos[(\phi + \phi')/2]}\right\} \tag{4.157}$$

with $a\ (\phi \pm \phi')$ given by (4.70). Note that, when the diffracted ray is not close to a shadow boundary, $F(kLa) \approx 1$ and (4.157) reduces to Keller's original diffraction coefficient as given in (4.23).

We now consider the curved screen shown in Figure 4.8(c), a special case of the curved wedges of Figure 4.18. Because $n = 2$, we find that $\hat{\mathbf{n}}_o = -\hat{\mathbf{n}}_n$ and $a_o = -a_n$. If we substitute these values into (4.82), we find that $\rho^{ro} = \rho^{rn}$. From (4.61) and (4.62), $L^{ro} = L^{rn}$. We can now substitute these parameters into (4.55)–(4.59) to yield the diffraction coefficients for a curved screen:

$$D_{s,h}(L^i, L^r, \phi, \phi') = \frac{-e^{-j\pi/4}}{2\sqrt{2\pi k}}\left\{\frac{F[kL^i a(\phi - \phi')]}{\cos[(\phi - \phi')/2]} \mp \frac{F[kL^r a(\phi + \phi')]}{\cos[(\phi + \phi')/2]}\right\} \tag{4.158}$$

with $a(\phi \pm \phi')$ given in (4.70). A more detailed discussion of diffraction by a curved screen can be found in [42].

4.4.7 Continuity across the Shadow Boundary: Grazing Incidence

The enforcement of continuity across the shadow boundary in the case of grazing incidence now will be investigated. Consider a magnetic line source to be located on the half-plane as shown in Figure 4.24. In this case, $\phi' = 0$. The total field thus is given by

$$H_z^t = \begin{cases} H_z^i + H_z^d, & 0 \leq \phi \leq \pi \\ H_z^d, & \pi \leq \phi \leq 2\pi \end{cases} \tag{4.159}$$

where

$$H_z^i = \frac{e^{-jkl}}{\sqrt{l}} \tag{4.160}$$

$$H_z^d = \frac{e^{-jks'}}{\sqrt{s'}}\frac{D_h}{2}\frac{e^{-jks}}{\sqrt{s}} \tag{4.161}$$

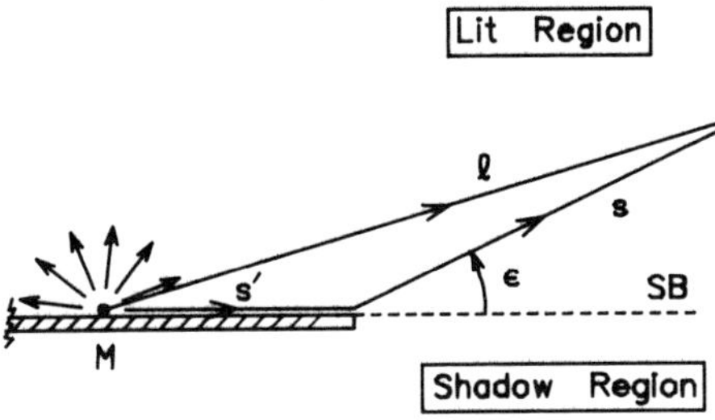

Figure 4.24 Continuity across the shadow boundary in the case of grazing incidence (M is a magnetic line source).

The diffraction coefficient is given in (4.157). In the case of grazing incidence, all diffraction coefficients are divided by two as explained in Section 4.4.5. Substituting $\phi' = 0$ into (4.157), we find that

$$D_{s,h}(L^i, \phi) = \frac{-e^{-j\pi/4}}{2\sqrt{2\pi k}} \left\{ \frac{F[kL^i a(\phi)]}{\cos(\phi/2)} \mp \frac{F[kL^i a(\phi)]}{\cos(\phi/2)} \right\} \tag{4.162}$$

so that

$$D_s = 0 \tag{4.163}$$

$$\frac{D_h(L^i, \phi)}{2} = \frac{-e^{-j\pi/4}}{2\sqrt{2\pi k}} \frac{F[kL^i a(\phi)]}{\cos(\phi/2)} \tag{4.164}$$

where L^i is given in (4.60). Following the arguments in Section 4.4.3, we find that at an angle $\phi = \pi - \epsilon$ (with ϵ very small and positive in the lit region), the diffracted field in the lit region is given by

$$H^d = -H^i \frac{|\epsilon|}{2\epsilon} \tag{4.165}$$

so that the total field in the region close to the shadow boundary is given by

$$H^t = \begin{cases} H^i - \dfrac{H^i}{2} = \dfrac{H^i}{2}, & \text{in the lit region } (\epsilon > 0) \\[2ex] \dfrac{H^i}{2}, & \text{in the shadow region } (\epsilon < 0) \end{cases} \tag{4.166}$$

We see that there are no residue in fields (U^{cd}), as is the case in (4.112) and (4.127), so that the diffracted field is exactly equal to one-half the incident field on the shadow boundary. In summary, this discussion shows that, in the case of grazing incidence, the reflected field need not be taken into account and the diffraction coefficient must then be divided by two.

The discussion in this section has dealt with the case where $n = 2$; that is, an infinitely thin half-plane. The case of diffraction from a thick, perfectly conducting half-plane, in which higher-order effects are taken into account, is investigated in [43].

Example 4.4: Radiation from a magnetic line source mounted on a conducting strip

The field radiated by a line source is given by (4.4). Examination of this equation shows that the line source has an omnidirectional far-zone radiation pattern. If the line source is mounted on a structure, however, the radiation pattern will change. A magnetic line source on a plane screen, for instance, can be used to model a slot antenna.

Let us consider the case where a magnetic line source is mounted in the center of a strip of width l as shown in Figure 4.25. We must determine the far-zone radiation pattern of the structure as a function of the angle θ measured with respect to the normal vector ($\hat{\mathbf{n}}$). For convenience, we assume that the line source has a unit amplitude. In view of the symmetry of the structure the pattern need be calculated only in the region $0° \leq \theta \leq 180°$.

The total field is given by

$$H^t = \begin{cases} H^i + H_1^d + H_2^d, & \text{if } 0° \leq \theta < 90° \\ H_1^d + H_2^d, & \text{if } 90° < \theta \leq 180° \end{cases} \tag{4.167}$$

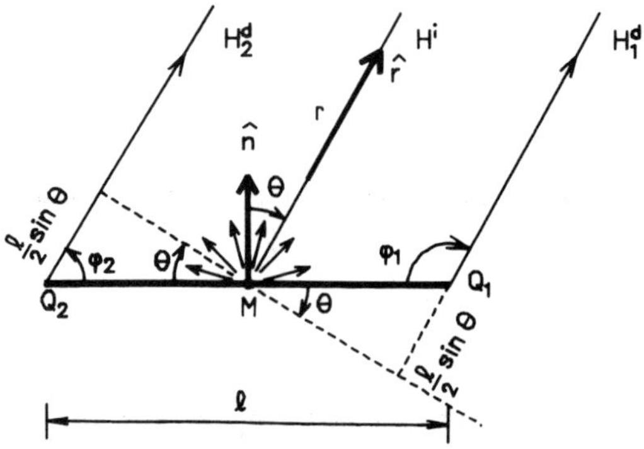

Figure 4.25 Magnetic line source (M) mounted on a conducting strip.

The phase reference is chosen to be at the line source. At a field point in the far-zone that is removed a distance r from the line source, the direct field thus is given by

$$H^i(\theta) = \begin{cases} \dfrac{e^{-jkr}}{\sqrt{r}}, & \text{if } \theta < 90° \\ 0, & \text{if } \theta > 90° \end{cases} \tag{4.168}$$

Because we are interested in determining the far-zone radiation pattern, we need to consider those diffracted rays that propagate in the direction $\hat{\mathbf{r}}$ where $\hat{\mathbf{r}} = \cos\theta$.

Note that the edges of the strip are illuminated at grazing incidence from the line source. Using (4.164), we find that the diffracted field from the right edge (Q_1) is given by

$$H_1^d(\theta) = H^i(Q_1)\frac{D_h(L^i, \phi_1)}{2}\frac{e^{-jk[r-(l/2)\sin\theta]}}{\sqrt{r - (l/2)\sin\theta}} \tag{4.169}$$

where

$$H^i(Q_1) = \frac{e^{-jkl/2}}{\sqrt{l/2}} \tag{4.170}$$

The distance parameter is found from (4.60) to be given by

$$L^i = \frac{(l/2)[r - (l/2)\sin\theta]}{(l/2) + r - (l/2)\sin\theta} \tag{4.171}$$

However, the fact that $r \gg l$ allows us to make the following simplifications:

$$L^i \approx \frac{l}{2} \tag{4.172}$$

$$\sqrt{r - (l/2)\sin\theta} \approx \sqrt{r} \tag{4.173}$$

where (4.173) is used in the amplitude terms of (4.169). As any textbook on antennas and radiation will tell us, a similar simplification cannot be made in the phase terms [6, 24, 25]. Furthermore, we find that

$$\phi_1 = 90° + \theta \tag{4.174}$$

so that

$$\cos\left(\frac{\phi_1}{2}\right) = \frac{\cos(\theta/2) - \sin(\theta/2)}{\sqrt{2}} \tag{4.175}$$

$$a(\phi_1) = 1 - \sin\theta \tag{4.176}$$

Substituting these parameters into (4.164) and (4.169), we find that

$$H_1^d(\theta) = \frac{-\mathrm{e}^{-j\pi[(l/\lambda)(1-\sin\theta)+1/4]}F[\pi(l/\lambda)(1 - \sin\theta)]}{\sqrt{l/\lambda}2\pi\,[\cos(\theta/2) - \sin(\theta/2)]}\frac{\mathrm{e}^{-jkr}}{\sqrt{r}} \tag{4.177}$$

We have expressed H_1^d as a function of l/λ, rather than k, to illustrate the effect that the width of the strip has on the pattern.

The field that is diffracted from the left edge (Q_2) is given by

$$H_2^d(\theta) = H^i(Q_2)\,\frac{D_h(L^i, \phi_2)}{2}\,\frac{\mathrm{e}^{-jk[r+(l/2)\sin\theta]}}{\sqrt{r + (l/2)\sin\theta}} \tag{4.178}$$

where

$$H^i(Q_2) = \frac{\mathrm{e}^{-jkl/2}}{\sqrt{l/2}} \tag{4.179}$$

Using the same arguments as above, we can make the following simplifications:

$$L^i \approx \frac{l}{2} \tag{4.180}$$

$$\sqrt{r + (l/2)\sin\theta} \approx \sqrt{r} \tag{4.181}$$

where (4.181) is used in the amplitude term of (4.178).

A casual inspection might lead us to believe that

$$\phi_2 = 90^\circ - \theta \tag{4.182}$$

can be used in (4.178). However, when $\theta > 90^\circ$, (4.182) will yield $\phi_2 < 0^\circ$. This is not allowed for the purposes of UTD calculations because (4.39) restricts ϕ_2 to the range $0^\circ \leqslant \phi_2 \leqslant 360^\circ$, as $n = 2$ in this case. In this example, the angle ϕ_2

therefore should be specified as follows:

$$\phi_2 = \begin{cases} 90° - \theta, & \text{if } 0° \leqslant \theta \leqslant 90° \quad (4.183a) \\ 450° - \theta, & \text{if } 90° < \theta \leqslant 180° \quad (4.183b) \end{cases}$$

Substitution gives

$$\cos\left(\frac{\phi_2}{2}\right) = \frac{\pm[\cos(\theta/2) + \sin(\theta/2)]}{\sqrt{2}} \tag{4.184}$$

where the plus sign applies in the case where $\theta \leqslant 90°$ and the minus sign in the case where $\theta > 90°$. Furthermore,

$$a(\phi_2) = 1 + \sin\theta \tag{4.185}$$

Substituting the parameters above into (4.178), we find that

$$H_2^d(\theta) = \frac{-e^{-j\pi[(l/\lambda)(1+\sin\theta)+1/4]}F[\pi(l/\lambda)(1 + \sin\theta)]}{\text{sgn}(90° - \theta)\sqrt{l/\lambda}\, 2\pi[\cos(\theta/2) + \sin(\theta/2)]} \frac{e^{-jkr}}{\sqrt{r}} \tag{4.186}$$

where

$$\text{sgn}(x) = \begin{cases} 1, & \text{if } x \geqslant 0 \\ -1, & \text{if } x < 0 \end{cases} \tag{4.187}$$

The factor $e^{-jkr}/\sqrt{r}$ is common to all three fields; that is, the direct as well as the two diffracted fields. Because this term is independent of angle, it usually is discarded when calculating the far-zone radiation pattern.

Figure 4.26 shows the pattern of a magnetic line source mounted in the center of a conducting strip of width $l = 5\lambda$, as expressed in (4.167). The constant line at 0 dB indicates the pattern of the magnetic line source in free space. We see from Figure 4.26 that there is a discontinuity in the pattern at $\theta = 90°$. Assuming that the discontinuity is not due to an error in our calculations or programming, such a discontinuity across a shadow boundary usually indicates that a scattering mechanism has not been taken into account. In this case, a higher-order diffraction term was not included.

To identify this higher-order term, we need to examine the continuity of the

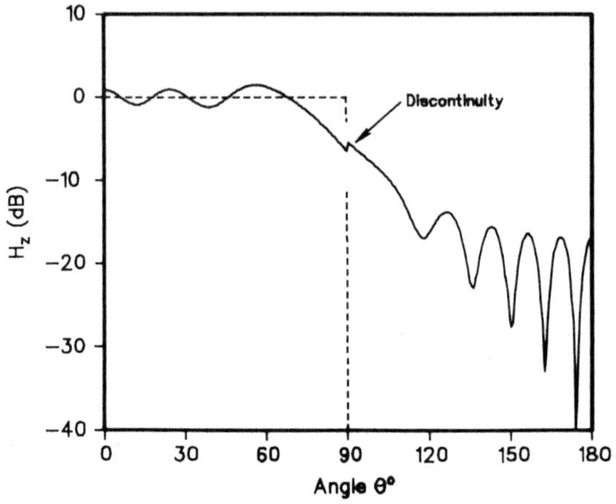

Figure 4.26 Radiation pattern of a magnetic line source mounted in the center of a conducting strip of width $l = 5\lambda$. The solid line indicates the direct and first-order diffracted fields, whereas the dashed line shows the pattern of a magnetic line source mounted on an infinite ground plane. Note the discontinuity at $\theta = 90°$.

terms in (4.167) across the shadow boundary at $\theta = 90°$. Let us start by considering $H^i + H_1^d$ at $\theta = 90° - \epsilon$. At this angle, we find that

$$1 - \sin(90° - \epsilon) = 1 - \cos\epsilon \approx \frac{\epsilon^2}{2} \tag{4.188}$$

$$\cos\frac{\theta}{2} - \sin\frac{\theta}{2} \approx \frac{\epsilon}{\sqrt{2}} \tag{4.189}$$

Hence,

$$H_1^d(\theta = 90° - \epsilon) = \frac{-e^{-j\pi/4}\sqrt{2}\, F\left(\pi(l/\lambda)\frac{\epsilon^2}{2}\right)}{\sqrt{l/\lambda}\; 2\pi\epsilon} \frac{e^{-jkr}}{\sqrt{r}} \tag{4.190}$$

where we have assumed that

$$e^{-j\pi}(l/\lambda)\epsilon^2 \approx 1 \tag{4.191}$$

Applying (4.73) to (4.190), we find that

$$H_1^d(\theta = 90° - \epsilon) = -\frac{|\epsilon|}{2\epsilon}\frac{e^{-jkr}}{\sqrt{r}} \tag{4.192}$$

so that

$$H^i + H_1^d(\theta = 90° - \epsilon, \epsilon > 0) = \frac{e^{-jkr}}{2\sqrt{r}} \tag{4.193}$$

$$H_1^d(\theta = 90° - \epsilon, \epsilon < 0) = \frac{e^{-jkr}}{2\sqrt{r}} \tag{4.194}$$

as $\epsilon \rightarrow 0$. Bear in mind that $H^i = 0$ for $\theta > 90°$. From (4.193) and (4.194), the field $H^i + H_1^d$ clearly is continuous across the shadow boundary at $\theta = 90°$.

Let us now examine H_2^d across the shadow boundary at $\theta = 90°$. From Figure 4.27, we see that two diffracted rays from the left edge actually propagate in the direction $\theta = 90°$; that is,

$$H_2^{d^-} = \lim_{\epsilon \rightarrow 0} H_2^d(\theta = 90° - \epsilon) \tag{4.195}$$

$$H_2^{d^+} = \lim_{\epsilon \rightarrow 0} H_2^d(\theta = 90° + \epsilon) \tag{4.196}$$

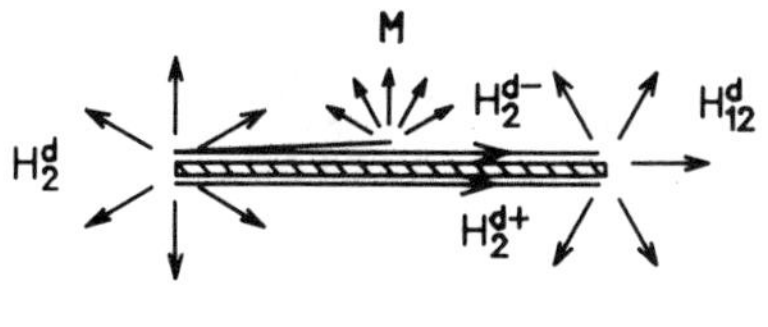

Figure 4.27 Second-order diffracted fields from a magnetic line source mounted on a conducting strip.

where $\epsilon > 0$ is assumed in both (4.195) and (4.196). Using (4.186), we find that the far-zone diffracted field from region $\theta \approx 90°$ is given by

$$H_2^{d^\pm} = \pm \frac{e^{-j\pi[(2l/\lambda) + 1/4]}}{\sqrt{l/\lambda}\, 2\pi\sqrt{2}}\frac{e^{-jkr}}{\sqrt{r}} \tag{4.197}$$

because

$$F[\pi(l/\lambda)(1 + \sin 90°)] \approx 1 \tag{4.198}$$

Although the magnitude of the amplitude of $H_2^{d^+}$ and $H_2^{d^-}$ are the same on their respective sides of the shadow boundary at $\theta \approx 90°$, the phase discontinuity between these two terms results in a discontinuity of magnitude $1/(\pi\sqrt{2l/\lambda})$ between $H_2^{d^+}$ and $H_2^{d^-}$ at $\theta = 90°$. Furthermore, this discontinuity causes the discontinuity in the pattern at $\theta = 90°$ in Figure 4.26. The problem can be solved by adding the second-order diffracted fields that will be excited at Q_1 by $H_2^{d^+}$ and $H_2^{d^-}$ to the solution for the total field.

The second-order diffracted field that emanates from Q_1 due to illumination from $H_2^{d^-}$ can be expressed as

$$H_{12}^{d^-} = H_2^{d^-}(Q_1)\frac{D_h(L^i, \phi_1)}{2}\frac{e^{-jk[r-(l/2)\sin\theta]}}{\sqrt{r - (l/2)\sin\theta}} \tag{4.199}$$

because this still is a case of grazing incidence. Note that (4.199) is similar to (4.169), the difference being that in the latter case the excitation is from a line source rather than from the field diffracted from Q_2. In a manner analogous to (4.171) and (4.172), we find that

$$L^i = l \tag{4.200}$$

in this case. Substituting (4.164), (4.175), (4.176), and (4.200) into (4.199), we find that

$$H_{12} = H_2(Q_1)\frac{-e^{j\pi[(l/\lambda)\sin\theta-(1/4)]}\sqrt{2\lambda}F[2\pi(l/\lambda)(1 - \sin\theta)]}{4\pi[\cos(\theta/2) - \sin(\theta/2)]}\frac{e^{-jkr}}{\sqrt{r}} \tag{4.201}$$

where

$$H_2^{d^-}(Q_1) = H^i(Q_2)\frac{D_h(L^i, 90°)}{2}\frac{e^{-jkl}}{\sqrt{l}} \tag{4.202}$$

with

$$L^i = \frac{(l/2)l}{(l/2) + l} = \frac{l}{3} \tag{4.203}$$

Because the shadow boundary of $H_2^{d^-}$ is in the direction $\theta = 270°$, we find from (4.185) that

$$a(\phi_2 = 0°) = 2 \tag{4.204}$$

and the transition function of $D_h/2$ in (4.202) becomes

$$F(kLa) \approx 1 \tag{4.205}$$

From (4.202)–(4.205), we find that

$$H_2^{d^-}(Q_1) = \frac{-\,e^{-j\,[(3kl/2)+\pi/4]}}{\sqrt{l/2}\;4\pi\;\sqrt{l/\lambda}} \tag{4.206}$$

Substituting (4.206) into (4.201), we then find that

$$H_{12}^{d^-} = \frac{-j\,e^{-j\pi(l/\lambda)(3-\sin\theta)}\,F[2\pi(l/\lambda)(1-\sin\theta)]}{(l/\lambda)8\pi^2\,[\cos(\theta/2)-\sin(\theta/2)]}\,\frac{e^{-jkr}}{\sqrt{r}} \tag{4.207}$$

In a similar manner the second-order diffracted field that emanates from Q_1 due to illumination from $H_2^{d^+}$ (i.e., $H_{12}^{d^+}$) is found to be equal to $H_{12}^{d^-}$. The total second-order field diffracted from Q_1 thus can be expressed as

$$H_{12}^{d} = H_{12}^{d^-} + H_{12}^{d^+} \tag{4.208}$$

or

$$H_{12}^{d} = \frac{-j\,e^{-j\pi(l/\lambda)(3-\sin\theta)}F[2\pi(l/\lambda)(1-\sin\theta)]}{(l/\lambda)4\pi^2[\cos(\theta/2)-\sin(\theta/2)]}\,\frac{e^{-jkr}}{\sqrt{r}} \tag{4.209}$$

Let us now examine the behavior of $H_2^d + H_{12}^d$ across the shadow boundary at $\theta = 90°$. Using (4.187) and the now-familiar small argument form of the transition function, we find that

$$H_{12}^{d}(\theta = 90° - \epsilon) = \frac{e^{-j\pi[2(l/\lambda)+1/2]}}{(l/\lambda)\,4\pi^2}\,\sqrt{\lambda}F\left(2\pi\frac{l}{\lambda}\frac{\epsilon^2}{2}\right)\frac{e^{-jkr}}{\sqrt{r}} \tag{4.210}$$

or

$$H_{12}^{d}(\theta = 90° - \epsilon) = \frac{e^{-j\pi[2(l/\lambda)+1/4]}}{\sqrt{l/\lambda}}\,\mathrm{sgn}(\epsilon)\frac{e^{-jkr}}{\sqrt{r}} \tag{4.211}$$

where $\epsilon > 0$ describes the field in the region smaller than 90° and $\epsilon < 0$ describes the field in the region greater than 90°. We thus find that

$$\lim_{\theta \to 90^\circ} H_{12}^{d} + H_{2}^{d^-} = \lim_{\theta \to 90^\circ} H_{12}^{d} + H_{2}^{d^+} = 0 \tag{4.212}$$

Equation (4.212) indicates that the second-order diffracted field H_{12}^d compensates for the discontinuity in the first-order diffracted field H_2^d across the shadow boundary at $\theta = 90°$ in that the total field consisting of $H_{12}^d + H_2^d$ is continuous across the shadow boundary.

Figure 4.28 shows the pattern of a magnetic line source mounted in the center of a conducting strip of width 5λ, where the pattern is given by

$$H^t = \begin{cases} H^i + H_1^d + H_2^d + H_{12}^d, & \text{if } 0° \leq \theta \leq 90° \\ H_1^d + H_2^d + H_{12}^d, & \text{if } 90° < \theta \leq 180° \end{cases} \tag{4.213}$$

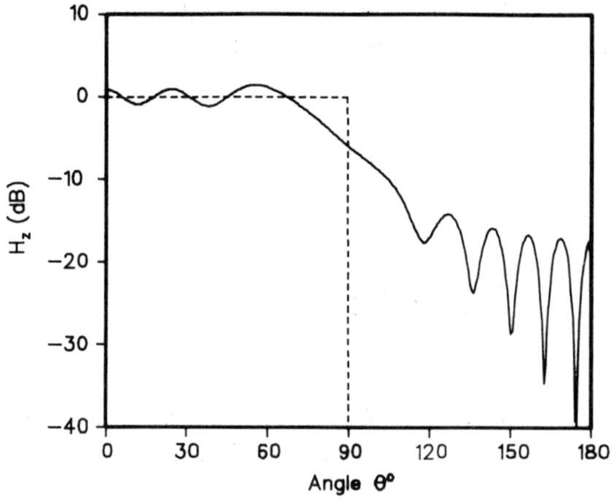

Figure 4.28 Radiation pattern of a magnetic line source mounted in the center of a conducting strip of width $l = 5\lambda$. The direct, first-, and second-order diffracted fields are shown by the solid line, whereas the dashed line shows the pattern of a magnetic line source mounted on an infinite ground plane.

Comparing Figure 4.26 to Figure 4.28, we notice that the incorporation of the second-order diffracted field ensures that fields in the latter figure are smooth and continuous at $\theta = 90°$. Figure 4.29 shows the relative amplitudes of the different component fields given in (4.213) for the case $l = 5\lambda$. Note that H_{12}^d is almost insignificant away from the direction $\theta \approx 90°$. The discontinuity in H_2^d across the shadow boundary as calculated in (4.197) is not reflected in Figure 4.29 because H_2^{d-} and H_2^{d+} have the same magnitude at $\theta = 90°$.

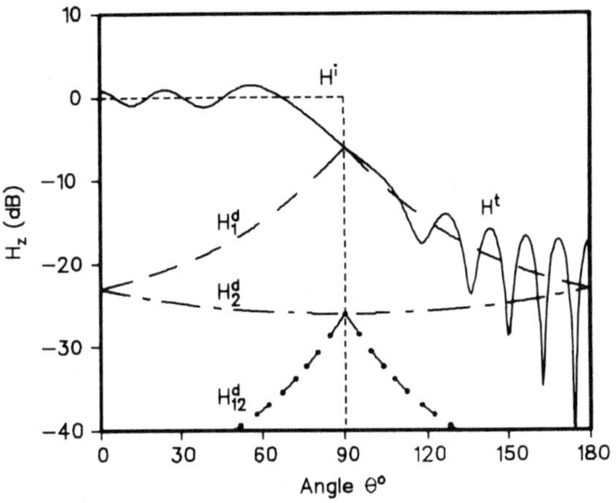

Figure 4.29 Component fields for a magnetic line source mounted on a strip of width $l = 5\lambda$.

The field H_{12}^d also can be expressed in terms of a diffraction coefficient. Referring to (4.209), we can write H_{12}^d as

$$H_{12}^d = H^i(Q_1)\, D(l, \theta)\, \frac{e^{-jk[r-(l/2)\sin\theta]}}{\sqrt{r}} \tag{4.214}$$

where $H^i\,(Q_1)$ is given in (4.170), and

$$D(l, \theta) = \frac{-j\, e^{-jkl}\, F[kl(1 - \sin\theta)]}{\sqrt{l}\, 2\pi k[\cos(\theta/2) - \sin(\theta/2)]} \tag{4.215}$$

One interesting problem involving a magnetic line source mounted on a strip is to obtain the front-to-back isolation. The *front-to-back ratio* (FBR) can be expressed as

$$\text{FBR} = \frac{|H^t(\theta = 0°)|}{|H^t(\theta = 180°)|} \tag{4.216}$$

Substituting (4.167) into (4.216), we find that the FBR can be expressed in decibels as

$$\text{FBR} = 20 \log \left| \frac{1 - e^{-j\pi[(l/\lambda) + 1/4]}/(\pi\sqrt{l/\lambda})}{1/(\pi\sqrt{l/\lambda})} \right| \tag{4.217}$$

or

$$\text{FBR} = 20 \log|\pi\sqrt{l/\lambda} - e^{-j\pi[(l/\lambda)+1/4]}| \tag{4.218}$$

Figure 4.30 shows the FBR as a function of l/λ. H_{12}^d need not be included in (4.211) because it is very small at $\theta = 0°$ and $\theta = 180°$.

When we calculate the pattern by computer, to evaluate the fields at exactly $\theta = 0°$, $\theta = 90°$, or $\theta = 180°$ is sometimes risky because that may lead to errors due to numerical overflow. Furthermore, we cannot evaluate the field exactly on a shadow boundary, whereas evaluation at a very small angle away from the shadow boundary presents no problem. In this case, we would evaluate the fields at $\theta = 0° + \epsilon$, $\theta = 90° \pm \epsilon$, and $\theta = 180° - \epsilon$.

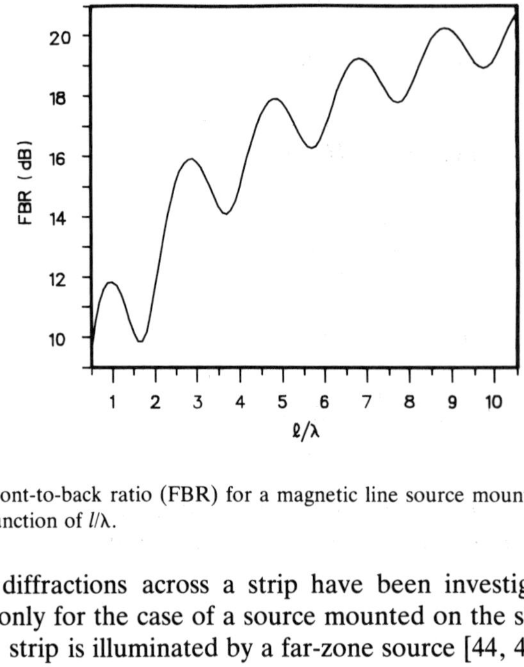

Figure 4.30 The front-to-back ratio (FBR) for a magnetic line source mounted on a conducting strip as a function of l/λ.

Multiple diffractions across a strip have been investigated by several researchers, not only for the case of a source mounted on the strip, but also for the case where the strip is illuminated by a far-zone source [44, 45]. A discussion and a modification of the GTD to compensate for low-frequency defects prevalent in transient scattering analyses is given in [46].

4.4.8 Full-Plane

We now consider the case of a full-plane; that is, where $\alpha = 180°$ ($n = 1$) as shown in Figure 4.8(e). We have seen in (4.27) that in this case Keller's diffraction coefficients always are zero for both polarizations: $D^k_{s,h} \equiv 0$. At first, the full-plane seems to be a trivial case and its analysis pointless. However, recall that Keller's diffraction coefficients do not take into account the radii of curvature of the *o*-face and *n*-face at the point of diffraction, which the UTD diffraction coefficients do. We find that diffraction occurs when $n = 1$ if the radii of curvature of the *o*-face and *n*-face differ [47].

Consider the full-plane in Figure 4.8(e). Substituting $n = 1$ into (4.56) and (4.57) yields

$$D_1 \equiv D_2 \equiv 0 \tag{4.219}$$

However, $D_3 \neq 0$ and $D_4 \neq 0$, because in general $L^{ro} \neq L^{rn}$. Bear in mind that D_1 and D_2 are the coefficients of diffracted fields that provide continuity across the incident shadow boundaries. In this case, however, there are no incident shadow boundaries so we expect $D_{1,2}$ to vanish. Furthermore, (4.71) degenerates to (4.70). Substituting $n = 1$ into (4.58) and (4.59), we find that the diffraction coefficients in (4.55) reduce to

$$D_{s,h}(L^{ro}, L^{rn}, \phi, \phi') = \frac{\pm\, e^{-j\pi/4}}{2\sqrt{2\pi k}} \tan\left(\frac{\phi + \phi'}{2}\right) \{F[kL^{ro}a(\phi + \phi')] - F[kL^{rn}a(\phi + \phi')]\} \tag{4.220}$$

If the radii of curvature of the *o*-face and *n*-face are equal at the point of diffraction, $L^{ro} = L^{rn}$, resulting in $D_{s,h} = 0$. Again, this is to be expected. If there is no discontinuity in the radius of curvature, there is no discontinuity in the reflected field and hence no diffracted field. Actually, we would be more correct to say that there is no first-order diffracted field. We have seen that, if there is a discontinuity in the derivative of the radius of curvature across an edge, there will be a diffracted field that is a function of the discontinuity in the derivative [48].

When the *o*-face of the wedge is flat, (4.220) reduces to

$$D_{s,h}(\infty, L^{rn}, \phi, \phi') = \frac{\pm\, e^{-j\pi/4}}{2\sqrt{2\pi k}} \tan\left(\frac{\phi + \phi'}{2}\right) \{1 - F[kL^{rn}a(\phi + \phi')]\} \tag{4.221}$$

A similar expression can be determined for the case where the *n*-face is flat.

Kouyoumjian and Pathak [20, 27] have developed an extended theory for the case of a perfectly conducting full-plane; that is, where $n = 1$ but there is a discontinuity in the radius of curvature. They found that in this case the diffraction coefficients were given by

$$D_{s,h}(L^{ro}, L^{rn}, \phi, \phi') = \frac{e^{-j\pi/4}}{\sqrt{2\pi k}} f_{s,h}$$

$$\{F[kL^{ro}a(\phi + \phi')] - F[kL^{rn}a(\phi + \phi')]\} \quad (4.222)$$

where

$$f_s = \frac{\sin\phi \sin\phi' \tan\left(\frac{\phi + \phi'}{2}\right)}{\left(1 + \frac{\sin\phi}{\sin\phi'}\right) \cos^2\left(\frac{\phi - \phi'}{2}\right)} \quad (4.223)$$

$$f_h = -\frac{(1 + \cos\phi \cos\phi') \tan\left(\frac{\phi + \phi'}{2}\right)}{\left(1 + \frac{\sin\phi}{\sin\phi'}\right) \cos^2\left(\frac{\phi - \phi'}{2}\right)} \quad (4.224)$$

Examples of the applications of edge diffraction from a full plane with a discontinuity in the radii of curvature will be presented in the next chapter.

4.5 SLOPE DIFFRACTION

Equations (4.53) and (4.54) express the first-order diffracted fields from a point Q_e on an edge. As is evident from these equations, the first-order diffracted fields are directly proportional to the field incident upon the edge. By considering these equations, if a null field were incident upon an edge, the edge diffracted field would seem to be zero. Consider, for instance, a dipole illuminating an edge so that the null of the dipole is directed at the edge, as in Figure 4.31. Substituting E_z^i $(Q_e) = 0$ into (4.53) we find that the first-order edge diffracted field is zero.

However, the total edge-diffracted field consists not only of the first-order diffracted field considered so far, but also of the so-called slope-diffracted fields [17, 27, 49, 50]. Whereas the first-order diffracted field is proportional to the amplitude of the incident field at Q_e, the slope-diffracted field is proportional to the derivative of the incident field at Q_e. We can express the two-dimensional slope-diffracted field as

$$U^d(s) = \frac{1}{jk} \frac{\partial D_{s,h}}{\partial \phi'} \frac{\partial U}{\partial n}\bigg|_{Q_e} \frac{e^{-jks}}{\sqrt{s}} \quad (4.225)$$

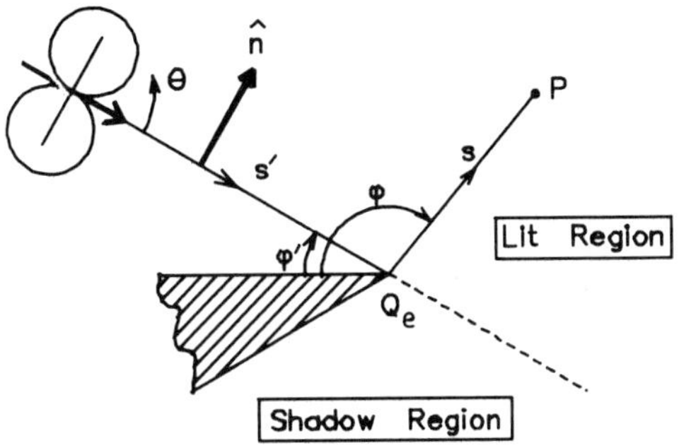

Figure 4.31 Dipole illuminating an edge.

where $\partial D_{s,h}/\partial \phi'$ are the soft and hard slope diffraction coefficients and $\partial U/\partial n$ is the directional derivative of the field in the direction $\hat{\mathbf{n}}$. Figure 4.31 shows that $\hat{\mathbf{n}}$ points from the shadow region into the lit region. The slope diffraction coefficients are given by

$$\frac{\partial D_{s,h}}{\partial \phi'}(L^i, L^{ro}, L^{rn}, \phi, \phi', n)\frac{\partial U(Q_e)}{\partial n} = D_1^s + D_2^s - R_{s,h}\,(D_3^s + D_4^s) \tag{4.226}$$

where

$$D_1^s = -\frac{e^{-j\pi/4}}{4n^2\sqrt{2\pi k}}\operatorname{cosec}^2\left(\frac{\pi + \beta^-}{2n}\right) F_s[kL^i a^+(\beta^-)]\,\frac{\partial U^i(Q_e)}{\partial n_n^i} \tag{4.227}$$

$$D_2^s = +\frac{e^{-j\pi/4}}{4n^2\sqrt{2\pi k}}\operatorname{cosec}^2\left(\frac{\pi - \beta^-}{2n}\right) F_s[kL^i a^-(\beta^-)]\,\frac{\partial U^i(Q_e)}{\partial n_o^i} \tag{4.228}$$

$$D_3^s = -\frac{e^{-j\pi/4}}{4n^2\sqrt{2\pi k}}\operatorname{cosec}^2\left(\frac{\pi + \beta^+}{2n}\right) F_s[kL^{rn} a^+(\beta^+)]\,\frac{\partial U^r(Q_e)}{\partial n_n^r} \tag{4.229}$$

$$D_4^s = +\frac{e^{-j\pi/4}}{4n^2\sqrt{2\pi k}}\operatorname{cosec}^2\left(\frac{\pi - \beta^+}{2n}\right) F_s[kL^{ro} a^-(\beta^+)]\,\frac{\partial U^r(Q_e)}{\partial n_o^r} \tag{4.230}$$

and $\beta^{\pm} = \phi \pm \phi'$. In (4.8), we found that $R_{s,h} = \mp 1$ in the case where the radius of curvature was not too small relative to a wavelength. The functions $a^{\pm}$ are defined in (4.66). Note that we can make the same approximations as in (4.70) and (4.71) for $a^{\pm}$. The function F_s is given in terms of the transition function F as

$$F_s(x) = 2jx[1 - F(x)] \tag{4.231}$$

or

$$F_s(x) = 2jx + 4x^{3/2} e^{jx} \int_{\sqrt{x}}^{\infty} e^{-j\tau^2} d\tau \tag{4.232}$$

The small argument form of $F_s(x)$ can be found by substituting (4.73) into (4.231). By substituting (4.74) into (4.232), the large argument form of $F_s(x)$ is found to be

$$\lim_{x \to \infty} F_s(x) = 1 \tag{4.233}$$

The parameters of the function F_s are the same as those of the transition function F.

The directional derivative $\partial U / \partial n$ of a field U in the direction $\hat{\mathbf{n}}$ can be expressed as

$$\frac{\partial U}{\partial n} = \nabla U \cdot \hat{\mathbf{n}} \tag{4.234}$$

where $\hat{\mathbf{n}}$ points from the shadow region into the lit region.

In the two-dimensional case $\partial U^i / \partial n$ also can be formulated in cylindrical coordinates. Consider a line source to be located at the origin of a cylindrical coordinate system as shown in Figure 4.32 (a). Because $\hat{\mathbf{n}} = \hat{\phi}_s$, we find that

$$\frac{\partial U^i}{\partial n} = \left(\frac{\partial U^i}{\partial \rho} \hat{\rho} + \frac{1}{\rho} \frac{\partial U^i}{\partial \phi_s} \hat{\phi}_s + \frac{\partial U^i}{\partial z} \hat{\mathbf{z}} \right) \cdot \hat{\phi}_s \tag{4.235}$$

so that

$$\frac{\partial U^i}{\partial n} = \frac{1}{\rho} \frac{\partial U^i}{\partial \phi_s} \tag{4.236}$$

$$\frac{\partial U^i(Q_e)}{\partial n} = \frac{1}{s'} \frac{\partial U^i}{\partial \phi_s} \tag{4.237}$$

as shown in Figure 4.32 (b). Note that ϕ_s increases in the direction of $\hat{\phi}_s$.

In the case of a half-plane, $n = 2$, so that

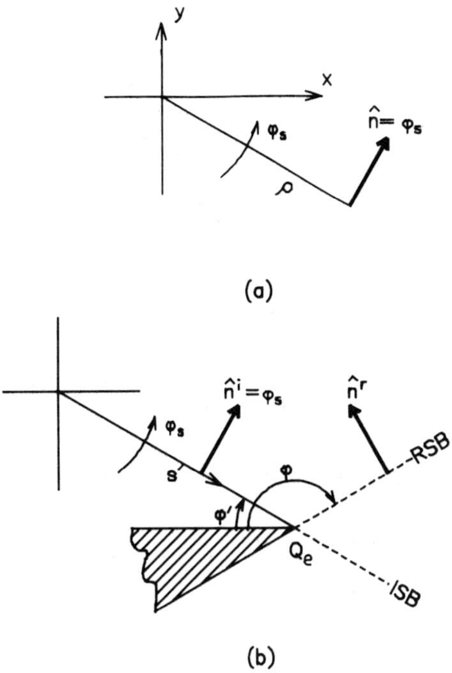

Figure 4.32 Line source (a) located at the origin of a cylindrical coordinate system, and (b) directional derivative.

$$\frac{\partial D_{s,h}}{\partial \phi'}\frac{\partial U}{\partial n} = \frac{e^{-j\pi/4}}{4\sqrt{2\pi k}}\left\{\frac{\sin[(\phi - \phi')/2]}{\cos^2[(\phi - \phi')/2]}F_s[kL^i a(\phi - \phi')] - R_{s,h}\frac{\sin[(\phi + \phi')/2]}{\cos^2[(\phi + \phi')/2]}F_s[kL^r a(\phi + \phi')]\right\}\frac{\partial U^i(Q_e)}{\partial n^i} \tag{4.238}$$

To illustrate how the slope-diffracted fields enforce continuity across a shadow boundary, consider the case where a line source with a pattern $f(\theta)$ illuminates the edge of a half-plane, as shown in Figure 4.33. We assume that $f(\theta) = 0$. Because the edge is not illuminated directly, no first-order diffracted field will emanate from it. There will be a slope-diffracted field, however. The incident field in the lit region will be that of the line source:

$$U^i(\theta) = f(\theta)\frac{e^{-jk\rho}}{\sqrt{\rho}} \tag{4.239}$$

whereas the incident field in the shadow region will be zero.

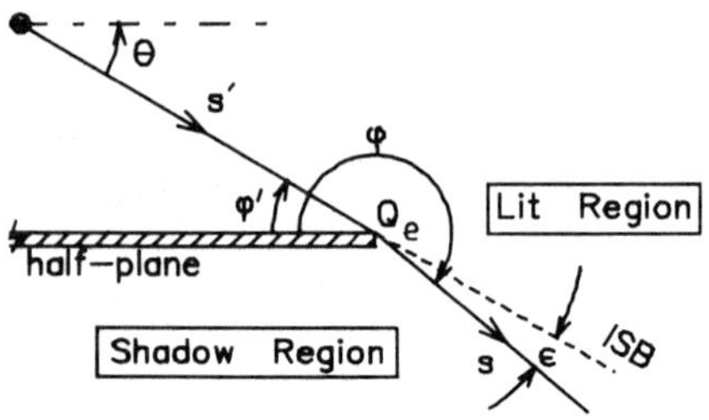

Figure 4.33 Continuity of a slope diffracted field across the ISB.

Let us now consider the field at a very small angle ϵ from the ISB; that is, at $\phi = \phi' + \pi + \epsilon$. If $\epsilon > 0$, the field point is in the shadow region; and if $\epsilon < 0$, the field point is in the lit region. If we substitute this value for ϕ into (4.238), we find that in this region the diffraction coefficient can be expressed as

$$\frac{\partial D_{s,h}}{\partial \phi'} = \frac{e^{-j\pi/4}}{4\sqrt{2\pi k}} \left\{ \frac{\sin[(\pi + \epsilon)/2]}{\cos^2[(\pi + \epsilon)/2]} F_s[kL^i a(\pi + \epsilon)] + C \right\} \tag{4.240}$$

where C is continuous, as the discontinuity is across the ISB in this case. Using (4.231) and (4.73), we find that the small argument form of F_s reduces to

$$F_s\left(\frac{kL^i\epsilon^2}{2}\right) = jkL^i\epsilon^2 - 2j\sqrt{\pi}\left(\frac{kL^i\epsilon^2}{2}\right)^{3/2} e^{j\pi/4} \tag{4.241}$$

Substituting (4.241) into (4.240) and (4.226) and ignoring C because it is continuous and small across the ISB, we find that the slope diffracted field is given by

$$U^d = \frac{e^{-j\pi/4}}{\sqrt{2\pi k}} \sqrt{\frac{s}{s'(s + s')}}\, f'(0) \frac{e^{-jk(s'+s)}}{\sqrt{s' + s}} - U^{dc} \tag{4.242}$$

where

$$U^{dc} = |\epsilon| \frac{f'(0)}{2} \left(\frac{s}{s' + s}\right) \frac{e^{-jk(s'+s)}}{\sqrt{s' + s}} \tag{4.243}$$

and $f'(0)$ is the derivative of $f(\theta)$ at $\theta = 0°$. We see that if the source is located reasonably close to the edge or $f'(0)$ is large, the slope diffracted field can be quite substantial.

In the region close to the ISB, we can expand the incident field in a Taylor series and express it as

$$U^i = \begin{cases} -\epsilon f'(0)\dfrac{e^{-jk(s'+s)}}{\sqrt{s'+s}}, & \text{lit side of the ISB} \\ 0, & \text{shadow side of the ISB} \end{cases} \tag{4.244}$$

Because the first term on the right-hand side of (4.242) is continuous across the ISB and both the incident field and U^{dc} go to zero as $\epsilon \rightarrow 0$, the total field also is continuous across the shadow boundary. The value of the slope diffracted field at the shadow boundary is given by the first term in (4.242).

Note that whereas the first-order diffracted fields are inversely proportional to $1/\sqrt{k}$ outside the transition region, the slope diffracted fields are inversely proportional to $1/k^{3/2}$. As the frequency increases, the effect of the slope diffracted fields will diminish with respect to the first-order fields, so that the slope diffracted fields can be referred to as second-order diffracted fields. Examples on the application of slope diffraction will be given in the next section and the following chapter.

4.6 GENERAL TWO-DIMENSIONAL EDGE DIFFRACTED FIELDS

In general, the UTD diffracted field from a point Q_e on an edge can be expressed as

$$U^d = \left[U^i(Q_e)\, D_{s,h} + \frac{1}{jk}\frac{\partial D_{s,h}}{\partial \phi'}\frac{\partial U^i(Q_e)}{\partial n}\right]\frac{e^{-jks}}{\sqrt{s}} \tag{4.245}$$

A logical extension would be the postulation of the existence of higher-order diffracted fields. Although the UTD diffraction coefficients for such fields have not been published, we would expect them to be inversely proportional to $1/k^n$, where $n = 5/2, 7/2, \ldots$

Example 4.5: Diffraction from a wedge illuminated by a nonisotropic line

To illustrate the effect of (4.245), consider the case where a line source with a pattern illuminates a half-plane as shown in Figure 4.34. Let us assume that the line source has a pattern

$$U^i(\theta) = \sin(\theta - \theta_0) \tag{4.246}$$

which has a null in the direction $\theta = \theta_0$. In this example we further assume that the frequency is 3 GHz, the line source is 5 m from the edge Q_e, and that $\phi' = 60°$. Using the expressions just derived, we calculate the field around the ISB. Two

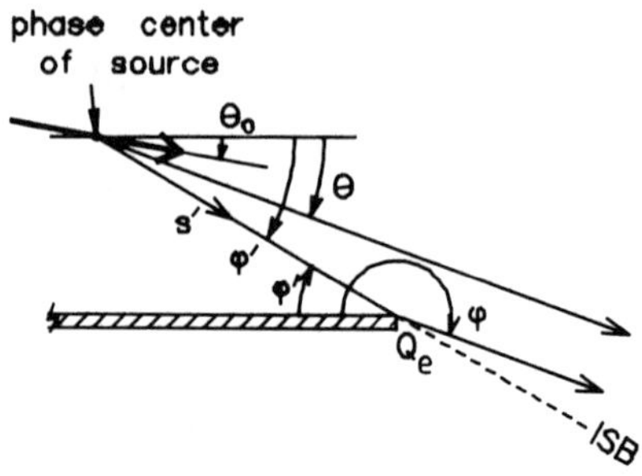

Figure 4.34 Scattering from a half-plane by a nonisotropic line source.

cases will be considered: the first where the edge is illuminated by the line source and the second where the null of the source's pattern is directed at the edge.

In the first case, the edge is illuminated by the source, so that a first-order discontinuity exists across the ISB. This discontinuity is corrected by the first-order diffracted field as shown in Figure 4.35. However, because the field illuminating the edge also has a finite slope, a slope diffracted field is also present. If we do not add the slope diffracted field to the total solution, we find a slope discontinuity

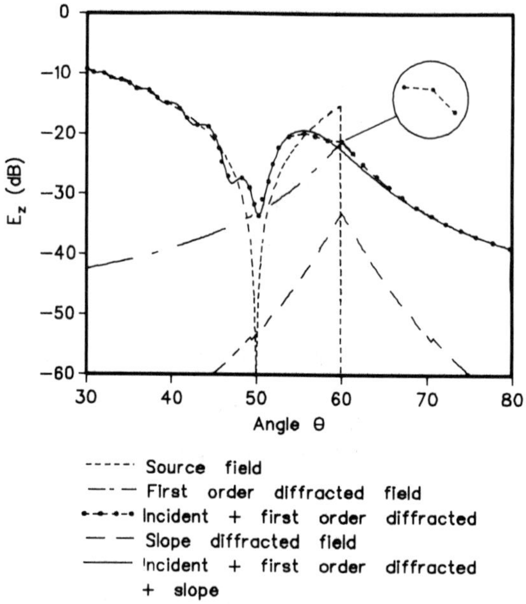

Figure 4.35 Fields scattered from a half-plane illuminated by a line source with a sinusoidal pattern as in Figure 4.34, with $s' = 5$ m, $\phi' = 60°$, $\theta_0 = 40°$ at a frequency of 3 GHz. Inset shows the slope discontinuity in the total field when slope diffraction is omitted.

in the field (comprising the incident and first-order diffracted fields), shown in the inset of Figure 4.35. Such slope discontinuities in the pattern are typical of slope diffracted fields that have not been taken into account, in contrast to the step discontinuities that result when first-order diffractions are omitted. When the slope diffracted field is added, we find that the pattern for the total field is smooth and continuous.

A second example was evaluated where $\theta_0 = \phi'$, so that the null of the source pattern points directly at the edge. Consequently the line source does not illuminate the edge directly and the first-order diffracted field will be zero. The slope diffracted field still exists and illuminates the shadow region. The calculated fields are shown in Figure 4.36. The irregularities in the slope-diffracted field in the region of -52 dB are caused by the computer routine FFCT (given in Appendix F), used for calculating the transition functions of the slope diffraction coefficients. The routine is prone to exhibit a slight discontinuity when used to calculate diffracted fields. This happens when the routine switches from an iterative to the small argument solutions.

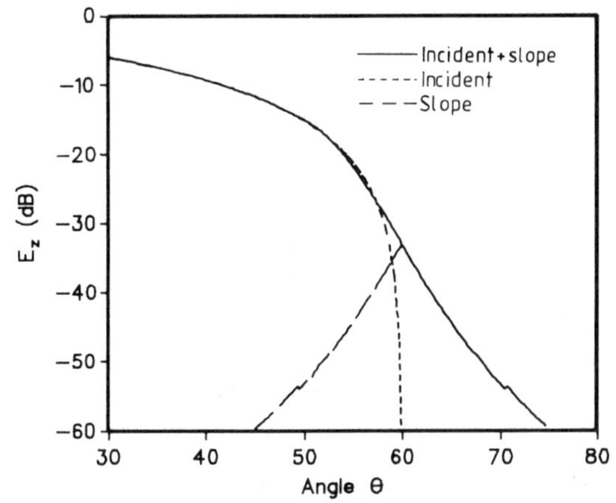

Figure 4.36 Fields scattered from a half-plane illuminated by a line source with a sinusoidal pattern as in Figure 4.34, with $s' = 5$ m, $\phi' = 60°$, $\theta_0 = 60°$ at a frequency of 3 GHz. In this case, the null of the source pattern is pointing at the edge.

4.7 DIELECTRIC AND IMPEDANCE WEDGES

In this chapter, we have considered only perfectly conducting structures. The UTD, however, can be extended to treat dielectric wedges as well as wedges that have surfaces described by impedance conditions; that is, where there is a specified relationship between the tangential electric and magnetic fields at the surfaces. Discussion will not be given on these topics here, but some references are included.

Dielectric wedges are treated in [51–53], and scattering from pyramidal and wedge absorbers is discussed in [54]. References [55–64] deal with wedges that have impedance boundary conditions and lossy wedges.

PROBLEMS

4.1 Derive (4.16).

4.2 Show that the lines demarcating the transition regions in Figure 4.20 are parabolas with their foci located at the edge.

4.3 Investigate the effect that an increase in frequency has on the transition region.

4.4 Consider Example 4.3. Investigate the effects that changes in α, ϕ', and s have on the diffracted and total fields.

4.5 In Example 4.4, we examined the radiation pattern of a magnetic line source mounted on a half-plane. The same geometry can be used to calculate the radiation pattern of an axial-TEM slot in a ground plane. The slot is modeled by using a source that radiates the field:

$$H^i = \frac{\sin(ka\sin\theta)/2}{(ka\sin\theta)/2}\frac{e^{-jkr}}{\sqrt{r}} \tag{4.247}$$

where a is the width of the slot and is assumed to be large [25]. Calculate and plot the radiated field for an axial-TEM slot on a strip as shown in Figure 4.26. Investigate the effect that the width a and the ratio l/a have on the pattern.

4.6 Complete Example 3.3 by adding the edge-diffracted fields.

4.7 The doubly curved, infinitely thin plate shown in Figure 4.37 is illuminated by a plane wave. Derive an expression for the backscattered field, taking into account first-order reflection and diffraction. Plot σ^w/λ for $0° \leq \theta \leq 90°$.

4.8 A compact range [65] is used to measure far-zone characteristics of antennas such as radiation patterns and also the RCS of targets. To do this, the antennas and targets need to be illuminated by a plane wave. In a compact range, a local plane is produced by illuminating a parabolic reflector with a feed antenna located at its focus, as shown in Figure 4.38.

Normally, the target zone is a volume in space, but for the purposes of this problem we assume the target zone to be on the vertical line at $z = z_p$. Furthermore, we assume that no direct illumination from the feed antenna at the focus illuminates the target zone. The feed antenna is modeled by a magnetic line source.

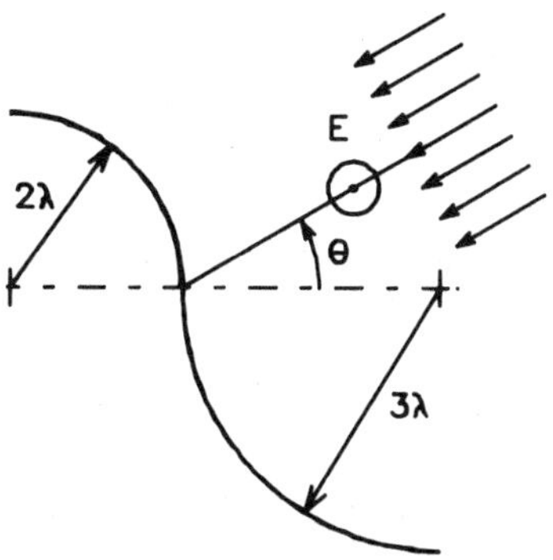

Figure 4.37 Doubly curved, infinitely thin plate illuminated by a plane wave.

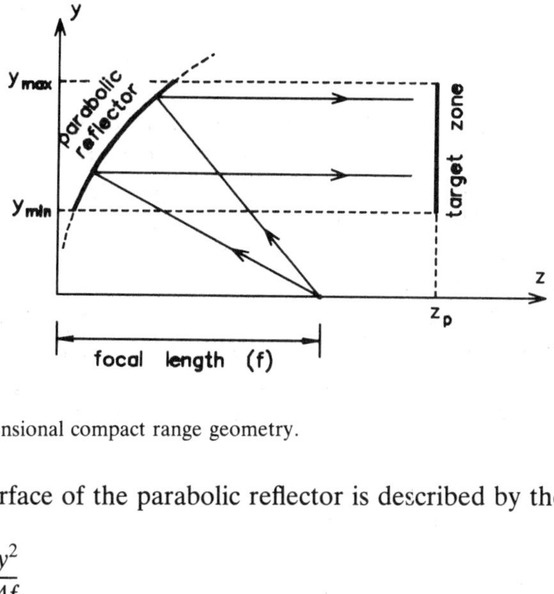

Figure 4.38 Two-dimensional compact range geometry.

The surface of the parabolic reflector is described by the equation

$$z = \frac{y^2}{4f} \tag{4.248}$$

where f is the focal length. We saw in Chapter 3 that the reflected rays illuminating the target zone are parallel and in phase at $z = z_p$. The reflected field illuminating the target zone is thus a plane wave.

(a) Show that the reflected field in the target zone has an amplitude taper in the vertical direction. Although the reflected field is a plane wave, it is not a uniform plane wave.

(b) Assume that $y_{min} = 1$ m, $y_{max} = 7$ m, and $f = 4$ m. Plot the amplitude (in dB) and phase of the reflected field in the target zone. If an amplitude taper of 1 dB or less [66] is acceptable in the target zone, how large is the usable target zone, assuming it is centered around $(y_{min} + y_{max})/2$?

(c) Derive expressions for the diffracted fields from the top and bottom edges of the parabolic reflector that illuminate a point in the target zone.

(d) Plot the amplitudes of the reflected field, diffracted fields, and total field as well as the phase of the total field in the target zone ($z_p =$ 8 m and $z_p = 10$ m) for the cases where the frequency is 2 GHz and 6 GHz.

(e) What effect do the diffracted fields have on the local plane wave in the target zone?

(f) The ripple in the amplitude of the total field can be defined as

$$R = -20 \log[1 - 10^{(E^d - E^r)/20}] \text{ dB} \tag{4.249}$$

where E^d is the diffracted field and E^r is the reflected field (in dB). Assuming that a ripple of less than 0.1 dB is required in the target zone [66], how big is the usable target zone?

(g) Explain the effect that an increase in frequency has on the quality of the local plane wave in the target zone (amplitude taper, phase uniformity, and ripple)?

4.9 The diffracted fields from the edges of the compact range reflector can be reduced by adding rolled edges [67]. Consider two semiellipses to be attached to the edges of the reflector as shown in Figure 4.39, with $a =$ 0.85 m and $b = 0.18$ m. Note that the ellipses have tangentially smooth joints with the parabola.

(a) Plot the amplitudes of the reflected field, diffracted fields and total field as well as the phase of the total field in the target zone for the cases where the frequency is 2 GHz and 6 GHz. Compare the results to those obtained in Problem 4.9.

(b) Explain why the rolled edges reduce the diffracted fields in the target zone.

The diffractions can be further reduced by using blended rolled edges [47].

4.10 In this problem we develop the slope diffraction solution. Let the edge be illuminated by two electric line sources of equal magnitude (I), shown in Figure 4.40. As $\epsilon \to 0°$, the two line sources represent a magnetic line dipole. Derive the four-term slope diffraction coefficient outside the transition region; that is, where the transition function is equal to unity. Start with Keller's solution for a single line source.

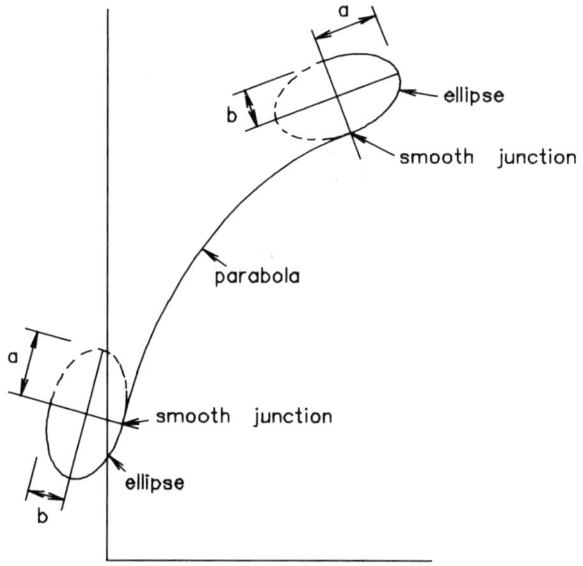

Figure 4.39 Two dimensional compact range reflector with elliptical rolled edges.

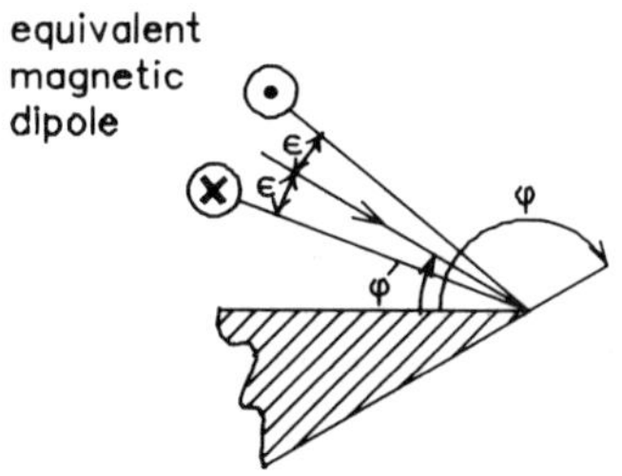

Figure 4.40 Geometry for the derivation of the slope diffraction coefficients.

REFERENCES

[1] A. Sommerfeld, "Mathematische Theorie der Diffraktion," *Math. Ann.*, Vol. 47, 1896, pp. 314–374.

[2] P.C. Clemmow, "A Method for the Exact Solution of a Class of a Two-Dimensional Diffraction Problems," *Proc. Roy. Soc.*, Vol. A205, 1951, pp. 286–308.

[3] G.L. James, *Geometrical Theory of Diffraction for Electromagnetic Waves,* 3rd Ed., Peter Peregrinus, London, 1986.

[4] E.V. Jull, *Aperture Antennas and Diffraction Theory,* Peter Peregrinus, London, 1981.

[5] M. Born and E. Wolf, *Principles of Optics,* 6th Ed., Pergamon Press, 1984.

[6] E. Jordan and K. Balmain, *Electromagnetic Waves and Radiating Systems,* 2nd Ed., Prentice-Hall, Englewood Cliffs, NJ, 1968.

[7] J.B. Keller, "The Geometrical Theory of Diffraction," *Proc. Symp. on Microwave Optics,* Eaton Electronics Research Laboratory, McGill University, Montreal, Canada, June 1953.
[8] I. Kay and J.B. Keller, "Asymptotic Evaluation of the Field at a Caustic," *J. Appl. Phys.,* Vol. 25, July 1954, pp. 876–883.
[9] J.B. Keller, "Diffraction by a Convex Cylinder," *IRE Trans. Antennas and Propagation,* Vol. AP-4, July 1956, pp. 312–321.
[10] J.B. Keller, "Diffraction by an Aperture", *J. Appl. Phys.,* Vol. 28, April 1957, pp. 426–444.
[11] S.N. Karp and J.B. Keller, "Multiple Diffraction by an Aperture in a Hard Screen," *Optica Acta,* Vol. 8, January 1961, pp. 61–72.
[12] J.B. Keller, "Geometrical Theory of Diffraction," *J. Opt. Soc. of America,* Vol. 52, No. 2, February 1962, pp. 116–130.
[13] T.B.A. Senior and P.L.E. Uslenghi, "Experimental Detection of the Edge Diffraction Cone," *Proc. IEEE,* Vol. 53, August 1965, pp. 877–882.
[14] D.S. Jones, *Methods in Electromagnetic Wave Propagation,* Vol. 2, Oxford Science Publications, 1987, pp. 583–782.
[15] J.B. Keller, "One Hundred Years of Diffraction Theory", *IEEE Trans. Antennas and Propagation,* Vol. AP-33, No. 2, February 1985, pp. 123–125.
[16] R.G. Kouyoumjian and P.H. Pathak, "A Uniform Geometrical Theory of Diffraction for an Edge in a Perfectly Conducting Surface," *Proc. IEEE,* Vol. 62, November 1974, pp. 1448–1461.
[17] R.G. Kouyoumjian, "The Geometrical Theory of Diffraction and Its Applications," in *Numerical and Asymptotic Techniques in Electromagnetics,* R. Mittra (ed.), Springer Verlag, New York, 1975.
[18] P.H. Pathak, "Techniques for High Frequency Problems," in *Antenna Handbook,* Y.T. Lo and S. W. Lee (Eds.), ITT-Howard W. Sams, Indianapolis, 1985.
[19] R.G. Kouyoumjian, P.H. Pathak, and W.D. Burnside, "A Uniform GTD for the Diffraction by Edges, Vertices and Convex Surfaces," in *Theoretical Methods for Determining the Interaction of Electromagnetic Waves with Structures,* J.K.S. Skwirzynski (ed.), Sijthoff and Nordhoff, Amsterdam 1979.
[20] R.G. Kouyoumjian and P.H. Pathak, "A Uniform GTD Approach to EM Scattering and Radiation," in *Handbook on Acoustic and Elastic Wave Scattering—High and Low Asymptotics,* Vol. II, V.J. Varadan and V.V. Varadan (Eds.), North Holland, Doderecht, 1986.
[21] R.C. Hansen (Ed.), *Geometric Theory of Diffraction,* IEEE Press, New York, 1981. This book is a collection of significant papers on GTD and UTD.
[22] F.A. Molinet, "Geometric Theory of Diffraction (GTD), Part I: Foundation and Theory," *IEEE Antennas and Propagation Soc. Newsletter,* Vol. 29, August 1987, pp. 6–17.
[23] F.A. Molinet, "Geometrical Theory of Diffraction, Part II: Extensions and Future Trends of the Theory," *IEEE Antennas and Propagation Soc. Newsletter,* Vol. 29, October 1987, pp. 5–16.
[24] W.L. Stutzman and G.A. Thiele, *Antenna Theory and Design,* John Wiley & Sons, New York, 1981.
[25] C.A. Balanis, *Antenna Theory,* Harper and Row, New York, 1982.
[26] C.A. Balanis, *Advanced Engineering Electromagnetics,* John Wiley and Sons, New York, 1989.
[27] Ohio State University ElectroScience Laboratory, Short Course Notes on The Modern Geometrical Theory of Diffraction. Authors include R.G. Kouyoumjian, P.H. Pathak, L. Peters, Jr., W.D. Burnside, R.C. Rudduck, N. Wang, and R.J. Marhefka.
[28] W.D. Burnside, unpublished class notes, Ohio State University.
[29] G.F. Herrmann, "Numerical Computation of Diffraction Coefficients," *IEEE Trans. Antennas and Propagation,* Vol. AP-35, No. 1, January 1987, pp. 53–61.

[30] G.F. Herrmann and S.M. Strain, "Numerical Diffraction Coefficients in the Shadow Transition Region," *IEEE Trans. Antennas and Propagation,* Vol. 36, No. 9, September 1988, pp. 1244–1251.

[31] "Symposia Cumulative Index (1963–1984)," *IEEE Trans. Antennas and Propagation,* Vol. AP-33, No. 4, April 1985.

[32] "Transactions Cumulative Index (1952–1984)," *IEEE Trans. Antennas and Propagation,* Vol. AP-33, No. 8, August 1985.

[33] A. Michaeli, "Contribution of a Single Face to the Wedge Diffracted Field," *IEEE Trans. Antennas and Propagation,* Vol. AP-33, No. 2, February 1985, pp. 221–223.

[34] J.D. Cashman, R.G. Kouyoumjian, and P.H. Pathak, "Comments on 'A Uniform Geometrical Theory of Diffraction from an Edge in a Perfectly Conducting Surface' and Authors' Reply," *IEEE Trans. Antennas and Propagation,* Vol. AP-25, No. 3, May 1977, pp. 447–451.

[35] K.R. Jacobsen, "An Alternative Diffraction Coefficient for the Wedge", *IEEE Trans. Antennas and Propagation,* Vol. AP-32, No. 2, February 1984, pp. 175–177.

[36] O.M. Buyukdura, *Radiation from Sources and Scatterers Near the Edge of a Perfectly Conducting Wedge,* Ph.D dissertation, Ohio State University, 1984.

[37] J.A. Aas, "On the Accuracy of the Uniform Geometrical Theory of Diffraction Close to a 90° Wedge," *IEEE Trans. Antennas and Propagation,* Vol. AP-27, No. 5, September 1979, pp. 704–705.

[38] R. Tiberio and R.G. Kouyoumjian, "A Uniform GTD Solution for the Diffraction by Strips Illuminated at Grazing Incidence," *Radio Science,* Vol. 14, 1979, pp. 933–941.

[39] R. Tiberio and R.G. Kouyoumjian, "An Analysis of Diffraction at Edges Illuminated by Transition Region Fields," *Radio Science,* Vol. 17, 1982, pp. 323–336.

[40] R. Tiberio and R.G. Kouyoumjian, "Calculation of the High-Frequency Diffraction by Two Nearby Edges Illuminated at Grazing Incidence," *IEEE Trans. Antennas and Propagation,* Vol. AP-32, No. 11, November 1984, pp. 1186–1196.

[41] A. Michaeli, R. Tiberio, and R.G. Kouyoumjian, "Comments on 'Calculation of the High-Frequency Diffraction by Two Nearby Edges Illuminated at Grazing Incidence' and Authors' Reply," *IEEE Trans. Antennas and Propagation,* Vol. AP-34, No. 1, January 1986, pp. 122–123.

[42] A.H. Serbest, "An Extension to GTD for an Edge on a Perfectly Conducting Surface," *IEEE Trans. Antennas and Propagation,* Vol. AP-34, No. 6, June 1986, pp. 837–842.

[43] J.L. Volakis and M.A. Ricoy, "Diffraction by a Thick Perfectly Conducting Half-Plane," *IEEE Trans. Antennas and Propagation,* Vol. AP-35, No. 1, January 1987, pp. 62–72.

[44] H. Shirai and L.B. Felsen, "Modified GTD for Generating Complex Resonances for Flat Strips and Disks," *IEEE Trans. Antennas and Propagation,* Vol. AP-34, No. 1, June 1986, pp. 779–790.

[45] H. Shirai and L.P. Felsen, "High-Frequency Multiple Diffraction by a Flat Strip: Higher Order Asymptotics," *IEEE Trans. Antennas and Propagation,* Vol. AP-34, No. 9, September 1987, pp. 1106–1112.

[46] H. Shirai, "Deemphasizing Low Frequency Defects in GTD Analysis of Pulsed Signal Scattering by a Perfectly Conducting Strip," *IEEE Trans. Antennas and Propagation,* Vol. AP-34, No. 10, October 1986, pp. 1261–1266.

[47] T.T. Chu, "First Order Uniform Geometrical Theory of Diffraction of the Scattering by Smooth Structures," Ph.D. dissertation, Ohio State University, 1982.

[48] C.W.I. Pistorius and W.D. Burnside, "An Improved Main Reflector Design for Compact Range Applications," *IEEE Trans. Antennas and Propagation,* Vol. AP-35, No. 3, March 1987, pp. 342–344.

[49] R.C. Rudduck and D.C.F. Wu, "Slope Diffraction Analysis of TEM Parallel-Plate Guide Radiation Patterns," IEEE *Trans. Antennas and Propagation,* Vol. AP-17, No. 6, November 1969, pp. 797–799.

[50] C.A. Mentzer, L. Peters, Jr., and R.C. Rudduck, "Slope Diffraction and Its Applications to Horns," *IEEE Trans. Antennas and Propagation,* Vol. AP-23, No. 2, March 1975, pp. 153–159.

[51] W.D. Burnside and K.W. Burgener, "High-Frequency Scattering by a Thin Lossless Dielectric Slab," *IEEE Trans. Antennas and Propagation,* Vol. AP-31, No. 1, January 1983, pp. 104–110.

[52] A. Chakrabarti, "Diffraction by a Dielectric Half-Plane," *IEEE Trans. Antennas and Propagation,* Vol. AP-34, No. 6, June 1986, pp. 830–833.

[53] J.L. Volakis and T.B.A. Senior, "Diffraction by a Thin Dielectric Half-Plane," *IEEE Trans. Antennas and Propagation,* Vol. AP-35, No. 12, December 1987, pp. 1483–1487.

[54] B.T. DeWitt and W.D. Burnside, "Electromagnetic Scattering by Pyramidal and Wedge Absorber," *IEEE Trans. Antennas and Propagation,* Vol. 36, No. 7, July 1988, pp. 956–970.

[55] R.G. Rojas, "Electromagnetic Diffraction of an Obliquely Incident Plane Wave Field by a Wedge with Impedance Faces," *IEEE Trans. Antennas and Propagation,* Vol. 36, No. 7, July 1988, pp. 956–970.

[56] J.L. Volakis, "Scattering by a Thick Impedance Half-Plane," *Radio Science,* Vol. 22, January–February 1987, pp. 13–25.

[57] M.I. Herman and J.L. Volakis, "High Frequency Scattering by a Double Impedance Wedge," *IEEE Trans. Antennas and Propagation,* Vol. 36, No. 5, May 1988, pp. 664–678.

[58] M.I. Herman and J.L. Volakis, "High Frequency Scattering from Polygonal Impedance Cylinders and Strips," *IEEE Trans. Antennas and Propagation,* Vol. 36, No. 5, May 1988, pp. 679–689.

[59] M.I. Herman and J.L. Volakis, "High Frequency Scattering by a Resistive Strip and Extensions to Conductive and Impedance Strips," *Radio Science,* Vol. 22, May–June 1987, pp. 335–349.

[60] R. Tiberio and G. Pelosi, "High Frequency Scattering from the Edges of Impedance Discontinuities on a Flat Plane," *IEEE Trans. Antennas and Propagation,* Vol. AP-31, July 1983, pp. 590–596.

[61] J.L. Volakis, "A Uniform Geometrical Theory of Diffraction for an Imperfectly Conducting Half-Plane," *IEEE Trans. Antennas and Propagation,* Vol. 34, No. 2, February 1986, pp. 172–180. "Revision of Figures in 'A Uniform Geometrical Theory of Diffraction from an Imperfectly Conducting Half-Plane'," *IEEE Trans. Antennas and Propagation,* Vol. AP-35, No. 6, June 1987, pp. 742–744.

[62] R. Tiberio, G. Pelosi, and G. Manara, "A Uniform GTD Formulation for the Diffraction by a Wedge with Impedance Faces," *IEEE Trans. Antennas and Propagation,* Vol. AP-33, No. 8, August 1985, pp. 867–873.

[63] R. Tiberio, G. Pelosi, G. Manara, and P.H. Pathak, "High-Frequency Scattering from a Wedge with Impedance Faces Illuminated by a Line Source, Part I: Diffraction," *IEEE Trans. Antennas and Propagation,* Vol. 37, No. 2, February 1989, pp. 212–218.

[64] R.J. Luebbers, "A Heuristic UTD Slope Diffraction Coefficient for Rough Lossy Wedges," *IEEE Trans. Antennas and Propagation,* Vol. 37, No. 2, February 1989, pp. 206–211.

[65] R.C. Johnson, H.A. Ecker, and R.A. Moore, "Compact Range Techniques and Measurements", *IEEE Trans. Antennas and Propagation,* Vol. AP-17, No. 5, September 1969, pp. 568–576.

[66] W.D. Burnside and L. Peters, Jr., "Target Illumination Requirements for Low RCS Target Measurements," *Proc. 1985 AMTA Symp.,* Melbourne, FLA.

[67] W.D. Burnside, M.C. Gilreath, B.M. Kent, and G.L. Clerici, "Curved Edge Modification of Compact Range Reflector," *IEEE Trans. Antennas and Propagation,* Vol. AP-35, No. 2, February 1987, pp. 176–182.

Chapter 5
Applications of Two-Dimensional Wedge Diffraction

"I am half sick of shadows," said The Lady of Shalott.

The Lady of Shalott,
Alfred Lord Tennyson

In this chapter, certain applications of two-dimensional wedge diffraction will be discussed as examples. The applications can be divided broadly into two classes: radiation and scattering. The intention certainly is not to give an exhaustive catalog of cookbook solutions, but rather to illustrate the use of the UTD in solving scattering and radiation problems through comprehensive examples.

5.1 RADIATION FROM A PARALLEL PLATE WAVEGUIDE WITH TEM MODE PROPAGATION, TERMINATED IN AN INFINITE GROUND PLANE

Consider the parallel plate waveguide with width a, terminated in an infinite ground plane as illustrated in Figure 5.1. Such a waveguide sometimes also is referred to as a *flange guide*. We consider the far-zone radiation from the waveguide when a TEM mode wave is propagating in the guide as shown. In this section, a first-order UTD analysis will be presented to illustrate the method by which we would attack problems of this nature. The reader is referred to references [1–15] for more detailed discussions on the application of the GTD and UTD to the radiation from waveguides. The case where a TE_{01} mode wave is propagating down the guide is left as an exercise.

In this case, the electric and magnetic fields in the guide can be expressed as

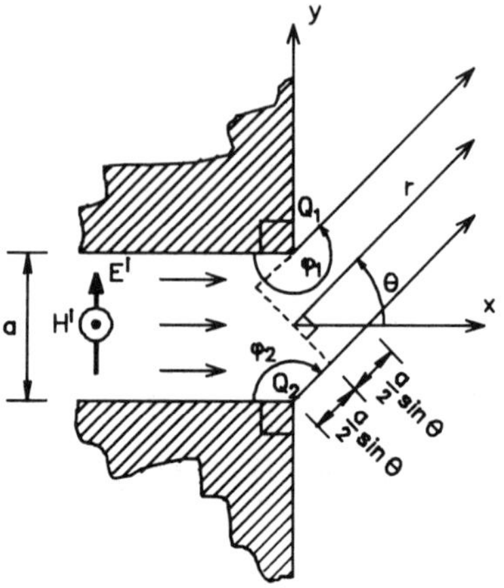

Figure 5.1 Radiation from a parallel plate waveguide with TEM mode propagation, terminated in an infinite ground plane.

$$\mathbf{E} = E_0 \, e^{-jkx}\hat{\mathbf{y}} \tag{5.1}$$

$$\mathbf{H} = H_0 \, e^{-jkx}\hat{\mathbf{z}} \tag{5.2}$$

with $E_0/H_0 = Z$, where Z is the free space impedance. For convenience's sake the phase reference is chosen to be in the center of the aperture. Because the magnetic fields are parallel to the edges and thus perpendicular to the x-y plane, we calculate the radiated fields in terms of the magnetic fields; that is, hard-polarization. The incident field in the aperture can be expressed as

$$H^i = H_0, \quad -\frac{a}{2} \leqslant y \leqslant \frac{a}{2} \tag{5.3}$$

Note that the illumination in the aperture has a uniform amplitude and phase distribution.

A field point in the far-zone located at a distance r and angle θ from the guide origin is illuminated by the fields diffracted from the edges at Q_1 and Q_2, respectively. In the far-zone, direct rays illuminate only field points located in the direction $\theta = 0°$; that is, on the x-axis. Because the x-axis is approached asymptotically in the far-zone, the direct radiation is not included in the expression for the total radiated field.

As a first-order approximation, the radiated far-zone field from the guide thus can be expressed as

$$H^t(\theta) = H_1^d(\theta) + H_2^d(\theta) \tag{5.4}$$

where

$$H_1^d(\theta) = \frac{1}{2} D_h(L^i, L^{ro}, L^{rn}, \phi_1, \phi', n)\, \mathrm{e}^{jk(a/2)\sin\theta} \frac{\mathrm{e}^{-jkr}}{\sqrt{r}} \tag{5.5}$$

We set $|H_0| = 1$, as we are interested in only the pattern and not the absolute value of the field. Generally the pattern also will be normalized to 0 dB at its peak. Note that the wedge faces at Q_1 and Q_2 in the guide are flat and that the phase front of the incident rays is planar. From (4.91) and (4.96) we find that $L^i = L^{rn} = L^{ro} = s$, with s large. Furthermore $\phi' = 0$ (indicating grazing incidence and hence the factor 1/2 in (5.5)), $\phi_1 = \pi + \theta$ and $n = 1.5$. Similarly,

$$H_2^d(\theta) = \frac{1}{2} D_h(L^i, L^{ro}, L^{rn}, \phi_2, 0, 1.5)\, \mathrm{e}^{-jk(a/2)\sin\theta} \frac{\mathrm{e}^{-jkr}}{\sqrt{r}} \tag{5.6}$$

with $L^i = L^{rn} = L^{ro} = s$ and $\phi_2 = \pi - \theta$. It is good practice to verify that $0 \leqslant \phi_{1,2} \leqslant 3\pi/2$ for $-90° \leqslant \theta \leqslant 90°$ as stated in (4.38) and (4.39).

Because $L \to \infty$ in the far-zone, $F \approx 1$, so that (5.5) and (5.6) can be expanded to yield, respectively,

$$H_1^d(\theta) = \frac{-\mathrm{e}^{-j\pi/4}}{3\sqrt{2\pi k}} \left[\cot\left(\frac{2\pi + \theta}{3}\right) - \cot\left(\frac{\theta}{3}\right)\right] \mathrm{e}^{jk(a/2)\sin\theta} \tag{5.7}$$

$$H_2^d(\theta) = \frac{-\mathrm{e}^{-j\pi/4}}{3\sqrt{2\pi k}} \left[\cot\left(\frac{2\pi - \theta}{3}\right) + \cot\left(\frac{\theta}{3}\right)\right] \mathrm{e}^{-jk(a/2)\sin\theta} \tag{5.8}$$

Because we are dealing with the far-zone and therefore interested mainly in the pattern, the common term $\mathrm{e}^{-jkr}/\sqrt{r}$ has been dropped in (5.7) and (5.8). Strictly speaking, (5.7) and (5.8) thus no longer represent magnetic fields, but rather pattern functions. Figure 5.2 shows the far-zone radiation pattern of the guide in Figure 5.1 for the cases $a = 2\lambda$ and $a = 5\lambda$, obtained by substituting (5.7) and (5.8) into (5.4).

The characteristics of a uniform illuminated aperture are clearly evident from Figure 5.2. As is well known from other analysis techniques, such as aperture integration [16–21], a uniform illuminated aperture has a $\sin(u)/u$ pattern where $u = (\pi a/\lambda)\sin\theta$. Note that in both cases illustrated in Figure 5.2 the first sidelobes are -13.2 dB down from the main beam peak and that the nulls occur at $\theta_n = \arcsin(n\lambda/a)$. As the aperture becomes wider, the main beam becomes narrower.

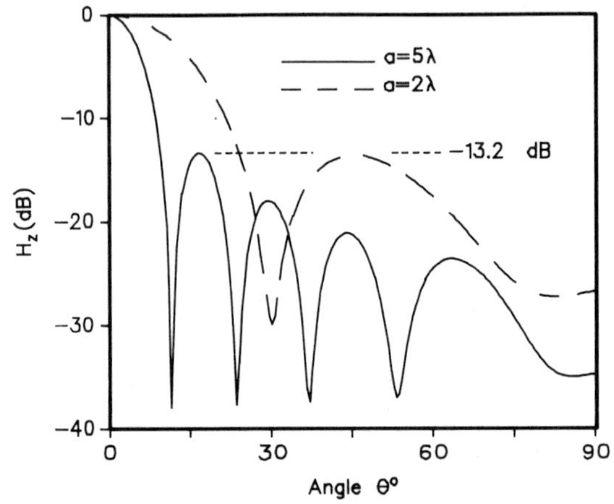

Figure 5.2 Far-zone radiation pattern from a parallel plate waveguide with TEM propagation terminated in an infinite ground plane.

A UTD analysis of the problem lends physical insight into the scattering mechanisms involved. The ripple in the pattern can be explained as due to the interference between the diffracted rays from Q_1 and Q_2. As the aperture size increases and hence the separation between Q_1 and Q_2 increases, we would expect the frequency of the ripple to increase.

5.2 ANTENNA GAIN

The directive power gain of an antenna $G(\theta, \phi)$ is the ratio of the power density radiated in the direction (θ, ϕ) relative to the power density radiated by an isotropic source. The maximum value of the directive gain is known simply as the *gain of the antenna*.

For the two-dimensional geometry, the directive gain can be expressed as [16, 18]

$$G = \frac{2\pi r\, S}{P_0} \tag{5.9}$$

where S is the radiated power density at distance r from the antenna, and P_0 is the total power incident in the waveguide feeding the antenna. Assuming no losses, P_0 also is the total power radiated by the antenna. The total radiated field can be expressed as

$$H^t = D_t \frac{e^{-jkr}}{\sqrt{r}} \tag{5.10}$$

In general, the radiated power density is given by

$$S = |\mathbf{E} \times \mathbf{H}^*| \tag{5.11}$$

where **E** and **H** are root mean square values. The asterisk denotes the complex conjugate. In this case,

$$S = \frac{Z}{r} |D_t|^2 \tag{5.12}$$

The directional gain can thus be expressed as

$$G = \frac{2\pi\, Z}{P_0} |D_t|^2 \tag{5.13}$$

Consider now the case of the parallel plate waveguide with TEM mode propagation described in Section 5.1. Assuming an incident magnetic field of unit strength, the power propagating down the guide is given by

$$P_0 = a\, Z \tag{5.14}$$

Substituting (5.14) into (5.13), we find that the directional gain for the guide is given by

$$G = \frac{2\pi}{a} |D_t|^2 \tag{5.15}$$

where $D_t = H_1^d(\theta) + H_2^d(\theta)$ with $H_1^d(\theta)$ and $H_2^d(\theta)$ given by (5.7) and (5.8), respectively. In the high frequency case where k is large, the gain of the guide in the direction $\theta = 0°$ is equal to ka. This is left as an exercise for the reader to verify that (5.15) yields this result.

In the two-dimensional case, the effective aperture of an antenna is related to the gain by the equation:

$$A = \frac{\lambda}{2\pi} G \tag{5.16}$$

so that the effective aperture of the parallel plate guide is given by

$$A = a \tag{5.17}$$

This is what we expect as the parallel plate guide has a uniform aperture distribution. In the receiving mode, the guide therefore would accept all the energy incident upon it.

5.3 RADIATION FROM AN *E*-PLANE HORN ANTENNA

The UTD is particularly well suited to the analysis of horn antennas [22–37]. The E-plane far-zone radiation pattern from the pyramidal horn antenna of Figure 5.3 can be calculated by considering the two-dimensional model shown in Figure 5.4. Due to considerations of symmetry, we need only calculate $0 \leq \theta \leq \pi$. However, to specify the relative shadow boundaries, we will allow θ to exist in the region $0 \leq \theta \leq 2\pi$ (rather than $-\pi \leq \theta \leq \pi$). We assume that the horn is fed by a magnetic line source of unit strength at the apex. The horn has a half-angle θ_e and length L_e. The phase reference is chosen at the apex of the horn.

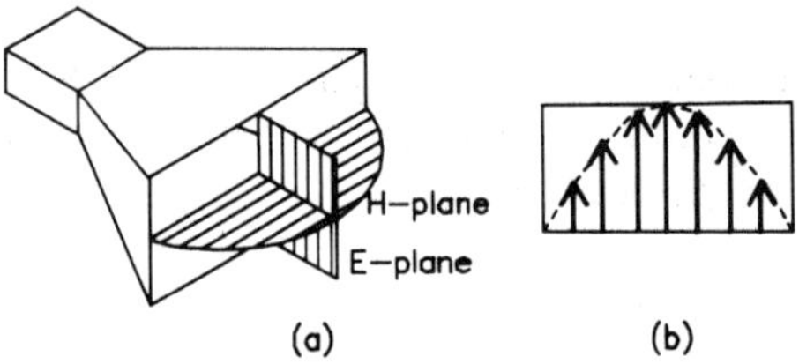

Figure 5.3 Pyramidal horn antenna: (a) 3D view; (b) E-field distribution in the aperture.

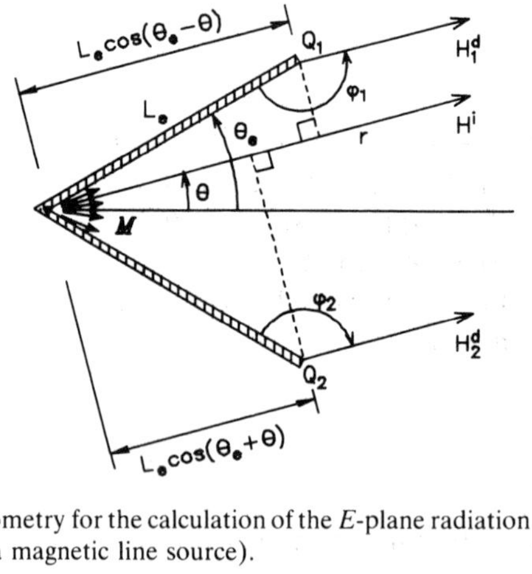

Figure 5.4 Two-dimensional geometry for the calculation of the E-plane radiation pattern of a pyramidal horn antenna (M is a magnetic line source).

As a first-order approximation, the far-zone pattern is given by

$$H^t(\theta) = H^i(\theta) + H_1^d(\theta) + H_2^d(\theta) \tag{5.18}$$

where H_1^d and H_2^d are the diffracted fields from Q_1 and Q_2, respectively; and H^i is the direct field from the line source at the apex.

The direct radiation is given by

$$H^i(\theta) = \begin{cases} \dfrac{e^{-jkr}}{\sqrt{r}}, & \text{for } -\theta_e \leqslant \theta \leqslant \theta_e \\ 0, & \text{elsewhere} \end{cases} \tag{5.19}$$

Note that H^i is included in the total field because it radiates in an angular sector $(-\theta_e \leqslant \theta \leqslant \theta_e)$ rather than just at a specific angle ($\theta = 0°$), as was the case with the parallel plate guide in Section 5.1. In that case, the direct radiation was not included in the expression for the total radiated field.

We find that Q_1 (and Q_2) is illuminated at grazing incidence; that is, $\phi' = 0$. Furthermore, $L^i = L^r = s' = L_e$ (because s is large), $n = 2$, and $\phi_1 = (\pi - \theta_e) + \theta$. As usual, we verify that $0 \leqslant \phi_1 \leqslant 2\pi$ for θ in the range of interest to satisfy (4.39). Substituting these parameters into the diffraction coefficient for grazing incidence on a half-plane given in (4.164), we find that the diffracted field from Q_1 can be expressed as

$$H_1^d(\theta) = \begin{cases} 0, & \text{for } \pi + \theta_e < \theta \leqslant \pi \\ \dfrac{e^{-jkL_e}}{\sqrt{L_e}} \dfrac{D_h(L_e, \pi - \theta_e + \theta)}{2} e^{jkL_e\cos(\theta_e - \theta)} \dfrac{e^{-jkr}}{\sqrt{r}}, & \text{elsewhere} \end{cases} \tag{5.20}$$

The diffracted field from Q_1 is obscured by the side of the horn for $\pi + \theta_e < \theta \leqslant 2\pi$. Similarly, the diffracted field from Q_2 can be expressed as

$$H_2^d(\theta) = \begin{cases} 0, & \text{for } \pi/2 < \theta \leqslant \pi - \theta_e \\ \dfrac{e^{-jkL_e}}{\sqrt{L_e}} \dfrac{D_h(L_e, \phi_2)}{2} e^{jkL_e\cos(\theta_e + \theta)} \dfrac{e^{-jkr}}{\sqrt{r}}, & \text{elsewhere} \end{cases} \tag{5.21}$$

The diffracted field from Q_2 is obscured by the side of the horn for $\pi/2 \leqslant \theta < \pi - \theta_e$.

Because of symmetry, we need to evaluate θ only in the interval $(0 \leqslant \theta \leqslant \pi)$, so that ϕ_2 could be calculated according to

$$\phi_2 = \begin{cases} (\pi - \theta_e) - \theta, & \text{for } 0 \leqslant \theta \leqslant \pi/2 \\ (3\pi - \theta_e) - \theta, & \text{for } \pi - \theta_e \leqslant \theta \leqslant \pi \end{cases} \tag{5.22}$$

to satisfy the requirement that $0 \leq \phi_2 \leq 2\pi$.

Figure 5.5 shows the pattern of an E-plane horn with $L_e = 13.5\lambda$ and $\theta_e = 17.5°$, using (5.18) to (5.22). Note that the pattern is continuous across the shadow boundary at $\theta = 17.5°$ (as it should be), but that there is a discontinuity at $\theta = 90°$. An examination of Figure 5.6 shows that the edge at Q_1 creates a shadow boundary for the diffracted field from Q_2 at $\theta = 90°$. A second-order diffracted

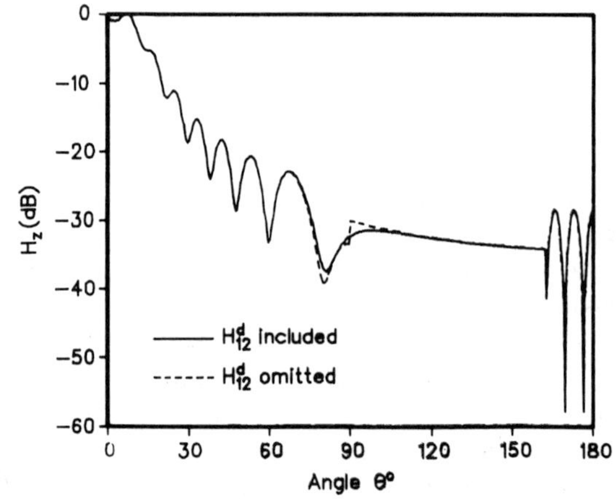

Figure 5.5 Far-zone radiation pattern of an E-plane horn antenna with $L_e = 13.5\lambda$ and $\theta_e = 17.5°$. Results are shown for the cases with and without second-order diffraction.

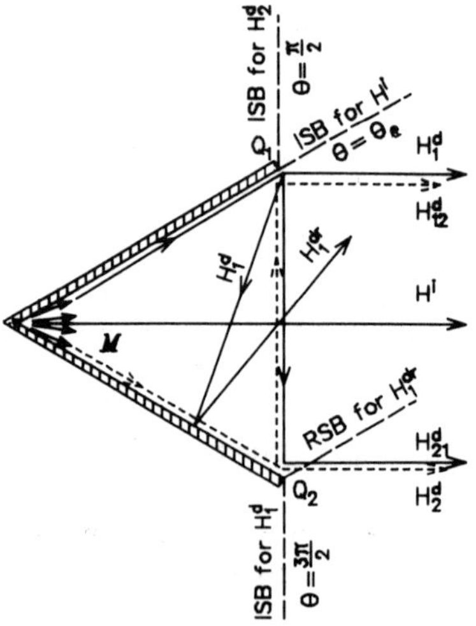

Figure 5.6 First- and second-order diffracted rays in the E-plane geometry of a horn antenna (M is a magnetic line source).

field emanating from Q_1 therefore should be included in the solution for the total radiated field to compensate for this discontinuity.

The second-order diffracted field from Q_1 due to illumination from Q_2 is given by

$$H_1^d(\theta) = \begin{cases} 0, & \text{for } \pi + \theta_e < \theta \leq \pi \\ H_2^d(Q_1) \dfrac{D_h(L_e, \pi - \theta_e + \theta)}{2} e^{jkL_e\cos(\theta_e - \theta)} \dfrac{e^{-jkr}}{\sqrt{r}}, & \text{elsewhere} \end{cases} \tag{5.23}$$

where D_h for a half-plane is given in (4.157). The angles are given by $\phi_1 = (\pi - \theta_e) - \theta$. Here, $\phi' = (\pi/2) - \theta_e$, so that this is not a case of grazing incidence. Because $s \gg s'$ we know from (4.97) that $L^i = s'$. In this case, the distance parameter hence is given by $L^i = 2L_e \sin\theta$; that is, the distance from Q_1 to Q_2. The diffracted field from Q_2 that illuminates Q_1 can be expressed as

$$H_2^d(Q_1) = \frac{e^{-jkL_e}}{\sqrt{L_e}} \frac{D_h(L^i, \pi/2 - \theta_e)}{2} \frac{e^{-j2kL_e\sin\theta_e}}{\sqrt{2L_e \sin\theta_e}} \tag{5.24}$$

where L^i in (5.24) is given by (4.95) as

$$L^i = \frac{2L_e^2 \sin\theta_e}{L_e(1 + 2\sin\theta_e)} \tag{5.25}$$

For the purpose of (5.25), $s' = L_e$ and $s = 2L_e \sin\theta_e$. The diffraction coefficient $D_h/2$ in (5.24) is given by (4.164).

The corrected total field therefore can be expressed as

$$H^t(\theta) = H^i(\theta) + H_1^d(\theta) + H_2^d(\theta) + H_{12}^d(\theta) \tag{5.26}$$

Figure 5.5 also shows the total field with H_{12}^d included. As expected, the pattern is continuous across the shadow boundary at $\theta = 90°$.

A more rigorous treatment would include the field H_{21}^d in addition to H_{12}^d. This is a second-order diffracted field from Q_2 due to illumination from Q_1. However, H_{21}^d will attain a maximum in the direction of its shadow boundary, located at $\theta = 3\pi/2$, and therefore its effect in the region around $\theta \approx \pi/2$ will be very small; consequently, this field is neglected when calculating the pattern in the region $0 \leq \theta \leq \pi$.

Apart from H_{12}^d and H_{21}^d a multitude of other second-order fields can be included; however, the effect of these fields on the total pattern would be small. Consider for example the ray H_1^{dr} that is diffracted from Q_1 and then reflected from the bottom face of the horn, as shown in Figure 5.6. From the previous chapter, we recall that diffracted rays attain a maximum amplitude in the direction of a shadow boundary. Because the ISB for H_1^d is in the direction $\theta = \theta_e$, the amplitude of the diffracted ray reflected from the bottom face would be small, because spatially it is far removed from the ISB. The field H_1^{dr} has an RSB in the direction $\theta = (\pi/2) - 2\theta_e$; the field H_{21}^d would have compensated for the discontinuity across this RSB.

We now proceed to calculate the on-axis gain of the two-dimensional E-plane horn. In order to use (5.13), it is necessary to calculate P_0. In the aperture of the horn, P_0 is given by

$$P_0 = \int_{-\theta_e}^{\theta_e} |\mathbf{E} \times \mathbf{H}^*| r \, d\theta \tag{5.27}$$

with

$$\mathbf{H} = \frac{e^{-jkL_e}}{\sqrt{L_e}} \hat{\mathbf{z}} \tag{5.28}$$

and $\mathbf{E} = Z (\mathbf{H} \times \hat{\mathbf{s}})$. We find that

$$P_0 = Z 2\theta_e \tag{5.29}$$

so that

$$S = \frac{Z}{r} \left| 1 - \frac{e^{-j[kL_e(1-\cos\theta_e) + \pi/4]}}{\sqrt{2\pi k} \sin(\theta_e/2) \sqrt{L_e}} \right|^2 \tag{5.30}$$

The on-axis gain therefore can be expressed as

$$G = \frac{\pi}{\theta_e} \left| 1 + \frac{e^{-j[kL_e(1-\cos\theta_e) - 3\pi/4]}}{2\pi\sqrt{L_e/\lambda} \sin(\theta_e/2)} \right|^2 \tag{5.31}$$

From (5.31) an optimum relation between L_e and θ_e can be determined to maximize the gain. This is left as an exercise for the reader.

5.4 RADIATION FROM AN *H*-PLANE HORN ANTENNA

The H-plane far-zone radiation pattern of a horn antenna can be calculated by considering the two-dimensional model shown in Figure 5.7. Because of the sym-

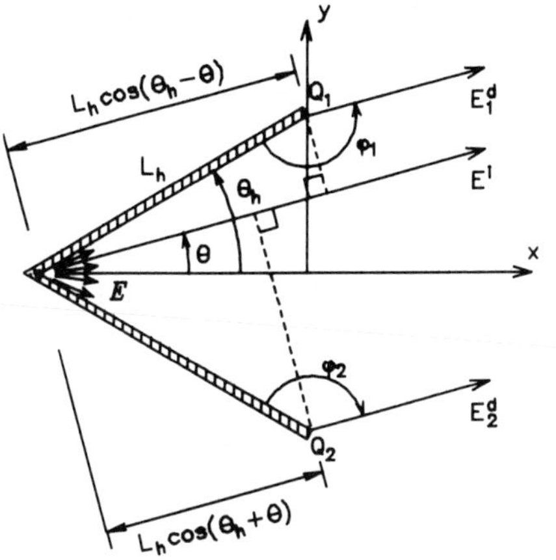

Figure 5.7 Two-dimensional geometry for the calculation of the H-plane radiation pattern of a pyramidal horn antenna (E is an electric line source).

metry of the problem, we need consider only $(0 \leqslant \theta \leqslant \pi)$. The horn has a half-angle θ_h and length L_h. The phase reference is again chosen at the apex of the horn.

In this case, the horn is fed by an electric line source. Because the tangential electric fields must vanish on the conducting surfaces of the horn, the line source must have a pattern, rather than being an isotropic line source as in the case of the E-plane horn. For the purposes of this problem we assume that the aperture distribution in the plane $x = L_h \cos\theta_h$ is given by

$$A(y) = \cos\left(\frac{\pi y}{2L_h \sin\theta_h}\right) \tag{5.32}$$

with $y = L_h \cos\theta_h \tan\theta$, so that

$$A(\theta) = \cos\left(\frac{\pi \tan\theta}{2 \tan\theta_h}\right) \tag{5.33}$$

and that this is the pattern of the line source. The incident field therefore can be expressed as

$$E^i(\theta) = \begin{cases} \cos\left(\dfrac{\pi \tan\theta}{2 \tan\theta_h}\right) \dfrac{e^{-jkr}}{\sqrt{r}}, & \text{for } -\theta_h \leqslant \theta \leqslant \theta_h \\ 0, & \text{elsewhere} \end{cases} \tag{5.34}$$

In contrast to the E-plane case, the aperture is cosinusoidally rather than uniformly illuminated.

As a first-order approximation, the far-zone radiation pattern is given by

$$E^t(\theta) = E^i(\theta) + E_1^d(\theta) + E_2^d(\theta) \tag{5.35}$$

where E_1^d and E_2^d are the slope diffracted fields from Q_1 and Q_2, respectively. Referring to Section 4.5, we find that the slope diffracted field from Q_1 is given by

$$E_1^d(\theta) = \begin{cases} 0, & \text{for } \pi + \theta_h \leqslant \theta \leqslant 2\pi \\ \dfrac{1}{2jk} \left.\dfrac{\partial E^i}{\partial n}\right|_{Q_1} \dfrac{\partial D_s}{\partial \phi'} e^{jkL_h \cos(\theta_h - \theta)} \dfrac{e^{-jkr}}{\sqrt{r}}, & \text{elsewhere} \end{cases} \tag{5.36}$$

The 1/2 term in (5.36) is to account for grazing incidence. The slope diffraction coefficient for a half-plane is given in (4.238), with $\phi_1 = (\pi - \theta_e) + \theta$, $\phi_1' = 0$ and $L^i = L^r = L_h$. Furthermore,

$$\left.\frac{\partial E^i}{\partial n}\right|_{Q_1} = \frac{-1}{L_h} \left.\frac{\partial E^i}{\partial \theta}\right|_{\theta = \theta_h} \tag{5.37}$$

as $\hat{\boldsymbol{\theta}} = -\hat{\mathbf{n}}$ at Q_1. We find that

$$\left.\frac{\partial E^i}{\partial n}\right|_{Q_1} = \frac{\pi}{L_h \sin 2\theta_h} \frac{e^{-jkL_h}}{\sqrt{L_h}} \tag{5.38}$$

Similarly, the diffracted field from Q_2 can be expressed as

$$E_2^d(\theta) = \begin{cases} 0, & \text{for } \pi/2 \leqslant \theta \leqslant \pi - \theta_e \\ \dfrac{1}{2jk} \left.\dfrac{\partial E^i}{\partial n}\right|_{Q_2} \dfrac{\partial D_s}{\partial \phi'} e^{jkL_h \cos(\theta_h + \theta)} \dfrac{e^{-jkr}}{\sqrt{r}}, & \text{elsewhere} \end{cases} \tag{5.39}$$

where ϕ_2 is given in (5.22). As earlier, $\phi_2' = 0$ and $L^i = L^r = L_h$. At Q_2, we find that

$$\left.\frac{\partial E^i}{\partial n}\right|_{Q_2} = \frac{1}{L_h} \left.\frac{\partial E^i}{\partial \theta}\right|_{\theta = -\theta_h} \tag{5.40}$$

so that

$$\left.\frac{\partial E^i}{\partial n}\right|_{Q_2} = \frac{\pi}{L_h \sin 2\theta_h} \frac{e^{-jkL_h}}{\sqrt{L_h}} \tag{5.41}$$

From our experience with the E-plane horn, we know that E_2^d will be shadowed by the edge at Q_1 and that a doubly diffracted ray E_{12}^d emanating from Q_1 needs to be added to make the pattern smooth and continuous around $\theta = \pi/2$. E_{12}^d will be a first-order diffracted field and can be expressed as

$$E_{12}^d(\theta) = \begin{cases} 0, & \text{for } \pi + \theta_h \leq \theta \leq 2\pi \\ E_2^d(Q_1) D_s(L^i, \phi_1, \phi_1')\, e^{jkL_h \cos(\theta_h - \theta)} \dfrac{e^{-jkr}}{\sqrt{r}}, & \text{elsewhere} \end{cases} \tag{5.42}$$

where D_s for a half-plane is given in (4.157). The angles are given by $\phi_1 = (\pi - \theta_e) - \theta$ and $\phi_1' = (\pi/2) - \theta_e$. Because $s \gg s'$ we know from (4.97) that $L^i = s'$, and therefore the distance parameter for (5.42) is given by $L^i = 2L_h \sin\theta$.

The diffracted field that illuminates Q_1 can be expressed as

$$E_2^d(Q_1) = \frac{1}{2jk} \left.\frac{\partial E^i}{\partial n}\right|_{Q_2} \frac{\partial D_s(L, \phi_2)}{\partial \phi'} \frac{e^{-jk2L_h \sin\theta_h}}{\sqrt{2L_h \sin\theta_h}} \tag{5.43}$$

where $\partial E^i/\partial n$ at Q_2 is given in (5.41). The distance parameter L is the same as that in (5.25), and $\phi_2 = (\pi/2) - \theta_h$.

The H-plane radiation pattern for a horn with $L_h = 13.1\lambda$ and $\theta_h = 16.1°$ is shown in Figure 5.8. This pattern is in good agreement with measured values

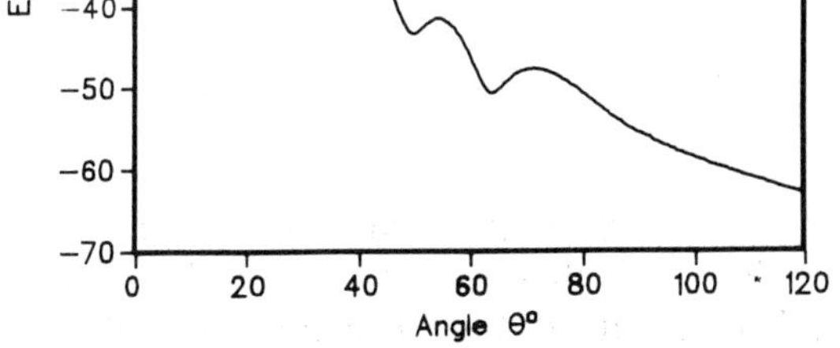

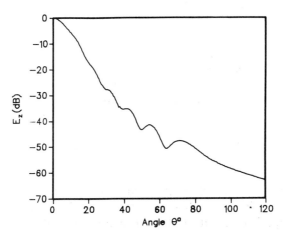

Figure 5.8 Far-zone radiation pattern of an H-plane horn antenna with $L_e = 13.1\lambda$ and $\theta_e = 16.1°$. Results are shown for the case where second-order diffraction is included.

in the region $0° \leq \theta \leq 120°$, for this particular geometry. Diffraction from the *E*-plane edges of a pyramidal horn influences the *H*-plane pattern of the *H*-plane horn in the region $120° \leq \theta \leq 180°$, so that the solution obtained here is not sufficient to model the *H*-plane radiation pattern of a real horn antenna [18]. It was shown [26, 28] that the complete *H*-plane pattern of a pyramidal horn can be obtained by combining a slope diffraction analysis with the *E*-plane diffracted rays.

5.5 RADAR WIDTH OF A TWO-DIMENSIONAL STRUCTURE

In this section we calculate the radar width of a two-dimensional structure. Although the structure to be considered has no obvious practical use, it illustrates how the radar width of a typical scattering body is calculated. The concept of radar width (σ^w) as well as the assumptions on which the calculation of σ^w (as given in (3.133)) depends, have been discussed in Section 3.3.9.

Consider now the rectangular cylinder capped by a half-circular cylinder as shown in Figure 5.9. We assume that the structure is illuminated by a plane wave incident at an angle θ, as shown, with the magnetic field perpendicular to the plane of incidence. Due to the symmetry of the structure, the radar width need be calculated only in the region $0 \leq \theta \leq \pi$.

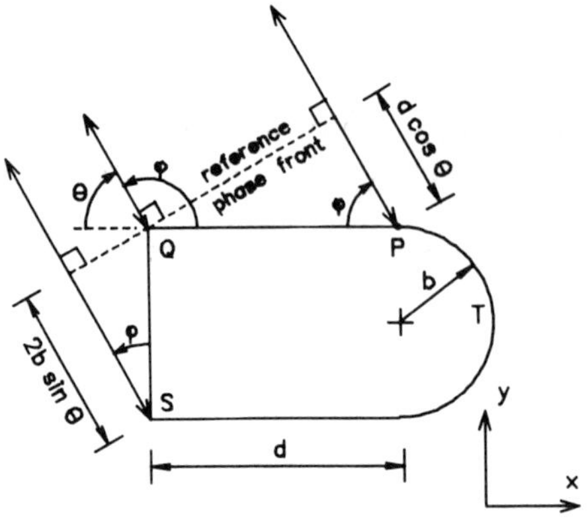

Figure 5.9 Two-dimensional rectangular scattering body with cylindrical cap (the phase origin lies at *Q*).

In the monostatic case, where the source and receiver are collocated, the problem is to identify those rays that are scattered back in the direction opposite to the incoming rays. With regard to the diffraction coefficients,

$$\hat{\mathbf{s}} = -\hat{\mathbf{s}}' \tag{5.44}$$

$$\phi = \phi' \tag{5.45}$$

so that $\phi - \phi' = 0$ and $\phi + \phi' = 2\,\phi$.

The phase reference of the structure is chosen to be at Q. The next step in the calculation process is to decompose the total angular region over which σ^w is to be determined into subregions. Every subregion is chosen so that σ^w in that region is determined from the same scattering centers. In our present problem there will be two subregions: $0 \leqslant \theta \leqslant \pi/2$ and $\pi/2 \leqslant \theta \leqslant \pi$.

The hard polarized, monostatic backscattered fields therefore can be expressed as

$$H^t = \begin{cases} H_P^d + H_Q^d + H_S^d, & \text{for } 0 \leqslant \theta < \pi/2 \quad \text{(Region 1)} \\ H_P^d + H_Q^d + H_T^r, & \text{for } \pi/2 < \theta \leqslant \pi \quad \text{(Region 2)} \end{cases} \tag{5.46}$$

where H_P^d, H_Q^d, and H_S^d are diffracted fields from P, Q, and S, respectively, and H_T^r is the reflected field from a point T on the cylindrical cap. At the boundary between two subregions ($\theta = \pi/2$ in this case), the scattered fields from the two regions should converge to the same value, as the total field should be smooth and continuous everywhere.

Let us now proceed to calculate σ^w in region I. We begin by finding an expression for the diffracted field from Q. This field can be expressed as

$$H_Q^d(\theta) = H^i(Q)\, D_h \frac{e^{-jks}}{\sqrt{s}} \tag{5.47}$$

where s is the distance from Q to the far-zone observation point. In this case the observation point is the far-zone phase reference, so that $s = r$. Although we know that $r \to \infty$, it is convenient to consider r as a very large distance rather than infinity, for the moment. From (3.133) we know that the total backscattered field will be normalized with respect to the amplitude of the incident field. To choose $H^i(Q) = 1$ therefore is convenient. The diffraction coefficient is given in (4.55), with parameters $n = 1.5$ and $\phi = \pi - \theta$. The horizontal face QP has been chosen as the o-face. As always, we should check that $0 \leqslant \phi \leqslant n\pi$. Furthermore, because $s' \to \infty$, it was shown in (4.91) and (4.96) that $L^i = L^r = s$. Because the observation point is in the far-zone, $F(kLa) \approx 1$.

Let us digress for a moment to consider the process whereby $F \to 1$. Although s' and s both tend to infinity, to know which one does so first is important. In this case, there is a plane wave incident, so that $s' \to \infty$. Therefore, the diffraction coefficient initially should be set up so that $s' \to \infty$ and s is finite. At that point, the solution is valid for any s. Because we are calculating σ^w, we need s also to be very large. Only then do we let $s \rightarrow \infty$. This results in the distance parameters L^i and L^r also tending to infinity with the result that $F \approx 1$. Now we can argue that although $L^{i,r}$ are very large, the function a in (4.70) and (4.71) will tend to zero as $\phi \pm \phi' \approx \pi$. We then might further argue that the small value of a would cancel the large value of $L^{i,r}$, leaving a finite value for the argument kLa of the transition function and hence a value of less than 1 for F.

However, this is not the case. First, although $\phi - \phi' = 0$, the backscattered diffracted rays are spatially far removed from the ISB, so that the transition functions in D_1 and D_2 will tend to unity in any case. Because $\phi = \phi'$, the RSB will be approached when $\phi \approx \pi/2$, and a will take on a small value. However, for an angle $\phi = (\pi/2) \pm \epsilon$ (with ϵ small), we may argue that, because the field point is in the far-zone, s is so much larger than a, that $F \approx 1$. If ϵ and hence a are decreased, we can increase s arbitrarily (as $s \rightarrow \infty$ in the far-zone) thereby ensuring that $ks \gg a$ and $F \approx 1$.

The diffraction coefficient in (5.47) can now be expressed as

$$D_h = \frac{-e^{-j\pi/4}\sqrt{\lambda}}{6\pi}\left[2\cot\frac{\pi}{3} + \left(\cot\frac{2\theta}{3} + \cot\frac{2\theta - \pi}{3}\right)\right] \tag{5.48}$$

Note that the diffraction coefficient in (5.48) has been expressed in terms of λ rather than k. This will be done throughout this example to illustrate that the radar width of the structure can be expressed as σ^w/λ, in which case all dimensions of the scattering body are normalized to a wavelength.

Substituting (5.48) into (5.47), we can express the backscattered diffracted field from Q as

$$H_Q^d(\theta) = \frac{-e^{-j\pi/4}}{6\pi}\left[2\cot\frac{\pi}{3} + \left(\cot\frac{2\theta}{3} + \cot\frac{2\theta - \pi}{3}\right)\right]\frac{e^{-jkr}}{\sqrt{r/\lambda}} \tag{5.49}$$

or

$$H_Q^d(\theta) = H_Q^{d\prime}\frac{e^{-jkr}}{\sqrt{r/\lambda}} \tag{5.50}$$

Note that the $\sqrt{\lambda}$ term in (5.48) has been included in the $1/\sqrt{r/\lambda}$ term in (5.50).

Similarly, the diffracted field from S can be expressed as

$$H_S^d(\theta) = H^i(S)D_h\frac{e^{-jks}}{\sqrt{s}} \tag{5.51}$$

where s is the distance from S to the observation point. In this case $s = r + 2b \sin\theta$, so that

$$H^i(S) = e^{-j2kb\sin\theta} \tag{5.52}$$

The parameters for the diffraction coefficient are $n = 1.5$ and $\phi = (\pi/2) - \theta$, where the vertical face SQ has been chosen as the o-face. As earlier, $L^i = L^r \to \infty$, and the transition functions in the diffraction coefficient tend to unity. Substituting (5.52) and the value of s into (5.51), the backscattered diffracted field from S is given by

$$H_S^d(\theta) = e^{-j4\pi(b/\lambda)\sin\theta} D_h \frac{e^{-jk(r+2b\sin\theta)}}{\sqrt{r + 2b\sin\theta}} \tag{5.53}$$

with

$$D_h = \frac{-e^{-j\pi/4}\sqrt{\lambda}}{6\pi}\left[2\cot\frac{\pi}{3} + \left(\cot\frac{2\theta}{3} + \cot\frac{2\theta - 2\pi}{3}\right)\right] \tag{5.54}$$

Because $r \gg 2b \sin\theta$,

$$H_S^d(\theta) = e^{-j8\pi(b/\lambda)\sin\theta} D_h' \frac{e^{-jkr}}{\sqrt{r/\lambda}} \tag{5.55}$$

where

$$D_h = D_h'\sqrt{\lambda} \tag{5.56}$$

As earlier, we define $H_S^{d\prime}$ so that

$$H_S^d(\theta) = H_S^{d\prime} \frac{e^{-jkr}}{\sqrt{r/\lambda}} \tag{5.57}$$

The junction at P is smooth; that is, it is a full-plane ($n = 1$). The diffracted field from P therefore can be expressed as

$$H_P^d(\theta) = H^i(P) D_h \frac{e^{-jks}}{\sqrt{s}} \tag{5.58}$$

The appropriate diffraction coefficient to be used here is the extended theory form given by (4.222). Note that the radius of curvature of the *o*-face is infinite, so that $F(kL^{rn}a) \rightarrow 1$. From (4.62) we find that, with $\phi = \theta$, as shown in Figure 5.9,

$$L^{rn} = \frac{\rho^{rn} s}{\rho^{rn} + s} \tag{5.59}$$

From (4.86) the caustic distance of reflection (ρ^{rn}) is found to be

$$\rho^{rn} = \frac{b \sin\theta}{2} \tag{5.60}$$

In this case, $s = r + d \cos\theta$ so that

$$H^i(P) = e^{-jkd\cos\theta} \tag{5.61}$$

Because $s \gg \rho^{rn}$, we thus have

$$L^{rn} = \frac{b \sin\theta}{2} \tag{5.62}$$

Substituting (5.62) and (4.222) into (5.58), we find after some manipulation that the backscattered diffracted field from P can be expressed as

$$H_P^d(\theta) = -e^{-j\pi[4(L/\lambda)\cos\theta + (1/4)]} \frac{f_h}{4\pi}(1 - F(kL^{rn}a))\frac{e^{-jkr}}{\sqrt{r/\lambda}} \tag{5.63}$$

where the argument in the transition function in (5.63) is given by

$$kL^{rn}a = 2\pi(L/\lambda) \sin\theta \cos^2\theta \tag{5.64}$$

As earlier, we define

$$H_P^d(\theta) = H_P^{d\prime} \frac{e^{-jkr}}{\sqrt{r/\lambda}} \tag{5.65}$$

The total backscattered field from the structure now can be expressed as

$$H^t = (H_P^{d\prime} + H_Q^{d\prime} + H_S^{d\prime})\frac{e^{-jkr}}{\sqrt{r/\lambda}} \tag{5.66}$$

Substituting (5.66) into (3.133), we find that

$$\frac{\sigma^w}{\lambda} = 2\pi \, |H_P^{d\prime} + H_Q^{d\prime} + H_S^{d\prime}|^2 \tag{5.67}$$

Recall that we have assumed that the amplitude of the incident plane wave is unity. Radar width usually is expressed in decibels; that is,

$$\sigma^w_{\mathrm{dB}/\lambda} = 10 \log\frac{\sigma^w}{\lambda} \tag{5.68}$$

Figure 5.10 shows the radar width in the region $0 \leqslant \theta \leqslant \pi/2$, obtained by considering only the diffracted fields $H_Q^d + H_S^{d'}$ for the case where $d = 2\lambda$ and $b = \lambda$. Figure 5.10 also shows the radar width due only to $H_P^{d'}$, as well as the radar width resulting from $H_Q^d + H_S^d + H_P^d$. Note that $H_Q^d + H_S^d$ converges to a finite value at $\theta = 0$. In fact, at $\theta = 0$ we find that

$$\sigma^w_{\mathrm{dB}/\lambda} = 10 \log 2\pi\left(\frac{l}{\lambda}\right)^2 \tag{5.69}$$

where $l = 2b$ so that $\sigma^w/\lambda = 14$ dB/λ. The expression for σ^w/λ in (5.69) yields the radar width perpendicular to a two-dimensional surface of width l. Equation (5.69) also can be obtained by other methods, such as physical optics.

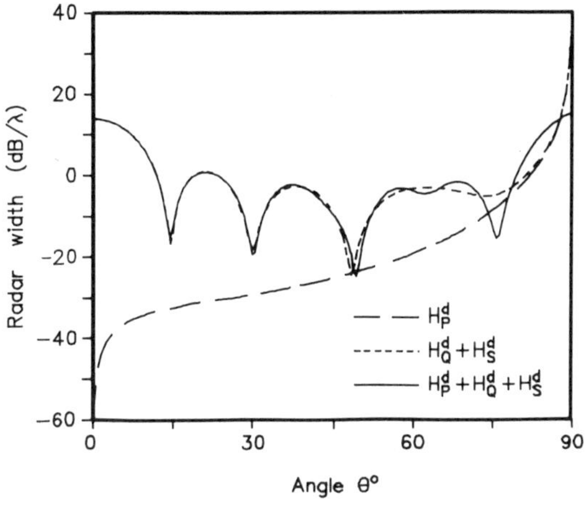

Figure 5.10 Radar width (σ^w) of the two-dimensional scattering body shown in Figure 5.9, in the angular sector $0 \leqslant \theta \leqslant \pi/2$. Hard polarized case, with $d = 2\lambda$ and $b = \lambda$.

As is evident from Figure 5.10, the contribution from H_P^d at $\theta \approx 0$ is very small so that σ^w in the angular region around $\theta \approx 0$ is determined principally by the diffraction from Q and S. Note that σ^w due to $H_Q^d + H_S^d$ becomes singular at $\theta = \pi/2$. In the case where only $H_Q^d + H_S^d$ are used to calculate σ^w in Figure 5.10, the incident field experiences the structure to be infinitely long; that is, $d \rightarrow \infty$. Hence, an infinite amount of energy is intercepted and scattered back at $\theta = \pi/2$, leading to $\sigma^w \rightarrow \infty$. However, once point P is taken into account by adding H_P^d to the total field, the incident fields "realize" that the structure is finite and σ^w converges to a finite value at $\theta = \pi/2$ as shown in Figure 5.10, where $H_Q^d + H_S^d + H_P^d$ is used. We can verify that this value also is approximately 14 dB/λ; the UTD solution actually yields a value of 14.6 dB/λ that can be expected because P is not a sharp edge and consequently the incident fields may experience the structure to be slightly larger than d. This will be true especially at lower frequencies where the wavelengths are longer.

To calculate σ^w in region II, we need to find an expression for the reflected field from T. The reflected field is given by (3.56) as

$$H^r(\theta) = H^i(T)\, R_h \sqrt{\frac{\rho^r}{s + \rho^r}}\, e^{-jks} \tag{5.70}$$

where s is the distance from T to the far-zone phase reference. The caustic distance of reflection (ρ^r) is found by substituting $s' \rightarrow \infty$, $\hat{\mathbf{s}}' \cdot \hat{\mathbf{n}} = -1$ and $a = b$ into (4.70) to yield

$$\rho^r = \frac{b}{2} \tag{5.71}$$

To express s in terms of r, let us consider the general case illustrated in Figure 5.11, where point C is the phase origin and a plane wave is incident. Consider a

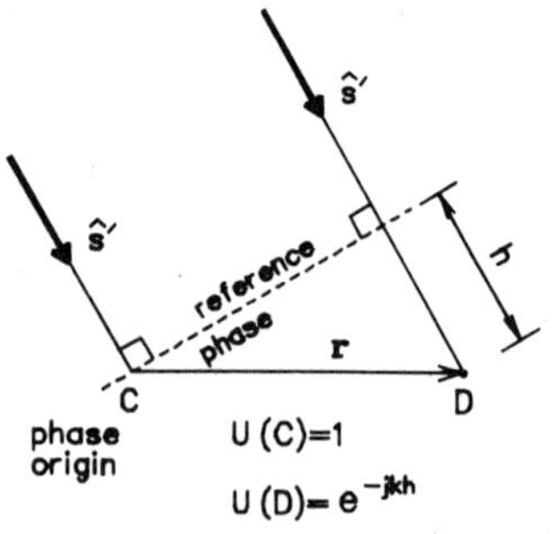

Figure 5.11 Generic case to illustrate phase delay due to spatial separation.

second point, D, and let h be the distance from the reference phase front (which passes through C and is perpendicular to the rays) to point D. The distance h is given by

$$h = \hat{\mathbf{s}}' \cdot \mathbf{r} \tag{5.72}$$

where $\mathbf{r}$ is the vector from C to D and $\hat{\mathbf{s}}'$ is a unit vector in the direction of the incident ray. If the incident field at the phase reference is unity, the field at D will be given by

$$U(D) = e^{-jkh} \tag{5.73}$$

If the incident field reaches D before C, (5.72) will yield a negative value for h. In Figure 5.9, the vector from the phase reference (Q) to S is given by $\mathbf{r} = -2b\,\hat{\mathbf{y}}$, so that $h = 2b \sin\theta$. The incident field at S was given in (5.54) as $e^{-j2kb\sin\theta}$, which is consistent with (5.73).

In the case of our scattering body, the point of specular reflection (T) moves as θ is increased, in contrast to P, Q, and S, which are stationary. We see from Figure 5.12 that

$$\mathbf{r} = (d - b\cos\theta)\hat{\mathbf{x}} - b(1 - \sin\theta)\hat{\mathbf{y}} \tag{5.74}$$

so that

$$h = d\cos\theta + b(\sin\theta - 1) \tag{5.75}$$

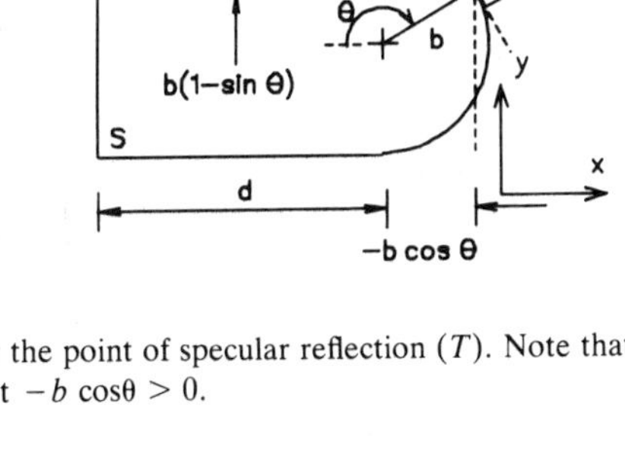

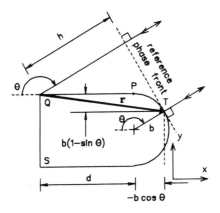

Figure 5.12 Phase delay for the point of specular reflection (T). Note that in the case where $\theta > \pi/2$, $\cos\theta < 0$ so that $-b\cos\theta > 0$.

The incident field at T thus can be expressed as

$$H^i(T) = e^{-j2\pi[(d/\lambda)\cos\theta + (b/\lambda)(\sin\theta - 1)]} \tag{5.76}$$

referred to the phase origin at Q. Substituting (5.71) and (5.76) into (5.70), the specular reflected field from T in the direction θ can be expressed as

$$H_T^r(\theta) = e^{-j4\pi[(d/\lambda)\cos\theta + (b/\lambda)(\sin\theta - 1)]}\sqrt{b/2}\,\frac{e^{-jkr}}{\sqrt{r}} \tag{5.77}$$

Note that the dimension b in the square root is not normalized to a wavelength. To be able to add the reflected field from T to the diffracted fields in region II, we write the reflected field as

$$H_T^r(\theta) = e^{-j4\pi[(d/\lambda)\cos\theta + (b/\lambda)(\sin\theta - 1)]}\sqrt{\frac{b/\lambda}{2}}\,\frac{e^{-jkr}}{\sqrt{r/\lambda}} \tag{5.78}$$

or

$$H_T^r(\theta) = H_T^{r\prime}\,\frac{e^{-jkr}}{\sqrt{r/\lambda}} \tag{5.79}$$

where

$$H_T^{r\prime}(\theta) = e^{-j4\pi[(d/\lambda)\cos\theta + (b/\lambda)(\sin\theta - 1)]}\sqrt{\frac{b/\lambda}{2}} \tag{5.80}$$

Be very careful not to interpret (5.80) as indicating that the amplitude of the reflected field is frequency dependent. It is not, because the $\sqrt{\lambda/\lambda}$ term was artificially introduced to conform to the format of (5.66). The diffracted fields from Q and P, however, are frequency dependent because the diffraction coefficients are proportional to $\sqrt{\lambda}$.

Figure 5.13 shows σ^w/λ as a function of θ in the region $0 \leq \theta \leq \pi$ for both hard and soft polarizations. Note that the pattern is smooth, continuous, and finite everywhere, including the region around $\theta = \pi/2$, which separates regions I and II. In region II, $H_P^{d\prime}$ shows behavior similar to that in region I. In the region around $\theta \approx \pi$, the principal contributor to σ^w is the reflection from T. Therefore, we would expect σ^w to oscillate around $10 \log 2\pi\,[\sqrt{(b/\lambda)/2}]^2$, or $\sigma^w/\lambda \approx 5$ dB. The ripple in the region $\theta \geq 100°$ is due to the interference of H_T^r with H_Q^d and H_P^d.

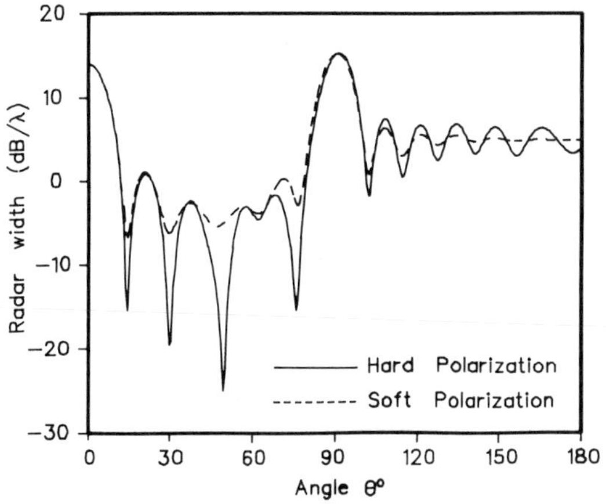

Figure 5.13 Radar width (σ^w) of the scattering body shown in Figure 5.9 in the angular sector $0 \leq \theta \leq \pi$, with $d = 2\lambda$ and $b = \lambda$.

PROBLEMS

5.1 Consider the parallel plate waveguide with TEM mode propagation discussed in Section 5.1.

(a) Show analytically that the radiation pattern has a $\sin(u)/u$ form, with $u = (\pi a/\lambda)\sin\theta$.

(b) Show that the gain of the parallel plate guide is equal to ka.

(c) Examine the scattered field at a point (R, θ) in the near-zone. Examine the far-zone case by letting $R \to \infty$; and pay particular attention to the on-axis case, the limit of $\theta = \epsilon$ as $\epsilon \to 0$.

(d) Examine the effect of higher-order diffraction terms.

5.2 Consider a TE_{01} mode plane wave to propagate down the parallel plate waveguide as shown in Figure 5.14. This field can be expressed as

$$\mathbf{E} = E_0 \cos\left(\frac{\pi y}{a}\right) e^{-jkx}\hat{\mathbf{z}} \tag{5.81}$$

(a) Show that the TE_{01} mode field can be represented by two crossing plane waves as shown, where

$$\sin\theta_c = \frac{\lambda}{2a} \tag{5.82}$$

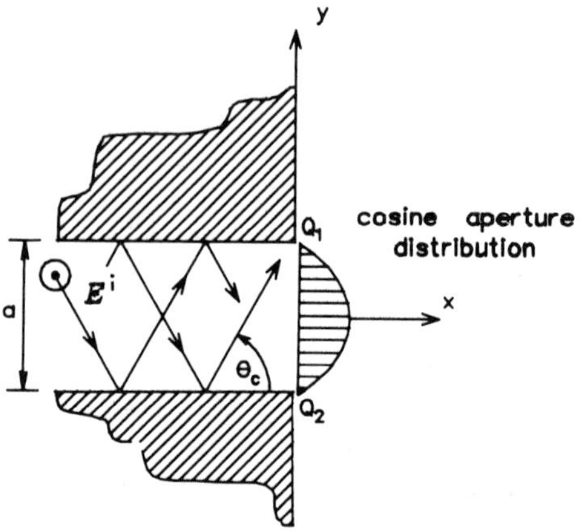

Figure 5.14 Parallel plate waveguide with TE_{01} mode propagation, showing crossing plane waves.

(b) Derive an expression for the far-zone radiated field from the guide using the crossing plane waves.

(c) Plot the far-zone radiated field from the guide for the cases where $a = 2\lambda$ and $a = 5\lambda$.

5.3 In this problem, analyze the parallel plate waveguide with TE_{01} mode plane wave propagation using slope diffraction. Note that the incident field is given by (5.81) and that it is zero at Q_1 and Q_2.

(a) Derive an expression for the far-zone radiated field from the guide using slope diffraction.

(b) Plot the far-zone radiated field from the guide for the cases where $a = 2\lambda$ and $a = 5\lambda$.

(c) Compare the results with those of Problem 5.1, and discuss the validity of both solutions.

5.4 Determine the values of L_e and θ_e that will maximize the gain of the E-plane horn antenna.

5.5 Include diffracted-reflected rays in the E-plane pattern of the horn antennas and investigate their effect.

5.6 Determine the front-to-back ratio for the E-plane horn.

5.7 The FBR for the E-plane horn can be improved by adding flanges to the horn as shown in Figure 5.15. For the case where $L_e = 11\lambda$ and $\theta_e = 15°$, determine the length d of the flanges required for an FBR of 60 dB. In the chapter on creeping waves we will examine a much more efficient way to improve the FBR.

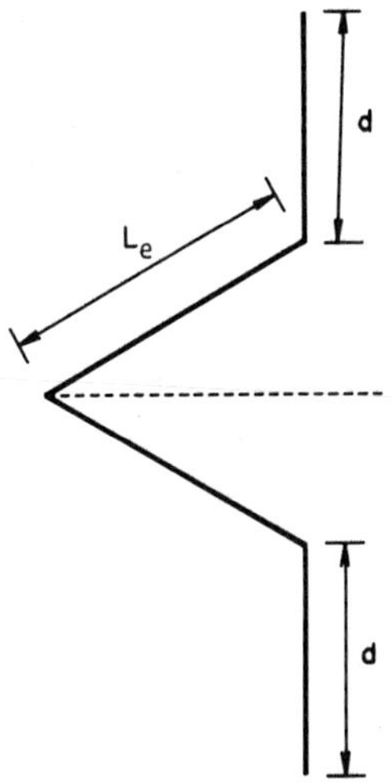

Figure 5.15 *E*-plane horn antenna with flanges.

5.8 In Example 3.4 the radar width of a conducting two-dimensional ellipse due to specular reflection was derived. In this problem calculate the radar width of a truncated two-dimensional ellipse.

(a) Derive the solution for σ^w of the semiellipse shown in Figure 5.16, using wedge diffraction and reflection.

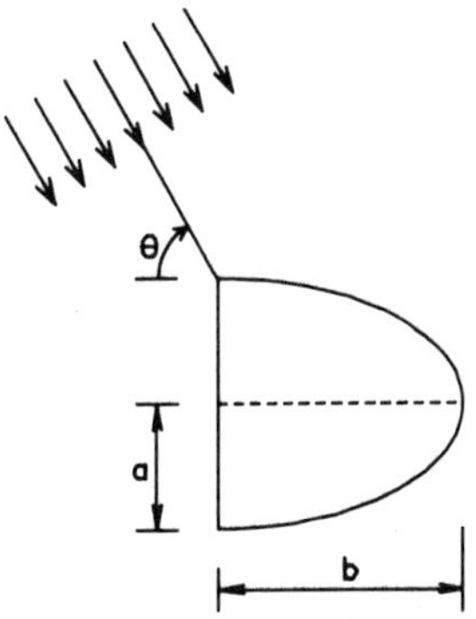

Figure 5.16 Two-dimensional semiellipse.

(b) Plot σ^w in dB/λ where $a = \lambda$ and $b = 2\lambda$, for the complete and truncated ellipses in the region $0 \leqslant \theta \leqslant \pi$.

5.9 Derive an expression for the radar width of the perfectly absorbing strip shown in Figure 5.17. Why does this strip have a backscattered field at all? Assume that the reflection and transmission coefficients both are zero.

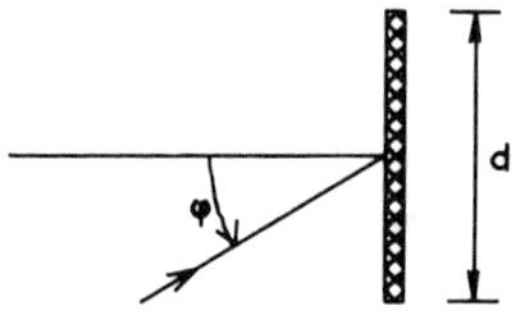

Figure 5.17 Backscatter from a perfectly absorbing strip.

REFERENCES

[1] R.B. Dybdal, R.C. Rudduck, and L.L. Tsai, "Mutual Coupling between TEM and TE_{01} Parallel-Plate Waveguide Apertures," *IEEE Trans. Antennas and Propagation,* Vol. AP-14, No. 5, September 1966, pp. 574–580.

[2] R.C. Rudduck and L.L. Tsai, "Aperture Reflection Coefficient of TEM and TE_{01} Mode Parallel-Plate Waveguides," *IEEE Trans. Antennas and Propagation*, Vol. AP-16, No. 1, January 1968, pp. 83–89.

[3] C.E. Ryan, Jr., and R.C. Rudduck, "A Wedge Diffraction Analysis of the Radiation Patterns of Parallel-Plate Waveguides," *IEEE Trans. Antennas and Propagation,* Vol. AP-16, No. 4, July 1968, pp. 490–491.

[4] J.E. Jones, L.L. Tsai, R.C. Rudduck, C.T. Swift, and W.D. Burnside, "The Admittance of a Parallel-Plate Waveguide Aperture Illuminating a Metal Sheet," *IEEE Trans. Antennas Propagation,* Vol. AP-16, No. 5, September 1968, pp. 528–535.

[5] R.C. Rudduck, L.L. Tsai, and W.D. Burnside, "Reflection Coefficient of a Parallel-Plate Waveguide Illuminating a Conducting Sheet," *IEEE Trans. Antennas Propagation,* Vol. AP-17, No. 2, March 1969, pp. 175–179.

[6] W.D. Burnside, R.C. Rudduck, and L.L. Tsai, "Reflection Coefficient of a TEM Mode Symmetric Parallel-Plate Waveguide Illuminating a Dielectric Layer," *Radio Science*, Vol. 4, No. 6, June 1969, pp. 545–556.

[7] S.W. Lee, "On Edge Diffracted Rays of an Open-Ended Waveguide," *Proc. IEEE,* Vol. 57, August 1969, pp. 1445–1446.

[8] R.C. Rudduck and D.C.F. Wu, "Slope Diffraction Analysis of TEM Parallel-Plate Guide Radiation Patterns," *IEEE Trans. Antennas Propagation,* Vol. AP-17, No. 6, November 1969, pp. 797–799.

[9] J.J. Bowman, "Comparison of Ray Theory with Exact Theory for Scattering by Open Waveguides," *Siam J. Appl. Math.,* Vol. 18, June 1970, pp. 818–829.

[10] W.D. Burnside, E.L. Pelton, and L. Peters, Jr., "Analysis of Finite Parallel-Plate Waveguide Arrays," *IEEE Trans. Antennas and Propagation,* Vol. AP-18, No. 5, September 1970, pp. 701–705.

[11] S.W. Lee, "Ray Theory of Diffraction by Open-Ended Waveguides. Part I: Field in Waveguides," *J. Math. Phys.,* Vol. 11, September 1970, pp. 2830–2850.

[12] S.W. Lee, "Ray Theory of Diffraction by Open-Ended Waveguides. Part II: Applications," *J. Math. Phys.*, Vol. 13, May 1972, pp. 657–664.

[13] S.W. Lee and J. Boersma, "Ray-Optical Analysis of Fields on Shadow Boundaries of Two-Parallel Plates," *J. Math. Phys.*, Vol. 16, September 1975, pp. 1746–1764.

[14] J. Boersma, "Ray-Optical Analysis of Reflection in an Open-Ended Parallel-Plate Waveguide I—TM Case," *Siam J. Appl. Math.*, Vol. 29, 1975, pp. 164–195.

[15] J. Boersma, "Diffraction by Two Parallel Half-Planes," *J. Mech and Appl. Math.*, Vol. 38, 1975, pp. 406–425.

[16] E.V. Jull, *Aperture Antennas and Diffraction Theory*, Peter Peregrinus, London, 1981.

[17] G.L. James, *Geometrical Theory of Diffraction for Electromagnetic Waves,* 3rd ed., Peter Peregrinus, London, 1986.

[18] R.C. Rudduck, "Application of GTD to antennas," short course notes on the modern geometrical theory of diffraction, Vol. 1, Ohio State University.

[19] S. Silver, *Microwave Antenna Theory and Design,* Peter Peregrinus, London, 1984.

[20] W.L. Stutzman and G.A. Thiele, *Antenna Theory and Design*, John Wiley and Sons, New York, 1984.

[21] C.A. Balanis, *Antenna Theory*, Harper and Row, New York, 1982.

[22] P.M. Russo, R.C. Rudduck and L. Peters, Jr., "A Method for Computing E-Plane Patterns of Horn Antennas," *IEEE Trans. Antennas and Propagation,* Vol. AP-13, No. 2, March 1965, pp. 219–224.

[23] J.S. Yu, R.C. Rudduck, and L. Peters, Jr., "Comprehensive Analysis for E-Plane of Horn Antennas by Edge Diffraction Theory," *IEEE Trans. Antennas and Propagation,* Vol. AP-14, No. 2, March 1966, pp. 138–149.

[24] R.E. Lawrie and L. Peters, Jr., "Modifications of Horn Antennas for Low Sidelobe Levels," *IEEE Trans. Antennas and Propagation,* Vol. AP-14, No. 5, September 1966, pp. 605–610.

[25] C.E. Ryan, Jr., and R.C. Rudduck, "Radiation Patterns of Rectangular Waveguides," *IEEE Trans. Antennas and Propagation*, Vol. AP-16, No. 4, July 1968, pp. 488–489.

[26] J.S. Yu and R.C. Rudduck, "H-Plane Pattern of a Pyramidal Horn," *IEEE Trans. Antennas and Propagation*, Vol. AP-17, No. 5, September 1969, pp. 651–652.

[27] E.V. Jull, "Reflection from the Aperture of a Long E-Plane Sectoral Horn," *IEEE Trans. Antennas and Propagation,* Vol. AP-20, January 1972, pp. 62–68.

[28] C.A. Mentzer, L. Peters, Jr., and R.C. Rudduck, "Slope Diffraction and Its Application to Horns," *IEEE Trans. Antennas and Propagation,* Vol. AP-23, March 1975, pp. 153–159.

[29] C.A. Mentzer and L. Peters, Jr., "Pattern Analysis of Corrugated Horn Antennas," *IEEE Trans. Antennas and Propagation,* Vol. AP-24, No. 3, May 1976, pp. 304–309.

[30] M.S. Narasimhan and M.S. Sheshadri, "GTD Analysis of the Radiation Patterns of Conical Horns," *IEEE Trans. Antennas and Propagation*, Vol. AP-26, No. 6, November 1978, pp. 774–778.

[31] M.S. Narasimham and M.S. Sheshadri, "GTD Analysis of Radiation Patterns of Wide-Flare Conical and E-Plane Corrugated Sectoral Horns," *IEEE Trans. Antennas and Propagation,* Vol. AP-27, No. 2, March 1979, pp. 276–279.

[32] M.S. Narasimhan, "GTD Analysis of the Radiation Pattern of Open-Ended Circular Cylindrical Waveguide Horns," *IEEE Trans. Antennas and Propagation,* Vol. AP-27, No. 3, May 1979, pp. 438–441.

[33] M.S. Narasimhan and K.S. Rao, "GTD Analysis of the Near-Field Patterns of Conical and Corrugated Conical Horns," *IEEE Trans. Antennas and Propagation*, Vol. AP-27, No. 5, September 1979, pp. 705–708. See also Bhattacharya's comments and authors' reply in *IEEE Trans. Antennas and Propagation*, Vol. AP-30, No. 5, September 1982, pp. 1042–1043.

[34] M.S. Narasimhan and K. Sudhakar, "GTD Analysis of the E-Plane Patterns of Conical Horns," *IEEE Trans. Antennas and Propagation,* Vol. AP-28, No. 5, September 1980, pp. 715–717.

[35] J.C. Mather, "Broad-Band Flared Horn with Low Sidelobes," *IEEE Trans. Antennas and Propagation,* Vol. AP-29, No. 6, November 1981, pp. 967–969.

[36] J. Huang, Y. Rahmat-Samii, and K. Woo, "A GTD Study of Pyramidal Horns for Offset Reflector Studies," *IEEE Trans. Antennas and Propagation,* Vol. AP-31, No. 2, March 1983, pp. 305–309.

[37] D.J. Heedy and W.D. Burnside, "An Aperture-Matched Compact Range Feed Horn Design," *IEEE Trans. Antennas and Propagation*, Vol. AP-33, No. 11, November 1985, pp. 1249–1255.

Chapter 6
Three-Dimensional Wedge Diffraction and Corner Diffraction

"Nevertheless it is interesting to see that Young's theory concerned with the illumination of the edge of the screen is not confirmed here just by coincidence. We shall actually show that the function . . . may be split by a transformation into an (in the sense of geometrical optics) incident light wave and into a diffraction wave originating at the edge of the screen."

A. Rubinowitz
Annalen der Physik, Vol. 53, 1917

6.1 INTRODUCTION

In this chapter the first-order two-dimensional UTD wedge diffraction concepts discussed in the previous chapters will be extended to three dimensions. Three-dimensional slope diffraction will not be covered in this text. Although the basic principles stay the same, addition of a third dimension brings with it certain complications. Before we start with a formal treatment of three-dimensional UTD wedge diffraction, some differences between two-dimensional and three-dimensional cases need to be indicated.

A two-dimensional wedge is shown in Figure 6.1(a). The wedge is located in the *x-y* plane. By the nature of the two-dimensional assumption, the *x-y* projection of the wedge extends to infinity in the $\pm z$ direction. In Figure 6.1 (a) the edge is located on the $\hat{\mathbf{z}}$-axis and is straight along its entire length. The wedge faces have finite radii of curvature a_o and a_n. Although a_o, $a_n > 0$ in Figure 6.1 (a), they can become infinite in the case of flat faces or negative in the case of concave

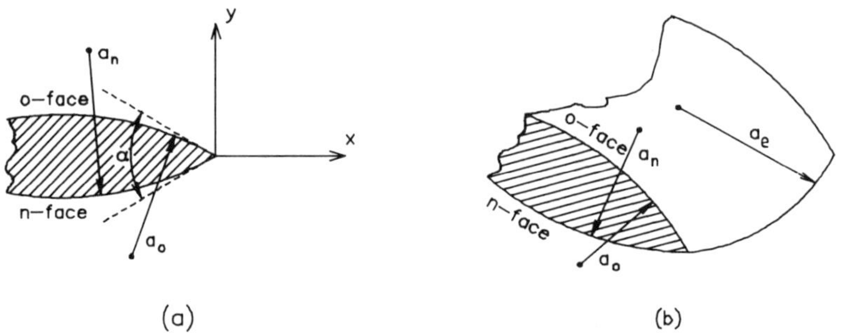

Figure 6.1 (a) Two-dimensional wedge and (b) three-dimensional wedge with edge curvature a_e.

faces. Because the incident and diffracted rays are contained in the *x-y* plane, these rays always are perpendicular to the edge, so that $\beta_0 = \pi/2$. In the three-dimensional case, the edge contour need not be straight nor infinitely long. In addition, the wedge faces can be curved or straight as in the two-dimensional case. Figure 6.1(b) shows a typical three-dimensional wedge where the edge has a curvature a_e.

In the three-dimensional case the incident and diffracted rays in general are not perpendicular to the edge. Consider a point Q_e on an edge, where edge diffraction occurs as shown in Figure 4.5. The unit vector that is tangential to the edge at Q_e is $\hat{\mathbf{e}}$. If an incident ray propagates in the direction defined by the unit vector $\hat{\mathbf{s}}'$, then the angle subtended between the incident ray and the edge is β_0'; where

$$\sin\beta_0' = |\hat{\mathbf{s}}' \times \hat{\mathbf{e}}| = |\hat{\mathbf{s}} \times \hat{\mathbf{e}}| \tag{6.1}$$

Equation (6.1) is an expression of the law of edge diffraction mentioned in Section 4.3. Recall from our discussion there that (6.1) is doubly valued and that (4.19) rather should be used to compute β_0. Diffracted rays propagate in directions along a diffraction cone, where the half-angle of the cone is β_0, so that $\beta_0 = \beta_0'$. Note that $0 \leqslant \beta_0, \beta_0' \leqslant \pi$. One ray incident upon an edge thus will cause an infinite number of diffracted rays. The unit vector $\hat{\mathbf{s}}$ in (6.1) denotes the direction of propagation along any one of the diffracted rays. Figures 4.5 and 4.6 illustrate the case where a ray is incident upon an edge with an arbitrary angle β_0' and the case where $\beta_0' = \pi/2$, respectively. Recall that, in the case where $\beta_0' = \pi/2$, the diffraction cone degenerates to a disk. In the two-dimensional case, we dealt only with the case where $\beta_0' = \pi/2$.

The fact that in the three-dimensional case the edges no longer need be straight means that the direction of $\hat{\mathbf{e}}$ will change as the edge curves through space and the edge may have a finite radius of curvature at Q_e, rather than the infinite radius of curvature of a straight edge. In the case of a curved edge, we find that the diffracted energy will spread out in a way similar to the reflected energy from a curved surface. We therefore surmise that a caustic distance will be associated with the edge just as caustic distances are associated with the curved surfaces making up the wedge faces. This indeed is the case.

In the two-dimensional case only two polarizations were possible: electric field perpendicular to the plane of incidence (E_z) and magnetic field perpendicular to the plane of incidence (H_z). This led us to use either the soft or hard diffraction coefficients. In three dimensions, an infinite number of polarizations are possible, the only restrictions being that the electric and magnetic fields are perpendicular to each other and to the direction of propagation. Subsequently, we find that, in the three-dimensional case, to deal with two perpendicular polarizations of the electric field is more convenient than the electric and magnetic fields, as has been done in the two-dimensional case.

6.2 EDGE-FIXED COORDINATE SYSTEM

To calculate the GO reflected fields from a surface, we use a ray-fixed coordinate system, as was discussed in Chapter 3. In such a ray-fixed coordinate system, the incident field is resolved in components parallel and perpendicular to the plane of incidence; that is, the plane containing the unit vector normal to the surface at the point of reflection as well as the incident and reflected rays. When this is done, the reflection coefficient reduces to a simple dyadic or a 2 × 2 diagonal matrix. However, the ray-fixed coordinate system is not suitable for diffracted rays, as the diffraction coefficients then would be composed of the sum of seven dyads [1]. In matrix form, this means that the diffraction coefficient is a 3 × 3 matrix with seven nonvanishing elements.

The soft and hard diffraction coefficients ($D_{s,h}$) used in the two-dimensional cases are valid only when the incident field is resolved into components parallel and perpendicular to the edge-fixed plane of incidence; that is, the plane containing $\hat{\mathbf{e}}$ and $\hat{\mathbf{s}}'$. The diffracted field components then are parallel and perpendicular to the edge-fixed plane of diffraction; that is, the plane containing $\hat{\mathbf{e}}$ and $\hat{\mathbf{s}}$. Note that every one of the diffracted rays will have its own edge-fixed plane of diffraction. We therefore could anticipate that the diffraction coefficient would have a much simpler form than the sum of seven dyads just mentioned, if the three-dimensional diffracted fields could be resolved into an edge-fixed coordinate system. In fact, we find that if this is done, the diffraction coefficient reduces to a diagonal 2 × 2 matrix, as will be shown later.

The edge-fixed coordinate system is shown in Figure 6.2. The incident field is resolved into $\hat{\beta}_0'$ and $\hat{\phi}'$ components and the diffracted field into $\hat{\beta}_0$ and $\hat{\phi}$ components [2], where

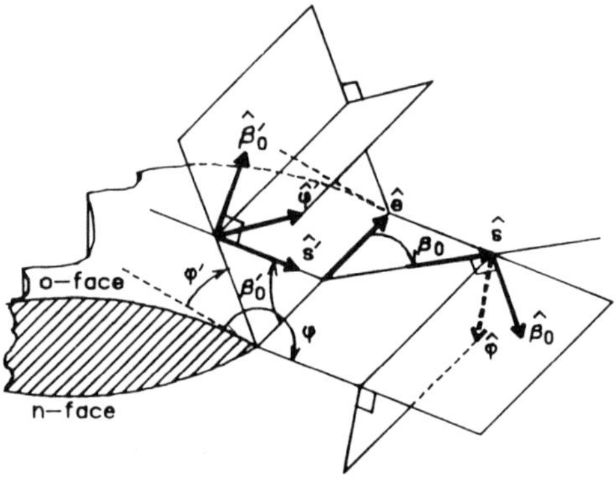

Figure 6.2 Edge-fixed coordinate system.

$$\hat{\phi}' = \frac{-\,\hat{\mathbf{e}} \times \hat{\mathbf{s}}'}{|\hat{\mathbf{e}} \times \hat{\mathbf{s}}'|} \tag{6.2}$$

$$\hat{\beta}_0' = \hat{\phi}' \times \hat{\mathbf{s}}' \tag{6.3}$$

$$\hat{\phi} = \frac{\hat{\mathbf{e}} \times \hat{\mathbf{s}}}{|\hat{\mathbf{e}} \times \hat{\mathbf{s}}|} \tag{6.4}$$

$$\hat{\beta}_0 = \hat{\phi} \times \hat{\mathbf{s}} \tag{6.5}$$

If (6.1) is substituted into these equations, we find that the unit vectors are not clearly defined when $\beta_0 \to 0$. This is not surprising because, when $\beta_0 \to 0$, the edge-fixed plane of incidence containing the incident ray and the edge is not defined. As we shall see, the situation where $\beta_0 \to 0$ leads to other problems, notably the violation of the asymptotic approximations upon which the UTD is based.

Note that although some other authors define these vectors differently [1], they yield the same results. Fields that lie in the edge-fixed planes of incidence and diffraction thus will be $\hat{\beta}_0'$ and $\hat{\beta}_0$ polarized, respectively, whereas fields perpendicular to the edge-fixed planes of incidence and diffraction are $\hat{\phi}'$ and $\hat{\phi}$ polarized, respectively. As will be seen, $\hat{\beta}_0'$ and $\hat{\beta}_0$ polarized fields are associated with the soft diffraction coefficient, whereas the $\hat{\phi}'$ and $\hat{\phi}$ polarized fields are associated with the hard diffraction coefficients. The edge-fixed coordinate system can be viewed as a spherical coordinate system with the diffraction point at the origin and $\hat{\mathbf{e}}$ along the z-axis (or $-$ z-axis). Note that β_0' and β_0 are measured from opposite sides of the z-axis; that is, if $\beta_0' = \pi - \theta$, then $\beta_0 = \theta$, because the law of edge diffraction requires that $\beta_0 = \beta_0'$; also, $\hat{\beta}_0'$ and $\hat{\beta}_0$ both will lie in θ-planes of the spherical coordinate system, albeit not necessarily in the same θ-plane.

Apart from the angles β_0' and β_0, the rays in a spherical coordinate system need to be specified by another angle in a plane perpendicular to the z-axis (edge). In the edge-fixed coordinate system, these angles are ϕ' and ϕ in the case of the incident and diffracted rays, respectively. The angles ϕ' and ϕ are the same angles used in the case of two-dimensional diffraction and measured with respect to the o-face, in a plane perpendicular to the edge ($\hat{\mathbf{e}}$).

The components of s' and s that lie in the plane perpendicular to the edge are given by

$$s_t' = |\hat{\mathbf{s}}' - (\hat{\mathbf{s}}' \cdot \hat{\mathbf{e}})\hat{\mathbf{e}}| \tag{6.6}$$

$$s_t = |\hat{\mathbf{s}} - (\hat{\mathbf{s}} \cdot \hat{\mathbf{e}})\hat{\mathbf{e}}| \tag{6.7}$$

We leave as an exercise to show that $s_t' = s' \sin\beta_0$ and $s_t = s \sin\beta_0$.

The angles ϕ' and ϕ can be determined analytically as follows. Let the unit vector tangential to the o-face be given by

$$\hat{\mathbf{t}}_o = \hat{\mathbf{n}}_o \times \hat{\mathbf{e}} \tag{6.8}$$

where $\hat{\mathbf{e}}$ is the unit vector tangential to the edge. Note that the direction of $\hat{\mathbf{e}}$ should be such that $\hat{\mathbf{t}}_o$ points toward the o-face as shown in Figure 6.3.

Following (6.6) and (6.7), we can define the unit vectors:

$$\hat{\mathbf{s}}_t' = \frac{\hat{\mathbf{s}}' - (\hat{\mathbf{s}}' \cdot \hat{\mathbf{e}})\hat{\mathbf{e}}}{|\hat{\mathbf{s}}' - (\hat{\mathbf{s}}' \cdot \hat{\mathbf{e}})\hat{\mathbf{e}}|} \tag{6.9}$$

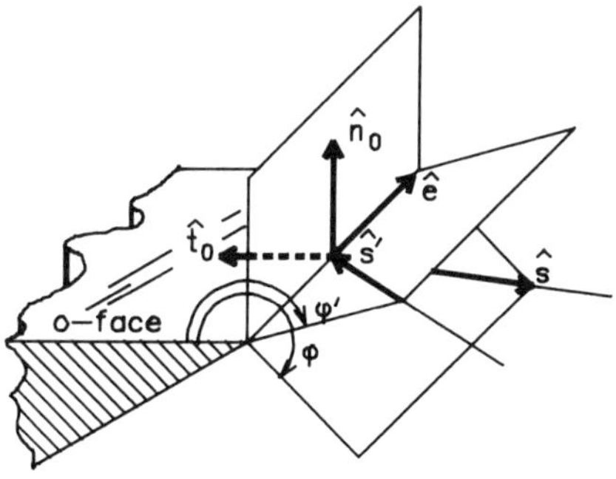

Figure 6.3 Normal and tangential unit vectors on the o-face of a wedge.

$$\hat{\mathbf{s}}_t = \frac{\hat{\mathbf{s}} - (\hat{\mathbf{s}} \cdot \hat{\mathbf{e}})\hat{\mathbf{e}}}{|\hat{\mathbf{s}} - (\hat{\mathbf{s}} \cdot \hat{\mathbf{e}})\hat{\mathbf{e}}|} \tag{6.10}$$

to lie in the plane perpendicular to the edge. The angle ϕ' then is given by

$$\phi' = \pi - [\pi - \arccos(-\hat{\mathbf{s}}_t' \cdot \hat{\mathbf{t}}_o)] \operatorname{sgn}(-\hat{\mathbf{s}}_t' \cdot \hat{\mathbf{n}}_o) \tag{6.11}$$

with $0 \leqslant \arccos(x) \leqslant \pi$. The function $\operatorname{sgn}(x)$ was defined in (4.187). Similarly, the angle ϕ can be expressed as

$$\phi = \pi - [\pi - \arccos(\hat{\mathbf{s}}_t \cdot \hat{\mathbf{t}}_o)] \operatorname{sgn}(\hat{\mathbf{s}}_t \cdot \hat{\mathbf{n}}_o) \tag{6.12}$$

Recall from Chapter 4 that $0 \leqslant \phi, \phi' \leqslant n\pi$ is required, where n is related to the wedge angle α by (4.25); that is, $\alpha = (2 - n)\pi$.

6.3 THREE-DIMENSIONAL UTD DIFFRACTION COEFFICIENTS

The three-dimensional first-order UTD diffracted fields from a general three-dimensional wedge now can be expressed in the edge fixed coordinate system as

$$\begin{bmatrix} E^d_{\beta_0} \\ E^d_{\phi} \end{bmatrix} = \begin{bmatrix} -D_s & 0 \\ 0 & -D_h \end{bmatrix} \begin{bmatrix} E^i_{\beta_0'}(Q_e) \\ E^i_{\phi'}(Q_e) \end{bmatrix} \sqrt{\frac{\rho}{s(s+\rho)}}\, e^{-jks} \tag{6.13}$$

where $D_{s,h}$ are soft and hard diffraction coefficients, ρ is the edge caustic distance, and s is the distance from the point of diffraction (Q_e) on the edge to the field point. The incident and diffracted field components are given by

$$\mathbf{E}^i = E^i_{\beta_0'}\hat{\beta}_0' + E^i_{\phi'}\hat{\phi}' \tag{6.14}$$

$$\mathbf{E}^d = E^d_{\beta_0}\hat{\beta}_0 + E^d_{\phi}\hat{\phi} \tag{6.15}$$

An arbitrary polarized incident field can be resolved into edge-fixed components as follows:

$$E^i_{\beta_0'} = \mathbf{E}^i \cdot \hat{\beta}_0' \tag{6.16}$$

$$E^i_{\phi'} = \mathbf{E}^i \cdot \hat{\phi}' \tag{6.17}$$

Equation (6.13) also can be expressed in dyadic form as

$$\mathbf{E}^d = \mathbf{E}^i \cdot \mathbf{D} \sqrt{\frac{\rho}{s(s+\rho)}}\, e^{-jks} \tag{6.18}$$

where the dyadic diffraction coefficient is given by

$$\mathbf{D} = -\hat{\beta}_0'\hat{\beta}_0 D_s - \hat{\phi}'\hat{\phi} D_h \tag{6.19}$$

Note that the minus signs in (6.13) and (6.19) follow from the way in which the unit vectors are defined.

Some authors [3] use the notation $E_{\parallel}^i$ and $E_{\perp}^i$ when referring to $E_{\beta_0'}^i$ and $E_{\phi'}^i$, respectively, and $E_{\parallel}^d$ and $E_{\perp}^d$ when referring to $E_{\beta_0}^d$ and E_{ϕ}^d, respectively.

The three-dimensional diffraction coefficients are given by

$$D_{s,h}(L^i, L^{ro}, L^{rn}, \phi, \phi', \beta_0, n) = D_1 + D_2 \mp (D_3 + D_4) \tag{6.20}$$

where ∓ 1 are the soft and hard reflection coefficients of the perfectly conducting surfaces of the wedge at the edge, respectively. The components of the diffraction coefficients are given by

$$D_1 = \frac{-e^{-j\pi/4}}{2n\sqrt{2\pi k}\,\sin\beta_0} \cot\left[\frac{\pi + (\phi - \phi')}{2n}\right] F[kL^i a^+(\phi - \phi')] \tag{6.21}$$

$$D_2 = \frac{-e^{-j\pi/4}}{2n\sqrt{2\pi k}\,\sin\beta_0} \cot\left[\frac{\pi - (\phi - \phi')}{2n}\right] F[kL^i a^-(\phi - \phi')] \tag{6.22}$$

$$D_3 = \frac{-e^{-j\pi/4}}{2n\sqrt{2\pi k}\,\sin\beta_0} \cot\left[\frac{\pi + (\phi + \phi')}{2n}\right] F[kL^{rn} a^+(\phi + \phi')] \tag{6.23}$$

$$D_4 = \frac{-e^{-j\pi/4}}{2n\sqrt{2\pi k}\,\sin\beta_0} \cot\left[\frac{\pi - (\phi + \phi')}{2n}\right] F[kL^{ro} a^-(\phi + \phi')] \tag{6.24}$$

where the transition function (F) and associated functions $a^{\pm}(\beta)$ are exactly the same as in the two-dimensional case. Comparison of (6.20)–(6.24) with (4.55)–(4.59) show that the two-dimensional coefficients are just special cases of the three-dimensional coefficients, because $\sin\beta_0 = 1$ in the two-dimensional case.

The distance parameters for three-dimensional diffraction are considerably more complicated than in the two-dimensional case. The distance parameter L^i associated with the incident shadow boundaries is given by

$$L^i = \frac{s(\rho_e^i + s)\rho_1^i \rho_2^i}{\rho_e^i(\rho_1^i + s)(\rho_2^i + s)} \sin^2\beta_0 \tag{6.25}$$

where ρ_1^i is the principal radius of curvature of the incident wavefront at Q_e in the plane of incidence (containing $\hat{\mathbf{s}}'$ and $\hat{\mathbf{n}}$)and ρ_2^i is the principal radius of curvature of the incident wavefront in the plane transverse to the plane of incidence. The radius of curvature of the incident wavefront in the edge-fixed plane of incidence is ρ_e^i. In the case where the incident field has a spherical wavefront, $\rho_{1,2}^i = \rho_e^i = s'$, where s' is the radius of curvature of the spherical wavefront. L^i then reduces to

$$L^i = \frac{ss'}{s + s'} \sin^2\beta_0 \tag{6.26}$$

In the case of plane wave incidence, L^i reduces to

$$L^i = s \sin^2\beta_0 \tag{6.27}$$

The distance parameters associated with the reflection shadow boundaries are given by

$$L^{ro,n} = \frac{s(\rho_e^{ro,n} + s)\rho_1^{ro,n}\rho_2^{ro,n}}{\rho_e^{ro,n}(\rho_1^{ro,n} + s)(\rho_2^{ro,n} + s)} \sin^2\beta_0 \tag{6.28}$$

where $\rho_1^{ro,n}$ and $\rho_2^{ro,n}$ are the principal radii of curvature of the reflected wavefront from the o- and n-faces, respectively.

The principal radius of curvature of the reflected wavefront in the plane of incidence was discussed in Section 3.5. For convenience, the relevant expressions are repeated here:

$$\frac{1}{\rho_1^r} = \frac{1}{\rho_1^i} + \frac{2}{a_1 \cos\theta^i} \tag{6.29}$$

where a_1 is the radius of curvature of the surface at Q_e in the plane of incidence. The principal radius of curvature of the reflected wavefront in the transverse plane is given by

$$\frac{1}{\rho_2^r} = \frac{1}{\rho_2^i} + \frac{2\cos\theta^i}{a_2} \tag{6.30}$$

where a_2 is the radius of curvature of the surface at Q_e in the plane transverse to the plane of incidence. In Section 4.4.2 we saw that the caustic distance of reflection in the plane of incidence can be expressed as

$$\frac{1}{\rho_1^r} = \frac{1}{\rho_1^i} - \frac{2}{a_1(\hat{\mathbf{s}}' \cdot \hat{\mathbf{n}})} \tag{6.31}$$

where $a_1 > 0$ if the o-, n-face is convex, and $a_1 < 0$ if it is concave. Similarly, we can express ρ_2^r as

$$\frac{1}{\rho_2^r} = \frac{1}{\rho_2^i} - \frac{2(\hat{\mathbf{s}}' \cdot \hat{\mathbf{n}})}{a_2} \tag{6.32}$$

Note that (6.31) and (6.32) may be applied to either the o- or n-face, so that $\hat{\mathbf{n}}$ is the unit vector perpendicular to the o- or n-face ($\hat{\mathbf{n}}_o$ or $\hat{\mathbf{n}}_n$), respectively. The case where the plane of incidence is different from the principal planes is discussed in [1].

The distances $\rho_e^{ro,n}$ are the radii of curvature of the reflected wavefront in the plane containing the reflected ray and the edge ($\hat{\mathbf{e}}$), so that

$$\frac{1}{\rho_e^{ro,n}} = \frac{1}{\rho_e^i} - \frac{2(\hat{\mathbf{n}}_e \cdot \hat{\mathbf{n}}_{o,n})(\hat{\mathbf{s}}' \cdot \hat{\mathbf{n}}_{o,n})}{|a_e| \sin^2\beta_0} \tag{6.33}$$

where $\hat{\mathbf{n}}_{o,n}$ are the unit vectors normal to the o- and n-faces at Q_e, respectively; a_e is the radius of curvature of the edge at Q_e; and $\hat{\mathbf{n}}_e$ is the unit vector normal to the edge at Q_e.

Consider for a moment the concept of $\hat{\mathbf{n}}_e$, the unit vector normal to the edge. The unit vector normal to a surface ($\hat{\mathbf{n}}_{o,n}$) is easy to visualize. An edge, however, is a curve in space, with an infinite number of possible unit vectors normal to it at any given point. The diffracted rays in the case where $\beta_0 = 90°$ as shown in Figure 4.6, illustrate the point. How do we determine which of these unit vectors is $\hat{\mathbf{n}}_e$? For the purpose of this section, a simple explanation is in order; a more complete treatment of the radius of curvature of a curve in space and the associated normal vector is given in Appendix C.

Consider the general case of a curve in space as shown in Figure C.1 in Appendix C. At any point (Q) on the curve, it will have a radius of curvature a_e, such that we can fit a circle (known as the *osculating circle*) with radius a_e "into" the curve at Q. The unit vector $\hat{\mathbf{n}}_e$ then is the radial unit vector of the circle at that point. Figure 6.4 (a) shows the case where a paraboloid is intercepted by a plane parallel to the x-y plane, so that the curve (C) where the plane and the paraboloid intercept is a circle. The center of curvature of C then is the center of the circle on the z-axis; the curve C being the osculating circle in this case. The unit vector $\hat{\mathbf{n}}_e$ at the point Q_e where C intercepts the y-z plane is $\hat{\mathbf{n}}_e = \hat{\mathbf{y}}$. In Figure 6.4 (b), a plane parallel to the x-z plane intercepts the paraboloid, so that $\hat{\mathbf{n}}_e = -\hat{\mathbf{z}}$ at Q_e. In Figure 6.4 (c), the paraboloid is intercepted by an oblique plane

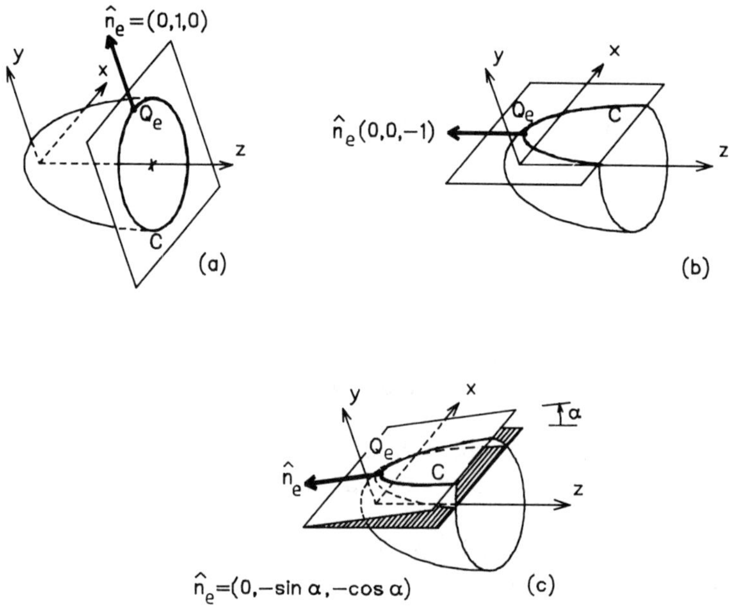

Figure 6.4 Paraboloid intercepted by various planes, showing $\hat{\mathbf{n}}_e$. Intercepting planes: (a) parallel to x-y plane; (b) parallel to x-z plane; and (c) tilted with respect to x-z plane.

tilted at an angle α relative to the x-z plane. In this case, $\hat{\mathbf{n}}_e = \sin\alpha\,\hat{\mathbf{y}} - \cos\alpha\,\hat{\mathbf{z}}$ at Q_e.

The edge caustic (ρ) contained in the spreading factor of (6.13) is given by

$$\frac{1}{\rho} = \frac{1}{\rho_e^i} - \frac{\hat{\mathbf{n}}_e \cdot (\hat{\mathbf{s}}' - \hat{\mathbf{s}})}{|a_e| \sin^2\beta_0} \tag{6.34}$$

where a_e is the radius of curvature of the edge at Q_e and ρ_e^i is the radius of curvature of the incident wavefront at Q_e in the edge-fixed plane of incidence; that is, the plane containing $\hat{\mathbf{s}}'$ and $\hat{\mathbf{e}}$.

Let us digress for a moment to consider the case where $s > \rho$ and $\rho < 0$. Following the discussion in Chapter 3, we should evaluate the spreading factor as

$$\sqrt{\frac{\rho}{s(\rho + s)}} = \pm j\sqrt{\frac{|\rho|}{s(s + \rho)}} \tag{6.35}$$

where the plus sign would be chosen if the propagation were in a direction through the caustic. Note that ρ actually is the second caustic distance related to the edge, as the edge itself also is a caustic.

By changing the radius of curvature of the edge at a point Q_e, the caustic distance ρ is affected, which, in turn, will affect the diffracted field from Q_e. This principle has been applied to reduce the diffracted fields from a parabolic compact range reflector [4].

Consider now the special case where the edge is straight; that is, $a_e \to \infty$. From (6.34), we thus have $\rho = \rho_e^i$, so that the spreading factor in (6.13) reduces to

$$\sqrt{\frac{\rho}{s(s + \rho)}} = \sqrt{\frac{\rho_e^i}{\rho_e^i + s}}\frac{1}{\sqrt{s}} \tag{6.36}$$

In the case where a cylindrical wave is incident upon the straight edge (as in the two-dimensional case), the radius of curvature of the incident wavefront in the plane containing $\hat{\mathbf{s}}'$ and $\hat{\mathbf{e}}$ (ρ_e^i) is infinite so that

$$\sqrt{\frac{\rho}{s(s + \rho)}} = \frac{e^{-jks}}{\sqrt{s}} \tag{6.37}$$

The diffracted field therefore also will have a cylindrical wavefront. Note that (6.37) is exactly the form described in the two-dimensional case in Chapter 4. We thus have shown that the two-dimensional case actually is a special case of the three-dimensional case.

Whenever $s \gg \rho$, such as when $|\rho| < \infty$ and $s \to \infty$, we find that

$$\sqrt{\frac{\rho}{s(s + \rho)}} \approx \frac{\sqrt{\rho}}{s} \tag{6.38}$$

In the three-dimensional case, the largeness parameter (as far as the asymptotic approximation is concerned) is

$$\kappa = kL \sin^2\beta_0 \tag{6.39}$$

In Section 4.4.4, we mentioned that $\kappa > 1$ for the UTD to be valid. This condition can be violated if the frequency is too low or if s' or s is too small; that is, the source point or field point is too close to the point of diffraction. In the three-dimensional case, a small angle β_0 also can result in κ being unacceptably small; this happens when the incident rays are in the paraxial region close to the edge. Additional coefficients that extend the validity of the UTD in these regions have been developed [5]. Incident rays in the paraxial region also give rise to edge waves [6].

6.4 EXAMPLES OF THREE-DIMENSIONAL WEDGE DIFFRACTION

To illustrate some aspects of three-dimensional edge diffraction, we now consider several examples.

Example 6.1: Monopole mounted on a thin circular disk

Consider a short monopole mounted on a thin circular disk with radius a [7–9], as shown in Figure 6.5(a). The field radiated by the monopole is assumed to be

$$\mathbf{E}^m(\theta) = \sin\theta \frac{e^{-jkr}}{r} \hat{\theta} \tag{6.40}$$

For this example, the short monopole is considered to be a point source with a $\sin\theta$ pattern, located at the center of the disk. The problem is to determine the far-zone radiation pattern of the monopole mounted on the disk. To avoid confusion, the symbol Φ will be used to indicate the angle phi of the spherical coordinate system, whereas the symbols ϕ' and ϕ will be used to indicate the angles of incidence and diffraction of the incident and diffracted rays, respectively, measured with respect to the *o*-face.

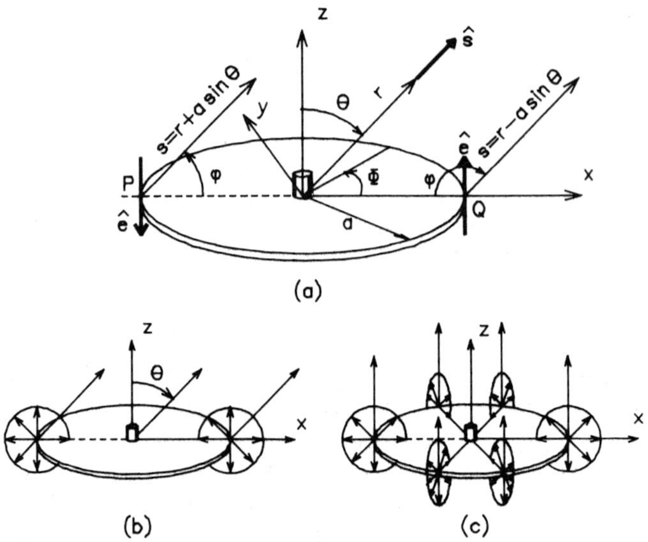

Figure 6.5 Short monopole mounted on a circular disk with radius a: (a) geometry; (b) radiation in direction away from the axial region; (c) radiation in the axial direction, resulting in the entire edge becoming a caustic.

Because the structure is symmetric with respect to Φ, it is sufficient to calculate the pattern only in the x-z plane for $0° \leq \theta \leq 180°$. From Figure 6.5 it is clear that only the diffracted fields from the two points P and Q on the edge and the radiation from the monopole itself will contribute to the far-zone field in any given direction, except when θ is close to 0° or 180°. Note that the incident ray $\hat{\mathbf{s}}'$ from the center of the disk to any point on the edge intercepts the edge at an angle of $\beta_0 = 90°$. Subsequently all the diffraction cones around the edge degenerate to disks as shown in Figure 6.5(b). Close to the angles $\theta = 0°$ and 180° the entire edge contour thus becomes a caustic as shown in Figure 6.5(c). Diffracted rays from the entire edge contribute in these directions instead of just two diffracted rays, as in the case when θ is removed from the axial direction, so that the UTD wedge diffraction fails in the axial region. In the next chapter, we shall see how the concept of equivalent currents can be used to determine the fields in the region of a caustic.

In the regions away from the axis of symmetry, the far-zone radiated fields from the structure can be expressed as

$$\mathbf{E}^t = \begin{cases} \mathbf{E}^m + \mathbf{E}_P^d + \mathbf{E}_Q^d + \mathbf{E}_{QP}^d, & 0 < \theta < \pi/2 \\ \mathbf{E}_P^d + \mathbf{E}_Q^d + \mathbf{E}_{QP}^d, & \pi/2 < \theta < \pi \end{cases} \tag{6.41}$$

where $\mathbf{E}_P^d$ and $\mathbf{E}_Q^d$ are the first-order diffracted fields from P and Q, respectively, and $\mathbf{E}_{QP}^d$ is the second-order diffracted field from Q due to illumination from the first-order diffracted field from P. From our experience in Example 4.4 we know that $\mathbf{E}_{QP}^d$ must be included to compensate for the discontinuity of $\mathbf{E}_P^d$ across the shadow boundary at $\theta = \pi/2$. The contribution of $\mathbf{E}_{PQ}^d$ in the angular sector considered here is negligible, and therefore disregarded.

The first step is to determine the unit vectors that will be used to resolve the fields in components compatible with the edge-fixed coordinate system. Consider first the point Q, located at ($\theta = 90°$, $\Phi = 0°$) in the x-z plane. As mentioned earlier, the incident field at Q is perpendicular to the edge, so that $\beta_0 = 90°$. Because the diffraction cone degenerates to a disk for diffraction points on the rim of the circular disk on which the monopole is mounted, the diffracted rays from Q all are in the x-z plane. Through inspection we find that $\hat{\mathbf{n}}_o = \hat{\mathbf{z}}$, $\hat{\mathbf{e}} = \hat{\mathbf{y}}$, $\hat{\mathbf{s}}' = \hat{\mathbf{x}}$. Furthermore, $\hat{\mathbf{s}} = \sin\theta\,\hat{\mathbf{x}} + \cos\theta\,\hat{\mathbf{z}}$, because we are considering radiation in the direction $\Phi = 0°$ as a function of θ. Substituting these values into (6.2)–(6.5), we find that the following holds at Q:

$$\hat{\phi}' = \hat{\mathbf{z}} = -\hat{\theta} \tag{6.42}$$

$$\hat{\beta}_0' = \hat{\mathbf{y}} \tag{6.43}$$

$$\hat{\phi} = -\sin\theta\hat{\mathbf{z}} + \cos\theta\hat{\mathbf{x}} = \hat{\theta} \tag{6.44}$$

$$\hat{\beta}_0 = -\hat{\mathbf{y}} \tag{6.45}$$

$$\hat{\mathbf{n}}_e = \hat{\mathbf{x}} \tag{6.46}$$

Substituting (6.40) and the preceding values into (6.16) and (6.17), the incident field at Q can be expressed as

$$\mathbf{E}^i(Q) = -\frac{e^{-jka}}{a}\hat{\phi}' \tag{6.47}$$

so that

$$E^i_{\phi'}(Q) = -\frac{e^{-jka}}{a} \tag{6.48}$$

$$E^i_{\beta_0'}(Q) = 0 \tag{6.49}$$

Following (6.18), the diffracted field from Q can be expressed as

$$\mathbf{E}^d_Q(\theta) = -E^i_{\phi'}(Q)\frac{D_h}{2}\sqrt{\frac{\rho}{s(\rho + s)}}\, e^{-jks}\,\hat{\phi} \tag{6.50}$$

where $s = r - a\sin\theta$. The diffraction coefficient is divided by 2 as we have grazing incidence. From (6.34) the edge caustic distance is found to be

$$\rho = \frac{a}{\sin\theta} \tag{6.51}$$

Note that $0 \leqslant \sin\theta \leqslant 1$, so that $\rho > 0$. Equation (6.38) can be used in this case because the field point is assumed to be in the far-zone, so that (6.50) reduces to

$$\mathbf{E}^d_Q(\theta) = \frac{e^{-jka(1-\sin\theta)}}{\sqrt{a\sin\theta}}\frac{D_h}{2}\frac{e^{-jkr}}{r}\hat{\theta} \tag{6.52}$$

Recall from (6.44) that $\hat{\phi} = \hat{\theta}$.

What now remains is to determine the parameters of the diffraction coefficient, $D_h(L^i, L^{ro}, L^{rn}, \phi, \phi', \beta_0, n)$. We already have established that $\beta_0 = \pi/2$

and $n = 2$. Although we can determine through inspection that $\phi' = 0$ and $\phi = \pi/2 + \theta$ in this case, to apply (6.11) and (6.12) to find these values nevertheless is instructive. Because

$$\hat{\mathbf{n}}_o \times \hat{\mathbf{e}} = -\hat{\mathbf{x}} \tag{6.53}$$

the correct sign has been chosen for $\hat{\mathbf{e}}$ to satisfy (6.8). Also,

$$-\hat{\mathbf{s}}_t' = \hat{\mathbf{x}} \tag{6.54}$$

$$\hat{\mathbf{t}}_o = -\hat{\mathbf{x}} \tag{6.55}$$

so that

$$-\hat{\mathbf{s}}_o' \cdot \hat{\mathbf{n}}_o = 0 \tag{6.56}$$

$$-\hat{\mathbf{s}}_o' \cdot \hat{\mathbf{t}}_o = 1 \tag{6.57}$$

Hence,

$$\phi' = \arccos(1) = 0 \tag{6.58}$$

Applying (6.12), we find that

$$\phi = \pi - [\pi - \arccos(-\sin\theta)] \operatorname{sgn}(\cos\theta) \tag{6.59}$$

We can see that (6.59) yields the following result:

$$\phi = \pi/2 + \theta \quad \text{for} \quad 0 \leqslant \theta \leqslant \pi \tag{6.60}$$

In other, more complicated problems, ϕ' and ϕ may not be so self-evident in terms of a system angle such as θ, in which case the use of (6.11) and (6.12) can considerably simplify matters.

Because the monopole illuminates Q with a spherical wavefront, $\rho_e = \rho_1 = \rho_2^i = s' = a$. Keeping in mind that $s \gg a$, L^i in (6.26) reduces to

$$L^i = a \tag{6.61}$$

Because both faces of the disk are flat, $a_1 = a_2 \to \infty$ so that $\rho_1^r = \rho_2^r = a$. Furthermore, because $\hat{\mathbf{n}}_{o,n} \cdot \hat{\mathbf{n}}_e = 0$ (and $\hat{\mathbf{n}}_{o,n} \cdot \hat{\mathbf{s}}' = 0$), from (6.33) we have

$$\rho_e^r = a \tag{6.62}$$

for both faces so that

$$L^{ro,n} = a \tag{6.63}$$

Consider now the point P located at ($\theta = 90°$, $\Phi = 180°$) in the x-z plane, where $\hat{\mathbf{e}} = -\hat{\mathbf{y}}$, $\hat{\mathbf{s}}' = -\hat{\mathbf{x}}$, $\hat{\mathbf{s}} = \sin\theta\ \hat{\mathbf{x}} + \cos\theta\ \hat{\mathbf{z}}$. Substituting these values into (6.2)–(6.5), we find that the following hold at P:

$$\hat{\phi}' = \hat{\theta} \tag{6.64}$$

$$\hat{\beta}_0' = \hat{\mathbf{y}} \tag{6.65}$$

$$\hat{\phi} = -\hat{\theta} \tag{6.66}$$

$$\hat{\beta}_0 = -\hat{\mathbf{x}} \tag{6.67}$$

$$\hat{\mathbf{n}}_e = -\hat{\mathbf{y}} \tag{6.68}$$

The incident field at P thus can be expressed as

$$\mathbf{E}^i(P) = \frac{e^{-jka}}{a}\,\hat{\phi}' \tag{6.69}$$

so that

$$E^i_{\phi'}(P) = \frac{e^{-jka}}{a} \tag{6.70}$$

$$E^i_{\beta_0'}(P) = 0 \tag{6.71}$$

The diffracted field from P therefore can be expressed as

$$\mathbf{E}^d_P(\theta) = -E^i_{\phi'}(P)\,\frac{D_h}{2}\sqrt{\frac{\rho}{s(\rho + s)}}\,e^{-jks}\,\hat{\phi} \tag{6.72}$$

where $s = r + a\sin\theta$. From (6.34), the edge caustic is given by

$$\rho = -\frac{a}{\sin\theta} \tag{6.73}$$

Note that $\rho < 0$. From (6.35) and (6.37), the spreading factor and phase term hence can be expressed as

$$\sqrt{\frac{\rho}{s(\rho + s)}}\, e^{-jks} = j\sqrt{\frac{a}{\sin\theta}}\, e^{-jka\sin\theta}\, \frac{e^{-jkr}}{r} \tag{6.74}$$

To see why $+j$ rather than $-j$ is chosen in (6.74), consider Figure 6.6. Recall from the discussions in the previous chapters that the caustic point can be found by extrapolating a distance ρ from the diffraction point in the direction opposite to the direction of propagation. We saw in (6.51) that $\rho > 0$ for the ray diffracted from Q, so this ray will not pass through the caustic point. However, for the ray diffracted from P, $\rho < 0$. This caustic point is found by extrapolating a distance $-a/\sin\theta$ in the direction opposite to that direction of propagation, or alternatively, a distance $a/\sin\theta$ in the direction of propagation as shown in Figure 6.6. This ray thus will propagate through the caustic. In Chapter 3 we learned that the field expression is multiplied by $+j$ when the ray propagates through the caustic.

The far-zone expression for the diffracted field from P thus is given by

$$\mathbf{E}_P^d(\theta) = \frac{-j\, e^{-jka(1+\sin\theta)}}{\sqrt{a\sin\theta}}\, \frac{D_h}{2}\, \frac{e^{-jkr}}{r}\, \hat{\theta} \tag{6.75}$$

because $\hat{\phi} = -\hat{\theta}$ in this case. As in the previous case, $\phi' = 0$, $\beta_0 = \pi/2$, $n = 2$, and $L^i = L^r = a$. We saw in (4.183) that

$$\phi = \begin{cases} 90^\circ - \theta, & \text{if} \quad 0^\circ \le \theta \le 90^\circ \\ 450^\circ - \theta & \text{if} \quad 90^\circ < \theta \le 180^\circ \end{cases} \tag{6.76}$$

Alternatively, (6.12) could have been used to yield

$$\phi = \pi \dot{-} [\pi - \arccos(\sin\theta)]\ \mathrm{sgn}(\cos\theta) \tag{6.77}$$

The derivation of an expression for $\mathbf{E}_{QP}^d$ is left as an exercise for the reader.

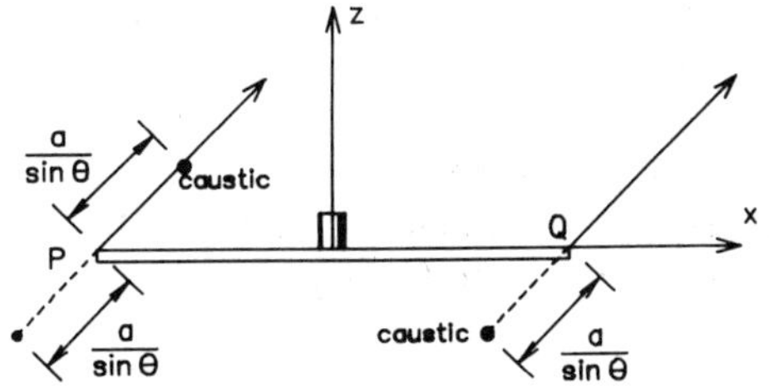

Figure 6.6 Projection in the x-z plane of a short monopole mounted on a disk to show the location of the caustic points for radiation in regions away from the axial direction.

Figure 6.7 shows the far-zone radiation pattern obtained for a short monopole on a disk, using (6.41). The sinθ pattern of the monopole on an infinite ground plane also is shown. The pattern of the monopole on the disk obviously fails in the regions around $\theta = 0°$ and $\theta = 180°$. As explained previously, the entire edge becomes a caustic in these regions, so the wedge diffraction solutions break down. In the next chapter, we will see how the concept of equivalent currents can be used to determine the radiation pattern in the caustic regions. Note that the pattern in Figure 6.7 is normalized to 0 dB at the peak of the lobe in the region around 60°.

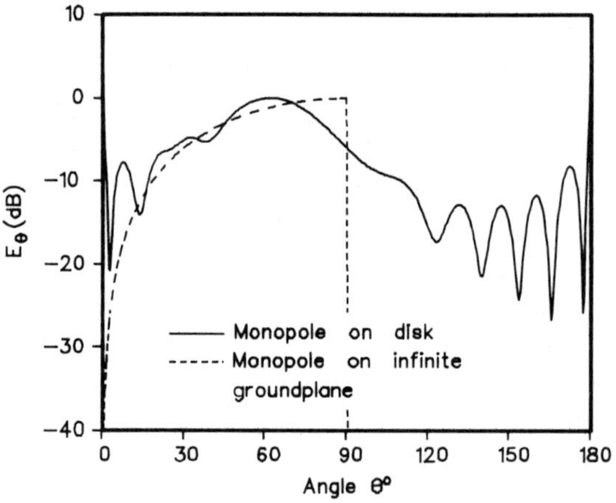

Figure 6.7 Normalized radiation patterns of a short monopole mounted on a disk of radius $a = 2.5\lambda$ and a short monopole mounted on an infinite ground plane. Note that the wedge diffraction solution fails in the axial regions.

Example 6.2: Scattered field from a half-ellipsoidal shell illuminated by a point source in the near-field

Consider the half-ellipsoidal shell with axes of lengths R_1, and R_2, and R_3 shown in Figure 6.8. The structure is illuminated by a point source located in the x-z plane at $x = x'$ and $z = z'$ with $x' > 0$ and $z' < 0$. In this example we examine the diffracted field from Q_e, as shown in Figure 6.9(b) and pay particular attention to the determination of the parameters.

We start by expressing the incident field at Q_e in the format of (6.14). At Q_e the unit vector tangential to the edge ($\hat{\mathbf{e}}$) is $\hat{\mathbf{y}}$, and

$$\hat{\mathbf{s}}' = \frac{(R_1 - x')}{s'}\,\hat{\mathbf{x}} - \frac{z'}{s'}\,\hat{\mathbf{z}} \tag{6.78}$$

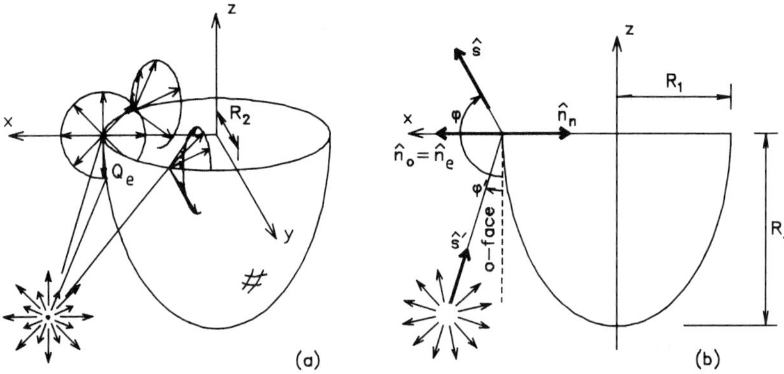

Figure 6.8 (a) Half-ellipsoidal shell illuminated by a point source in the near-field; (b) *x-z* projection of the geometry.

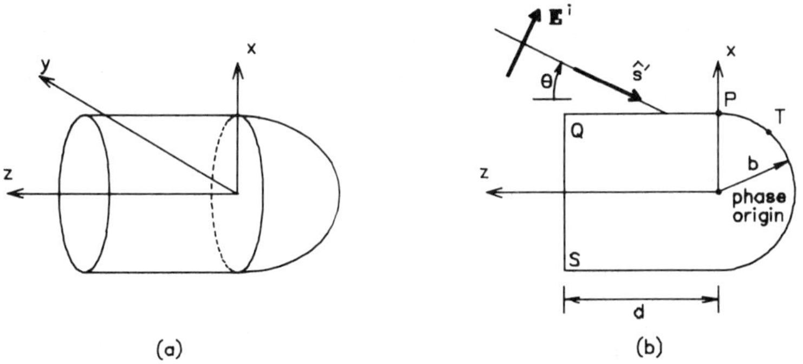

Figure 6.9 Spherically capped cylinder: (a) oblique view; (b) side view.

with

$$s' = \sqrt{(R_1 - x')^2 + z'^2} \tag{6.79}$$

From (6.2) and (6.3),

$$\hat{\phi}' = -\frac{z'}{s'}\,\hat{\mathbf{x}} - \frac{(R_1 - x')}{s'}\,\hat{\mathbf{z}} \tag{6.80}$$

$$\hat{\beta}_0' = \hat{\mathbf{y}} \tag{6.81}$$

Because $\hat{\mathbf{s}} = \sin\phi\,\hat{\mathbf{x}} - \cos\phi\,\hat{\mathbf{z}}$, we find from (6.4) and (6.5) that

$$\hat{\phi} = -\cos\phi\,\hat{\mathbf{x}} - \sin\phi\,\hat{\mathbf{z}} \tag{6.82}$$

$$\hat{\beta}_0 - \hat{\mathbf{y}} \tag{6.83}$$

Because the shell is infinitely thin at Q_e, $n = 2$. The diffraction coefficients therefore can be expressed as

$$D_{s,h} = \frac{-e^{-j\pi/4}}{2\sqrt{2\pi k}\,\sin\beta_0}\left\{\frac{F[kL^i a(\phi - \phi')]}{\cos[(\phi - \phi')/2]} \mp \frac{F[kL^r a(\phi + \phi')]}{\cos[(\phi + \phi')/2]}\right\} \tag{6.84}$$

Note that $\beta_0 = \pi/2$ at Q_e. As the incident wavefront at Q_e is spherical, L^i is given by (6.26); and L^r is given by (6.28). Note that the discussion in Section 4.4.6 that shows why $L^{ro} = L^{rn}$ also is applicable here. It therefore is sufficient to evaluate only L^{ro}. To find L^{ro}, we have to find expressions for ρ_1^{ro}, ρ_2^{ro}, and ρ_e^{ro}. The expressions for ρ_1^{ro} is given by (6.31), where $\rho_1^i = s'$ and $\hat{\mathbf{n}}_o = \hat{\mathbf{x}}$. Note that a_1 is the radius of curvature of the o-face at Q_e in the plane of incidence (containing $\hat{\mathbf{s}}'$, $\hat{\mathbf{n}}_o$, and $\hat{\mathbf{s}}$); that is, the x-z plane. From (C.29), we thus have

$$a_1 = \frac{R_3^2}{R_1} \tag{6.85}$$

The caustic distance ρ_2^{ro} is given in (6.32) where $\rho_2^i = s'$, $\hat{\mathbf{n}}_o = \hat{\mathbf{x}}$, and $a_2 = R_2^2/R_1$. Note that a_2 is measured in a plane transverse to the plane of incidence; that is, the x-y plane. Because the o-face is convex at Q_e, $a_1 > 0$ and $a_2 > 0$. To evaluate ρ_e^{ro} in (6.33), we find that $\rho_e^i = s'$, $\hat{\mathbf{n}}_e = \hat{\mathbf{x}}$ and

$$a_e = \frac{R_2^2}{R_1} \tag{6.86}$$

Note that $a_2 = a_e$ here. This is coincidental and not always the case. In our previous example $a_1 = a_2 \to \infty$, whereas a_e was finite. All the parameters of the edge caustic distance (ρ) given in (6.34) already have been defined.

Example 6.3: Radar cross section of a spherically capped cylinder

In this section, we calculate the *radar cross section* (RCS) of a three-dimensional structure. As in the two-dimensional case, the particular structure we consider here has no obvious practical use, but illustrates how the RCS of a structure can be calculated. From Chapter 3, we know that the RCS is defined by

$$\sigma = \lim_{r \to \infty} 4\pi r^2 \frac{|\mathbf{E}^s|^2}{|\mathbf{E}^i|^2} \tag{6.87}$$

where r is the distance from the far-zone phase reference to the phase origin on the structure, $\mathbf{E}^i$ is the field of the incident electric plane wave at the phase origin of the structure, and $\mathbf{E}^s$ is the scattered electric field at the far-zone phase reference. Alternatively, the magnetic incident and scattered fields also can be used. $\mathbf{E}^s$ and $\mathbf{E}^i$ usually are copolarized. In the three-dimensional case the phase reference on the structure is a plane perpendicular to the incident rays. This plane passes through the phase origin, so that it is an equiphase plane for the incident plane wave. Recall from (5.45) and (5.46) that $\hat{\mathbf{s}} = -\hat{\mathbf{s}}'$ and $\phi = \phi'$ in the case of a monostatic RCS calculation.

Consider now a cylinder with length d and a spherical cap that has a radius b, as shown in Figure 6.9. Assume that the cylinder is orientated coaxially along the z-axis, with the incident field $\hat{\theta}$ polarized in the x-z plane, as shown. We must determine the radar cross section as a function of angle θ, for $0 \leq \theta \leq \pi$. Note that this three-dimensional problem is analogous to the two-dimensional problem considered in Section 5.5. In fact, we will use many of the results obtained in Section 5.5 here.

The incident field at a point $\mathbf{r}$ can be expressed as

$$\mathbf{E}^i = e^{-jk(\hat{\mathbf{s}}'\mathbf{r})}\,\hat{\theta} \tag{6.88}$$

where the direction of propagation of the incident field is given by

$$\hat{\mathbf{s}}' = -\sin\theta\,\hat{\mathbf{x}} - \cos\theta\,\hat{\mathbf{z}} \tag{6.89}$$

$$\mathbf{r} = x\hat{\mathbf{x}} + z\hat{\mathbf{z}} \tag{6.90}$$

so that

$$\mathbf{E}^i = e^{jk(x\sin\theta + z\cos\theta)}\,\hat{\theta} \tag{6.91}$$

The origin of the coordinate system and phase center of the structure now has been chosen to be at the center of the spherical cap, rather than at Q as in the example in Section 5.5. In anticipation of the amplitude normalization associated with the calculation of the RCS, the amplitude of the incident field has been set equal to unity.

Following (5.47), the scattered field can be expressed as

$$\mathbf{E}^s = \begin{cases} \mathbf{E}_P^d + \mathbf{E}_Q^d + \mathbf{E}_S^d, & 0 \leq \theta < \pi/2 \quad \text{(region I)} \\ \mathbf{E}_P^d + \mathbf{E}_Q^d + \mathbf{E}_T^r, & \pi/2 < \theta < \pi \quad \text{(region II)} \end{cases} \tag{6.92}$$

We start by calculating the diffraction from Q. Following the convention in (6.8) we choose the tangential vector to the edge at Q to be $\hat{\mathbf{e}} = -\hat{\mathbf{y}}$. From (4.19), hence $\beta_0 = \pi/2$; this also holds for the diffracted fields from P and S. From (6.2)–(6.5), we thus have

$$\hat{\phi}' = -\hat{\theta} \tag{6.93}$$

$$\hat{\beta}_0' = -\hat{\mathbf{y}} \tag{6.94}$$

$$\hat{\phi} = -\hat{\theta} \tag{6.95}$$

$$\hat{\beta}_0 = \hat{\mathbf{y}} \tag{6.96}$$

Because the incident field is $\hat{\theta}$-polarized, as is evident from (6.19), the diffracted fields from Q are hard polarized. In this example, we can determine through inspection that the diffracted fields from P and S also are hard polarized, because the magnetic field is tangential to the edges at P, Q, and S. In the general case, however, the incident field is to be decomposed as in (6.14), and the soft and hard diffracted fields calculated separately. The diffracted fields are calculated according to (6.18) in edge-fixed coordinates. The total vector-diffracted fields then are resolved in the required components; for example, rectangular or spherical.

Following (6.17), (6.18), and (6.19), the diffracted field from Q can be expressed as

$$\mathbf{E}^d_\phi(\theta) = E^i_{\phi'}(Q)\, D_h \sqrt{\frac{\rho}{s(\rho + s)}}\, e^{-jks}\, \hat{\phi} \tag{6.97}$$

or

$$\mathbf{E}^d_Q(\theta) = -E^i_\theta(Q)\, D_h \sqrt{\frac{\rho}{s(\rho + s)}}\, e^{-jks}\, \hat{\theta} \tag{6.98}$$

because $E^i_{\phi'} = -E^i_\theta$ and $\hat{\phi}' = -\hat{\theta}$. Substituting $\mathbf{r}(Q) = b\hat{\mathbf{x}} + d\hat{\mathbf{z}}$ into (6.88), we find that

$$E^i_\theta(Q) = e^{jk(b\sin\theta + d\cos\theta)} \tag{6.99}$$

If we choose the horizontal face QP as the o-face, $\phi = \pi - \theta$. Furthermore, $n = 1.5$ and

$$s = r - (b \sin\theta + d \cos\theta) \tag{6.100}$$

As in the two-dimensional example of Section 5.5, we find that the transition functions in the diffraction coefficient tend to unity.

The edge caustic distance is given by (6.34). In this case, $\rho_e^i \to \infty$ because the incident field is a plane wave. Furthermore, $a_e = b$ and $\hat{\mathbf{n}}_e = \hat{\mathbf{x}}$ so that

$$\rho(Q) = \frac{b}{2 \sin\theta} \tag{6.101}$$

The diffracted field from Q thus can be expressed as

$$\mathbf{E}_Q^d(\theta) = -e^{j2k(b\sin\theta + d\cos\theta)} D_h \sqrt{\frac{b}{2 \sin\theta}} \frac{e^{-jkr}}{r} \hat{\theta} \tag{6.102}$$

where r is the distance from the phase origin on the structure to the far-zone phase reference. D_h is given in (6.20).

Similarly, the diffracted field from S can be expressed as

$$\mathbf{E}_S^d(\theta) = -j\, e^{j2k(-b\sin\theta + d\cos\theta)} D_h \sqrt{\frac{b}{2 \sin\theta}} \frac{e^{-jkr}}{r} \hat{\theta} \tag{6.103}$$

where $\phi = \pi/2 - \theta$ and $n = 1.5$. The j term is due to the fact that $\rho < 0$.

At P, $n = 1$ and $\phi = \theta$. Combining (4.222) and (5.63), the diffraction coefficient can be as expressed as

$$D_h = \frac{e^{-j\pi/4}}{\sqrt{2\pi k} \sin\beta_0} f_h \{1 - F[kL^{rn}a(\phi + \phi')]\} \tag{6.104}$$

with f_h given by (4.224). In this case, $\beta_0 = \pi/2$.

As in Section 5.5, $L^{ro} \to \infty$. The distance parameter L^{rn} is given by (6.28). Because the incident illumination is a plane wave, we have from (6.31) and (6.32):

$$\rho_1^{rn} = \frac{b \sin\theta}{2} \tag{6.105}$$

$$\rho_2^{rn} = \frac{b}{2 \sin\theta} \tag{6.106}$$

To evaluate ρ_e^{rn} note that $\rho_e^i \to \infty$, $a_e = b$, $\hat{\mathbf{n}}_e = \hat{\mathbf{n}}_n = \hat{\mathbf{x}}$. The unit vector $\hat{\mathbf{s}}'$ is given in (6.89). Substituting these values into (6.33), we have

$$\rho_e^{rn} = \frac{b}{2\sin\theta} \tag{6.107}$$

so that

$$L^{rn} = \frac{b\sin\theta}{2} \tag{6.108}$$

Because the edge contours at Q and P are the same, the edge caustic distances are also the same; that is,

$$\rho(P) = \frac{b}{2\sin\theta} \tag{6.109}$$

The diffracted field from P can thus be expressed as

$$\mathbf{E}_P^d(\theta) = -\mathrm{e}^{j2kb\sin\theta} D_h \sqrt{\frac{b}{2\sin\theta}} \frac{\mathrm{e}^{-jkr}}{r} \hat{\theta} \tag{6.110}$$

The total scattered field in region I can be determined from (6.92), (6.102), (6.103), and (6.110). To determine the total scattered field in region II, we need to find an expression for the reflected field from T. From (3.243) and (3.245),

$$\mathbf{E}_T^r = (-\mathbf{E}^i + 2(\hat{\mathbf{n}} \cdot \mathbf{E}^i)\hat{\mathbf{n}}) \sqrt{\frac{\rho_1^r \rho_2^r}{(\rho_1^r + s)(\rho_2^r + s)}} \mathrm{e}^{-jks} \tag{6.111}$$

with $s = r - b$. Substituting $\rho_1^i = \rho_2^i = \rightarrow \infty$ and $a_1 = a_2 = b$ into (6.29) and (6.30) and noting that the reflection is specular (i.e., $\theta^i = 0$), we have

$$\rho_1^r = \rho_2^r = \frac{b}{2} \tag{6.112}$$

At T the incident field can be expressed as

$$\mathbf{E}^i(T) = \mathrm{e}^{j2kb}\,\hat{\theta} \tag{6.113}$$

with

$$\hat{\theta} = \cos\theta\,\hat{\mathbf{x}} - \sin\theta\,\hat{\mathbf{z}} \tag{6.114}$$

in the x-z plane. Because the unit vector normal to the surface at T is given by

$$\hat{\mathbf{n}} = \sin\theta\, \hat{\mathbf{x}} + \cos\theta\, \hat{\mathbf{y}} \tag{6.115}$$

we thus have $\hat{\mathbf{n}} \cdot \mathbf{E}^i = 0$. Because $s \gg \rho'$, the reflected field from T therefore can be expressed as

$$\mathbf{E}_T^r = -\, e^{j2kb}\left(\frac{b}{2}\right)\frac{e^{-jkr}}{r}\,\hat{\theta} \tag{6.116}$$

The total scattered field in region II is found by substituting (6.102), (6.110), and (6.116) into (6.92). Once the total scattered field has been found, we then can substitute it into (6.87) to yield the RCS. Recall that the amplitude of the incident field has been set equal to unity so that $|\mathbf{E}^i| = 1$ in (6.87).

In Section 5.5, all the dimensions of the structure were normalized relative to a wavelength, and the radar width was subsequently expressed in dB relative to a wavelength. In this example, the lengths are expressed in meters and the RCS (σ) is expressed in decibel relative to a square meter; that is,

$$\sigma_{\mathrm{dB/m^2}} = 10 \log \sigma \tag{6.117}$$

Figure 6.10 shows the RCS of the structure for θ-polarized electric fields calculated from (6.92) as well as measured values from [10]. A more accurate

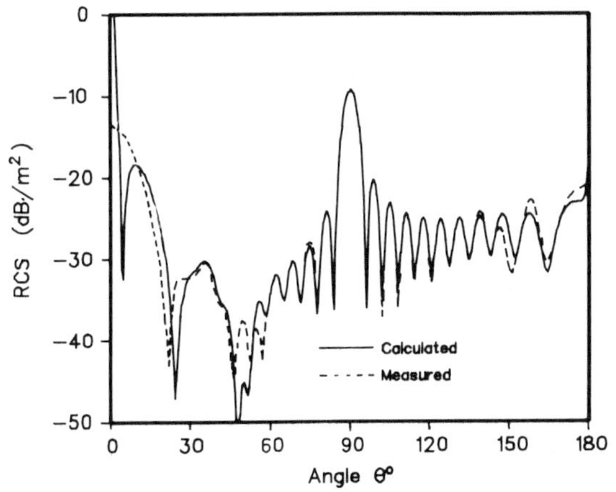

Figure 6.10 E-plane RCS of a spherically capped cylinder with d = 15.2 cm (4.57λ) and b = 2.5 cm (0.75λ) at 9 GHz (measured values from Chu [10]). Note that the wedge diffraction solution fails in the axial region.

UTD model would include higher-order diffracted fields as well as creeping waves (which will be discussed in Chapter 8). The wedge diffraction solution shown in Figure 6.10 fails in the axial regions, due to the caustic effects from the rim that passes through Q and S and the ring discontinuity in radius of curvature passing through P. In the next chapter, we will see that the method of equivalent currents can be used to overcome this sort of problem.

Good practice is to verify the RCS calculation with various spot checks. At $\theta = 90°$, the RCS should be approximately equal to that of cylinder of length d and radius b. Using physical optics, we find that the broadside RCS of such a cylinder at $\theta = 90°$ is given by [11]

$$\sigma = kb\, d^2 \tag{6.118}$$

In this case, (6.118) yields $\sigma = -9.6$ dB/m^2, which corresponds very well with the results obtained with the UTD. Note that the calculated curve is smooth and continuous at $\theta = 90°$, where the expression for the total scattered field changes from region I to region II. The pattern cannot be calculated at exactly $\theta = 90°$ because this is the direction of the reflection shadow boundary. However, the results at $\theta = 90° - \epsilon$ and $\theta = 90° + \epsilon$ converge to the correct value at $\theta = 90°$ as $\epsilon \rightarrow 0$. At $\theta = 0°$ the RCS should be approximately equal to that of a disk of radius b at $\theta = 90°$ [11]; that is,

$$\sigma = \frac{4\pi^3 b^4}{\lambda^2} \tag{6.119}$$

In this case, (6.119) yields $\sigma = -13.6$ dB/m^2, which corresponds well to the measured value. The wedge diffraction solution of (6.92) fails in the region around $\theta = 180°$ due to the caustic effect. The RCS due to the specular reflection from T alone is given by

$$\sigma = \pi b^2 \tag{6.120}$$

or -27 dB/m^2. The measured value is somewhat higher than this value, which indicates that, in addition to the specular reflection, other effects play a significant role in the direction $\theta = 180°$.

6.5 CORNER DIFFRACTION

6.5.1 Corner Diffraction from a Flat Plate

Consider the edge of finite length shown in Figure 6.11. For this discussion, let the edge be orientated along the z-axis, with Q_a located at $z = 0$ and Q_b located

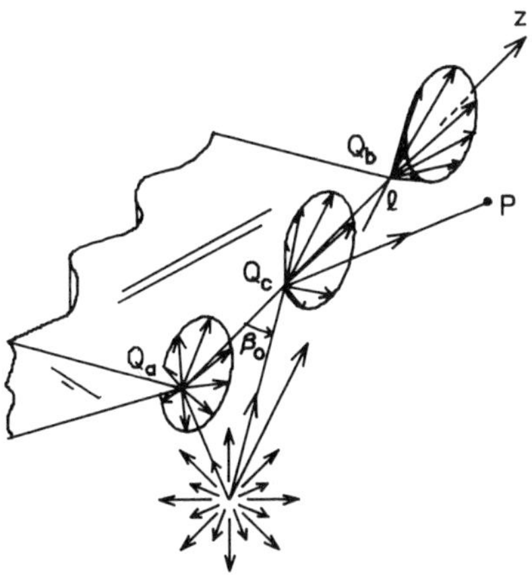

Figure 6.11 Discontinuity of diffracted fields across the vertices at the corners.

at $z = l$. From our previous discussions on wedge diffraction, we know that edge-diffracted fields will emanate from all points $Q_e(z)$ on the edge, with $0 \leqslant z \leqslant l$. A field point will be illuminated if the law of edge diffraction is satisfied; that is, given that the incident ray at Q_e subtends an angle β_0 with the edge, the field point is intercepted by a diffracted ray also subtending an angle β_0 with the edge.

However, because the diffraction occurs only from $Q_e(z)$, with $0 \leqslant z \leqslant l$, there will be a discontinuity in diffracted fields at Q_a and Q_b. The diffraction cones at Q_a and Q_b can be viewed as shadow boundaries for the edge-diffracted rays. As far as the UTD is concerned, these discontinuities can be treated by using corner diffraction and equivalent currents. Corner diffraction will be discussed in this section; equivalent currents will be treated in the next chapter.

In Chapter 4, we saw how edge-diffracted fields complemented GO fields so that the total field across GO shadow boundaries is smooth and continuous. In a similar fashion corner, diffracted fields complement edge-diffracted fields so that the total field across the shadow boundaries of the edge-diffracted fields, caused by the abrupt termination of the edge, is smooth and continuous.

This section is designed to give the reader an understanding of the way in which corner diffraction coefficients work. Note that the corner diffraction coefficients presented here [12, 13] certainly have limitations [14]; however, as the UTD matures further, better coefficients are bound to be developed. Having worked through this section, we should be able to understand and apply the new corner diffraction coefficients that surely will be forthcoming.

A corner diffraction coefficient was derived for corners on a flat plate, based on an asymptotic evaluation of the radiation integral containing equivalent currents [12]. In a flat plate (where $n = 2$), the far-zone corner diffracted fields associated with one corner and one edge, with spherical field incidence, are given by

$$\begin{bmatrix} E^c_{\beta 0} \\ E^c_{\phi} \end{bmatrix} = \begin{bmatrix} I & Z \\ M & Y \end{bmatrix} \frac{\sqrt{\sin\beta_c \sin\beta_{0c}}}{\cos\beta_{0c} - \cos\beta_c} F[kL_c a(\pi + \beta_{0c} - \beta_c)] \frac{e^{-jks}}{4\pi s} \tag{6.121}$$

where Z and Y are the free space impedance and admittance, respectively. The terms I and M refer to the equivalent currents used in the derivation, given by

$$\begin{bmatrix} I \\ M \end{bmatrix} = -\begin{bmatrix} E^i_{\beta_0'} (Q_c) \\ E^i_{\phi_i} (Q_c) \end{bmatrix} \begin{bmatrix} C_s(Q_e)\, Y \\ C_h(Q_e)\, Z \end{bmatrix} \sqrt{\frac{8\pi}{k}}\, e^{-j\pi/4} \tag{6.122}$$

where Q_c is the corner and

$$\begin{aligned} C_{s,h} = \frac{-e^{-j\pi/4}}{2\sqrt{2\pi k}\sin\beta_0} \Bigg\{ & \frac{F[kLa(\phi - \phi')]}{\cos\left(\frac{\phi - \phi'}{2}\right)} \left| F\left(\frac{[La(\phi - \phi')/\lambda]}{kL_c a(\pi + \beta_{0c} - \beta_c)}\right) \right| \\ & \mp \frac{F[kLa(\phi + \phi')]}{\cos\left(\frac{\phi + \phi'}{2}\right)} \left| F\left(\frac{[La(\phi - \phi')/\lambda]}{kL_c a(\pi + \beta_{0c} - \beta_c)}\right) \right| \Bigg\} \end{aligned} \tag{6.123}$$

The geometry for the corner diffraction coefficients is shown in Figure 6.12. Note that the incident fields in (6.122) are evaluated at the corner (Q_c) whereas the coefficients $C_{s,h}$ in (6.123) are evaluated at Q_e. More will be said about Q_e later. The distance parameters are given by

$$L = \frac{s's''}{s' + s''} \sin^2\beta_0 \tag{6.124}$$

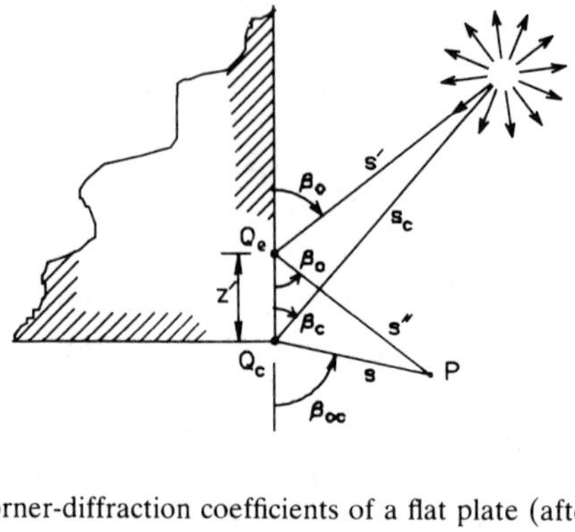

Figure 6.12 Geometry for corner-diffraction coefficients of a flat plate (after [12]).

$$L_c = \frac{\rho s}{\rho + s} \tag{6.125}$$

where s' is the distance from the source point to the diffraction point Q_e, and s'' is the distance from Q_e to the field point. In (6.125), ρ is the edge caustic, given in (6.34) with $\rho_e^i = s_c$ where s_c is the distance from the source point to the corner. The distance s is measured from the corner to the field point. When the edge is straight, we have

$$L_c = \frac{s_c s}{s_c + s} \tag{6.126}$$

From (4.70), we write

$$a(\phi \mp \phi') = 2\cos^2\left(\frac{\phi \mp \phi'}{2}\right) \tag{6.127}$$

Let us now return to Q_e. The point on the edge satisfies the law of edge diffraction for the field point to be intercepted by a diffracted ray. Q_e need not necessarily lie on the edge, but can be located on the edge extension, too. The angle β_0 is the angle measured from the edge to a ray that will be diffracted from Q_e to the field point, whether Q_e lies on the edge or not. If Q_e lies on the edge rather than on its extension, there will be an edge-diffracted ray that illuminates the field point in addition to the corner-diffracted ray; if Q_e is located on the edge extension rather than on the edge itself, there of course will not be an edge-diffracted ray intercepting the field point. In this case, Q_e still serves a mathematical purpose because it is needed to calculate s' and s''.

The angle β_c is measured from the edge at the corner to the incident ray illuminating the edge, whereas β_{0c} is the angle measured from the edge extension to the diffracted ray. Figure 6.13 [15] illustrates the case where corner-diffracted fields from both edges and only one edge-diffracted field illuminates the receiver. Note the $Q_e(b)$ is located on the edge whereas $Q_e(a)$ is located on the edge extension. The diffracted ray from edge (b) thus will illuminate the field point. Because $Q_e(a)$ is not located on the edge, no diffracted ray from edge (a) satisfies the law of edge diffraction to illuminate the field point. Figure 6.14 [15] illustrates the case where only corner-diffracted fields illuminate the field point. The case where edge-diffracted fields from both edges as well as two corner-diffracted fields illuminate the field point is illustrated in Figure 6.15 [15].

The factor:

$$\left| F\left\{\frac{[La(\phi \mp \phi')/\lambda]}{kL_c a(\pi + \beta_{0c} - \beta_c)}\right\}\right|$$

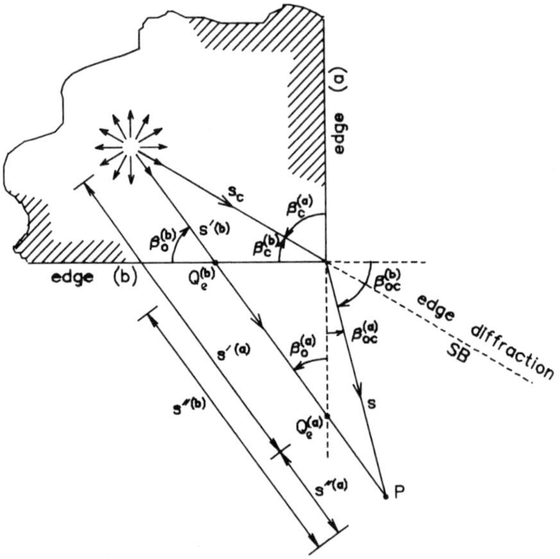

Figure 6.13 Geometry for corner-diffraction coefficients, when two corner-diffracted fields and only one edge-diffracted field intercept a field point (after [12]).

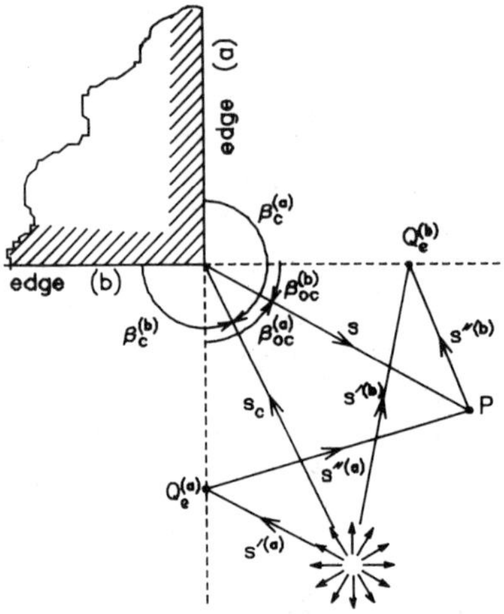

Figure 6.14 Geometry for corner-diffraction coefficients, when only two corner-diffracted fields intercept a field point (after [12]).

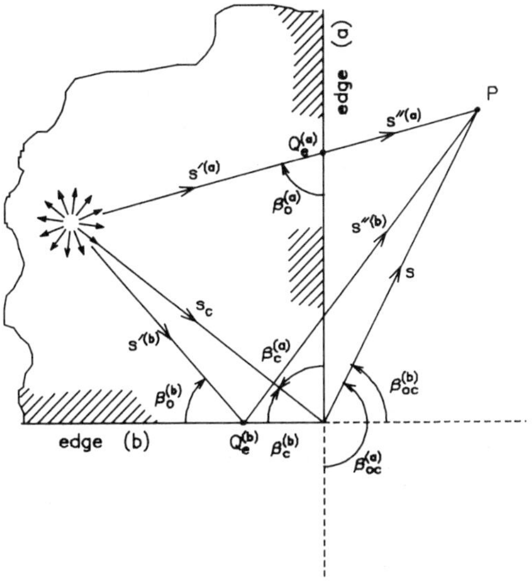

Figure 6.15 Geometry for corner diffraction coefficients, when two corner-diffracted fields and two edge-diffracted fields intercept a field point (after [12]).

in (6.123) is an empirically derived term which ensures that the corner-diffraction coefficient does not abruptly change sign when it passes through the shadow boundaries of the edge [12].

Note that the far-zone corner-diffracted fields discussed here have a $1/s$ amplitude dependence, indicating that these diffracted fields have spherical phase fronts. The corner thus acts as a point source, in contrast to an edge that acts more or less as a line source. Furthermore, the corner-diffracted fields have a $1/k$ amplitude dependence, whereas the edge-diffracted fields are proportional to $1/\sqrt{k}$. Because we assume that k is large, corner-diffracted fields therefore decay faster than edge-diffracted fields as the frequency increases. We leave as an exercise for the reader to show that corner diffracted field enforces continuity across a shadow boundary.

Equation (6.121) needs to be applied for every edge terminating in the corner; in the case of a flat plate there will be two edges. To be able to treat a more general case, equation (6.121) later will be extended for more general structures, where there may be more than two edges terminating in a corner.

Example 6.4: Monopole mounted on a thin square plate

Consider the case of a short monopole mounted on a thin square plate of side length $2d$, as shown on Figure 6.16. We shall calculate the θ-polarized radiation pattern of the structure in the plane $\Phi = 45°$. Note the similarities and differences between this case and Example 6.1, where the short monopole was mounted on a circular disk with radius a. The radiation pattern of the monopole is given in (6.40).

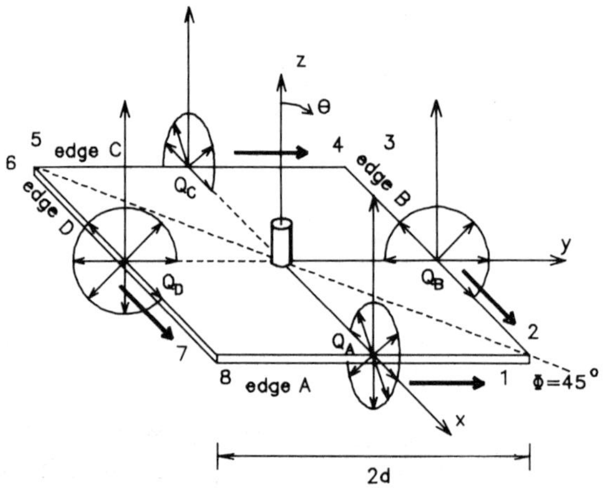

Figure 6.16 Short monopole mounted on a thin square plate, with diffracted rays in the axial region shown pronounced (the thick arrows along the edges show movement of the points of edge diffraction as $\theta \to 90°$ for radiation in the $\theta = 45°$ plane).

We have designated edges A–D and associated corners 1–8 on the plate shown in Figure 6.16. Although there are only four physical vertices, for the purposes of corner diffraction we need to apply a corner diffraction term for every edge terminating in a vertex. Because eight corner-diffracted terms will be required we identify eight corners.

The diffracted fields from the edges that contribute in the direction $\theta = 0°$ are shown in Figure 6.16. Although the monopole itself radiates no fields in this direction, there will be diffraction from each of the edges. The diffraction points are designated as Q_A, Q_B, Q_C, and Q_D. The four edge-diffracted fields cancel each other out in this direction so that the radiation pattern of the structure has a null in the direction $\theta = 0$.

However, as θ increases toward 90°, Q_A, Q_B, Q_C, and Q_D move toward the corners, as shown in Figure 6.17, to satisfy the law of edge diffraction. Consider, for instance, Q_B. The unit vector in the direction of radiation is given by (A.1) as

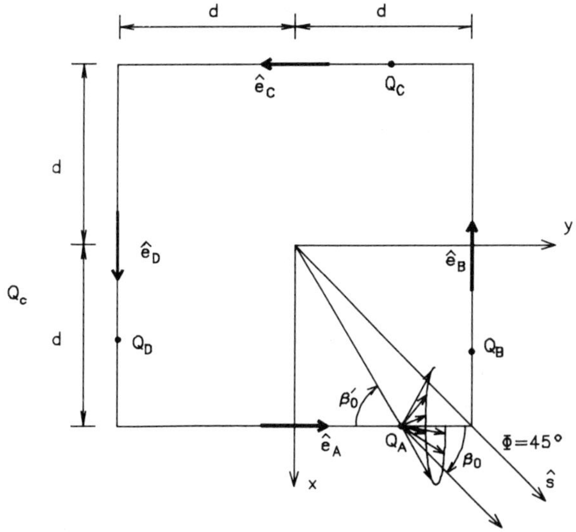

Figure 6.17 Diffracted rays from edge A in the $\Phi = 45°$ plane.

$$\hat{\mathbf{s}} = \sin\theta \cos\Phi \, \hat{\mathbf{x}} + \sin\theta \sin\Phi \, \hat{\mathbf{y}} + \cos\theta \, \hat{\mathbf{z}} \tag{6.128}$$

with $\Phi = 45°$. Because the edge vector of edge B is

$$\hat{\mathbf{e}}_B = -\hat{\mathbf{x}} \tag{6.129}$$

the angle between $\hat{\mathbf{e}}_B$ and $\hat{\mathbf{s}}$ is given by

$$\beta_0(\theta) = \arccos (\hat{\mathbf{e}}_B \cdot \hat{\mathbf{s}}) = \arccos (-\sin\theta \cos\Phi) \tag{6.130}$$

Applying the law of edge diffraction given in (6.1), we find that the position of Q_B is given by

$$\mathbf{r}_B(\theta) = \frac{-d}{\tan\beta_0(\theta)} \, \hat{\mathbf{x}} + d \, \hat{\mathbf{y}} \tag{6.131}$$

Table 6.1 gives several values for β_0 and positions of Q_B as functions of θ. We see that Q_B moves toward corner 2 and then back toward the y-axis as θ varies between 0° and 180°, as shown in Figure 6.16. Q_B is at corner 2 when $\theta = 45°$. Similar expressions can be derived for Q_A, Q_C, and Q_D.

Table 6.1
Monopole Mounted on a Thin Square Plate: Angle β_0 and Position of Q_B for Various Values of θ in the $\Phi = 45°$ Plane

θ	β_0	$\mathbf{r}_B(\theta)$
0°	90°	$y = d, \quad x = 0$
45°	120°	$y = d, \quad x = 0.58d$
90°	135°	$y = d, \quad x = d$
135°	120°	$y = d, \quad x = 0.58d$
180°	90°	$y = d, \quad x = 0$

To evaluate the edge-diffracted field from Q_B, we first have to find the unit polarization vectors. It follows from (6.2)–(6.5) that $\hat{\phi}' = \hat{\mathbf{z}}$ whereas β_0' lies in the x-y plane. From (A.2) we know that

$$\hat{\theta} = \cos\theta \cos\Phi \,\hat{\mathbf{x}} + \cos\theta \sin\Phi \,\hat{\mathbf{y}} - \sin\theta \,\hat{\mathbf{z}} \tag{6.132}$$

so that $\hat{\theta} = -\hat{\mathbf{z}}$ when $\theta = 90°$. We find that $E^i_{\beta_0'} = 0$, and hence the diffracted field from Q_B can be expressed as

$$\mathbf{E}^d_Q = -\frac{1}{2} E^i_{\phi'} D_h \, e^{jk(\mathbf{r}_B \cdot \hat{\mathbf{s}})} \frac{e^{-jkr}}{r} \hat{\phi} \tag{6.133}$$

or

$$\mathbf{E}^d_Q = \frac{1}{2} \frac{e^{-jks'}}{s'} D_h \, e^{jk(\mathbf{r}_B \cdot \hat{\mathbf{s}})} \frac{e^{-jkr}}{r} \hat{\phi} \tag{6.134}$$

where r is the distance from the monopole at the origin to the far-zone point and $s' = |\mathbf{r}_B|$. The factor 1/2 once again is due to the grazing incidence. The unit vector $\hat{\phi}$ is found by substituting $\hat{\mathbf{e}}_B$ and $\hat{\mathbf{s}}$ into (6.4), so that the θ-polarized edge-diffracted field from edge B is given by

$$E^d_{Q\theta} = \frac{1}{2} \frac{e^{-jks'}}{s'} D_h \, e^{jk(\mathbf{r}_B \cdot \hat{\mathbf{s}})} \frac{e^{-jkr}}{r} (\hat{\theta} \cdot \hat{\phi}) \tag{6.135}$$

This expression is valid in the region $0 \leqslant \theta \leqslant 180°$. To evaluate the diffraction coefficient, note that $n = 2$, $L = s'$, and $\phi' = 0$. The angle ϕ can be found from (6.12) and β_0 from (6.130). Similar expressions can be derived for the edge-diffracted fields from the other three edges. In the special case considered here, where the plate is square and $\Phi = 45°$, the symmetry of the problem can be exploited;

the edge-diffracted fields from Q_A and Q_B are the same, whereas the edge-diffracted fields from Q_C and Q_D are the same.

Figure 6.18 shows the radiation pattern of a short monopole mounted on a square plate when only the direct field from the monopole and edge-diffracted fields are taken into account. The pattern of a short monopole mounted on a infinite ground plane also is shown. The serious discontinuity in the radiation pattern of the monopole on the square plate at $\theta = 90°$ should not surprise us, because we argued earlier in this section that a discontinuity in edge-diffracted fields would occur at vertices. Table 6.1 showed that the points where edge diffraction occurred would reach the vertices when $\theta = 90°$. In the remainder of this example, we show how the use of corner diffraction terms compensate for the discontinuity and cause the total radiation pattern to be smooth and continuous.

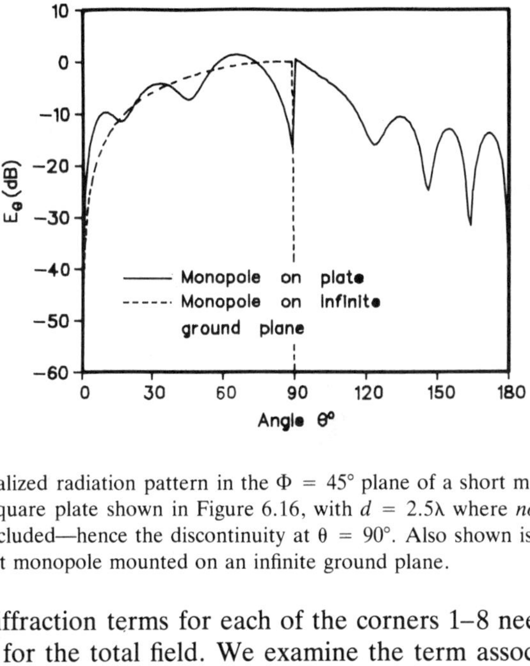

Figure 6.18 Normalized radiation pattern in the $\Phi = 45°$ plane of a short monopole mounted on the thin square plate shown in Figure 6.16, with $d = 2.5\lambda$ where *no* corner-diffracted terms are included—hence the discontinuity at $\theta = 90°$. Also shown is the radiation pattern of a short monopole mounted on an infinite ground plane.

Corner diffraction terms for each of the corners 1–8 need to be included in the expression for the total field. We examine the term associated with corner 2 in detail. Because $E^i_{\beta_0'} = 0$, it follows from (6.121) that the corner diffracted field from corner 2 is given by

$$E^c_\phi = (M\ Y)\frac{\sqrt{\sin\beta_c \sin\beta_{0c}}}{\cos\beta_{0c} - \cos\beta_c} F[kL_ca(\pi + \beta_{0c} - \beta_c)]\ e^{jk\hat{\mathbf{s}}\cdot\mathbf{r}_2}\frac{e^{-jkr}}{4\pi r} \tag{6.136}$$

where M is given in (6.122) and $\mathbf{r}_2$ is the location of corner 2; that is,

$$\mathbf{r}_2 = \sqrt{2}d\,\hat{\mathbf{x}} + \sqrt{2}d\,\hat{\mathbf{y}} \tag{6.137}$$

The θ-polarized component of E^c_ϕ is given by

$$E^c_\theta = E^c_\phi\,(\hat{\phi}\cdot\hat{\theta}) \tag{6.138}$$

This field is present in the region $0 \leqslant \theta \leqslant 180°$.

To evaluate C_h, keep in mind that the point Q_e in (6.122) actually is Q_B in this example, so that the parameters n, ϕ, ϕ', L, and β_0 are as for the wedge-diffraction terms and $s' = |\mathbf{r}_B|$. Because the edge is straight and $s \to \infty$ it follows that $L_c = s_c$, with $s_c = \sqrt{2}d$. Furthermore, $\beta_c = \pi/4$ and $\beta_{0c} = \arccos(-\hat{\mathbf{e}}_B \cdot \hat{\mathbf{s}})$ with $\hat{\mathbf{e}}_B = -\hat{\mathbf{x}}$. The incident field at the corner is given by

$$E^i_{\phi'} = -\frac{e^{-jk\sqrt{2}d}}{\sqrt{2}d} \tag{6.139}$$

The negative sign in (6.137) is because $\hat{\phi} = -\hat{\theta}$ on the perimeter of the plate and hence also at corner 2. As with the edge-diffracted fields we can exploit the symmetry of the problem in this case. The corner-diffracted fields from the following pairs of corners are the same for the particular pattern in which we are interested: 1 and 2, 3 and 8, 4 and 7 as well as 5 and 6.

Figure 6.19 shows the total θ-polarized radiated field where edge-diffracted as well as wedge-diffracted fields have been included in addition to the radiated field from the monopole itself. These results show good agreement with measured values [2]. Note that the total field is smooth and continuous across the shadow boundary at $\theta = 90°$, in contrast to the case where corner-diffracted fields had been omitted as shown in Figure 6.19. The small discontinuity at $\theta = 90°$ is due to second-order effects, as discussed previously. The amplitude of the combined corner-diffracted fields also are shown in Figure 6.19. As in the case of wedge-diffracted fields, we find that the corner-diffracted fields attain a maximum in the direction of the shadow boundary. Because the diffraction points Q_A, Q_B, Q_C, and Q_D move toward corners 1, 2, 4, and 7, respectively, as θ approaches 90°, we also find that the contributions from these corners are much more significant than those from corners 3, 5, 6, and 8. Note that the total field is bound in the axial region.

6.5.2 Corner Diffraction from a Vertex in Which Wedges with Arbitrary Wedge Angles Are Terminated

In the case of a general wedge with $n \neq 2$, the corner diffraction coefficients are given by [16]

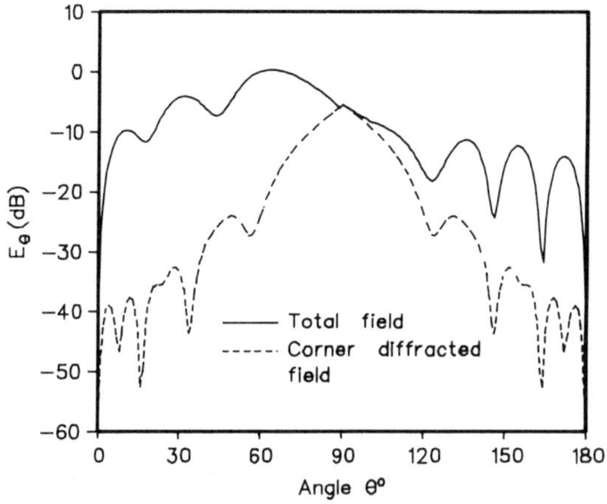

Figure 6.19 Normalized radiation pattern of a short monopole mounted on the thin square plate with $d = 2.5\lambda$ shown in Figure 6.16, where the corner-diffracted terms are included. Also shown is the pattern of all the corner-diffracted fields combined.

$$
\begin{aligned}
C_{s,h} = \frac{-e^{-j\pi/4}}{2n\sqrt{2\pi k}\,\sin\beta_0}\Bigg[&\cot\left(\frac{\pi + \beta^-}{2n}\right) F[kL^i a(\beta^-)] \\
\cdot &\left| F\left\{\frac{[L^i a(\beta^-)/\lambda}{kL_c a(\pi + \beta_{0c} - B_c)}\right\}\right| + \cot\left(\frac{\pi - \beta^-}{2n}\right) F[kL^i a(\beta^-)] \\
\cdot &\left| F\left\{\frac{[L^i a(\beta^-)/\lambda]}{kL_c a(\pi + \beta_{0c} - \beta_c)}\right\}\right| \mp \cot\left(\frac{\pi - \beta^+}{2n}\right) F[kL^{rn} a(\beta^+)] \\
\cdot &\left| F\left\{\frac{[L^{rn} a(\beta^+)/\lambda]}{kL_c a(\pi + \beta_{0c} - \beta_c)}\right\}\right| + \cot\left(\frac{\pi - \beta^+}{2n}\right) F[kL^{ro} a(\beta^+)] \\
\cdot &\left| F\left\{\frac{[L^{ro} a(\beta^+)/\lambda]}{kL_c a(\pi + \beta_{0c} - \beta_c)}\right\}\right|\Bigg]
\end{aligned}
\qquad (6.140)
$$

where $\beta^{\mp} = \phi \mp \phi'$. The functions $a(\beta^{\pm})$ are given in (4.70) and (4.71). To find the corner-diffracted field from a wedge with arbitrary wedge angle terminated in a vertex, (6.140) is substituted into (6.122). Consider for example the structure shown in Figure 6.20. There are three edges (A–C) terminating in the vertex (Q_c), so that the total corner diffracted-field from the vertex will contain three terms of the form of (6.140).

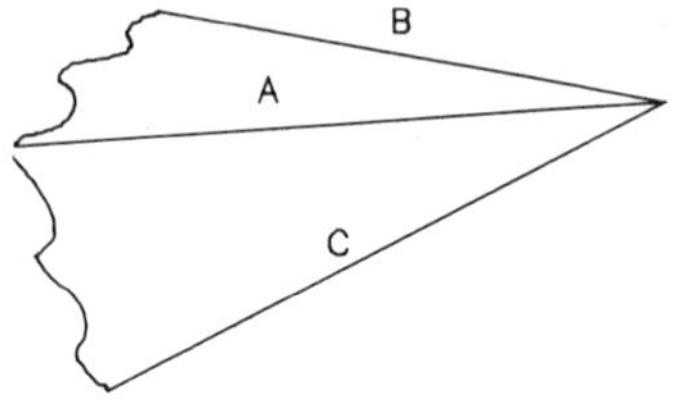

Figure 6.20 Three edges (A, B, and C) terminated in a vertex (Q_c). Each edge will give rise to a corner-diffracted field.

6.6 ALTERNATIVE FORMS OF THE DIFFRACTION COEFFICIENTS

Alternative forms of the diffraction coefficients often are found in the literature. A short summary of some of these expressions is given in this section.

The v_B form, common in older publications, is related to the diffraction coefficients when the receiver is very far away from the point of diffraction, that is, $\rho \rightarrow \infty$, and describes the total diffracted field [2]. It is given by [17]

$$v_B = [D_1 + D_2 \mp (D_3 + D_4)] \frac{e^{-jks'}}{\sqrt{s'}} \tag{6.141}$$

where D_1, D_2, D_3, and D_4 are given in (6.21)–(6.24). Because a cylindrical incident wavefront is incorporated into the v_B form, v_B is convenient for two-dimensional analysis but not for the three-dimensional case.

The function $G(n, \Psi)$ is defined [18] as

$$G(n, \Psi) = \frac{\frac{1}{n} \sin \frac{\pi}{n}}{\cos\frac{\pi}{n} - \cos\frac{\Psi}{n}} \tag{6.142}$$

The XY-coefficients are just special cases of G, given by [19]

$$X = G(n, \phi - \phi') \tag{6.143}$$

$$Y = G(n, \phi + \phi') \tag{6.144}$$

Comparing (6.143) and (6.144) with (6.20)–(6.24), and using the identity

$$\cot\left(\frac{\pi + \beta}{2n}\right) + \cot\left(\frac{\pi - \beta}{2n}\right) = \frac{-2\sin\dfrac{\pi}{n}}{\cos\dfrac{\pi}{n} - \cos\dfrac{\beta}{n}} \tag{6.145}$$

we find that

$$D_{s,h} = \frac{e^{-j\pi/4}}{\sqrt{2\pi k}\,\sin\beta_0}(X \mp Y) \tag{6.146}$$

in regions far removed from the shadow boundaries; that is, where $F \approx 1$. Note that (6.146) contains Keller's original coefficients given in (4.24).

PROBLEMS

6.1 Show that $s_t' = s' \sin\beta_0$ and $s_t = s \sin\beta_0$.

6.2 Investigate the continuity of E_P^d and E_Q^d across the shadow boundary in Example 6.1. Develop an expression for the higher-order diffraction term E_{QP}^d and show that this field enforces continuity across the shadow boundary.

6.3 In this problem we investigate the effect of the curvature of an edge on scattering by considering the compact range reflector (see Problem 4.7). The xy-projections of the top sections of three reflectors are shown in Figure 6.21. Although this is a three-dimensional problem, we consider only the diffraction from Q in the y-z plane.

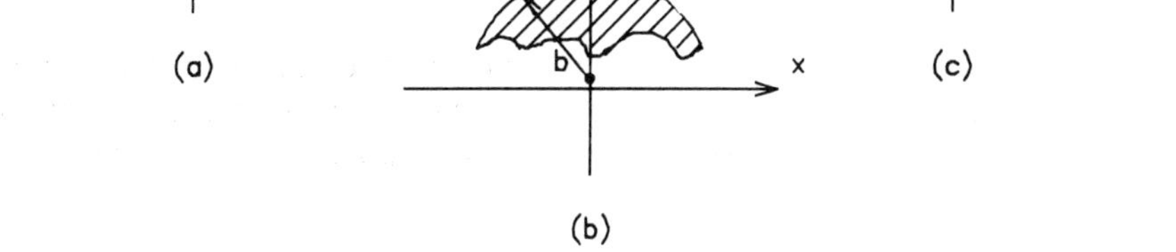

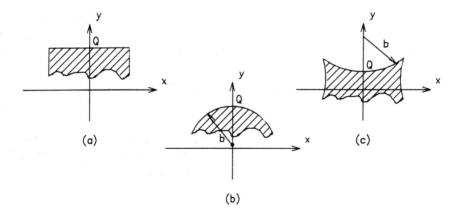

Figure 6.21 The xy-projection of the top sections of compact range reflectors with contoured edges: (a) straight edge; (b) convex edge; (c) concave edge.

Assume the same parameters for the range as in Problem 4.7. In the y-z plane the source is assumed to radiate the field

$$\mathbf{E} = \frac{e^{-jkr}}{r}\,\hat{\mathbf{y}} \tag{6.147}$$

Calculate and plot the total and diffracted fields in the target zone for various values of b and frequencies. Discuss the effect of the curvature of the edge on the diffracted fields.

6.4 Consider the spherically capped cylinder in Example 6.3. Assume that the cylinder is capped with a spheroid rather than a sphere, as shown in Figure 6.22. Plot the RCS with $d = 15.2$ cm, $b = 2$ cm, $f = 9$ GHz, and various cases of $c > b$.

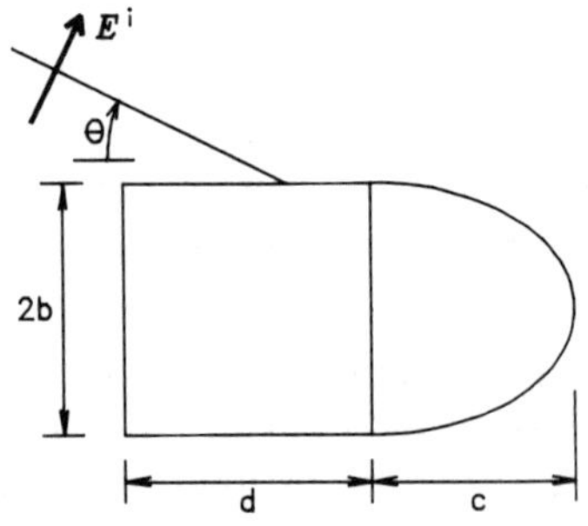

Figure 6.22 Circular cylinder with spheroidal cap.

6.5 Show that the corner diffraction coefficient enforces continuity across the edge-diffraction shadow boundary, as shown in Figure 6.23. Note that the edge shadow boundary forms a cone around the edge. The center of the cone is at the corner.

6.6 Determine an expression for the backscattered field from the ogival plate shown in Figure 6.24. The incident ray and field is in the plane of the plate.

6.7 Develop a first-order UTD solution for the backscattered field along the z-axis from the infinite pyramidal absorber cone shown in Figure 6.25. Assume that the material has a reflection coefficient of -20 dB.

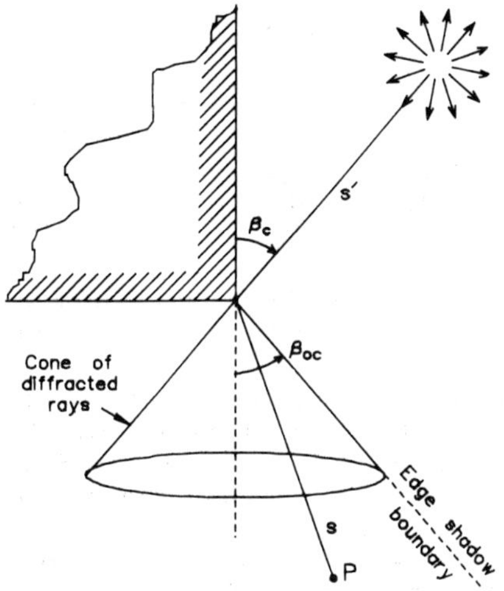

Figure 6.23 Continuity across the shadow boundary created by a corner on a flat plate.

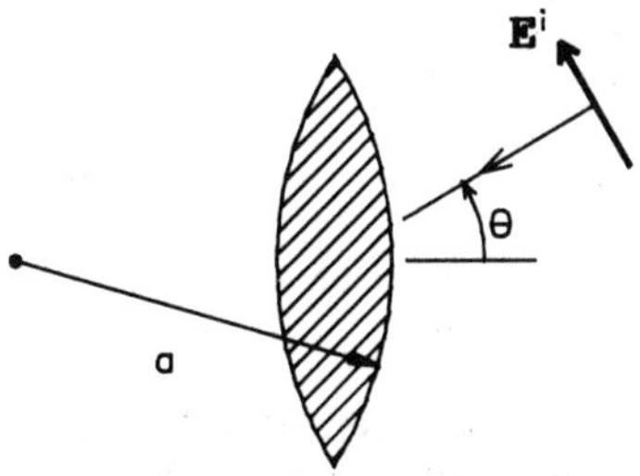

Figure 6.24 Backscatter from an ogival plate.

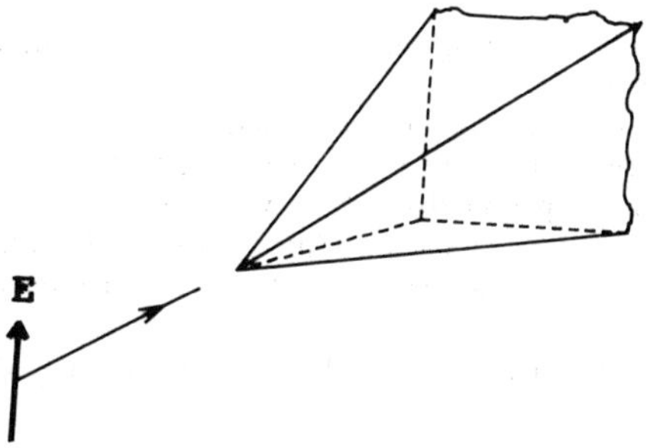

Figure 6.25 Backscatter from an infinite pyramidal absorber cone (normal incidence).

REFERENCES

[1] R.G. Kouyoumjian and P.H. Pathak, "A Uniform Geometrical Theory of Diffraction for an Edge in a Perfectly Conducting Surface," *Proc. IEEE,* Vol. 62, November 1974, pp. 1448–1461.

[2] W.D. Burnside and R.J. Marhefka, "Applications of Edge Diffraction," in short course notes on the modern geometrical theory of diffraction," Ohio State University.

[3] W.L. Stutzman and G.A. Thiele, *Antenna Theory and Design,* John Wiley & Sons, New York, 1981.

[4] C.W.I. Pistorius and W.D. Burnside, "Improved Main Reflector Design for Compact Range Applications," *IEEE Trans. Antennas and Propagation,* Vol. AP-35, No. 3, March 1987, pp. 342–347.

[5] O.M. Buyukdura, "Radiation from Sources and Scatterers Near the Edge of a Perfectly Conducting Wedge," Ph.D. dissertation, Ohio State University, 1984.

[6] L.W. Pearson, "Electromagnetic Edge Wave Due to a Point Source of Current Radiating in the Presence of a Conducting Wedge," *IEEE Trans. Antennas and Propagation,* Vol. AP-34, No. 9, September 1986, pp. 1125–1132.

[7] C.T. Tai, "Dipoles and Monopoles," in *Antenna Engineering Handbook,* R.C. Johnson and H. Jasik (eds.), 2nd ed., McGraw-Hill, New York, 1984.

[8] A.R. Lopez, "The Geometrical Theory of Diffraction Applied to Antenna Pattern and Impedance Calculations," *IEEE Trans. Antennas and Propagation,* Vol. AP-14, No. 1, January 1966, pp. 40–45.

[9] C.A. Balanis and D. de Carlo, "Monopole Antenna Patterns on Finite Sized Composite Ground Planes," *IEEE Trans. Antennas and Propagation,* Vol. AP-30, No. 3, July 1982, pp. 764–768.

[10] T.T. Chu, "First Order Uniform Theory of Diffraction Analysis of the Scattering by Smooth Structures," Ph.D. dissertation, Ohio State University, 1982.

[11] G.T. Ruck, D.E. Barrick, W.D. Stuart, and C.K. Krichbaum, *Radar Cross Section Handbook,* Plenum Press, New York, 1970.

[12] F.A. Sikta, W.D. Burnside, T.T. Chu, and L. Peters, Jr., "First-Order Equivalent Current and Corner Diffraction from Flat Plate Structures," *IEEE Trans. Antennas and Propagation,* Vol. AP-31, No. 4, July 1983, pp. 584–589.

[13] W.D. Burnside, N. Wang, and E.L. Pelton, "Near-Field Pattern Analysis of Airborne Antennas," *IEEE Transactions on Antennas and Propagation,* Vol. AP-28, No. 3, May 1980, pp. 318–327.

[14] A. Michaeli, "Comments on 'First-Order Equivalent Current and Corner Diffraction from Flat Plate Structures'," *IEEE Trans. Antennas and Propagation,* Vol. AP-32, No. 9, September 1984, p. 1011.

[15] F.A. Sikta, "UTD Analysis of Electromagnetic Scattering by Flat Plate Structures," Ph.D. dissertation, Ohio State University, 1981.

[16] B.T. de Witt, "Analysis and Measurement of Electromagnetic Scattering by Pyramidal and Wedge Absorbers," Ph.D. dissertation, Ohio State University, 1986.

[17] C.A. Balanis, *Antenna Theory,* Harper and Row, New York, 1982.

[18] C.E. Ryan and L. Peters, Jr., "Evaluation of Edge-diffracted Fields Including Equivalent Currents for the Caustic Regions," *IEEE Transactions on Antennas and Propagation,* Vol. AP-17, No. 3, May 1969, pp. 292–299.

[19] E.F. Knott, "A Progression of High-frequency RCS Prediction Techniques," *Proc. IEEE,* Vol. 73, No. 2, February 1985, pp. 252–264.

Chapter 7
Equivalent Currents

"Everyone has come across the sort of problem which seems impossible to solve until suddenly a surprisingly simple solution is revealed. Once it has been thought of, the solution is so obvious that one cannot understand why it was ever so difficult to find."

Edward de Bono,
The Use of Lateral Thinking (Penguin Books, 1977)

7.1 INTRODUCTION

The method of equivalent currents is a powerful technique in the analysis of the scattering of electromagnetic fields. The term *equivalent* refers to the fact that these currents are not physical currents, but fictitious currents used as an extremely convenient analytical tool. In addition to predicting the scattered fields in the direction of a caustic, the method allows us to calculate the diffracted fields from an edge of finite length and for observation angles away from the cone of diffracted rays. The corner-diffraction coefficients introduced in the previous chapter were derived from equivalent currents. No doubt the reader will recognize some of the terms in the corner-diffraction coefficients while working through the derivation of the equivalent current formulation. Apart from its applicability to diffracted fields, the method of equivalent currents also can be applied to find the reflected fields from curved bodies of finite length.

Equivalent currents have been the subject of many heated debates in the past, and the last word has certainly not been spoken. The intention of this chapter therefore is to introduce the reader to the concept and basic principles of equivalent currents, rather than to be an exhaustive treatise. The reader is encouraged to review the literature and keep abreast of new publications that certainly will keep appearing.

The concept of equivalent currents is not new. Millar [1–3] used equivalent currents in the 1950s, mainly to analyze the scattering from apertures. Ryan and Rudduck [4] first combined the concept of equivalent currents with the GTD in 1968. The work was extended by Ryan and Peters in 1969 [5], in a paper where the singularity at the base of a cone is treated. The controversy surrounding the initial proposal of equivalent currents in [5] can be followed in [6–10]. Burnside and Peters [7] used equivalent currents to analyze the axial radar cross section of cones, including the effects of double diffraction, as well as [10] edge-diffracted caustic fields.

Knott and Senior discuss equivalent currents as a viable technique to analyze high-frequency diffraction problems in [11]. A recent book by Knott, Schaeffer, and Tuley [12] also includes a discussion of equivalent currents. Mentzer, Peters, and Rudduck [13] showed that equivalent currents can be used to treat slope diffracted terms. More recently, in a paper by Sikta *et al.* [14], equivalent currents and corner diffraction were used to analyze the scattering from flat plate structures. Michaeli derived further forms for equivalent currents [15–18]. Comments on [15] by Knott [19] compared Michaeli's currents to Mitzner's *incremental length diffraction coefficients* (ILDC) [20], the latter being an extension of Umfimtsev's *physical theory of diffraction* (PTD). Volakis and Peters showed that equivalent currents also can be used to evaluate reflected fields in caustic regions [21, 22].

7.2 EQUIVALENT CURRENTS FOR EDGE DIFFRACTION

The derivation of the equivalent currents starts with the calculation of the diffracted field from an infinite wedge at a field point, using the UTD. A line source of unknown amplitude then is placed along the edge. Using the potential theory, the current is integrated along the infinite length of the edge to yield the value of the resulting radiated field at the field point, still in terms of the unknown amplitude. The field values at the field point obtained by UTD and the integration of the equivalent current then are equated, to express the amplitude of the equivalent current in terms of the UTD diffraction coefficient. To calculate the diffracted field from a wedge of finite length, the equivalent current then is integrated only over the finite length of the edge. We assume that the amplitude of the equivalent current that flows on the wedge with finite length is the same as that of the current that flows on the infinite wedge. Integration of the equivalent current also yields a finite value for the diffracted field at a caustic.

Consider now a wedge of infinite length as shown in Figure 7.1. Let the wedge be orientated along the z-axis of a spherical coordinate system as shown. The incident electric field is a plane wave of the form:

$$\mathbf{E}^i(\mathbf{r}) = E^i \, \mathrm{e}^{-jk(\hat{\mathbf{s}}'\cdot\mathbf{r})} \, \hat{\mathbf{e}}^i \qquad (7.1)$$

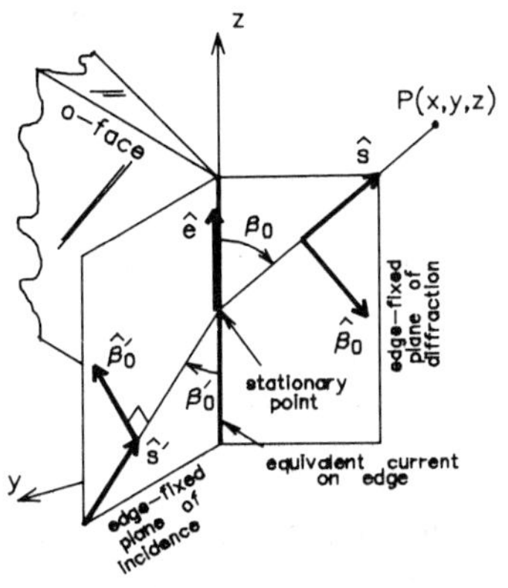

Figure 7.1 Equivalent current on the edge of a wedge.

where E^i is the constant complex amplitude of the incident electric field at the origin, $\hat{\mathbf{s}}'$ is the unit vector in the direction of propagation, $\mathbf{r}$ is the position vector of a point in space and $\hat{\mathbf{e}}^i$ is a unit vector indicating the polarization of the incident electric field. The incident field at a point $z = z'$ on the edge can thus be expressed as

$$\mathbf{E}^i(z') = E^i \, e^{-jkz'\cos\beta_0'} \hat{\mathbf{e}}^i \tag{7.2}$$

Note that $\theta = \pi - \beta_0'$ for the case illustrated in Figure 7.1. We know from previous chapters that a diffracted ray will emanate from the diffraction point located at $z = z'$ and intercept a field point $P(\mathbf{r})$.

Let us now assume that an electric current of the form:

$$\mathbf{I}^e(z') = I^e(z') \, e^{-jkz'\cos\beta_0'} \hat{\mathbf{e}} \tag{7.3}$$

flows along the edge, where $\hat{\mathbf{e}}$ is the unit vector along the edge. Note that the exponential term in (7.3) relates to the phase of the incident field along the edge and has been deliberately excluded from I^e. The problem is now to find I^e in terms of known quantities so that it will radiate the correct field at P.

The magnetic vector potential at the field point due to the current is given by [23]

$$\mathbf{A}(\mathbf{r}) = \int_{-\infty}^{\infty} \mathbf{I}^e(z') \frac{e^{-jks}}{4\pi s} \, dz' \tag{7.4}$$

where the distance from a current element of incremental length dz' at $z = z'$ to the field point is given by

$$s = \sqrt{(z' - z)^2 + \rho^2} \tag{7.5}$$

and $\rho^2 = x^2 + y^2$. Substituting (7.3) and (7.5) into (7.4), we find that

$$A_z(\mathbf{r}) = \int_{-\infty}^{\infty} \frac{I^e(z')\, e^{-jk[z'\cos\beta_0' + \sqrt{(z'-z)^2+\rho^2}]}}{4\pi\sqrt{(z' - z)^2 + \rho^2}} dz' \tag{7.6}$$

because in this case we assume that $\hat{\mathbf{e}} = \hat{\mathbf{z}}$. Equation (7.6) can be evaluated by the method of stationary phase described in Appendix D, assuming that $I^e(z')$ is slowly varying and k is large. The phase term in (7.6) can be expressed as

$$\Psi(z') = -[z' \cos\beta_0' + \sqrt{(z' - z)^2 + \rho^2}] \tag{7.7}$$

so that the derivative of the phase term is

$$\Psi'(z') = -\left[\cos\beta_0' + \frac{(z' - z)}{\sqrt{(z' - z)^2 + \rho^2}}\right] \tag{7.8}$$

At the stationary point, $\Psi'(z_s') = 0$, so that

$$\cos\beta_0' = \frac{z - z'}{\sqrt{(z' - z)^2 + \rho^2}} \tag{7.9}$$

An examination of the geometry in Figure 7.1 shows that the right-hand side of (7.9) also is equal to $\cos\beta_0$. This once again confirms the law of edge diffraction: $\beta_0' = \beta_0$. The stationary point thus also is the point where the diffracted ray intercepting the field point will diffract from the edge.

Furthermore,

$$\Psi''(z_s') = \frac{-\sin^2\beta_0}{s} \tag{7.10}$$

so that

$$\operatorname{sgn}\psi_s''(z') = -1 \tag{7.11}$$

Substituting these values into (7.10), we find that (7.6) can be asymptotically approximated as

$$A_z(\mathbf{r}) \sim \frac{I^e(z_s') \, e^{-j\pi/4}}{\sqrt{8\pi k} \, \sin\beta_0} \, e^{-jkz_s'\cos\beta_0} \, \frac{e^{-jks}}{\sqrt{s}} \tag{7.12}$$

The far-zone scattered field caused by A_z can be expressed as [24]

$$\mathbf{E}(\mathbf{r}) \approx -jkZ \, \mathbf{A}_t \tag{7.13}$$

where $\mathbf{A}_t$ is the transverse component of $\mathbf{A}$ relative to the direction of propagation of the scattered field. Hence,

$$\mathbf{A}_t = \mathbf{A} - (\mathbf{A} \cdot \hat{\mathbf{s}}) \, \hat{\mathbf{s}} \tag{7.14}$$

where $\hat{\mathbf{s}}$ is a unit vector in the direction of propagation of the scattered field. In the spherical coordinate system of Figure 7.1, $\theta = \beta_0$ and $\hat{\theta} = \hat{\beta}_0$. The unit vector $\hat{\mathbf{s}}$ is given by (A.1) as

$$\hat{\mathbf{s}} = \sin\theta \, \cos\Phi \, \hat{\mathbf{x}} + \sin\theta \, \sin\Phi \, \hat{\mathbf{z}} + \cos\theta \, \hat{\mathbf{z}} \tag{7.15}$$

As earlier, the symbol Φ is used to indicate the angle phi in spherical coordinates whereas the symbol ϕ is used to indicate the angle measured from the *o*-face in the diffraction coefficients. Substituting these values and (7.14) into (7.13) and keeping in mind that

$$\hat{\theta} = \cos\theta \, \cos\Phi \, \hat{\mathbf{x}} + \cos\theta \, \sin\Phi \, \hat{\mathbf{y}} - \sin\theta \, \hat{\mathbf{z}} \tag{7.16}$$

the radiated field from the current can be expressed as

$$E_{\beta_0} = jkZ \, \sin\beta_0 \, A_z \tag{7.17}$$

However, the $\hat{\beta}_0$-polarized component of the scattered field also can be found from our knowledge of the UTD. The field diffracted from the stationary point satisfying (7.9) can be expressed as

$$E_{\beta_0} = -E^i_{\beta_0'} \, e^{-jkz_s'\cos\beta_0} \, D_s \, \frac{e^{-jks}}{\sqrt{s}} \tag{7.18}$$

where $E^i_{\beta_0'}$ is the complex amplitude of the $\hat{\beta}_0'$-polarized incident field at the phase origin and D_s is the UTD soft diffraction coefficient at the point of diffraction, given by (6.20). We now substitute (7.12) into (7.17) and equate this to (7.18) to find

$$I^e(z') = -\frac{E^i_{\beta_0'}(z')}{Z} \sqrt{\frac{8\pi}{k}}\, e^{-j\pi/4}\, D_s(z') \tag{7.19}$$

We leave as an exercise to show that

$$E^i_{\beta_0'} = \frac{\hat{\mathbf{e}}^i \cdot \hat{\mathbf{e}}}{\sin\beta_0}\, E^i \tag{7.20}$$

The equivalent electric current is obtained by substituting (7.20) into (7.19):

$$I^e(\zeta) = -\frac{(\hat{\mathbf{e}}^i \cdot \hat{\mathbf{e}})}{Z \sin\beta_0}\, E^i(\zeta) \sqrt{\frac{8\pi}{k}}\, e^{-j\pi/4}\, D_s(\zeta) \tag{7.21}$$

$$\mathbf{I}^e(\zeta) = I^e(\zeta)\, \hat{\mathbf{e}} \tag{7.22}$$

where ζ is the length parameter measured along the current.

Comparison of (7.22) with (7.3) shows that the term $e^{-jkz'\cos\beta_0}$ is missing from (7.22). To make these two equations compatible, $E^i(\zeta)$ in the general expression given in (7.21) should be interpreted as having the correct phase at ζ, rather than just being a constant amplitude term. Note that (7.22) applies not only for an edge orientated along the z-axis, but also for an edge orientated along in the arbitrary direction $\hat{\mathbf{e}}$; hence, we have the change from z' to ζ. The parameters of D_s will be those applicable at ζ.

The substitution of (7.19) and (7.12) into (7.17) to find the scattered field may seem to be a pointless exercise of going around in circles. However, it enabled us to decompose the diffraction process, originally defined as an infinitely long process, into an infinitesimal phenomenon. Consequently, for an edge of finite length, the integration in (7.6) is accomplished only over the finite length of the edge to yield the diffracted field, where I^e given in (7.21) is assumed to flow along the edge. In this case, the method of equivalent currents is used as an alternative to wedge and corner diffraction. Equivalent currents also can be used to evaluate fields in caustic regions where wedge diffraction fails.

The reader needs to realize that these currents are artifices and not real. This is illustrated by the fact that D_s depends on the observation angle, and consequently so does the amplitude of the equivalent current. Real currents would be independent of observation angle.

A procedure similar to the preceding one can be followed to find the equivalent magnetic current. Let the magnetic line current have the form:

$$\mathbf{I}^m(z') = I^m(z')\, e^{-jkz'\cos\beta_0'}\, \hat{\mathbf{e}} \tag{7.23}$$

The electric vector potential at the field point due to the current is given by [23]:

$$\mathbf{F}(\mathbf{r}) = \int_{-\infty}^{\infty} \mathbf{I}^m(z') \frac{e^{-jks}}{4\pi s}\, dz' \tag{7.24}$$

Following (7.12), we find that

$$\mathbf{F}(\mathbf{r}) \sim \frac{I_s^m(z')\, e^{-j\pi/4}}{\sqrt{8\pi k}\,\sin\beta_0}\, e^{-jkz_s'\cos\beta_0} \frac{e^{-jks}}{\sqrt{s}}\, \hat{\mathbf{e}} \tag{7.25}$$

In the far-zone, the scattered magnetic field can be expressed as

$$\mathbf{H}(\mathbf{r}) \approx -jkY\, \mathbf{F}_t \tag{7.26}$$

The corresponding scattered electric field is given by

$$\mathbf{E} = Z(\mathbf{H} \times \hat{\mathbf{s}}) \tag{7.27}$$

so that

$$\mathbf{E} = -jk(\mathbf{F} \times \hat{\mathbf{s}}) \tag{7.28}$$

Because the vector cross product between $\mathbf{F}$ and $\hat{\mathbf{s}}$ is taken, there is no need to calculate the transverse component of $\mathbf{F}$, as the cross product operation eliminates all components of $\mathbf{F}$ in the direction of $\hat{\mathbf{s}}$. In this case $\mathbf{F} = F_Z\, \hat{\mathbf{z}}$ and $\hat{\mathbf{s}}$ is given in (7.15), so that

$$E_\phi = -jk\, \sin\beta_0\, F_Z \tag{7.29}$$

Note that $\hat{\Phi} = \hat{\phi}$ in this case.

The $\hat{\phi}$-polarized component of the scattered field also can be expressed as

$$E_\phi = -E^i_{\phi'}\, e^{-jkz_s'\cos\beta_0}\, D_h \frac{e^{-jks}}{\sqrt{s}} \tag{7.30}$$

where $E^i_{\phi'}$ is the complex amplitude of the $\hat{\phi}'$-polarized incident field at the phase origin. D_h is the UTD hard diffraction coefficient at the point of diffraction, given by (6.20). Equating (7.29) and (7.30), we find that

$$I^m(z') = -E^i_{\phi'} \sqrt{\frac{8\pi}{k}}\, e^{-j\pi/4}\, D_h(z') \tag{7.31}$$

or

$$I^m(\zeta) = -\frac{[\hat{\mathbf{e}}^i \cdot (\hat{\mathbf{e}} \times \hat{\mathbf{s}}')]}{\sin\beta_0} E^i(\zeta) \sqrt{\frac{8\pi}{k}}\, e^{-j\pi/4}\, D_h(\zeta) \tag{7.32}$$

so that

$$\mathbf{I}^m(\zeta) = I^m(\zeta)\, \hat{\mathbf{e}} \tag{7.33}$$

For (7.33) and (7.23) to be the same, $E^i(\zeta)$ in (7.32) should be interpreted as having the correct phase at ζ and not merely being a constant amplitude term. We leave as an exercise to show that

$$E^i_{\phi'} = \frac{-\hat{\mathbf{e}}^i \cdot (\hat{\mathbf{e}} \times \hat{\mathbf{s}}')}{\sin\beta_0} E^i \tag{7.34}$$

7.3 RADIATION FROM EQUIVALENT CURRENTS

Consider now a straight wedge of finite length l orientated in the direction $\hat{\mathbf{e}}$ in space. To calculate the diffraction from the wedge we assume an equivalent current to flow along the entire length of the wedge. For the moment we consider only the equivalent electric currents, which are used instead of wedge diffraction and corner diffraction.

The magnetic vector potential at a field point $P(\mathbf{r})$ due to the equivalent electric current in (7.22) is given by

$$\mathbf{A}(\mathbf{r}) = \int_{-l/2}^{l/2} \mathbf{I}(\mathbf{r}') \frac{e^{-jk(r-\hat{\mathbf{s}}\cdot\mathbf{r}')}}{4\pi r}\, dr' \tag{7.35}$$

where the local phase reference is chosen to be at the center of the wedge. The distance r is measured from the phase center to the field point, $\mathbf{r}'$ is the position vector of a current filament on the wedge, r' is a linear distance along the wedge, and $\hat{\mathbf{s}}$ is the direction of propagation of the radiated field. Substituting (7.1) and (7.22) into (7.35) leads to

$$\mathbf{A} = \hat{\mathbf{e}} \frac{e^{-jkr}}{4\pi r} \int_{-l/2}^{l/2} I^e(\zeta)\, e^{jkg\zeta}\, d\zeta \tag{7.36}$$

where

$$g = \hat{\mathbf{e}} \cdot (\hat{\mathbf{s}} - \hat{\mathbf{s}}') \tag{7.37}$$

In the case where the scattered field propagates in a direction that lies on the diffraction cone, we can see that $g = 0$ (see Problem 7.2). However, equivalent currents for arbitrary observation angles have been proposed, in which case $g \neq 0$. Because this chapter essentially is tutorial in nature, we do not summarily set $g = 0$, but instead evaluate the integral for the case where $g \neq 0$.

In the case where I^e is constant over the length of the wedge, (7.36) reduces to

$$\mathbf{A} = \hat{\mathbf{e}} \frac{e^{-jkr}}{4\pi r} I^e \int_{-l/2}^{l/2} e^{jkg\zeta}\, d\zeta \tag{7.38}$$

or

$$\mathbf{A} = A_e\, \hat{\mathbf{e}} \tag{7.39}$$

with

$$A_e = I^e\, l\, \mathrm{sinc}\left(\frac{kgl}{2}\right) \frac{e^{-jkr}}{4\pi r} \tag{7.40}$$

I^e is given in (7.21) and $\mathrm{sinc}(x) = \sin(x)/x$. From (7.40) and (7.13), the scattered field from the edge can be expressed as

$$\mathbf{E}_e = -jkZ\, A_e\, [\hat{\mathbf{e}} - (\hat{\mathbf{e}} \cdot \hat{\mathbf{s}})\hat{\mathbf{s}}] \tag{7.41}$$

Similarly, we can find the diffracted field due to the magnetic equivalent current. The electric vector potential at a field point $P(\mathbf{r})$ due to the equivalent current in (7.33) is given by

$$\mathbf{F}(\mathbf{r}) = \int_{-l/2}^{l/2} \mathbf{I}^m(\mathbf{r}') \frac{e^{-jk(r - \hat{\mathbf{s}}\cdot\mathbf{r}')}}{4\pi r}\, dr' \tag{7.42}$$

Substituting (7.33) into (7.42) leads to

$$\mathbf{F} = \hat{\mathbf{e}} \frac{e^{-jkr}}{4\pi r} \int_{-l/2}^{l/2} I^m(\zeta)\, e^{jkg\zeta}\, d\zeta \tag{7.43}$$

In the case where I^m is constant over the length of the wedge, (7.43) reduces to

$$\mathbf{F} = \hat{\mathbf{e}} \frac{e^{-jkr}}{4\pi r} I^m \int_{-l/2}^{l/2} e^{jkg\zeta}\, d\zeta \tag{7.44}$$

or

$$\mathbf{F} = F_e\, \hat{\mathbf{e}} \tag{7.45}$$

with

$$F_e = I^m l \,\mathrm{sinc}(kgl/2) \frac{e^{-jkr}}{4\pi r} \tag{7.46}$$

and I^m given in (7.32). From (7.28), we thus have

$$\mathbf{E}_m = -jk\, F_e\, (\hat{\mathbf{e}} \times \hat{\mathbf{s}}) \tag{7.47}$$

In general, the total field diffracted from the edge can be expressed as

$$\mathbf{E} = \mathbf{E}_e + \mathbf{E}_m \tag{7.48}$$

or

$$\mathbf{E} = -jkl \,\mathrm{sinc}\left(\frac{kgl}{2}\right) [Z\, I^e[\hat{\mathbf{e}} - (\hat{\mathbf{e}} \cdot \hat{\mathbf{s}})\hat{\mathbf{s}}] + I^m(\hat{\mathbf{e}} \times \hat{\mathbf{s}})] \frac{e^{-jkr}}{4\pi r} \tag{7.49}$$

When applying (7.49) take note of the assumptions upon which it is based, particularly the plane wave incidence, far-zone radiation condition, and the assumption that $I^{e,m}$ and $D_{s,h}$ are constant over the length of the edge. If these cases are not met, we can readily derive equations that hold for a particular situation. Integrals for which closed form solutions cannot be found often can be solved numerically.

Keep in mind that the equivalent currents, as derived here, yield the diffracted field in the direction of the diffraction cones. As mentioned previously, $g = 0$ in this case, so that $\operatorname{sinc}(kgl/2) = 1$ in (7.49). However, equivalent currents also can be applied to find the scattered fields in other directions. The reader is referred specifically to [11] and [15]. Equivalent currents have been successfully applied to calculate RCS in directions away from the diffraction cone. In [14] the RCS from flat plate structures is calculated, and a stripping technique is applied to the plates in which the inner product of the incident field is taken with a factor that decomposes the plate into strips with edges either perpendicular or parallel to the direction of incidence.

Example 7.1: Backscatter from an edge on a thin plate

In this example we calculate the far-zone backscattered field from an edge on a thin plate with the incident ray perpendicular to the edge of the plate and the incident field polarized in the plane of the plate; that is, parallel to the edge of the plate. Consider edge *A* of length *l* on the thin plate shown in Figure 7.2. Assume that this edge is located along the *z*-axis and that the incident field is a θ-polarized plane wave in the plane of the plate. The edge thus is a caustic, because an infinite number of diffracted rays will scatter back toward the source, and therefore regular wedge diffraction cannot be used to determine the backscattered field from this edge for normal incidence.

The scattered field is given by (7.49) and $\hat{\mathbf{s}}$ is given by (7.15), with $\phi = \pi$, $\theta = \pi/2$, and $\hat{\mathbf{e}} = \hat{\mathbf{z}}$. It follows through inspection that $\beta_0 = \pi/2$. Because we are dealing with the backscatter case, $\hat{\mathbf{s}}' = -\hat{\mathbf{s}}$ and $\phi = \phi'$. The incident field can be expressed as

$$\mathbf{E}^i = -E^i \, e^{-jkx} \, \hat{\mathbf{z}} \tag{7.50}$$

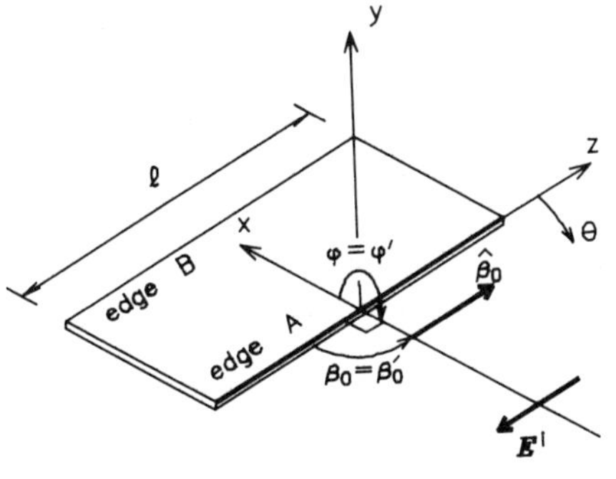

Figure 7.2 Backscatter from an edge of finite length: normal incidence.

where $x = 0$ on the edge. Substituting (7.50) into (7.20) and (7.34), leads to $E^i_{\beta 0} = -E^i$ and $E^i_{\phi'} = 0$. From (7.31), we know that $I^m = 0$. Because $s', s \to \infty$, the transition functions in the diffraction coefficient in the expression for the equivalent current tend to unity so that the diffraction coefficient reduces to

$$D_s = \frac{-e^{-j\pi/4}}{\sqrt{2\pi k}} \tag{7.51}$$

Note that D_s is a constant, which is one of the assumptions upon which the derivation of I^e in (7.19) is based. The equivalent electric current therefore is given by

$$I^e = j\frac{2E^i}{kZ} \tag{7.52}$$

Furthermore, from (7.15) and (7.37), we know that $g = 0$. We also can see that $[\hat{\mathbf{e}} - (\hat{\mathbf{e}} \cdot \hat{\mathbf{s}})\hat{\mathbf{s}}] = \hat{\mathbf{z}}$. Substituting these values into (7.49), we find that the scattered field from the edge can be expressed as

$$\mathbf{E} = E^i \frac{l}{2\pi} \frac{e^{-jkr}}{r} \hat{\mathbf{z}} \tag{7.53}$$

where r is the distance from the phase origin on the edge to the field point. The broadside ($\theta = \pi/2$) RCS of the edge for a θ-polarized incident field therefore is

$$\sigma = \frac{l^2}{\pi} \tag{7.54}$$

For this polarization and at grazing incidence, edge B does not scatter back toward the observer because the incident field is zero, having been shorted out by the plate.

Consider now the case where the incident field is polarized perpendicular to the edge; that is, Φ-polarized. This corresponds to the hard case, where it can be shown that $D_h = 0$ at edge A, so that zero field is scattered from this edge. However, we find that there is a finite backscattered field component from the back edge (edge B in Figure 7.2) for this polarization (see Problem 7.3).

Example 7.2: Backscatter from the ring caustic at the joint between base of a cone and a cylinder

Consider the geometry of a right circular cylinder of radius a capped by a circular cone as shown in Figure 7.3. We need to calculate the backscatter from the body

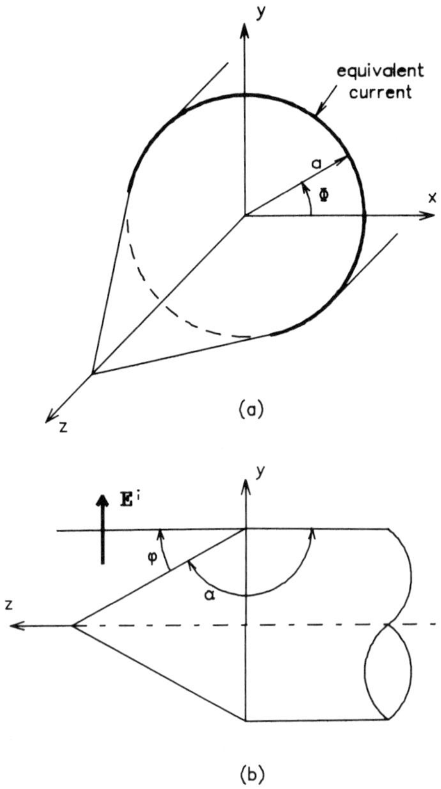

Figure 7.3 Equivalent current on the joint between a cone and a cylinder: (a) oblique view; (b) side view.

when it is illuminated with a plane wave propagating in the axial direction (the negative z-axis).

In this example, we consider only diffraction from the joint between the two bodies. Clearly, the joint is a caustic because rays from the entire joint will diffract back in the z-direction. This problem is similar to that described in Example 6.1, where a short monopole was mounted on a circular disk. There we found that the rim of the disk was a caustic when the radiation was in the axial direction. Because the joint was a caustic, wedge diffraction could not be used to calculate the backscatter from it, and therefore we had to resort to equivalent currents.

Without loss of generality we can assume that the incident plane wave is given by

$$\mathbf{E} = E^i \, e^{jkz} \, \hat{\mathbf{y}} \tag{7.55}$$

In this case,

$$\hat{\mathbf{e}} = \hat{\Phi} = -\sin\Phi\,\hat{\mathbf{x}} + \cos\Phi\,\hat{\mathbf{y}} \tag{7.56}$$

and $\beta_0 = \pi/2$. Substituting these values into (7.22), we find that the electrical equivalent current is given by

$$\mathbf{I}^e = -\frac{E^i}{Z}\cos\Phi\sqrt{\frac{8\pi}{k}}\,e^{-j\pi/4}\,D_s\,\hat{\mathbf{e}} \tag{7.57}$$

The diffraction coefficient D_s is not a function of the angle Φ, so that we find from (7.4) and (7.57) that

$$\mathbf{A} = -\frac{E^i}{Z}\sqrt{\frac{8\pi}{k}}\,e^{-j\pi/4}\,\frac{e^{-jkr}}{r}\,D_s\int_0^{2\pi}(-\hat{\mathbf{x}}\sin\Phi\cos\Phi + \hat{\mathbf{y}}\cos^2\Phi)\,a\,d\Phi \tag{7.58}$$

or

$$\mathbf{A} = -\frac{E^i}{Z}\,D_s\,e^{-j\pi/4}\sqrt{\frac{8\pi}{k}}\,a\,\frac{e^{-jkr}}{r}\,\hat{\mathbf{y}} \tag{7.59}$$

Substituting (7.59) into (7.13), we find that the far-zone backscattered field due to the equivalent electric current at a distance r from the origin can be expressed as

$$\mathbf{E}_e = ja\,E^i\sqrt{\frac{\pi k}{2}}\,D_s\,e^{-j\pi/4}\,\frac{e^{-jkr}}{r}\,\hat{\mathbf{y}} \tag{7.60}$$

Similarly, we find that the magnetic equivalent current is given by

$$\mathbf{I}^m = E^i\sin\Phi\sqrt{\frac{8\pi}{k}}\,e^{-j\pi/4}\,D_h\,\hat{\mathbf{e}} \tag{7.61}$$

so that the electric vector potential can be expressed as

$$\mathbf{F} = -E^i\,D_h\,e^{-j\pi/4}\sqrt{\frac{8\pi}{k}}\,a\,\frac{e^{-jkr}}{r}\,\hat{\mathbf{x}} \tag{7.62}$$

Substituting (7.62) into (7.28) we find that the far-zone backscattered field due to the equivalent magnetic current at a distance r from the origin can be expressed as

$$\mathbf{E}_m = -ja\, E^i \sqrt{\frac{\pi k}{2}}\, D_h\, e^{-j\pi/4} \frac{e^{-jkr}}{r}\, \hat{\mathbf{y}} \tag{7.63}$$

The total backscattered field from the rim is therefore given by

$$\mathbf{E} = \mathbf{E}_e + \mathbf{E}_m \tag{7.64}$$

or

$$\mathbf{E} = -ja\, E^i \sqrt{\frac{\pi k}{2}}\, (D_h - D_s)\, e^{-j\pi/4} \frac{e^{-jkr}}{r}\, \hat{\mathbf{y}} \tag{7.65}$$

We can show that this result can be expressed as

$$\mathbf{E} = -a\, E^i\, G(n, \beta^+) \frac{e^{-jkr}}{r} \tag{7.66}$$

where $\beta^+ = 2\phi$. Substituting this value into (6.142), we find that

$$G(n, \beta^+) = \frac{\dfrac{\sin(\pi/n)}{n}}{\cos(\pi/n) - \cos(2\phi/n)} \tag{7.67}$$

In this case,

$$2\phi/n = \pi\,(1 - 3/2n) \tag{7.68}$$

where $\alpha = (2 - n)\pi$ is the wedge angle at the joint as shown in Figure 7.3. The expression for the scattered field in (7.66) is in the form given in [5].

The more general case, where the backscattered field is found when the incident ray subtends a small angle θ with the z-axis, also is discussed in [5] and offered as an advanced problem for the reader. In the case where $\alpha = \pi/2$, the cone degenerates into a disk. In this case, there also will be specular reflection from the disk in the z direction. Ryan and Peters [5] showed that equivalent currents then still can be used to find the backscattered field. Burnside and Peters [7] showed that second-order diffractions across the base of a truncated cone also can be taken into account by an equivalent current to improve the solution in the axial region. The reader is referred to [6–10] for additional discussions on the diffraction from cones.

Example 7.3: Axial field of a short monopole mounted on a circular disk

The radiated fields from a short monopole mounted on a circular disk of radius a was discussed in Example 6.1. The geometry was shown in Figure 6.5 with the radiated field shown in Figure 6.7. We saw that the wedge-diffracted fields became singular in the axial regions where $\theta \approx 0°$ and $\theta \approx 180°$. This is to be expected as the rim of the disk is a caustic at these aspect angles; that is, diffracted rays seem to emanate from the entire rim rather than just from two discrete diffraction points P and Q. In this example, we see how equivalent currents can be used to find a bounded solution for the field in the axial region. The field radiated by the short monopole is given by (6.40).

Let us start by considering the equivalent electric current given by (7.22). In this case, $\hat{\mathbf{e}}^i = \hat{\theta}$ and $\hat{\mathbf{e}} = \Phi$ so that $\hat{\mathbf{e}}^i \cdot \hat{\mathbf{e}} = 0$. It therefore follows from (7.21) that $I^e = 0$. This is to be expected because I^e is associated with D_s, and from Example 6.1 we know that the diffraction coefficient applicable in this problem is D_h.

By substituting

$$\hat{\mathbf{s}}' = \cos\Phi\ \hat{\mathbf{x}} + \sin\Phi\ \hat{\mathbf{y}} \tag{7.69}$$

and the values of $\hat{\mathbf{e}}^i$ and $\hat{\mathbf{e}}$ into (7.33), we find that

$$\mathbf{I}^m = -\sqrt{\frac{8\pi}{\mathbf{k}}}\, e^{-j\pi/4}\, D_h\, \frac{e^{-jka}}{a}\, \hat{\Phi} \tag{7.70}$$

where

$$\hat{\Phi} = -\sin\Phi\hat{\mathbf{x}} + \cos\Phi\hat{\mathbf{y}} \tag{7.71}$$

The electric vector potential is given by

$$\mathbf{F} = \int_0^{2\pi} \mathbf{I}^m \frac{e^{-jks}}{4\pi s}\, a\, d\Phi \tag{7.72}$$

with

$$s = r - \hat{\mathbf{s}} \cdot \mathbf{r}' \tag{7.73}$$

where

$$\hat{\mathbf{s}} = \sin\theta\ \hat{\mathbf{x}} + \cos\theta\ \hat{\mathbf{z}} \tag{7.74}$$

$$\mathbf{r}' = a\cos\Phi\ \hat{\mathbf{x}} + a\sin\Phi\ \hat{\mathbf{y}} \tag{7.75}$$

We calculate the radiated field from the structure not only at the angles $\theta = 0°$ and $\theta = 180°$, but also in a small angular region around these angles. To be compatible with Example 6.1, we shall stay in the x-z plane. As an approximation, we assume that D_h is a constant and equal to its value at $\theta = 0°$ and $\theta = 180°$, respectively. Consequently,

$$\mathbf{F} = -\frac{e^{-j\pi/4}}{\sqrt{2\pi k}} D_h \, e^{-jka} \frac{e^{-jkr}}{r} \mathbf{C} \tag{7.76}$$

where

$$\mathbf{C} = \int_0^{2\pi} e^{ju\cos\Phi} \, \hat{\Phi} \, \mathbf{d}\Phi \tag{7.77}$$

$$u = ka \sin\theta \tag{7.78}$$

Applying (7.28) to (7.76), we find that

$$E_\theta^d(\theta) = j \sqrt{\frac{k}{2\pi}} \, e^{-j\pi/4} D_h \, e^{-jka} \frac{e^{-jkr}}{r} B(u) \tag{7.79}$$

where

$$B(u) = \int_0^{2\pi} e^{ju\cos\Phi} \cos\Phi \, d\Phi \tag{7.80}$$

We can show that

$$B(u) = 2\pi j \, J_1(u) \tag{7.81}$$

where J_1 is the Bessel function of the first kind of order one (see Problem 7.5). Keeping in mind that we have grazing incidence, $\beta_0 = \pi/2$, and that the disk is flat, we find that the diffraction coefficient can be expressed as

$$D_h = \frac{-e^{-j\pi/4}}{2\sqrt{2\pi k} \cos(\xi/2)} \tag{7.82}$$

where $\xi = \pi/2$ when $\theta \approx 0°$ and $\xi = 3\pi/2$ when $\theta \approx 180°$. Substituting (7.80) and (7.82) into (7.79), we find that the θ-component of the diffracted field from the structure in the axial region is given by

$$E_\theta^d(\theta) = j \, e^{-jka} J_1(ka \sin\theta) \, G(2, \xi) \frac{e^{-jkr}}{r} \tag{7.83}$$

where the function $G(2, \xi)$ is given by (6.142) as

$$G(2, \xi) = \frac{-1}{2\cos(\xi/2)} \tag{7.84}$$

The total radiated field from the structure in the region $\theta \approx 0°$ is the sum of (6.40) and (7.83). Because the lower half space is not illuminated by the monopole, (7.83) represents the total field in the axial region around $\theta \approx 180°$. Figure 7.4 shows the axial solution obtained by using equivalent currents as well as the wedge-diffraction solution calculated in Example 6.1. Normally we would use the axial solution for θ not too far removed from 0°, and then switch to the wedge-diffraction solution. The bottom of the first null would be a sensible place to switch between the two solutions. Similarly, we can switch to the axial solution in the last null before $\theta = 180°$.

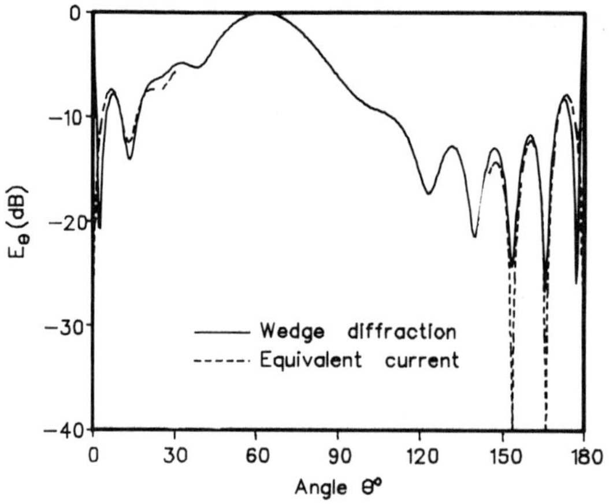

Figure 7.4 Normalized radiation pattern from a short monopole mounted on a circular disk of radius 2.5λ, showing equivalent current solution in the axial region as well as the two-point wedge diffraction solution.

7.4 REFLECTED FIELDS USING EQUIVALENT CURRENTS

In the previous sections we saw how equivalent currents could be used to evaluate diffracted fields in caustic regions. However, caustic regions also can occur when evaluating reflected fields, in which case geometrical optics cannot be used to calculate the fields. A typical example would be that of a finite cylinder illuminated by a plane wave. Volakis and Peters [21] have shown that a set of equivalent currents similar to those used for edge diffraction can be used to evaluate reflected fields in caustic regions. These currents produce the same result as geometrical optics, but remain valid when the GO solution fails.

Consider the conducting cylinder of infinite length and radius a shown in Figure 7.5. To simplify the mathematics somewhat, the z-axis has been placed on the specular line rather than along the axis of the cylinder. Let the incident field be a plane wave polarized parallel to the axis of the cylinder that propagates from a direction perpendicular to the axis of the cylinder. The incident field can then be expressed as

$$\mathbf{E}^i(\mathbf{r}) = E^i \, e^{-jk(\hat{\mathbf{s}}' \cdot \mathbf{r})} \, \hat{\mathbf{z}} \tag{7.85}$$

where $\hat{\mathbf{s}}'$ is the direction of propagation of the incident field.

Fields will be reflected from the entire specular line. The direction of propagation of the reflected rays, of course, will be such that the angle of incidence is equal to angle of reflection, given by (3.23) as

$$\hat{\mathbf{s}} = \hat{\mathbf{s}}' - 2(\hat{\mathbf{s}}' \cdot \hat{\mathbf{n}})\hat{\mathbf{n}}$$

where $\hat{\mathbf{n}}$ is the unit vector normal to the surface at the point of reflection. The field along a reflected ray can be expressed as

$$\mathbf{E}(s) = -E^i \sqrt{\rho_1^r} \, \frac{e^{-jks}}{\sqrt{s}} \, \hat{\mathbf{z}} \tag{7.86}$$

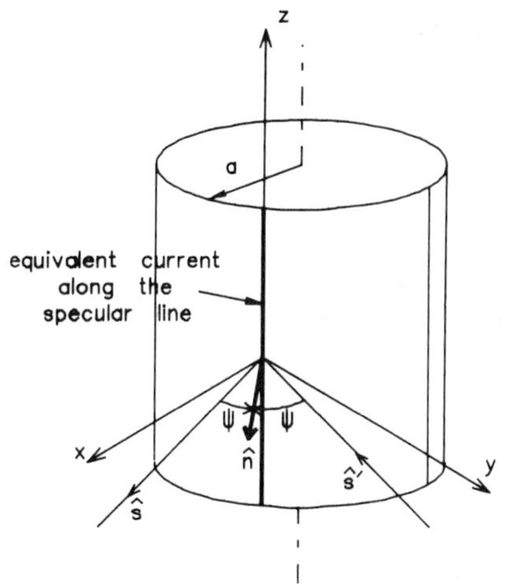

Figure 7.5 Equivalent current due to reflection from a circular cylinder.

Equation (7.86) is found by setting $\rho_2^r \to \infty$ in (3.245). Recall that ρ_1^r is the principal radius of curvature of the reflected wavefront in the plane of incidence, and ρ_2^r is the principal radius of curvature of the reflected wavefront in the plane perpendicular to the plane of incidence. In this case, the plane of incidence is parallel to the x-y plane, and hence

$$\rho_1^r = \frac{a \cos\psi}{2} \tag{7.87}$$

where

$$\cos\psi = -\hat{\mathbf{s}}' \cdot \hat{\mathbf{n}} \tag{7.88}$$

Let us assume that an equivalent current I^e flows along the specular line, and that I^e produces the same scattered field as (7.86). Setting $\beta_0 = \pi/2$ in (7.12), we know that the radiated field from I^e for normal incidence can be expressed as

$$\mathbf{E} = -jZI^e \sqrt{\frac{k}{8\pi}}\, e^{-j\pi/4} \frac{e^{-jks}}{\sqrt{s}}\, \hat{\mathbf{z}} \tag{7.89}$$

Equating (7.86) and (7.89), we find that

$$I^e = E^i \sqrt{\rho_1^r}\, \frac{e^{-j\pi/4}}{Z} \sqrt{\frac{8\pi}{k}} \tag{7.90}$$

In general, the equivalent current can be modified to include the case of oblique incidence. We then find [21] that

$$\mathbf{I}^e = \hat{\mathbf{e}} \frac{\hat{\mathbf{e}} \cdot \mathbf{E}^i}{Z \sin\beta_0} \sqrt{\rho^r} \sqrt{\frac{8\pi}{k}}\, e^{-j\pi/4} \tag{7.91}$$

where $\hat{\mathbf{e}}$ is the unit vector tangent to the specular line, ρ^r is the radius of curvature of the reflected wavefront at each path point perpendicular to $\hat{\mathbf{e}}$ [22], and

$$\cos\beta_0 = -\hat{\mathbf{s}}' \cdot \hat{\mathbf{e}} \tag{7.92}$$

Similarly, the equivalent magnetic current can be expressed as [21]

$$\mathbf{I}^m = -\hat{\mathbf{e}} \frac{(\hat{\mathbf{e}} \times \hat{\mathbf{s}}') \cdot \mathbf{E}^i}{\sin\beta_0} \sqrt{\rho^r} \sqrt{\frac{8\pi}{k}}\, e^{-j\pi/4} \tag{7.93}$$

As in (7.21) and (7.32), $\mathbf{E}^i$ in (7.91) and (7.93) is interpreted to have the correct phase at the point on the surface where $\mathbf{I}^e$ and $\mathbf{I}^m$ are evaluated, rather than merely being a constant amplitude term.

Having found the equivalent currents in (7.91) and (7.93), the scattered fields from the curved surfaces can be found using the potential integrals discussed in Section 7.3.

The equivalent currents for edge diffraction in Section 7.2 were derived for directions on the Keller diffraction cone and, as such, were particularly useful to calculate diffracted fields in the caustic regions where regular wedge diffraction fails. However, we have mentioned that equivalent currents have been applied successfully to calculate RCS. The point here is that, in the case of RCS, the backscattered rays are not necessarily on the diffraction cone. Similarly, the equivalent currents derived in this section were derived for, and thus give the correct reflected fields in, the direction of the specularly reflected ray.

Relying on the power of potential theory, we also can build a case to use the equivalent currents to yield the scattered field in other directions. To illustrate the point we calculate the RCS of a circular cylinder.

Example 7.4: RCS of a cylinder

In this example, we use the equivalent currents derived earlier to calculate the RCS of the circular cylinder of radius a and length l illuminated by a θ-polarized field as shown in Figure 7.6. Without loss of generality, we assume the incident ray to be in the x-z plane. The incident field on the specular line therefore can be expressed as

$$\mathbf{E}^i = e^{jk(a\sin\theta + z\cos\theta)}\,\hat{\theta} \tag{7.94}$$

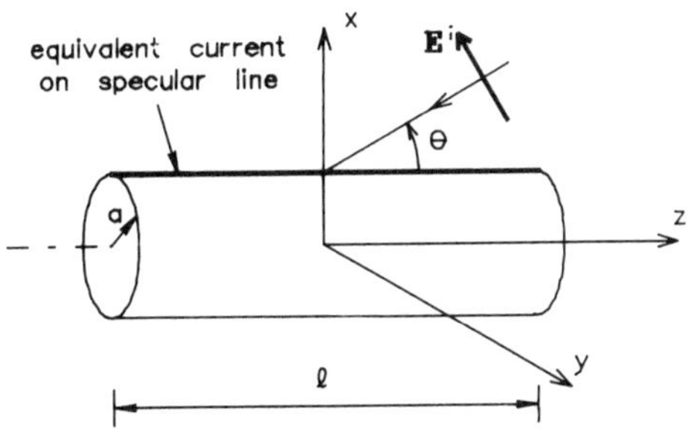

Figure 7.6 Scattering from a circular cylinder.

The amplitude of the incident field has been set equal to unity in anticipation of the normalization that is required in the calculation of RCS.

The axes once again are aligned with the equivalent current, so that $\hat{\mathbf{e}} = \hat{\mathbf{z}}$. Furthermore, because

$$\hat{\mathbf{s}}' = -\sin\theta\,\hat{\mathbf{x}} - \cos\theta\hat{\mathbf{z}} \tag{7.95}$$

we find that $(\hat{\mathbf{e}} \times \hat{\mathbf{s}}') \cdot \mathbf{E}^i = 0$, so that $I^m = 0$.

To evaluate the current in (7.91), we must determine ρ^r, the radius of curvature of the reflected wavefront perpendicular to $\hat{\mathbf{z}}$. From Figure 7.6, ρ^r in this case clearly is the radius of curvature of the reflected wavefront in the plane perpendicular to the plane of incidence. From (6.30), we know that

$$\rho^r = \frac{a}{2\,\sin\theta} \tag{7.96}$$

Substituting these values into (7.91), we find that

$$\mathbf{I}^e(z) = -\mathrm{e}^{jka\sin\theta}\sqrt{\frac{a}{4\pi k\,\sin\theta}}\,\mathrm{e}^{-j\pi/4}\,\frac{\mathrm{e}^{-jkr}}{r}\,W\,\hat{\mathbf{z}} \tag{7.97}$$

where r is the distance from the origin to the far-zone phase reference, and

$$W = \int_{-l/2}^{l/2} \mathrm{e}^{j2kz\cos\theta}\,dz \tag{7.98}$$

Evaluation of (7.98) and substitution of (7.97) into (7.13) yields

$$\mathbf{E}(\theta) = \mathrm{e}^{j2ka\sin\theta}\sqrt{\frac{a\,\sin\theta}{4\pi k}}\,\mathrm{e}^{j\pi/4}\,\sin(kl\,\cos\theta)\,\frac{\mathrm{e}^{-jkr}}{r}\,\hat{\theta} \tag{7.99}$$

The RCS is found by substituting (7.99) into (6.81); that is,

$$\sigma = \frac{a\,\lambda\,\sin\theta}{2\pi\,\cos^2\theta}\,\sin^2(kl\,\cos\theta) \tag{7.100}$$

The result in (7.100) is the same as that given in [25] and is valid when the wavelength is small in comparison to the radius and length of the cylinder. When $\theta = \pi/2$, we find from (7.100) that

$$\sigma = \frac{ka\,l^2}{\pi} \tag{7.101}$$

The broadside RCS given in (7.101) is in the direction where the specular line on the cylinder is a caustic, and we would expect the equivalent current formulation to give the correct result. As θ deviates from $\pi/2$, the RCS is not due to specular reflection anymore, so that we can expect (7.100) to break down as we move closer to the axial direction. Note that the contribution of the end faces of the cylinder has not been taken into account in the formulation just discussed.

PROBLEMS

7.1 Show that

$$E^i_{\beta_0'} = \frac{\hat{\mathbf{e}}^i \cdot \hat{\mathbf{e}}}{\sin\beta_0} E^i \qquad (7.102)$$

$$E^i_{\phi'} = -\frac{\hat{\mathbf{e}}^i \cdot (\hat{\mathbf{e}} \times \hat{\mathbf{s}}')}{\sin\beta_0} E^i \qquad (7.103)$$

7.2 Show that $g = 0$ in the case where the diffracted rays propagate in the direction of the diffraction cone.

7.3 Consider the thin flat plate of Example 7.3 shown in Figure 7.2.

(a) Show that there is no backscattered field from edge A for a Φ-polarized incident field at grazing incidence.
(b) Investigate the polarization dependence of the backscattered field from edge B at grazing incidence.
(c) Investigate the backscatter from the plate when the incident rays are perpendicular to edges A and B, but the incidence is no longer at grazing.

7.4 *Advanced problem.* Consider Example 7.4. Derive an expression for the backscattered field from the ring caustic at the joint between the base of a cone and a cylinder in the angular region $\theta \leq \theta_c$, where θ_c is a small angle. (See [5]).

7.5 Show that

$$J_1(u) = \frac{1}{2\pi j}\int_0^{2\pi} e^{ju\cos\Phi} \cos\Phi \, d\Phi \qquad (7.104)$$

where J_1 is the Bessel function of the first kind of order one.

7.6 Consider an axially symmetric parabolic reflector to be illuminated by a feed horn at its focus. Derive an expression for the edge-diffracted fields in the axial region using equivalent currents. Assume that the feed horn is a Huygens' source, which radiates the field

$$\mathbf{E}^f = (\sin\phi_f \hat{\theta}_f + \cos\phi_f \hat{\phi}_f) \cos^2\left(\frac{\theta_f}{2}\right) \frac{e^{-jkr}}{r} \tag{7.105}$$

where the subscript f denotes feed coordinates.

7.7 Derive an expression for the broadside backscattered field from a toroidal ring using equivalent currents. Assume the ring is illuminated by a plane wave as shown in Figure 7.7.

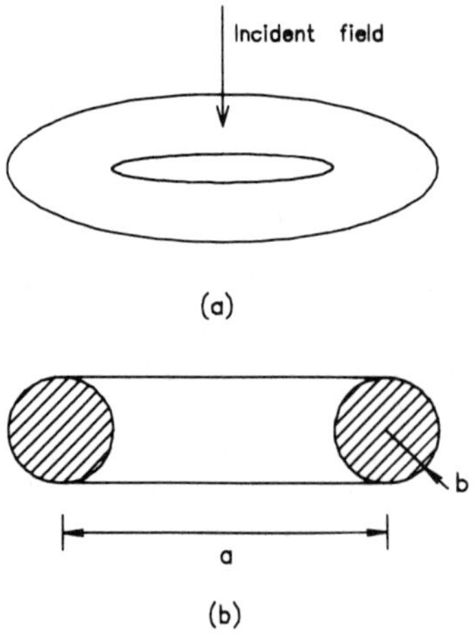

Figure 7.7 Toroidal ring: (a) oblique view showing normal incident field; (b) cross-sectional cut.

REFERENCES

[1] R.F. Millar, "An Approximate Theory of the Diffraction of an Electromagnetic Wave by an Aperture in a Plane Screen," *Proc. IEE,* Vol. 103, March 1956, pp. 177–185.

[2] R.F. Millar, "The Diffraction of an Electromagnetic Wave by a Circular Aperture," *Proc. IEE,* Vol. 104, March 1957, pp. 87–95.

[3] R.F. Millar, "The Diffraction of an Electromagnetic Wave by a Large Aperture," *Proc. IEE,* Vol. 104, September 1957, pp. 240–250.

[4] C.E. Ryan, Jr., and R.C. Rudduck, "Radiation Patterns of Rectangular Waveguides," *IEEE Trans. Antennas and Propagation,* Vol. AP-10, No. 4, July 1968, p. 488.

[5] C.E. Ryan, Jr., and L. Peters, Jr., "Evaluation of Edge Diffracted Fields Including Equivalent Currents for the Caustic Regions," *IEEE Trans. Antennas and Propagation,* Vol. AP-17, May 1969, pp. 292–299. Correction in *IEEE Trans. Antennas and Propagation,* Vol. AP-18, No. 2, March 1970, p. 275.
[6] T.B.A. Senior and P.L.E. Uslenghi, "High Frequency Back Scattering from a Finite Cone," *Radio Science,* Vol. 6, 1971, pp. 393–406.
[7] W.D. Burnside and L. Peters, Jr., "Axial-Radar Cross Section of Finite Cones by the Equivalent Current Concept with Higher Order Diffraction," *Radio Science,* Vol. 7, No. 10, October 1972, pp. 943–948.
[8] E.F. Knott and T.B.A. Senior, "Equivalent Currents for a Ring Discontinuity," *IEEE Trans. Antennas and Propagation,* Vol. AP-21, No. 5, September 1973, pp. 693–695.
[9] W.D. Burnside and L. Peters, Jr., "Comments on 'Equivalent Currents for a Ring Discontinuity'," *IEEE Trans. Antennas and Propagation,* Vol. AP-22, No. 3, May 1974, pp. 509–510.
[10] W.D. Burnside and L. Peters, Jr., "Edge Diffracted Caustic Field," *IEEE Trans. Antennas and Propagation,* Vol. AP-22, No. 4, July 1974, pp. 620–623.
[11] E.F. Knott and T.B.A. Senior, "Comparison of Three High Frequency Diffraction Techniques," *Proc. IEEE,* Vol. 62, No. 11, November 1974, pp. 1468–1474.
[12] E.F. Knott, J.F. Schaeffer, and M.T. Tuley, *Radar Cross Section,* Artech House, Dedham, MA, 1985.
[13] C.A. Mentzer, L. Peters, Jr., and R.C. Rudduck, "Slope Diffraction and Its Applications to Horns," *IEEE Trans. Antennas and Propagation,* Vol. AP-23, No. 2, March 1975, pp. 153–159.
[14] F.A. Sikta, W.D. Burnside, T.T. Chu, and L. Peters, Jr., "First-Order Equivalent Current and Corner Diffraction Scattering from Flat Plate Structures," *IEEE Trans. Antennas and Propagation,* Vol. AP-31, No. 4, July 1983, pp. 584–589.
[15] A. Michaeli, "Equivalent Currents for Arbitrary Aspects of Observation," *IEEE Trans. Antennas and Propagation,* Vol. AP-32, No. 3, March 1984, pp. 252–258. See also correction in *IEEE Trans. Antennas and Propagation,* Vol. AP-33, No. 1, January 1985, p. 227.
[16] A. Michaeli, "Elimination of Infinities in Equivalent Edge Currents, Part I: Fringe Current Components," *IEEE Trans. Antennas and Propagation,* Vol. AP-34, No. 7, July 1986, pp. 912–918.
[17] A. Michaeli, "Elimination of Infinities in Equivalent Edge Currents, Part II: Physical Optics Components," *IEEE Trans. Antennas and Propagation,* Vol. AP-34, No. 8, August 1986, pp. 1034–1037.
[18] A. Michaeli, "Equivalent Currents for Second-Order Diffraction by the Edges of a Perfectly Conducting Polygonal Surfaces," *IEEE Trans. Antennas and Propagation,* Vol. AP-35, No. 2, February 1987, pp. 183–190.
[19] E.F. Knott, "The Relationship between Mitzner's ILDC and Michaeli's Equivalent Currents," *IEEE Trans. Antennas and Propagation,* Vol. AP-33, No. 1, January 1985, pp. 112–185.
[20] K.M. Mitzner, "Incremental Length Diffraction Coefficients," Northrop Corp., Aircraft Division, Tech. Rep. AFAL-TR-73-296, April 1974.
[21] J.L. Volakis and L. Peters, Jr., "Evaluation of Reflected Fields at the Caustic Regions Using a Set of GO Equivalent Line Currents," *IEEE Trans. Antennas and Propagation,* Vol. AP-33, No. 8, August 1985, pp. 860–866.
[22] J.L. Volakis, "Electromagnetic Scattering from Inlets and Plates Mounted on Arbitrary Smooth Surfaces," Ph.D. dissertation, Ohio State University, 1982.
[23] R.F. Harrington, *Time Harmonic Electromagnetic Fields,* McGraw-Hill, New York, 1961.
[24] W.L. Stutzman and G.A. Thiele, *Antenna Theory and Design,* John Wiley and Sons, New York, 1981.
[25] J.W. Crispin, Jr., and K.M. Siegel (Eds.), *Methods of Radar Cross-Section Analysis,* Academic Press, New York, 1968.

Chapter 8
Diffraction at a Smooth Convex Conducting Surface

Diffraction shedding's also norm
When objects have a convex form

The ray fields here with sorrow weep
They travel not, but only creep

It sheds, alas, along its length
And thereby saps its vital strength

L.B. Felsen,
IEEE Antennas and Propagation Society Newsletter,
Vol. 21, No. 3, June 1979

8.1 THE PHENOMENON OF CREEPING WAVES, OR CURVED SURFACE DIFFRACTION

8.1.1 Introduction

Although GO predicts that the fields in the shadow region of an object are zero, experience indicates that this is not so even when the object is smooth and without sharp edges. Figure 8.1 shows an electric line source of amplitude I^e in the vicinity of an infinitely long conducting circular cylinder. We now examine this canonical problem. The classic eigenfunction solution[1] for the *total field* (scattered plus incident field *everywhere*) for this TM (soft) case is given, for $r > r_i$, by [1, 2]

$$E_z(r, \phi) = -\frac{k^2 I^e}{4\omega\epsilon}\{H_o^{(2)}[k\sqrt{r^2 + r_i^2 - 2rr_i\cos(\phi - \phi_i)}] + \sum_{n=-\infty}^{\infty} \frac{J_n(ka)}{H_o^{(2)}(ka)} H_n^{(2)}(kr_i)\, H_n^{(2)}(kr)\, e^{jn(\phi - \phi_i)} \tag{8.1}$$

[1]If the line source is allowed to recede to infinity, we obtain the solution given in Section 3.3.10.

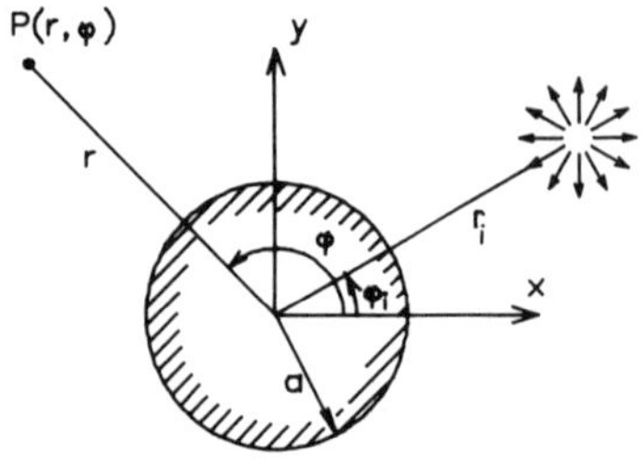

Figure 8.1 Line source radiating in the presence of a conducting circular cylinder. Notation for the eigenfunction solution.

The first term within the brackets is the field of the line source alone, an incident field in the sense of rigorous electromagnetic theory, as discussed in Section 2.9.

If this problem is considered from a GTD point of view (see Exercise 3.7), using only those ray mechanisms discussed up to this stage in the text, the solution for the *total* field would be the sum of the GO incident ray field and the GO reflected ray field, both being nonzero only in the lit region. The eigenfunction solution (8.1), however, predicts nonzero fields in the shadow zone as well. We therefore can conclude that the identically zero fields predicted by the GO in the shadow zone are in error; clearly, there are ray terms that have not been properly taken into account. However, because there are no sharp edges, these terms cannot be edge-diffracted ray contributions of any kind, and it is necessary to introduce the additional concept of surface-diffracted rays [1] to rectify matters.

8.1.2 Asymptotic Evaluation of Eigenfunction Solutions for Line Source Illumination of a Conducting Circular Cylinder

The purpose of the continued discussion of the solution for the circular cylinder scattering problem in the next few sections is fourfold:

(a) To immediately place the high-frequency curved-surface–diffraction principles within the framework of fundamental electromagnetic theory

(b) To indicate how Keller [1] could interpret the solution in terms of rays using an asymptotic evaluation of the eigenfunction solution and, thus, within a GTD framework generalize the result to arbitrarily curved surfaces.

(c) To point out where this early GTD result is inaccurate.

(d) Hence, to clearly show the advantage and significance of the UTD solutions that form the subject of this chapter.

At the outset we assume that ka is large, and the line source is sufficiently distant from any point on the cylinder that its field at the cylinder (e.g., at distance s' from the source itself) may be represented in the GO form (2.87):

$$U^i(s') = C_1 \frac{e^{-jks'}}{\sqrt{s'}} \tag{8.2}$$

with the constant C_1 given by (2.88). Furthermore, we assume that the observation points are well removed from the cylinder surface (that is, $kr > ka$) and these observation points are in the shadow zone, well away from the shadow boundaries. Subject to these assumptions (and only then), with reference to Figure 8.2, standard asymptotic techniques may be used to convert the series (8.1) to the following form [1, 2]:

$$\begin{aligned} U^d(P) \sim\ & U^i(Q_1') \frac{e^{-jks^d}}{\sqrt{s^d}} e^{-jkt_1} \sum_{n=1}^{N} (D_n)^2 e^{-\alpha_n t_1} \\ & + U^i(Q_2') \frac{e^{-jks^d}}{\sqrt{s^d}} e^{-jkt_2} \sum_{n=1}^{N} (D_n)^2 e^{-\alpha_n t_2} \end{aligned} \tag{8.3}$$

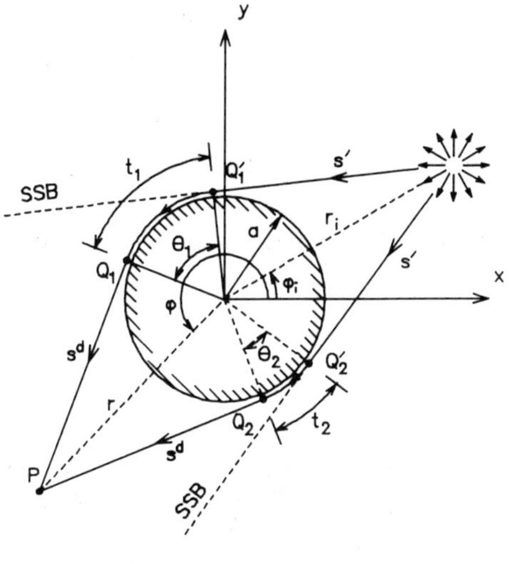

Figure 8.2 Geometry for the ray interpretation of the problem in Figure 8.1.

Only a few terms of the series are required for reasonable accuracy [1, 3]. Point P is the observation point, and the quantities $U^i(Q_1')$ and $U^i(Q_2')$ are the fields of the line source incident on the surface at points Q_1' and Q_2', as obtained from expression (8.2). The term:

$$(D_n)^2 = \sqrt{\frac{1}{2\pi k}} \frac{m\, e^{-j\pi/12}}{[Ai'(-q_n)]^2} \tag{8.4}$$

and

$$\alpha_n = \frac{q_n}{a} m\, e^{j\pi/6} \tag{8.5}$$

with

$$m = \left(\frac{ka}{2}\right)^{1/3} \tag{8.6}$$

the so-called curvature parameter. The assumption of a large ka therefore is equivalent to assumption of a large m. Note that D_n and α_n depend on the normalized radius of curvature ka of the cylindrical surface. The quantities $(-q_n)$ are the zeros of the Airy function $Ai(x)$, and thus the solutions of the equation $Ai(-q_n) = 0$. These are discussed in Appendix B. The terms t_1 and t_2 in the exponents are

$$t_1 = a[\phi - \phi_i - \sec^{-1}(r/a) - \sec^{-1}(r_i/a)] \tag{8.7}$$

$$t_2 = a\,[2\pi - \phi + \phi_i - \sec^{-1}(r/a) - \sec^{-1}(r_i/a)] \tag{8.8}$$

which are the arc lengths $Q_1'\,Q_1$ and $Q_2'\,Q_2$, respectively. The lengths s' and s^d are

$$s' = \sqrt{r_i^2 - a^2} \tag{8.9}$$

$$s^d = \sqrt{r^2 - a^2} \tag{8.10}$$

Readable accounts of the detailed derivation of (8.3) from (8.2) can be gleaned from the text by Jones [2, pp. 719–726], the book chapter by Christiansen [4], or the original papers by Keller and Levy [1, 5].

A similar asymptotic result can be found for the TE (hard) case. The form of the result is identical to (8.3), but with C_1 in (8.2) replaced by C_1/Z, and more important [1]:

$$(D_n)^2 = \sqrt{\frac{1}{2\pi k}} \frac{m\, e^{-j\pi/12}}{q_n'[Ai(-q_n')]^2} \tag{8.11}$$

$$\alpha_n = \frac{q'_n}{a} m \, e^{j\pi/6} \tag{8.12}$$

Here the quantities $(-q'_n)$ are the zeros of the derivative of the Airy function, $Ai'(x)$, also discussed in Appendix B.

8.1.3 Interpretation of the Asymptotic Solution in Terms of Surface Rays

An important task is to interpret geometrically the various terms in the asymptotic solution (8.3), as did Keller [1]. We refer again to Figure 8.2, and recall that asymptotic solutions (8.3) are valid only for observation points within the deep shadow region, which was one of the basic assumptions.

Expression (8.3) is the sum of two separate terms, each of which were interpreted by Keller [1] as a *surface-diffracted ray.* Note that each ray term in turn consists of a summation of terms with respect to index n. Keller called D_n the *diffraction coefficient* of the nth *surface ray mode* and α_n its *decay* or *attenuation constant.* (Note that these are *not* the diffraction coefficients for the two surface rays. There are two surface-diffracted rays, each one of which is composed of a sum of surface ray modes.) The surface diffracted rays become "attached" to the cylinder at points of grazing incidence (*attachment points*) Q'_1 and Q'_2, at which they are tangential to the surface, and then travel along the surface of the cylinder distances t_1 and t_2, respectively. The factor e^{-jkt_1} represents a phase change that would be associated with the one surface ray if it were traveling through free space between Q'_1 and Q_1. A similar statement can be made for the second ray. As they travel over the surface, the rays are attenuated exponentially due to the combined effect of the factors α_n. However, because these latter factors are complex, they also contribute a phase change to the surface ray over and above that caused by the e^{-jkt_1} and e^{-jkt_2} terms.

The rays leave the surface tangentially at the *shedding points* Q_1 and Q_2. Thereafter, they propagate to the observation point P as cylindrical GO wavefronts that appear to emanate from the shedding points Q_1 and Q_2, as is recognized from the factor:

$$\frac{e^{-jks^d}}{\sqrt{s^d}} \tag{8.13}$$

which is characteristic of cylindrical wave tubes.

This two-ray description is valid for *any* observation point P in the deep shadow region away from the *surface shadow boundary* (SSB). This implies that rays are shed continuously as the surface ray travels around the surface into the shadow zone, as illustrated in Figure 8.3. The continuous shedding of energy accounts for the attenuation experienced. It is clear from Figure 8.3 that the

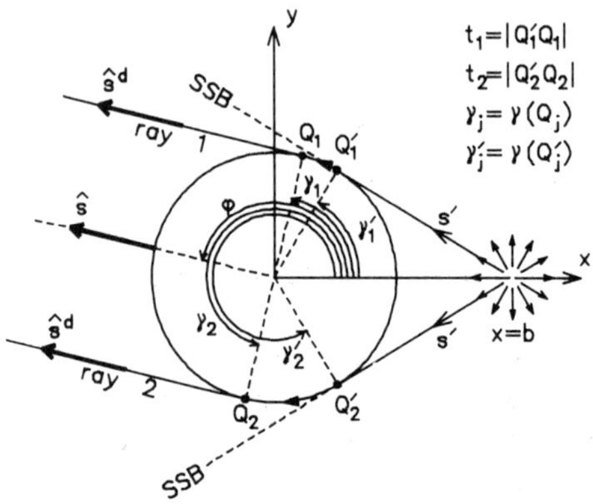

Figure 8.3 Tangential shedding of the surface diffracted rays.

constituent rays of any diffracted ray tube intersect at the surface of the scatterer. Thus, the surface is a caustic of the diffracted rays.

Remark 1: Asymptotic expansions similarly can be obtained for the current over the shadowed portion of the cylinder surface [6]. Franz and Depperman [7, 8] referred to this current that travels beyond the shadow boundaries as a *creeping wave* because of the way that it is attenuated as it propagates into the shadowed portion of the surface. The terms *creeping wave* and *surface-diffracted ray* often are used synonymously.

8.1.4 Invocation of Locality and the Generalized Fermat Principle

Insight rather than rigor is the watchword in the treatment here. In the previous section, we extracted from the asymptotic solution for the total fields a ray description that is valid in the deep shadow region of a smooth convex circular cylinder. With this as the canonical problem, we invoke the GTD principle of exploiting the local nature of the high-frequency diffraction effects and extend the solution (8.3) so that it can be used for more complicated shapes, for which exact solutions are not readily available. This is accomplished by approximating the vicinity of a point on a variable-curvature cylindrical surface (e.g., elliptic cylinder) by a circular cylinder whose radius is identical with the local radius of curvature [1].

The GTD approach is to consider a single ray at a time and then piece together an overall solution as a sum of separate direct, reflected, and diffracted rays. Let us therefore consider only the top surface-diffracted ray of Figure 8.2. The postulate that diffraction is a local phenomenon enables us to confine our attention to the portion of the geometry that determines the field properties along this particular ray, and we therefore consider only the top portion of the geometry of Figure 8.2, redrawn as Figure 8.4. The subscripts are superfluous when only a single ray is under consideration, and so have been discarded. The appropriate ray term in (8.3), rearranged into a slightly different form, for reasons that will become clear in the discussion, is

$$U^d(P) \sim \sum_{n=1}^{N} U^i(Q')\, D_n\, \mathrm{e}^{-\alpha_n t}\, \mathrm{e}^{-jkt} D_n \frac{\mathrm{e}^{-jks^d}}{\sqrt{s^d}} \tag{8.14}$$

The result (8.14) can be construed in the following manner [1, 3]. We use the word *construed* so as not to imply that all the intermediate quantities referred to are physically observable [9, 10], although the final answers are always so. We consider the contribution of the nth surface ray mode to the field $U^d(P)$. In this section, by the term *amplitude* we mean complex amplitude (that is, magnitude and phase). The term:

$$A_n(Q') = U^i(Q') D_n \tag{8.15}$$

is the amplitude (excitation function) at Q' of the nth mode, dependent on the incident field at Q' and the factor D_n. The quantity D_n in turn is a function of the radius of curvature of the surface at Q' (at present still simply a) and found from (8.4) or (8.11). The factor:

$$\mathrm{e}^{-\alpha_n t}\, \mathrm{e}^{-jkt}$$

describes the amplitude change of the nth surface ray mode between Q' and Q.

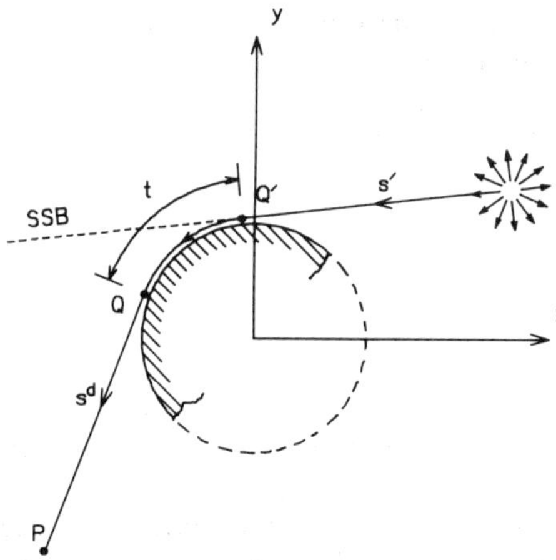

Figure 8.4 Single surface diffracted ray nomenclature.

At Q the amplitude of the nth surface ray mode is

$$A_n(Q) = A_n(Q')\, e^{-\alpha_n t}\, e^{-jkt} \tag{8.16}$$

Finally, the second D_n term serves to relate at Q the amplitude of the nth surface ray mode back into the amplitude contribution it makes to the cylindrical ray tube that appears to be shed from Q toward observation point P. The amplitude of this contribution is

$$A_n(Q)D_n = U^i(Q')\, D_n\, e^{-\alpha_n t}\, e^{-jkt} D_n \tag{8.17}$$

The summed contribution of each of the surface ray modes to the complex amplitude of the ray shed at Q is

$$U^d(Q) = \sum_{n=1}^{N} A_n(Q)D_n \tag{8.18}$$

This serves as the amplitude of the field of the cylindrical ray tube at its reference plane (taken to be the point Q), and thus the ray field at P is given by

$$U^d(P) = U^d(Q) \frac{e^{-jks^d}}{\sqrt{s^d}} \tag{8.19}$$

Suppose now that the curvature of the cylindrical surface in Figure 8.4 is not a constant at all points between Q' and Q, but that the assumptions about the largeness of the surface, and so on, are retained. We also assume that the radius of curvature of the surface is smoothly varying (with respect to wavelength) between Q' and Q, given at some general point Q_τ on the surface by $a_0(Q_\tau)$. Thus, the curvature parameter in (8.6) also will be a function of position; namely, $m(Q_\tau)$. In addition, let the surface between Q' and Q be parameterized in terms of an arc length parameter τ (see Appendix C), so that

$$t = \int_{\tau(Q')}^{\tau(Q)} d\tau \tag{8.20}$$

is distance along the surface from Q' to Q. The question now is this: What do we need to alter in (8.14) to make it applicable to the more general geometry just described?

Let us acknowledge that curved surface diffraction is a local phenomenon, and work through the previous physical interpretation again, rephrasing it into more general terms as we go along. We first see that (8.15) can remain as is except that the factor D_n used there will have to be evaluated from (8.4) or (8.11) using the particular value of the curvature parameter at Q'; namely, $m(Q')$, from (8.6). We write this as $D_n(Q')$. Second, the term:

$$e^{-\alpha_n t} \tag{8.21}$$

in (8.14) has to be generalized to the following form:

$$\exp\left[-\int_{\tau(Q')}^{\tau(Q)} \alpha_n(\tau)\, d\tau\right] \tag{8.22}$$

because α_n will not be a constant throughout the arc $|Q'Q|$, but a continuous function of the surface radius of curvature $a_0(Q_\tau)$ and curvature parameter $m(Q_\tau)$, according to (8.5) or (8.12). Expression (8.22) clearly simplifies to (8.21) when the surface radius of curvature, and hence α_n, is constant between Q' and Q. Finally, expression (8.18) also is unchanged, except that we must evaluate D_n using (8.4) or (8.11), but using the correct value of the curvature parameter at Q; namely, $m(Q)$. We thus write it $D_n(Q)$. What we have done amounts to locally approximating the characteristics of the arbitrarily curved cylinder at any point by a circular cylinder of the same radius of curvature placed about that point.

If this information is assembled, we find that for the cylindrical surface of variable curvature (8.14) becomes

$$U^d(P) \sim U^i(Q')\, T_{s,h} \frac{e^{-jks^d}}{\sqrt{s^d}} \tag{8.23}$$

where

$$T_{s,h} = \sum_{n=1}^{N} D_n^{s,h}(Q')\, e^{-jkt} \exp\left[-\int_{\tau(Q')}^{\tau(Q)} \alpha_n^{s,h}(\tau)\, d\tau\right] D_n^{s,h}(Q) \tag{8.24}$$

and the subscripts and superscripts, s, h, have been added to indicate that the appropriate forms must be used for the TM and TE cases, respectively. Expression (8.23) is the GTD curved surface diffraction result of Keller [1], albeit with a slight notational change and a less rigorous derivation. The coefficient $T_{s,h}$ essentially is a transmission function between Q' and Q.

This generalization to arbitrary cylinders, of the circular cylinder canonical problem, tacitly has taken for granted that the ray paths for the general case can be described along the same lines as the ray interpretation for the circular cylinder described in Section 8.1.3. As with direct rays, reflected rays, and edge-diffracted rays, in the GTD this is taken to result from the fact that ray paths satisfy a *generalized Fermat principle* (accepted as a postulate), which for the curved-surface scattering mechanisms being treated can be stated as follows (refer to Figure 8.4):

> In traveling from the source point (say P_s) to observation point P, the surface-diffracted ray path is such as to make the path $|P_sQ'QP|$ an extremum (although not necessarily a minimum).

This immediately provides us with the *law of convex curved-surface diffraction;* namely, that the paths P_sQ' and QP, being ray trajectories in free space, will be straight lines. These rays will be tangent (at grazing incidence) to the surface at Q' and Q, respectively. For that part of its trajectory $Q'Q$ on the surface itself, the ray follows a geodesic of the surface. These considerations apply to both the 2D problems considered in this text and the more general 3D cases [11–13]. In 2D problems, the surface geodesic simply is the cross-sectional curve of the scattering object between Q' and Q, a most convenient situation. This law of diffraction enables us to perform the necessary ray tracing, given the source and observation point locations and the geometry of the diffracting surface.

Remark 2: Closed Bodies. Consider the situation of the closed arbitrary cylinder in Figure 8.5. Because the generalized Fermat principle requires only that path lengths be extrema, and not necessarily minima (although the latter very often are the most significant), there will be two curved surface diffracted rays that can reach the point P. Each of these would have to be taken into account separately using the GTD result, then added together to obtain the complete solution for the field at P. The first of the rays has Q_1' and Q_1 as attachment and shedding points, respectively. For the second ray, these are Q_2' and Q_2. But these two curved-surface diffracted rays are not the only rays reaching observation point P. They simply are an additional kind of diffracted ray we, in this chapter, will be adding to the existing menu of ray types built up in the preceding chapters. Thus, if P does not lie in the shadow zone there will be a reflected ray and a direct ray as well, as illustrated.

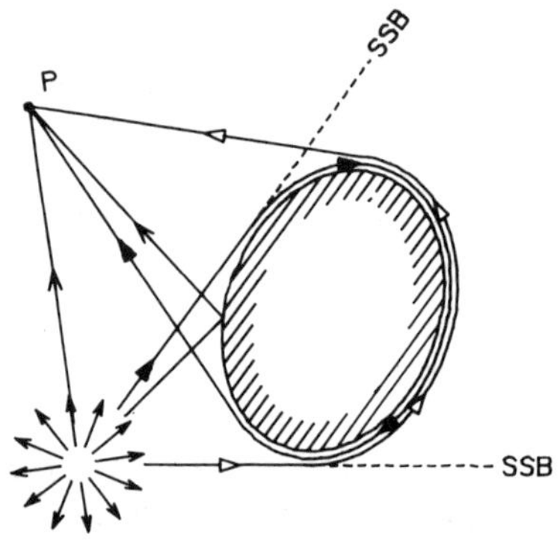

Figure 8.5 Ray mechanisms for a closed, smooth, convex surface.

In fact, each of these two surface-diffracted rays encircles the cylinder an infinite number of times. These multiple encircling rays may be summed to contribute [6, p. 874] a factor:

$$\left[1 - e^{-(jk+\alpha_n^{s,h})2\pi a}\right] \tag{8.25}$$

to the denominator of each of the coefficients $D_n^{s,h}$. This term can be retained for the large m assumption upon which the entire discussion depends, but the effect of the encircling rays is negligible [6; 2, p. 724].

8.1.5 The Significance of the UTD Results for Diffraction by Smooth Convex Surfaces

Asymptotic solutions are not unique. It is no surprise therefore that a number of alternative asymptotic expressions (which do not necessarily give identical answers at *all* observation points) exist in the literature for the solution to the canonical problem of Figure 8.1. However, in 1956 Keller [1] presented an asymptotic solution along with the ray interpretation of Sections 8.1.3 and 8.1.4.

Figure 8.6 illustrates regions of the boundary layer in which GTD is not valid. Recall that the results given up to now have been based on the assumption that observation points P are in the deep shadow zone, outside the transition region about the surface shadow boundary, and off the surface. Also, as noted at the end of Section 8.1.3, the shadowed part of the surface forms a caustic of the surface diffracted rays. Regions 3 and 4 of the shaded area close to the surface in Figure 8.6 are in the vicinity of this surface diffracted ray caustic. Region 5, on the other hand, is in the proximity of point Q', which is a caustic of the reflected ray for grazing incidence. Regions 3 to 5 commonly are referred to as the *surface* or *caustic boundary layer regions* [11]. Earlier discussions of caustics and the invalidity of

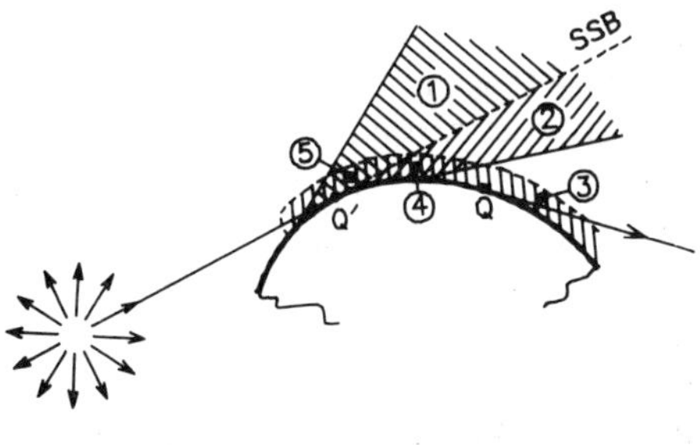

Figure 8.6 Transition and surface boundary layer regions in which the original GTD solution is invalid.

GO solutions there would seem to suggest that special care must be exercised with asymptotic solutions in such boundary layers. And this indeed proves to be the case. The GTD solution (8.24) is not uniform in the sense that its domain of validity excludes these transition regions. In addition the GTD curved surface diffraction solution fails to be accurate in region 2 (the shadow side) of the penumbral region about the SSB. This is readily established by simple numerical evaluation. In like manner, we noted in Chapter 3 that for near grazing incidence the GO reflected field solution is less accurate. Here, this means that the GO reflected field is not a good representation of the true field in region 1 of the transition region about the SSB. Furthermore, these GTD and GO solutions are discontinuous at the SSB, and the predicted fields are discontinuous there.

The angular extent of the transition region about the SSB is on the order $[2/ka_0(Q')]^{1/3}$, where $a_0(Q')$ is the radius of curvature of the cylinder at the attachment point Q' [14].

An asymptotic solution that removes these restrictions by being continuous at SSBs, valid in the transition regions about them, and perhaps allows P to be close to or even on the surface, is called a *uniform solution*. Although more than one such uniform asymptotic solution exists for the present canonical problem, the drawback is that they do not all necessarily have a ray description, and thus either cannot be incorporated easily into the GTD formalism or "are too complicated for numerical computation of fields in transition regions" [14]; and computational efficiency is a strain that runs throughout the UTD. The novelty of the UTD solutions to be discussed in the remainder of this chapter (and the origin of the title *uniform*) is that they, as for the UTD edge diffraction case,

(a) have been found in terms of a ray description,
(b) are continuous across the surface shadow boundaries and are valid in the transition regions about the shadow boundaries,
(c) outside the transition regions reduce smoothly to available GO or GTD solutions, and
(d) provide ray solutions for certain types of problems not available in the original GTD.

These UTD solutions, like those for edge diffraction, on the whole have not been derived through a purely mechanical application of asymptotic techniques, but combined with the mathematics certain heuristic arguments of much ingenuity. Although these derivations will not be presented in this text, references to the original papers will be given in the appropriate sections.

Lest there be any misunderstanding, we note here that in the list of UTD achievements, we do not mean to imply that each UTD curved-surface–diffraction solution (there are three, for three different situations) is valid for all source or observation points within the surface boundary layer. What we mean is that by meaningfully distinguishing three particular problem classes, the UTD covers any such requirement. These problem classes are discussed in the section that follows.

Remark 3: In the final UTD expressions, the surface ray modal terms do not appear explicitly, these effects having been "lumped" into the special Fock functions that have been studied in the literature and are presented in Appendix B.

8.1.6 Problem Classes for Curved-Surface Diffraction

The discussion in Section 8.1.2 concerned the scattering of a given incident GO field by a surface, with both the source and observation points off the surface. However, the phenomenon of diffraction by a smooth convex body embraces those situations where source or observation point or both are on the surface itself. We follow convention [15] by dividing the treatment into three separate situations: the scattering problem; the radiation or antenna problem; and the coupling problem.

These three cases are illustrated in Figure 8.7 to give the reader some qualitative idea of the differences between them. More detailed illustrations will be provided with the quantitative development of each class in the sections that follow.

For the scattering problem, both the source and observation points are off the surface. The radiation or antenna problem has the source point on, and the observation point off, the surface. Both the source and observation points are on the surface for the coupling problem.

In this text, our study of diffraction at smooth convex conducting surfaces will be restricted to problems of a 2D nature. As with reflection and edge diffraction, such 2D problems teach most of the essential ideas. Bibliographical remarks relevant to the 3D case are given at the end of the chapter. In the sections to follow we will be giving the general UTD solution associated with a *single ray* of

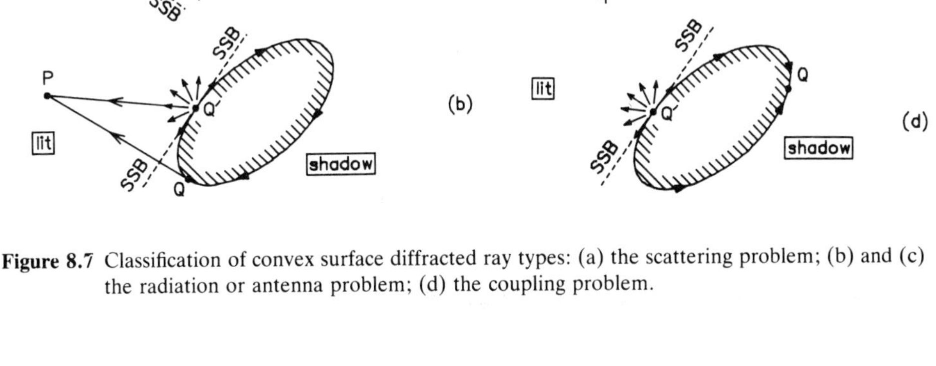

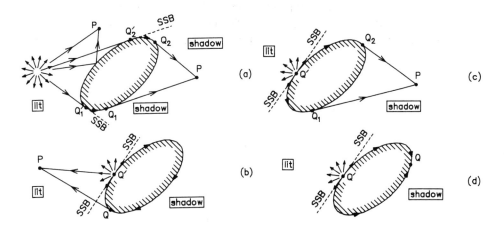

Figure 8.7 Classification of convex surface diffracted ray types: (a) the scattering problem; (b) and (c) the radiation or antenna problem; (d) the coupling problem.

each curved-surface–diffracted-ray class. In applying these results to particular geometries, the user must first identify exactly which surface-diffracted rays will be present, and then apply the general result to *each* such ray.

8.1.7 Differential Geometry for 2D Curved-Surface Diffraction

All that need concern us in 2D problems are the unit normal $\hat{\mathbf{n}}(Q_\tau)$, unit tangent $\hat{\mathbf{t}}(Q_\tau)$, and radius of curvature $a_0(Q_\tau)$ at any point Q_τ on the surface. A unit vector orthogonal to both $\hat{\mathbf{n}}$ and $\hat{\mathbf{t}}$ at all points Q_τ on the 2D surface will be

$$\hat{\mathbf{b}} = \hat{\mathbf{t}} \times \hat{\mathbf{n}} = \pm \hat{\mathbf{z}}$$

the plus (or minus) sign is used when $\hat{\mathbf{t}}$ is taken to be clockwise (or counter-clockwise). Further details are given in Appendix C.

8.2 THE TWO-DIMENSIONAL SCATTERING FORMULATION

8.2.1 The Scattering Problem Geometry

The relevant geometrical arrangement is shown in Figure 8.8, which also shows the location of the SSB, which for the scattering case is both an ISB and a RSB. The UTD offers a solution for the field both in the lit and shadow regions. The observation point P has been relabeled P_l if it falls in a lit zone, and P_d if it is situtated in the shadow zone. The distance s' is that between the source position (say P_s) and an attachment point Q'. The field in the lit zone is a reflected field, with Q_r the point of reflection and s^r the distance from Q_r to P_l. The distance along the surface between Q' and shedding point Q is t, and s^d is the distance from Q to P_d.

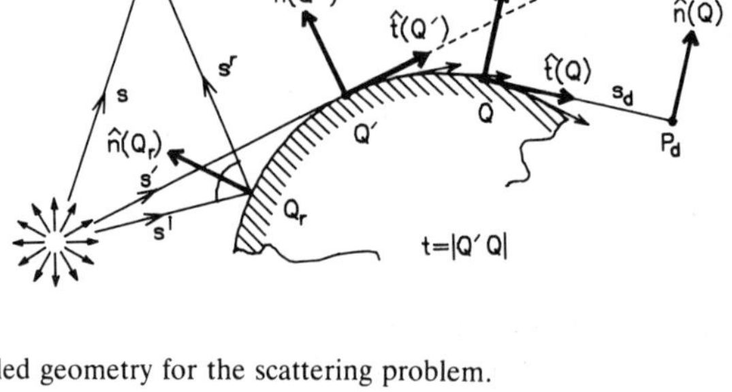

Figure 8.8 Detailed geometry for the scattering problem.

Note that, although we consider line source excitation, the directional properties of any source simply could be included in the expression for the incident field as a multiplicative factor with an angular dependence, as suggested in Section 2.4.

8.2.2 UTD Scattering Solution in the Lit Region

For observation points in the lit region, the surface diffracted field is a reflected field and the UTD expression for this reflected field at any observation point P_l is

$$U^r(P_l) = U^i(Q_r)R_{s,h}\sqrt{\frac{\rho^r}{\rho^r + s^r}}\, e^{-jks^r} \tag{8.26}$$

$U^i(Q_r)$ is the incident field at the reflection point Q_r, and of course must be of the GO type. Distances s' and s^r were defined in Section 8.2.1. For specified source and observation points (which *must* be off the surface), the location of Q_r can be determined from the law of reflection as used throughout Chapter 3. The reflected ray caustic distance ρ^r is computed from expression (3.77), repeated here for convenience:

$$\frac{1}{\rho^r} = \frac{1}{\rho^i(Q_r)} + \frac{2}{a_0(Q_r)\cos\theta^i} \tag{8.27}$$

with $\rho^i(Q_r) = s^i$ in this instance. The *generalized reflection coefficient* (or *UTD reflection coefficient*) is given by [11, 14]

$$R_{s,h} = -\sqrt{\frac{-4}{\xi_p}}\, e^{-j(\xi_p)^3/12}\left\{\frac{e^{-j\pi/4}}{2\xi_p\sqrt{\pi}}[1 - F(X_p)] + \hat{P}_{s,h}(\xi_p)\right\} \tag{8.28}$$

the factors of which are defined as given below.

(a) $F(X)$ is the *transition function* already used in connection with edge diffraction in Chapter 4 and defined in Appendix B.

(b) The *distance parameter* L_p is given by

$$L_p = \frac{s^r s^i}{s^r + s^i} \tag{8.29}$$

(c) The argument of the transition function is

$$X_p = 2kL_p\cos^2(\theta^i) \tag{8.30}$$

where

$$\cos\theta^i = \hat{\mathbf{n}}(Q_r) \cdot \hat{\mathbf{s}}^r \tag{8.31}$$

with $\hat{\mathbf{n}}(Q_r)$ the unit normal to the surface at Q_r. Because the cosine-squared term in (8.30) is positive, and parameter L_p is positive, being a combination of distances, we have $X_p \geqslant 0$ and always real.

(d) The quantity

$$\xi_p = -2m(Q_r)\cos\theta^i \tag{8.32}$$

is the *Fock parameter* associated with the reflected field in the lit region. Physical considerations restrict the angle θ^i to the range $-\pi/2 \leqslant \theta^i \leqslant \pi/2$, and so $\cos(\theta^i) \geqslant 0$. Hence, as $m(Q_r)$ always is positive for the convex surfaces under consideration, $\xi_p \leqslant 0$ and always real; hence, the reason for writing the first term in (8.28) as $\sqrt{-4/\xi_p}$. Far from the SSB both X_p and ξ_p are large, whereas on the SSB they both are zero.

(e)

$$m(Q_r) = \left[\frac{ka_0(Q_r)}{2}\right]^{1/3} \tag{8.33}$$

is the value of the *curvature parameter* at the reflection point Q_r, with $a_0(Q_r)$ the radius of curvature of the surface at Q_r. For the solution (8.26) to be valid, the parameter m must be large at, and in the vicinity of, the point Q_r.

(f) The function $\hat{P}_{s,h}(x)$ is the *Pekeris caret function* discussed in Appendix B. We have denoted it thus to maintain notational consistency with the key UTD papers on scattering by convex surfaces [11, 14]. Note that the caret does not imply that the function is a vector quantity. The function $\hat{P}_{s,h}$ can be written in terms of the *Fock scattering functions* $p^*(x)$ and $q^*(x)$ of Appendix B as

$$\hat{P}_s(x) = p^*(x)\,\mathrm{e}^{-j\pi/4} - \frac{\mathrm{e}^{-j\pi/4}}{2x\sqrt{\pi}} \tag{8.34}$$

for the TM (soft) case, and as

$$\hat{P}_h(x) = q^*(x)\,\mathrm{e}^{-j\pi/4} - \frac{\mathrm{e}^{-j\pi/4}}{2x\sqrt{\pi}} \tag{8.35}$$

for the TE (hard) case. If we substitute (8.34) or (8.35) into (8.28), we thus find

$$R_{s,h}(\xi_p, X_p) = -\sqrt{\frac{-4}{\xi_p}}\, e^{-j(\xi_p)^3/12}\, e^{-j\pi/4}\left[\frac{-1}{2\xi_p\sqrt{\pi}}F(X_p) + \begin{Bmatrix} p^*(\xi_p) \\ q^*(\xi_p) \end{Bmatrix}\right] \tag{8.36}$$

This is the form that ought to be used for computational purposes; routines for the numerical computation of $p^*(x)$ and $q^*(x)$ are given in Appendix F. However, expression (8.28) often is more convenient for certain analytical manipulations.

Total Field in the Lit Region

Expression (8.26) gives the UTD solution for the reflected field at point P_l in the lit zone. The total field at this point is the sum of the direct and reflected ray fields.

$$U(P_l) = U^i(P_l) + U^r(P_l) \tag{8.37}$$

where $U^i(P_l)$ is the direct ray field at P_l, of the form (8.1), but with the s' replaced by the s of Figure 8.8. Note that (8.37) is valid for points P_l on the lit side of the SSB transition region and on the SSB itself.

Special Case: Observation Point in Deep Lit Region

As the observation point P_l moves into the deep lit zone exterior to the SSB transition region, angle $\theta^i \to 0$ and thus $\cos\theta^i \to 1$. Because $m(Q_r)$ is by assumption large, we see from (8.32) that ξ_p will assume a large negative value; that is, $\xi_p \ll 0$. But if this is so, we can use the large argument form for $\hat{P}_{s,h}(\xi_p)$ given by equation (B.34) of Appendix B:

$$\hat{P}_{s,h}(\xi_p) \approx \pm\sqrt{\frac{-\xi_p}{4}}\, e^{j(\xi_p)^3/12} \tag{8.38}$$

Furthermore, as $\cos\theta^i \to 1$, parameter X_p also becomes large, which from equation (B.7) of Appendix B implies that $F(X_p) \approx 1$. This means that

$$[1 - F(X_p)] \approx 0 \tag{8.39}$$

If (8.38) and (8.39) are substituted into (8.28), it is easily seen to reduce to

$$R_{s,h} \approx \mp 1 \tag{8.40}$$

which is the GO reflection coefficient given by (3.38) of Chapter 3. The UTD result (8.26) then reduces to the GO reflected field solution, which is most reassuring. Note, however, that the generalized UTD form of $R_{s,h}$ in (8.28) or (8.36) applies only to convex surfaces, whereas the simple form (8.40) is used (see Example 3.7 for instance) even for concave surfaces.

Example 8.1: Line source in the presence of a conducting circular cylinder—lit region solution

This is the same problem as that in Example 3.1, and we refer to Figure 3.19, using the same symbols as we did there. Recall that the results presented in Example 3.1 were satisfactory except for a small sector of the lit region adjacent to the shadow boundary. We now make use of the UTD generalized reflection coefficient to see how significantly it alters the numerical results of Example 3.1.

All of the results in Example 3.1 are used, the only change is that the UTD reflection coefficient (8.36) is used in (3.104) for the reflected field $U^r(s, \phi)$ instead of simply ∓ 1. But before this can be achieved the parameters forming the arguments of the special functions in (8.36) need to be determined.

Because the radius of curvature $a_0(Q)$ is simply the constant a at all points Q on the cylinder, the curvature parameter (8.33) simplifies to

$$m = \left[\frac{ka}{2}\right]^{1/3} \tag{8.41}$$

The observation point P_l is in the far zone, so that $s^r \rightarrow \infty$ and the distance parameter L_p in (8.29) becomes

$$L_p = s^i$$

For any observation direction ϕ the reflection point Q_r is found as the solution of (3.94), and then from (3.95) $\cos\theta^i$ is known. Thus, the quantity X_p can be evaluated using (8.30), and so, too, the Fock parameter ξ_p in (8.32) using (8.41) and the preceding $\cos\theta^i$ information. This completely defines the UTD reflection coefficient (8.36). Note that it has to be recomputed for each observation direction ϕ. The sum of this reflected field and the direct field is shown in Figures 8.9 and 8.10 for the soft and hard cases, respectively. In this and subsequent examples the special functions needed all have been computed using the routines given in Appendix F. Note that the UTD results are noticeably different from those obtained using the GO reflection coefficient, except for observation directions in the deep lit zone. This is a consequence of (8.40).

The predicted ray field in the shadow regions of Figures 8.10 and 8.11 still

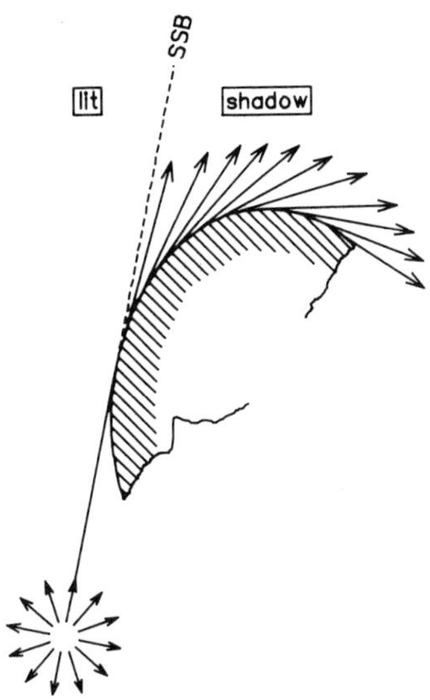

Figure 8.9 Total far-zone (direct + reflected) field for a circular cylinder illuminated by an electric line source, with $a = \lambda$ and $b = 2\lambda$. The solution obtained using the UTD reflection coefficient is compared to that for the GO reflection coefficient $R_{s,h} = \pm 1$.

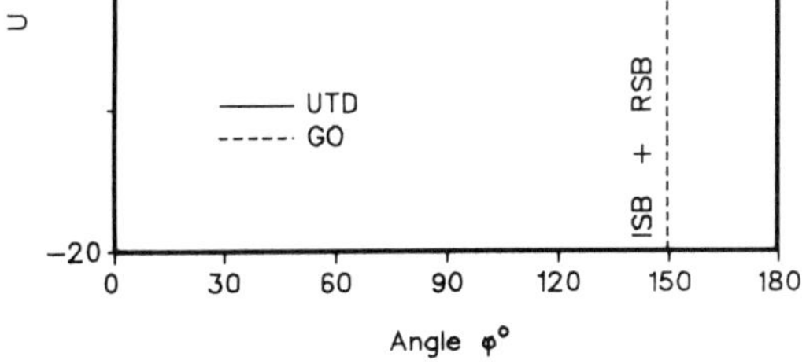

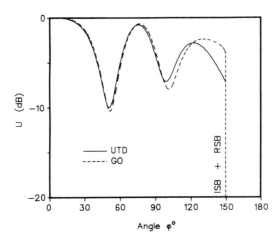

Figure 8.10 Total far-zone (direct + reflected) field for a circular cylinder illuminated by a magnetic line source, with $a = \lambda$ and $b = 2\lambda$. The solution obtained using the UTD reflection coefficient is compared to that for the GO reflection coefficient $R_{s,h} = \pm 1$.

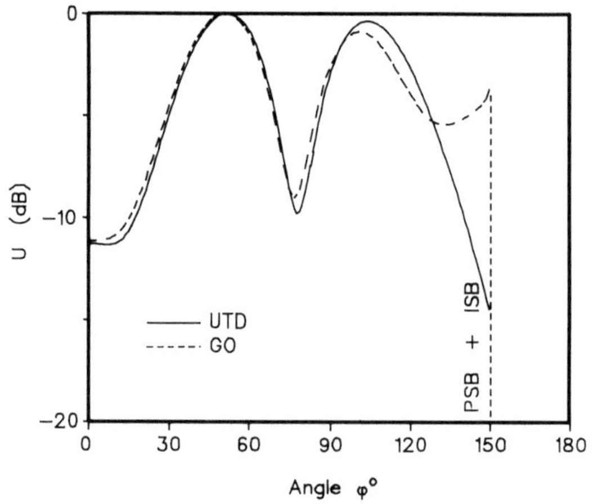

Figure 8.11 Circular cylinder of radius a, illuminated by a line source, showing the surface diffracted rays.

is zero, because we have not yet included the UTD expressions for the curved-surface–diffracted fields in the region beyond the SSBs. This is discussed in the following section.

8.2.3 UTD Scattering Solution in the Shadow Region

The surface diffracted field at any observation point P_d in the shadow zone is given by

$$U^d(P_d) = U^i(Q')\, T_{s,h} \frac{e^{-jks^d}}{\sqrt{s^d}} \tag{8.42}$$

$U^i(Q')$ is a GO type incident field at attachment point Q'. Observation point P_d may be in the shadow side of the SSB transition region and even on the SSB itself; however, it must be some distance off the scattering surface.

Let $\hat{\mathbf{n}}(Q')$ and $\hat{\mathbf{n}}(Q)$ be the unit normal to the surface at points Q' and Q, respectively, and $\hat{\mathbf{t}}(Q')$ and $\hat{\mathbf{t}}(Q)$ are the unit tangents to the surface there. According to the generalized Fermat principle, the path length $|P_sQ'QP_d|$ will be an extremum. Thus, the rays are tangent to the surface at Q' and Q. We therefore have

$$\hat{\mathbf{n}}(Q') \cdot \hat{\mathbf{s}}' = 0 \tag{8.43}$$

$$\hat{\mathbf{t}}(Q') \cdot \hat{\mathbf{s}}' = 1 \tag{8.44}$$

$$\hat{\mathbf{n}}(Q) \cdot \hat{\mathbf{s}}^d = 0 \tag{8.45}$$

$$\hat{\mathbf{t}}(Q) \cdot \hat{\mathbf{s}}^d = 1 \tag{8.46}$$

Given P_s and P_d, one or more of these forms of the law of diffraction may be employed to locate Q' and Q.

The *UTD surface diffraction coefficient* $T_{s,h}$ is [11, 14]

$$T_{s,h} = -\sqrt{m(Q')m(Q)}\sqrt{\frac{2}{k}}\,\mathrm{e}^{-jkt}\left\{\frac{\mathrm{e}^{-j\pi/4}}{2\xi_d\sqrt{\pi}}[1 - F(X_d)] + \hat{P}_{s,h}(\xi_d)\right\} \tag{8.47}$$

$T_{s,h}$ includes the integrated effect of the surface curvature over the entire path $Q'Q$ through its argument ξ_d, the diffracted field *Fock parameter* associated with the shadow zone. We have

$$\xi_d = \int_{\tau(Q')}^{\tau(Q)} \frac{m(\tau)}{a_0(\tau)}\,d\tau \tag{8.48}$$

with

$$m(\tau) = \left[\frac{ka_0(\tau)}{2}\right]^{1/3} \tag{8.49}$$

As earlier, $a_0(\tau)$ is the radius of curvature at some general point located by the parameter τ. The quantity

$$t = \int_{\tau(Q')}^{\tau(Q)} d\tau \tag{8.50}$$

is the path (or arc) length traversed by the surface ray between Q' and Q, and effects $T_{s,h}$ directly through the phase factor e^{-jkt}. The surface ray is not exactly like one that travels through free space; this is seen by the fact that e^{-jkt} is not the only phase shift added by $T_{s,h}$. For the convex surfaces to which these solutions apply, $a_0(\tau)$ and $m(\tau)$ always are positive, and so $\xi_d \geqslant 0$ and always is real. Clearly, the path length $t \geqslant 0$ as well.

A *distance parameter* L_d is defined as

$$L_d = \frac{s^d s'}{s^d + s'} \tag{8.51}$$

and the quantity:

$$X_d = \frac{kL_d\,(\xi_d)^2}{2m(Q')m(Q)} \tag{8.52}$$

with $L_d \geqslant 0$ and $X_d \geqslant 0$ always. Note that for each pair of locations P_s and P_d of the source and observation points, it is necessary to first perform the ray tracing procedure of locating Q' and Q before the coefficient $T_{s,h}$ can be evaluated. Remember that (8.42) is the total field in the shadow zone, unless the scattering surface is closed and a second curved-surface–diffracted ray creeps round in the counterclockwise direction similar to that shown in Figure 8.5.

Computational Aspects

If we substitute (8.34) or (8.35) into (8.47), we thus have

$$T_{s,h}(\xi_d, X_d, t) = -\sqrt{m(Q')m(Q)}\sqrt{\frac{2}{k}}\,e^{-jkt}\,e^{-j\pi/4}\left\{\frac{-F(X_d)}{2\xi_d\sqrt{\pi}} + \begin{bmatrix} p^*(\xi_d) \\ q^*(\xi_d) \end{bmatrix}\right\} \tag{8.53}$$

where p^* or q^* are chosen for the soft or hard cases, respectively. The form (8.53) should be used for computational purposes rather than the defining equation (8.47), but the latter sometimes is more convenient for analytical manipulation, as will be seen in the considerations of the special case that follows.

Special Case: Observation Point in Deep Shadow Region

As the observation point P_d moves into the deep shadow region exterior to the SSB transition region, the appropriate shedding point Q will move further into the shadowed region of the surface. Hence, with the integration interval from Q' to Q large, the Fock parameter ξ_d will become large, and from (8.52) so will X_d. Using equation (B.7) of Appendix B we then have $F(X_d) \approx 1$, and so $[1 - F(X_d)] \approx 0$ in (8.47). Furthermore, expressions (B.32) and (B.33) can be used for the function $\hat{P}_{s,h}(\xi_d)$. If these are substituted into (8.47), some cumbersome but straightforward algebraic manipulation reduces $T_{s,h}$ to the original GTD form (8.24) derived by Keller [1]. In other words, because $[1 - F(X_d)] \rightarrow 0$ in the deep shadow region, $F(X_d)$ has little influence there, and the function $\hat{P}_{s,h}(\xi_d)$ plays the dominant role in reducing the UTD solution uniformly to the GTD result.

Remark 4: Surface Diffracted Field in the Lit Zone: Although we have presented the diffracted field result (8.40) for observation points in the shadow region, remember that, if we have a closed object such as that in Figure 8.5, the surface diffracted ray that propagates counterclockwise around the object and then toward the lit region observation point P_l indeed is of the form $U^d(P_l)$ in the lit region. Similarly, for the surface-diffracted ray traveling clockwise. The form of the expressions for these diffracted fields are the same as when the observing is done in the shadow zone, and so nothing need be altered. This is the advantage of using the ray-based coordinates to express the UTD solutions.

Note that in the deep lit zone these surface diffracted rays usually are much smaller than the direct and reflected ray components shown in Figure 8.5, and in some cases will be negligible. But close to the SSB they are usually significant, as will be seen in Example 8.2.

Example 8.2: Line source radiating in the presence of a conducting circular cylinder—complete solution

We now add the UTD shadow region result to the solution of the problem whose geometry was given in Figure 3.19, and whose UTD lit region solution was discussed in Example 8.1. Figure 8.11 shows some of the additional nomenclature we will use for the shadow region analysis. Although we could write down many of the UTD solution parameters by inspection for this case of a circular cylinder, we will apply the definitions fully to become familiar with their use.

The observation point is in the far zone, and we therefore refer to the observation direction ϕ rather than an observation point as such. Observe from Figure 8.11 two surface-diffracted rays proceed in a given direction ϕ. Because of symmetry, only the upper half ($\phi < \pi$) of the far-zone radiation pattern need be computed. Points on the cylinder are located by the parameter γ, in terms of which the unit normal is given by (3.90). The diffracted ray direction $^{d\,\hat{}}$is simply the observation direction, which from (3.85) is

$$\hat{\mathbf{s}}^d = \cos\phi\hat{\mathbf{x}} + \sin\phi\hat{\mathbf{y}} \tag{8.54}$$

and is the same for both ray 1 and ray 2. The point Q_1', at which (8.43) must hold already has been found in Example 3.1 to be given by

$$\gamma_1' = \cos^{-1}(a/b) \tag{8.55}$$

where $\gamma_1' = \gamma(Q_1')$. Symmetry leads to

$$\gamma_2' = 2\pi - \cos^{-1}(a/b) \tag{8.56}$$

Points Q_1' and Q_2' thus are independent of the observation angle, and we always have

$$s' = \sqrt{b^2 - a^2} \tag{8.57}$$

and the GO incident fields at both Q_1' and Q_2' therefore are

$$U^i(Q_1') = U^i(Q_2') = \frac{e^{-jk\sqrt{b^2 - a^2}}}{(b^2 - a^2)^{1/4}} \tag{8.58}$$

Shedding points Q_1 and Q_2 are next located according to (8.45). For point Q_j, index $j = 1, 2$ identifying ray 1 or ray 2, we have

$$\hat{\mathbf{n}}(Q_j') \cdot \hat{\mathbf{s}}^d(Q_j) = \cos\phi \ \cos\gamma_j + \sin\phi \ \sin\gamma_j = 0 \tag{8.59}$$

which implies that

$$\cos[\phi - \gamma_j] = 0 \tag{8.60}$$

and has as solution:

$$\phi - \gamma_j = \pm\pi/2 \tag{8.61}$$

and hence that the position of the shedding points for a given observation direction, ϕ, are located by the parameters:

$$\gamma_1 = \phi - \pi/2 \tag{8.62}$$

$$\gamma_2 = \phi + \pi/2 \tag{8.63}$$

The element of arc length parameter $d\tau$ is related (see Appendix C) to the parameter γ as

$$d\tau = \pm\left|\frac{d\mathbf{r}}{d\gamma}\right| d\gamma \tag{8.64}$$

and the plus (or minus) sign applies if the arc length is measured counterclockwise (or clockwise). For the circular cylinder, $d\mathbf{r}(\gamma)/d\gamma$ was supplied in equation (3.88), and thus

$$d\tau = \pm\, a\, d\gamma \tag{8.65}$$

From (8.50), we then have

$$t_j = \int_{\tau(Q_j')}^{\tau(Q_j)} d\tau = \pm \int_{\gamma_j}^{\gamma_j'} a \, d\gamma = \pm a [\gamma_j - \gamma_j'] \tag{8.66}$$

which from (8.55) and (8.59) becomes

$$\begin{aligned} t_1 &= a [\phi - \pi/2 - \cos^{-1}(a/b)] \\ t_2 &= -a \{\phi + \pi/2 - [2\pi - \cos^{-1}(a/b)]\} \end{aligned} \tag{8.67}$$

or, in other words,

$$t_2 = a [3\pi/2 - \phi - \cos^{-1}(a/b)] \tag{8.68}$$

Because we are considering a circular cylinder with constant radius a:

$$m(Q_j') = m(Q_j) = (ka/2)^{1/3} \tag{8.69}$$

for both ray 1 and ray 2, and the Fock parameter (8.48) hence becomes

$$\xi_{dj} = \int_{\tau(Q_j')}^{\tau(Q_j)} \frac{(ka/2)^{1/3}}{a} d\tau = \left[\frac{k}{2a^2}\right]^{1/3} t_j \tag{8.70}$$

with t_j in (8.70) given by (8.67) or (8.68) for $j = 1$ or 2, respectively. Note that ξ_{dj} is a function of the observation angle ϕ. Because for the far-zone observation $s^d \to \infty$, (8.51) simplifies to

$$L_d = \sqrt{b^2 - a^2} \tag{8.71}$$

for both rays, and so the parameter X_{dj} is found by combining (8.69), (8.70), and (8.71) to yield

$$X_{dj} = \frac{k \sqrt{b^2 - a^2} \, t_j^2}{2a^2} \tag{8.72}$$

At this point, we have all the parameters necessary to compute the UTD diffracted field in the shadow zone. For a given frequency, source location, and geometry dimensions, a procedure for this computation can be defined:

(a) Compute the incident GO field at Q_j' using (8.58).
(b) For *each* ϕ compute t_j from (8.67) or (8.68), ξ_{dj} from (8.70), and X_{dj} from (8.72).

(c) Substitute these into (8.53) and find the diffracted field from (8.42) for j = 1 and j = 2 in turn, utilizing the special function routines of Appendix F.

The term $e^{-jks^d}/\sqrt{s^d}$ of course is factored out as earlier, and does not form part of the computation, as in all far-zone calculations. Figure 8.12 shows the total field (for magnetic line source excitation) obtained by combining the lit zone solution of Example 8.1 with the surface-diffracted field found in this example, but with the diffracted field included only in the shadow region. Note the discontinuity at the SSB. If the surface-diffracted field is included in the lit zone beyond the SSB, the solution shown in Figure 8.13 is obtained. The total field now is continuous across the SSB. This emphasizes the point made in Remark 4 that, although the UTD solution in (8.42) is known as the *shadow zone solution,* it also exists in the lit zone if the scattering object is closed. However, the effect of the solution is greatest in the vicinity of the SSB. This example completes the problem first begun as Example 3.1

8.2.4 Field Continuity at the SSB

Observe in Figure 8.13 that the UTD lit region and shadow region solutions blend both continuously and smoothly (that is, derivatives are continuous) at the SSB. The continuity property, ensured by the $F(X)$ terms in (8.28) and (8.47), in fact can be shown analytically to be true in general. To do this, we evaluate the lit region and shadow region field solutions separately, and then let the observation points approach the shadow boundary from either side. This exercise demonstrates the continuity property and also provides increased familiarity with the various terms in the UTD solutions. We refer to the geometry in Figure 8.8 and assume the line source that illuminates the surface has strength U_0.

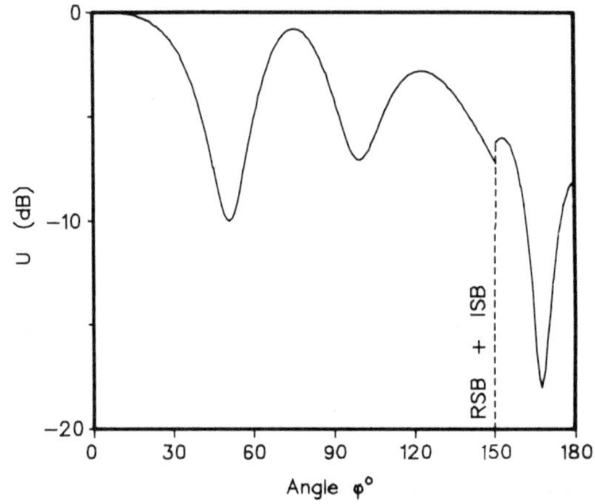

Figure 8.12 UTD solution for the problem depicted in Figure 8.11, but with the surface-diffracted ray contributions omitted from the lit zone.

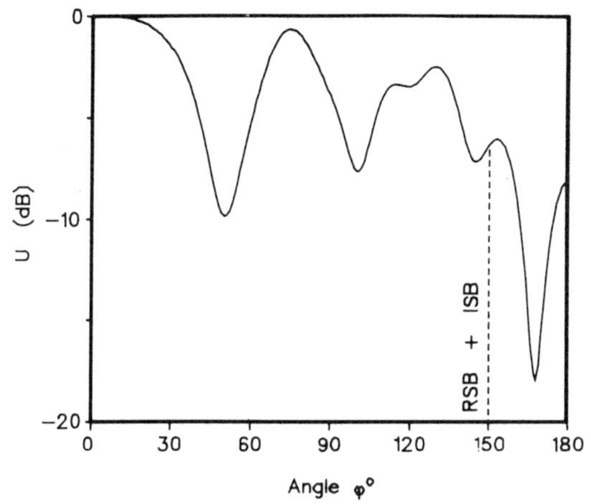

Figure 8.13 UTD solution for the problem depicted in Figure 8.11, with the surface-diffracted ray contributions included in the lit zone.

Approaching the SSB from the Lit Side

We examine the lit zone solution (8.37) first:

$$U(P_l) = U^i(P_l) + U^r(P_l) \tag{8.73}$$

When P_l is on the SSB we have $\theta^i = \pi/2$, and $Q = Q' = Q_r$. So we let P_l approach the SSB by setting $\theta^i = \pi/2 - \delta$, where $\delta \ll 1$, find the total UTD field at this point, and then take the limit as $\delta \to 0$. Therefore,

$$\cos\theta^i = \cos(\pi/2 - \delta) = \sin\delta \approx \delta \tag{8.74}$$

because δ is very small. From (8.32), we then have

$$\epsilon_p \approx -2m\delta \tag{8.75}$$

where m is understood to be $m(Q_r) = m(Q') = m(Q)$. The distance parameter L_p is found from (8.29) as

$$L_p = \frac{s^r s^i}{s^r + s^i} \tag{8.76}$$

and this together with (8.75) yields

$$X_p = 2kL_p\delta^2 \tag{8.77}$$

We next substitute (8.75) and (8.77) into (8.36) piece by piece. Because X_p is small, we may use the small argument form of $F(x)$ given in equation (B.7) of Appendix B, to yield

$$F(X_p) \approx \left[\sqrt{2\pi kL_p}\delta - 4k\delta^2 L_p\, e^{j\pi/4}\right] e^{j\pi/4}\, e^{j2k\delta^2 L_p} \tag{8.78}$$

Also

$$\frac{-1}{2\epsilon_p\sqrt{\pi}} \approx \frac{1}{4m\delta\sqrt{\pi}} \tag{8.79}$$

From (8.78) and (8.79), the first part of the bracketed term in (8.36) thus reduces to

$$f_1(\delta) = -\frac{F(X_d)}{2\epsilon_p\sqrt{\pi}} \approx \left[\frac{\sqrt{2kL_p}}{4m} - \frac{kL_p\delta}{m\sqrt{\pi}}\, e^{j\pi/4}\right] e^{j\pi/4}\, e^{j2k\delta^2 L_p} \tag{8.80}$$

The second part of the bracketed term in (8.36) simply becomes

$$f_2(\delta) = \begin{cases} p^*(-2m\delta) \\ q^*(-2m\delta) \end{cases} \tag{8.81}$$

Furthermore, the term outside the brackets in (8.36) is

$$-\sqrt{\frac{-4}{\epsilon_p}}\, e^{-j(\epsilon_p)^3/12}\, e^{-j\pi/4} \approx -\sqrt{\frac{2}{m\delta}}\, e^{jm^3\delta^3/12}\, e^{-j\pi/4} \tag{8.82}$$

From (8.27), because $\cos\theta^i \approx \delta$, we have the caustic distance for the reflected field:

$$\rho^r \approx \frac{a_0 s^i \delta}{a_0\delta + 2s^i} \tag{8.83}$$

where a_0 is understood to be $a_0(Q_r) = a_0(Q') = a_0(Q)$. This means that the spreading factor in (8.26) becomes

$$A(s^r) = \sqrt{\frac{\rho^r}{\rho^r + s^r}} \approx \sqrt{\frac{a_0 s^i \delta}{a_0\delta(s^i + s^r) + 2s^i s^r}} \tag{8.84}$$

Multiplication of (8.82) by (8.84) yields

$$f_3(\delta) = \sqrt{\frac{2(a_0/m)s^i}{a_0\delta(s^i + s^r) + 2s^is^r}}\ \mathrm{e}^{j8m^3\delta^3/12}\ \mathrm{e}^{-j\pi/4} \tag{8.85}$$

Assembly of the above results into (8.26) gives the reflected field at P_l as

$$\begin{aligned} U^r(P_l) &= U^i(Q_r)R_{s,h}A(s^r)\ \mathrm{e}^{-jks^r} \\ &= U^i(Q_r)f_3(\delta)[f_1(\delta) + f_2(\delta)]\ \mathrm{e}^{-jks^r} \end{aligned} \tag{8.86}$$

which at a point P_l on the SSB (that is, for $\delta \to 0$) becomes

$$U^r(P_l) = U^i(Q_r)\ \mathrm{e}^{-jks^r}\sqrt{\frac{a_0}{ms^r}}\ \mathrm{e}^{-j\pi/4}\left[\frac{\sqrt{2kL_p}}{4m}\ \mathrm{e}^{j\pi/4} + \begin{Bmatrix} p^*(0) \\ q^*(0) \end{Bmatrix}\right] \tag{8.87}$$

with $p^*(0)$ for the soft and $q^*(0)$ for the hard case. Note that we can write

$$\sqrt{\frac{a_0}{ms^r}} = m\sqrt{\frac{2}{k}} \tag{8.88}$$

Using this result and substituting for L_p from (8.76), we obtain the first term in (8.87) as

$$U^i(Q_r)\ \mathrm{e}^{-jks^r}\sqrt{\frac{a_0}{ms^r}}\ \mathrm{e}^{-j\pi/4}\ \frac{\sqrt{2kL_p}}{4m}\ \mathrm{e}^{j\pi/4} = \frac{U^i(Q_r)}{2}\ \frac{\mathrm{e}^{-jks^r}}{\sqrt{s^r}}\ \frac{\sqrt{s^rs^i}}{\sqrt{s^r + s^i}} \tag{8.89}$$

However, because the incident field at Q_r is

$$U^i(Q_r) = U_0\ \frac{\mathrm{e}^{-jks^i}}{\sqrt{s^i}} \tag{8.90}$$

the right-hand side of expression (8.89) reduces to

$$\frac{U_0}{2}\ \frac{\mathrm{e}^{-jk(s^i+s^r)}}{\sqrt{s^i + s^r}} \tag{8.91}$$

which is one-half times the incident field $U^i(P_l)$ at P_l. Using (8.88) leads to the second term in (8.79) being

$$U^i(Q_r)\frac{e^{-jks^r}}{\sqrt{s^r}}m\sqrt{\frac{2}{k}}\,e^{-j\pi/4}\begin{Bmatrix}p^*(0)\\q^*(0)\end{Bmatrix} \tag{8.92}$$

and so, if (8.91) and (8.92) are reassembled into (8.87), the reflected field at P_l on the SSB is found to be

$$U^r(P_l) = \frac{U_0}{2}\frac{e^{-jk(s^i+s^r)}}{\sqrt{s^i + s^r}} + U^i(Q_r)\frac{e^{-jks^r}}{\sqrt{s^r}}m\sqrt{\frac{2}{k}}\,e^{-j\pi/4}\begin{Bmatrix}p^*(0)\\q^*(0)\end{Bmatrix} \tag{8.93}$$

The first term in (8.93) is independent of the properties of the surface. The second term can be considered a correction term that indeed is a function of the geometrical properties of the surface and the particular polarization.

Approaching the SSB from the Shadow Side

We simply outline the procedure to be followed in this instance, as it is very similar to the one just presented. As the observation point P_d in (8.42) approaches the SSB, the point Q in Figure 8.8 moves closer toward Q', until, when P_d is on the SSB, they are collocated. Thus $m(Q') = m(Q) = m$, and

$$-\sqrt{m(Q')m(Q)} \rightarrow -m$$

As Q' approaches Q, so, too, does $t \rightarrow 0$ and $\xi_d \rightarrow 0$, and hence $Xd \rightarrow 0$. Thus,

$$e^{-jkt} \rightarrow 1$$

$$\begin{matrix}p^*(\xi_d)\\q^*(\xi_d)\end{matrix} \rightarrow \begin{matrix}p^*(0)\\q^*(0)\end{matrix}$$

In the limit $\xi_d \rightarrow 0$, the term

$$-\frac{F(X_d)}{2\xi_d\sqrt{\pi}}$$

reduces in a manner similar to (8.80). Finally, recognizing that as Q_i merges with Q we have $s^d = s^r$ because $P_d = P_l$, we find that $U^d(P_d)$ in (8.42) reduces precisely to that in (8.93). Thus, the UTD solution always is continuous at the SSB.

Example 8.3: Radar width of a conducting elliptical cylinder

Although Example 8.2 shows how the various UTD parameters are computed, the example in effect represents the application of the UTD to its own canonical problem. In this example, we therefore consider the case of an elliptical cylinder illuminated by a plane wave. The elliptical cross-section is a good intermediate step between the circular cross section and one that can be specified only numerically, for example. It also very often is used in practice [16–19].

We work through the detail somewhat stoically and comprehensively. Although for this particular elliptical cylinder problem (plane wave incidence and far-zone observation) we are able to write many of the required quantities by inspection, we will not do so. Not only will we thus gain increased familiarity with UTD practice, but we hope to make the detailed workings of this example easier to transfer to Example 8.4 in Section 8.3, as well as to some of the problem exercises at the end of this chapter.

The radar width of an elliptical cylinder was first considered as Example 3.4. For each incident angle, there was a single specular point that sent a single reflected ray back in the direction of incidence. We now wish to include the effects of the two surface-diffracted rays shown in Figure 8.14.

Incident Field

The phase of all fields is referenced to the origin. From Section 2.4.2, we know that at position (r, ϕ) a unit amplitude plane wave incident at angle ϕ_i is described by

$$U^i(r, \phi) = e^{jkr\cos(\phi - \phi_i)} \tag{8.94}$$

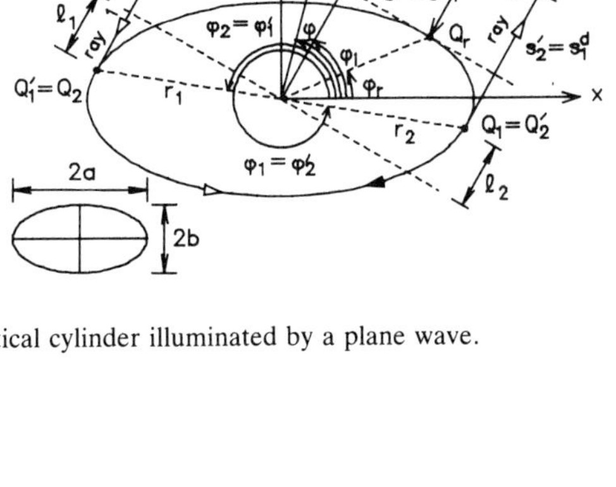

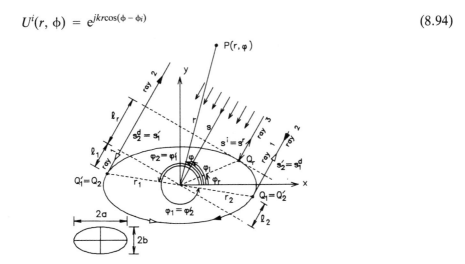

Figure 8.14 Elliptical cylinder illuminated by a plane wave.

Geometry

Although the geometry of the elliptical cylinder has been described previously in Example 3.4, to repeat some of this information here will be convenient. Recall that the elliptical curve is defined by

$$\mathbf{r}(\gamma) = a \cos\gamma \, \hat{\mathbf{x}} + b \sin\gamma \, \hat{\mathbf{y}} \tag{8.95}$$

where γ is a parameter ($0 \leq \gamma \leq 2\pi$) related to angle ϕ as

$$\tan\phi = \frac{b}{a} \tan\gamma \tag{8.96}$$

From (8.95), we thus have

$$\frac{d\mathbf{r}}{d\gamma} = -a \sin\gamma \, \hat{\mathbf{x}} + b \cos\gamma \, \hat{\mathbf{y}} \tag{8.97}$$

a result which we will use later. Furthermore, the radius of curvature at any point γ on the elliptical surface is

$$a_0(\gamma) = \frac{(a^2 \sin^2\gamma + b^2 \cos^2\gamma)^{3/2}}{ab} \tag{8.98}$$

Reflected Field

Because the plane wave is incident on the elliptical cylinder from a direction ϕ_i, it follows that the direction of the incident ray at the point of reflection Q_r is

$$\hat{\mathbf{s}}^i = -\cos\phi_i \, \hat{\mathbf{x}} - \sin\phi_i \, \hat{\mathbf{y}} \tag{8.99}$$

The point Q_r already has been determined in Example 3.4 to be specified by parameter:

$$\gamma_r = \tan^{-1}\left[\frac{b}{a} \tan\phi_i\right] \tag{8.100}$$

or, in other words, by the coordinates:

$$x_r = a \cos\gamma_r \tag{8.101}$$

$$y_r = b \sin\gamma_r \tag{8.102}$$

Note that (8.100) is not the same as (8.97). According to (8.94) the incident field at Q_r, with its phase referred to the origin according to (8.94), is given by

$$U^i(Q_r) = e^{jkl_r} \tag{8.103}$$

where

$$l_r = \sqrt{x_r^2 + y_r^2}\cos(\phi_r - \phi_i) \tag{8.104}$$

with

$$\phi_r = \tan^{-1}\left[\frac{b}{a}\tan\gamma_r\right] \tag{8.105}$$

from (8.97). Because for the plane wave incidence $s^i \gg 1$, and for the far-zone observation of the reflected field $s^r \gg 1$, we have $L_p \rightarrow \infty$. Thus,

$$X_p \rightarrow \infty \tag{8.106}$$

and because for backscatter $\theta^i = 0$, the Fock parameter for the lit region is

$$\xi_p = -2\,m(\gamma_r) \tag{8.107}$$

where

$$m(\gamma_r) = \left[\frac{ka_0(\gamma_r)}{2}\right]^{1/3} \tag{8.108}$$

This enables us to compute $R_{s,h}(\xi_p, X_p)$ for any incidence angle ϕ_i. Because the observation point is in the far zone:

$$\sqrt{\frac{\rho^r}{\rho^r + s^r}} \approx \frac{\sqrt{\rho^r}}{\sqrt{s^r}}$$

Furthermore, from (3.148) the caustic distance of the reflected field is

$$\rho^r = a_0(\gamma_r)/2$$

and hence the reflected ray component of the backscattered field is

$$U^r(s, \phi_i) = U^i(Q_r)\sqrt{a_0(\gamma_r)/2}\,R_{s,h}(\xi_p, X_p)\frac{e^{-jks^r}}{\sqrt{s^r}} \tag{8.109}$$

Because

$$s^r = s - l_r \tag{8.110}$$

we have

$$e^{-jks^r} = e^{-jks}\, e^{jkl_r} \tag{8.111}$$

Because $s \gg l_r$, we may write

$$1/\sqrt{s^r} \approx 1/\sqrt{s} \tag{8.112}$$

Using (8.103), (8.111), and (8.112), expression (8.109) becomes

$$U^r(s, \phi_i) = R_{s,h}(\xi_p, X_p)\, \sqrt{a_0(\gamma_r)/2}\; e^{j2kl_r}\, \frac{e^{-jks}}{\sqrt{s}} \tag{8.113}$$

Surface-Diffracted Fields

Let the directions of the incident rays at attachment points Q_1' and Q_2' of the two surface-diffracted rays (ray 1 and ray 2) be $\hat{\mathbf{s}}_1'$ and $\mathbf{s}_2'$, respectively. Clearly, for the plane wave incidence case under consideration these directions are the same, and both are equal to $\hat{\mathbf{s}}^i$ in (8.99). Symmetry considerations require that $Q_1' = Q_2$ and $Q_2' = Q_1$. Thus, the diffracted ray direction vectors at Q_1 and Q_2 are given by $\hat{\mathbf{s}}_1^d = \hat{\mathbf{s}}_2^d = -\hat{\mathbf{s}}^i$. In that which follows we adopt the following notation:

$$Q_1' = (x_1', y_1') = (a \cos\gamma_1', b \sin\gamma_1') \tag{8.114}$$

$$Q_1 = (x_1, y_1) = (a \cos\gamma_1, b \sin\gamma_1) \tag{8.115}$$

$$Q_2' = (x_2', y_2') = (a \cos\gamma_2', b \sin\gamma_2') \tag{8.116}$$

$$Q_2 = (x_2, y_2) = (a \cos\gamma_2, b \sin\gamma_2) \tag{8.117}$$

If for a given ϕ_i we locate (x_1', y_1'), we can exploit symmetry to provide immediately

$$(x_1, y_1) = (-x_1', -y_1') \tag{8.118}$$

$$(x_2', y_2') = (x_1, y_1) \tag{8.119}$$

$$(x_2, y_2) = (x_1', y_1') \tag{8.120}$$

To locate attachment point (x_1', y_1'), we apply the rule (8.43), namely that at Q_1'

we have $\hat{\mathbf{s}}_1' \cdot \hat{\mathbf{n}}(Q_1') = 0$. We then obtain

$$x_1' = -\frac{a^2 \sin\phi_i}{\sqrt{a^2 \sin^2\phi_i + b^2 \cos^2\phi_i}} \tag{8.121}$$

$$y_1' = \frac{b^2 \cos\phi_i}{\sqrt{a^2 \sin^2\phi_i + b^2 \cos^2\phi_i}} \tag{8.122}$$

We will sometimes find more convenient to specify the location of these attachment and shedding points by the value of parameter γ. For $j = 1, 2$, we have from (8.95):

$$\gamma_j' = \tan^{-1}(y_j'/x_j') \tag{8.123}$$

$$\gamma_j = \tan^{-1}(y_j/x_j) \tag{8.124}$$

The total arc length over which ray 1 travels on the surface from Q_1' to Q_1 is

$$t_1 = \int_{\tau(Q_1')}^{\tau(Q_1)} d\tau \tag{8.125}$$

The element of arc length parameter $d\tau$ is related to parameter γ as (see Appendix C)

$$d\tau = \left|\frac{d\mathbf{r}}{d\gamma}\right| d\gamma \tag{8.126}$$

From (8.90), we clearly have

$$\left|\frac{d\mathbf{r}}{d\gamma}\right| = \sqrt{a^2 \sin^2\gamma + b \cos^2\gamma} \tag{8.127}$$

and, thus,

$$d\tau = \sqrt{a^2 \sin^2\gamma + b^2 \cos^2\gamma}\, d\gamma \tag{8.128}$$

which means that

$$t_1 = \int_{\gamma_1'}^{\gamma_1} \sqrt{a^2 \sin^2\gamma + b^2 \cos^2\gamma}\, d\gamma \tag{8.129}$$

Symmetry then requires that

$$t_2 = t_1 \tag{8.130}$$

The Fock parameter for surface diffracted ray 1 is

$$\xi_{d1} = \int_{\tau(Q_1')}^{\tau(Q_1)} \frac{m(\tau)}{a_0(\tau)} \, d\tau \tag{8.131}$$

$$= \int_{\gamma_1'}^{\gamma_1} \frac{m(\gamma)}{a_0(\gamma)} \sqrt{a^2 \sin^2\gamma + b^2 \cos^2\gamma} \, d\gamma \tag{8.132}$$

which, from the definition of $m(\gamma)$ in (8.49), becomes

$$\xi_{d1} = \left[\frac{ka^2b^2}{2}\right]^{1/3} \int_{\gamma_1'}^{\gamma_1} \frac{d\gamma}{\sqrt{a^2 \sin^2\gamma + b^2 \cos^2\gamma}} \tag{8.133}$$

Again, symmetry demands that

$$\xi_{d1} = \xi_{d2} \tag{8.134}$$

For each incidence angle ϕ_i, (8.129) and (8.133) are to be evaluated numerically, and then (8.130) and (8.134) also provide the parameters for the second ray.

Because $s_1' \rightarrow \infty$ and $s_1^d \gg 1$, hence $L_d \gg 1$ for ray 1, and thus we take

$$X_{d1} \rightarrow \infty \tag{8.135}$$

because

$$m(\gamma_1') = \left[\frac{ka_0(\gamma_1')}{2}\right]^{1/3} \tag{8.136}$$

$$m(\gamma_1) = \left[\frac{ka_0(\gamma_1)}{2}\right]^{1/3} \tag{8.137}$$

in (8.52) are both finite. Again,

$$X_{d1} = X_{d2} \tag{8.138}$$

With this completed, we have all the parameters necessary to evaluate the surface diffraction coefficient $T_{s,h}$ for both ray 1 and ray 2.

The incident field at Q_1', with its phase referenced to the origin, from (8.94), is given by

$$U^i(Q_1') = e^{jkr_1\cos(\phi_1' - \phi_i)} \tag{8.139}$$

where

$$r_1 = \sqrt{(x_1')^2 + (y_1')^2} \tag{8.140}$$

$$\phi_1' = \tan^{-1}\left[\frac{b}{a}\tan\gamma_1'\right] \tag{8.141}$$

This can be rewritten in the form:

$$U^i(Q_1') = e^{-jkl_1} \tag{8.142}$$

with l_1 the distance shown in Figure 8.14, given by

$$l_1 = r_1 \sin(\phi_1' - \phi_i - \pi/2) = -r_1\cos(\phi_1' - \phi_i) \tag{8.143}$$

Similarly, the incident field at Q_2' is

$$U^i(Q_2') = e^{jkr_2\cos(\phi_2' - \phi_i)} \tag{8.144}$$

where

$$r_2 = \sqrt{(x_2')^2 + (y_2')^2} \tag{8.145}$$

$$\phi_2' = \tan^{-1}\left[\frac{b}{a}\tan\gamma_2'\right] \tag{8.146}$$

This, in turn, can be rewritten in the form:

$$U^i(Q_2') = e^{jkl_2} \tag{8.147}$$

with distance l_2 also shown in Figure 8.14, and given by

$$l_2 = r_2\sin(\phi_2' - \phi_i - 3\pi/2) = r_2\cos(\phi_2' - \phi_i) \tag{8.148}$$

From the symmetry made visible by equations (8.118) to (8.120), we clearly have

$$l_1 = l_2 \tag{8.149}$$

The diffracted field shed from Q_1 then is

$$U_1^d(s_1^d, \phi_i) = U^i(Q_1') \, T_{s,h}(\xi_{d1}, X_{d1}, t_1) \, \frac{e^{jks_1^d}}{\sqrt{s_1^d}} \tag{8.150}$$

From Figure 8.14, we see that

$$s_1^d = s - l_2 \tag{8.151}$$

Because $s \gg l_2$, we can write

$$1/\sqrt{s_1^d} \approx 1/\sqrt{s} \tag{8.152}$$

Using (8.142), (8.151), and (8.152), but recognizing (8.149), expression (8.150) can be rewritten as

$$U_1^d(s, \phi_i) = T_{s,h}(\xi_{d1}, X_{d1}, t_1) \frac{e^{-jks}}{\sqrt{s}} \tag{8.153}$$

The diffracted field shed from Q_2 is

$$U_1^d(s_2^d, \phi_i) = U^i(Q_2') \, T_{s,h}(\xi_{d2}, X_{d2}, t_2) \frac{e^{-jks_2^d}}{\sqrt{s_2^d}} \tag{8.154}$$

From Figure 8.14, we see that

$$s_2^d = s + l_1 \tag{8.155}$$

Because $s \gg l_1$, we can write

$$1/\sqrt{s_2^d} \approx 1/\sqrt{s} \tag{8.156}$$

Using (8.147), (8.155), and (8.156), but recognizing (8.149), expression (8.154) can be rewritten as

$$U_2^d(s, \phi_i) = T_{s,h}(\xi_{d2}, X_{d2}, t_2) \frac{e^{-jks}}{\sqrt{s}} \tag{8.157}$$

Next (8.153) and (8.157) can be summed to give the total diffracted field. Because

of (8.134), (8.138), and (8.130), we simply write the UTD parameters as ξ_d, X_d, and t, and the resultant diffracted field as

$$U^d(s, \phi_i) = 2T_{s,h}(\xi_d, X_d, t)\frac{e^{-jks}}{\sqrt{s}} \tag{8.158}$$

Radar Width

As defined in Chapter 3, the radar width for incidence angle ϕ_i is

$$\sigma^W = \lim_{s \to \infty} 2\pi s \left[\frac{|U^r(s, \phi_i) + U^d(s, \phi_i)|^2}{|U^i(s, \phi_i)|^2} \right] \tag{8.159}$$

Inserting (8.113) and (8.158) into (8.159), and remembering that $|U^i(s, \phi_i)|^2$ is unity, we finally obtain

$$\sigma^W = 2\pi |R_{s,h}(\xi_p, X_p)\sqrt{a_0(\gamma_r)/2}\, e^{j2kl^r} + 2T_{s,h}(\xi_d, X_d, t)|^2 \tag{8.160}$$

Reminder: The user of the UTD scattering solutions must always keep in mind that they are valid only under the following conditions:

(a) The curvature parameter m is large *everywhere* on the surface ray path between Q' and Q and in the vicinity (on either side) of Q' and Q.
(b) The surface curvature is slowly varying with respect to wavelength.
(c) The normalized distance parameters kL_p and kL_d are large, though practice has shown the solutions to be accurate even for values of these quantities as low as $kL_{p,d} = 3$, and even unity in special cases [15].
(d) The incident field U^i must be of the GO type, and the source must be off the surface.
(e) The observation point must be off the surface. The UTD scattering solutions are not valid in the boundary layer at the surface, which is a caustic of the surface diffracted rays.

Remark 5: How is it possible, in the TM (soft) case, for the electric field E_z of the ray that supposedly travels over the surface not to be exactly zero at all points on the surface between attachment and shedding points? This in fact is what the boundary conditions at conducting surfaces require. In answering this question, we remind ourselves that, although this indeed must be true, an observed fact also is that an E_z field is perceived at the observation point. Now, although for the physical phenomena the UTD scattering solution presents a paradigm in which we think in terms of a ray that travels over the surface, remember that it does not permit us actually to "observe" the field at any point on the surface. In other words, as far as its application is concerned, what happens to the field at the surface really does not matter. All we need to know is the precise location of the geodesic path followed between the attachment and shedding points and the geometrical properties of the surface at each point along this path; the UTD expressions will take care of the complex process occurring between points Q' and Q.

In summary, the surface ray field does not necessarily represent the actual field at the surface, but merely an artifice to relate the field at Q' to that at Q. The modal diffraction coefficients $D_n^{s,h}$ are used to relate these surface ray fields (boundary layer fields [11, 12]) to the physical fields that impinge on, or are shed from, the surface [9]. In the radiation and coupling problem classes of the UTD, which are discussed in Sections 8.3 and 8.4, we briefly indicate how further modal coefficients called *launching and attachment coefficients* [9, 12] are used for this purpose.

In the application of the UTD, we need never concern ourselves with the detailed workings of such modal coefficients though. Because of the dilemma of the TM case mentioned earlier, the surface ray often is considered to travel at a small but finite distance above the surface [10, p. 589].

Remark 6: The 2D UTD results presented in this section are obtained from the more general 3D case discussed in [11, 14] by letting one of the principal radii of curvature of the scattering surface become infinite and assuming 2D ray tubes as sources.

8.2.5 UTD Scattering Solution in the Surface-Based Ray Coordinate System

Thus far we have written the UTD solutions in terms of the scalar $U(P)$, which represents either a $\hat{\mathbf{z}}$-directed electric (TM) or magnetic (TE) field. This was done in view of the fact that a 2D field of arbitrary polarization can always be resolved as the composition of a TE and a TM part.

Suppose that we have a line source of some general polarization, providing an incident GO field $\mathbf{E}^i(Q')$, for example, at Q':

$$\mathbf{E}^i(Q') = \mathbf{E}^i_{\mathrm{TE}}(Q') + \mathbf{E}^i_{\mathrm{TM}}(Q') \tag{8.161}$$

Then, we wish to apply the UTD solutions to find the field at some observation point P. We determine in this section the form that a single compact expression for the field will take. This not only introduces ideas that are used in Sections 8.3 and 8.4, but also makes easier proceeding to the 3D situations for which bibliographic guidance is given at the end of the chapter.

We can dispense quickly with the case of observation points in the lit region. The reflected field at P_l then is given by the expressions in Section 3.2.4 in terms of the ray-fixed coordinate system, but with R_s and R_h now the UTD reflection coefficients of (8.36).

We next refer to Figure 8.8 and discuss the situation in which the observation point is P_d in the shadow zone. The TM and TE results are considered separately. The UTD solutions for each of these contexts is written in terms of ray-fixed coordinates, and these results finally are combined into a single expression. The diffracted field at P_d can be written as

$$\mathbf{E}^d(P_d) = \mathbf{E}^d_{\text{TE}}(P_d) + \mathbf{E}^d_{\text{TM}}(P_d) \tag{8.162}$$

TM Part

The TM part of the electric field $\mathbf{E}^i(Q')$ is $\hat{\mathbf{z}}$-directed (the only $\hat{\mathbf{z}}$-component of the incident field):

$$\mathbf{E}^i_{\text{TM}}(Q') \cdot \hat{\mathbf{z}} = \mathbf{E}^i(Q') \cdot \hat{\mathbf{z}} = E^i_z(Q') \tag{8.163}$$

and so

$$\mathbf{E}^i_{\text{TM}}(Q') = E^i_z(Q')\hat{\mathbf{z}} \tag{8.164}$$

For this component of the incident field, the associated surface diffracted field at P_d also is $\hat{\mathbf{z}}$-directed and according to (8.42) is given by

$$E^d_z(P_d) = E^i_z(Q')\, T_s \frac{e^{-jks^d}}{\sqrt{s^d}} \tag{8.165}$$

Thus, we can write

$$\begin{aligned}\mathbf{E}^d_{\text{TM}}(P_d) &= E^i_z(P_d)\hat{\mathbf{z}} \\ &= E^i_z(Q')\hat{\mathbf{z}}\, T_s \frac{e^{-jks^d}}{\sqrt{s^d}} \\ &= [\mathbf{E}^i_{\text{TM}}(Q') \cdot \hat{\mathbf{z}}]\hat{\mathbf{z}}\, T_s \frac{e^{-jks^d}}{\sqrt{s^d}}\end{aligned} \tag{8.166}$$

which from (8.163) is

$$\mathbf{E}^d_{\text{TM}}(P_d) = [\mathbf{E}^i(Q') \cdot \hat{\mathbf{z}}]\hat{\mathbf{z}}\, T_s \frac{e^{-jks^d}}{\sqrt{s^d}} \tag{8.167}$$

which is a $\hat{\mathbf{z}}$-polarized field.

TE Part

The TE part of the incident magnetic field is entirely $\hat{\mathbf{z}}$-directed, for example, $H_z^i(Q')$, and, of course, is the only $\hat{\mathbf{z}}$-directed magnetic field component of the arbitrary incident field. The incident field is of the GO type, and therefore the TE part of the incident electric field at Q' is given in terms of $H_z^i(Q')$ as

$$\mathbf{E}_{\text{TE}}^i(Q') = -Z\,\hat{\mathbf{s}}' \times \hat{\mathbf{z}} H_z^i(Q') = Z\,\hat{\mathbf{z}} \times \hat{\mathbf{s}}'\, H_z^i(Q') \tag{8.168}$$

We know that the polarization of this TE ray does not alter as it travels from the source point to Q'. If we can determine $\hat{\mathbf{z}} \times \hat{\mathbf{s}}'$ at *any* point on the ray, the value is known *everywhere* on that ray. Let this point be the attachment point Q'. Now at Q' the incident ray vector $\hat{\mathbf{s}}'$ always is tangent to the surface and in the surface ray direction $\hat{\mathbf{t}}\,(Q')$. In other words, $\hat{\mathbf{s}}' \cdot \hat{\mathbf{n}}(Q') = 0$, vector $\hat{\mathbf{n}}(Q')$ being the unit normal to the surface at Q'. Therefore,

$$\hat{\mathbf{z}} \times \hat{\mathbf{s}}' = \hat{\mathbf{n}}(Q') \tag{8.169}$$

and (8.168) becomes

$$\mathbf{E}_{\text{TE}}^i(Q') = \hat{\mathbf{n}}(Q')\, Z\, H_z^i(Q') \tag{8.170}$$

So, we have

$$\mathbf{E}_{\text{TE}}^i(Q') \cdot \hat{\mathbf{n}}(Q') = \mathbf{E}^i(Q') \cdot \hat{\mathbf{n}}(Q') = Z\, H_z^i(Q') \tag{8.171}$$

Now, from (8.42), the TE part of the surface-diffracted magnetic field at P_d is

$$H_z^d(P_d) = H_z^i(Q')\, T_h \frac{e^{-jks^d}}{\sqrt{s^d}} \tag{8.172}$$

The diffracted fields being of the GO type enables us to write

$$\mathbf{E}_{\text{TE}}^i(P_d) = -Z\,\hat{\mathbf{s}}^d \times \hat{\mathbf{z}} H_z^d(P_d) = Z\,\hat{\mathbf{z}} \times \hat{\mathbf{s}}^d\, H_z^d(P_d) \tag{8.173}$$

Again, the polarization of this TE surface-diffracted ray does not alter as it travels through free space from the shedding point Q to observation point P_d. If we have $\hat{\mathbf{z}} \times \hat{\mathbf{s}}^d$ at *any* point on the ray, then we know the value *everywhere* on that ray. Let this point be the shedding point Q. At Q, the diffracted ray vector $\hat{\mathbf{s}}^d$ always is tangent to the surface and in the surface ray direction $\hat{\mathbf{t}}(Q)$. In other words, $\hat{\mathbf{s}}^d \cdot \hat{\mathbf{n}}(Q) = 0$, vector $\hat{\mathbf{n}}(Q)$ being the unit normal to the surface at Q (*not* at Q').

Therefore,

$$\hat{\mathbf{z}} \times \hat{\mathbf{s}}^d = \hat{\mathbf{n}}(Q) \tag{8.174}$$

and (8.173) becomes

$$\mathbf{E}^i_{\mathrm{TE}}(P_d) = Z\, H^d_z(P_d)\, \hat{\mathbf{n}}(Q) \tag{8.175}$$

Substituting (8.172) into (8.175), we obtain

$$\mathbf{E}^i_{\mathrm{TE}}(P_d) = ZH^i_z(Q')\hat{\mathbf{n}}(Q)T_h \frac{e^{-jks^d}}{\sqrt{s^d}} \tag{8.176}$$

which from (8.171) can be written as

$$\mathbf{E}^i_{\mathrm{TE}}(P_d) = [\mathbf{E}^i(Q') \cdot \hat{\mathbf{n}}(Q')]\, \hat{\mathbf{n}}(Q)\, T_h \frac{e^{-jks^d}}{\sqrt{s^d}} \tag{8.177}$$

which clearly is polarized in the $\hat{\mathbf{n}}(Q)$ direction. That is, the polarization of a particular ray is determined by the surface normal at its shedding point.

Combining (8.167) and (8.177) according to (8.162), we have

$$\mathbf{E}^d(P_d) = \{[\mathbf{E}^i(Q') \cdot \hat{\mathbf{z}}]\hat{\mathbf{z}}T_s + [\mathbf{E}^i(Q') \cdot \hat{\mathbf{n}}(Q')]\hat{\mathbf{n}}(Q)T_h\} \frac{e^{-jks^d}}{\sqrt{s^d}} \tag{8.178}$$

This result is correct if the rays propagate in the clockwise direction shown in Figure 8.8. Our reasoning has been based on this premise. However, if the source position is moved to the right and the surface diffracted rays travel counterclockwise, the vectors $\hat{\mathbf{t}}(Q')$ and $\hat{\mathbf{t}}(Q)$, which always are chosen to be in the ray direction, also will be directed counterclockwise, though, of course, they will remain tangent to the surface. However, this means that the ray vectors $\hat{\mathbf{s}}' = \hat{\mathbf{t}}(Q')$ and $\hat{\mathbf{s}}^d = \hat{\mathbf{t}}(Q)$ in (8.169) and (8.174) will change sign, but units normal $\hat{\mathbf{n}}(Q')$ and $\hat{\mathbf{n}}(Q)$ are a property of the surface that will be unaffected by the alteration in ray direction. So, for (8.169) and (8.174) to remain correct, we must replace the $\hat{\mathbf{z}}$ there by $-\hat{\mathbf{z}}$. If we defined a vector $\hat{\mathbf{b}} = \pm\hat{\mathbf{z}}$, the plus sign for clockwise rays, and the minus sign for counterclockwise rays, then (8.178) would hold for both ray directions were we to replace the vectors $\hat{\mathbf{z}}$ there by $\hat{\mathbf{b}}$ and to write

$$\mathbf{E}^d(P_d) = \{[\mathbf{E}^i(Q') \cdot \hat{\mathbf{b}}]\hat{\mathbf{b}}T_s + [\mathbf{E}^i(Q') \cdot \hat{\mathbf{n}}(Q')]\hat{\mathbf{n}}(Q)T_h\} \frac{e^{-jks^d}}{\sqrt{s^d}} \tag{8.179}$$

The vector $\hat{\mathbf{b}} = \hat{\mathbf{t}}(Q') \times \hat{\mathbf{n}}(Q') = \hat{\mathbf{t}}(Q) \times \hat{\mathbf{n}}(Q)$, of course, is the binormal vector, and its use here is consistent with that introduced in Chapter 3.

Note: The result (8.179) is sometimes written in dyadic notation as

$$\mathbf{E}^d(P_d) = \mathbf{E}^i(Q') \cdot [\hat{\mathbf{b}}\hat{\mathbf{b}}T_s + \hat{\mathbf{n}}(Q')\hat{\mathbf{n}}(Q)T_h] \frac{e^{-jks^d}}{\sqrt{s^d}} \tag{8.180}$$

although (8.179) may be preferred because it elucidates the role of the various unit vectors.

8.3 THE RADIATION PROBLEM FOR A SOURCE MOUNTED ON A SMOOTH CONVEX CONDUCTING SURFACE

8.3.1 The Radiation Problem Geometry

The UTD solutions of the previous section involved incident fields arriving from sources located off the convex surface. We next discuss the UTD results for sources mounted on the surface itself, but with observation points off the surface. (For observation points on the surface the coupling class of solutions described in Section 8.4 should be used.)

The relevant geometrical arrangement is shown in Figure 8.15, which also shows the location of the SSBs. The source is located at point Q' on the surface; this point often is called the *launching point*. The observation point P has been relabeled P_l if it lies in the lit zone and P_d if it is situated in the shadow zone. A direct ray is shown from Q' to a point P_l, which satisfies the Fermat principle by being a straight line. The ray reaching P_d does so in such a way that the extended

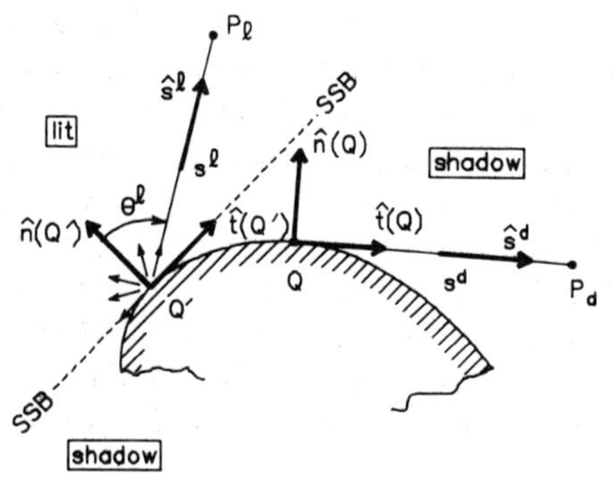

Figure 8.15 Detailed geometry for the radiation problem.

Fermat principle—namely, that $|Q'QP_d|$ be an extremum—is satisfied. Hence, it travels across the geodesic $Q'Q$ on the surface and then sheds tangentially to the surface at Q, propagating as a straight line toward P_d. Thus,

$$\hat{\mathbf{n}}(Q) \cdot \hat{\mathbf{s}}^d = 0 \tag{8.181}$$

where $\hat{\mathbf{n}}(Q)$ is the unit normal to the surface at the shedding point Q, and $\hat{\mathbf{s}}^d$ the diffracted ray direction. Distance s^l is that between the source position Q' and P_l, whereas s^d is that between Q and P_d. The distance along the surface between Q' and Q is t.

The UTD has two separate solutions for the radiation problem: one valid in the lit zone and the other for the shadow zone. These were derived in [9] and [12] for the general 3D case and have been specialized here to the 2D situation. These solutions are valid in the SSB transition regions as well as the deep lit and deep shadow zones. However, observation points must be at least a few wavelengths off the surface, although they may be in the near-zone of the surface.

8.3.2 Sources of the Radiated Fields

To express the UTD radiation solutions, we use a ray-coordinate vector notation similar to that in Section 8.2.5.

An electric current source placed onto and tangential to any conducting surface will not radiate at all, whereas an electric current on and normal to the surface will. Conversely, a magnetic line source on and tangential to the surface radiates, whereas a magnetic current on the surface but normal to it does not. Hence, we need consider only tangential magnetic and normal electric current sources at the radiating surface.

Magnetic Current Sources

To see how these are obtained in practice, consider the three types in Figure 8.16 (a) to (e), each treated as an infinitesimal source of either width $dt' \rightarrow 0$ or height $dl' \rightarrow 0$; their sizes relative to the surface are exaggerated in the sketches for clarity.

In that which follows, let Q_T denote a general point on the surface, and $\delta(Q_T - Q')$ a Dirac delta function that is non-zero only at $Q_T = Q'$. The source in Figure 8.16 (a) is an infinitesimal aperture in which there is a $\hat{\mathbf{z}}$-directed electric field:

$$E_z(Q')\hat{\mathbf{z}} \tag{8.182}$$

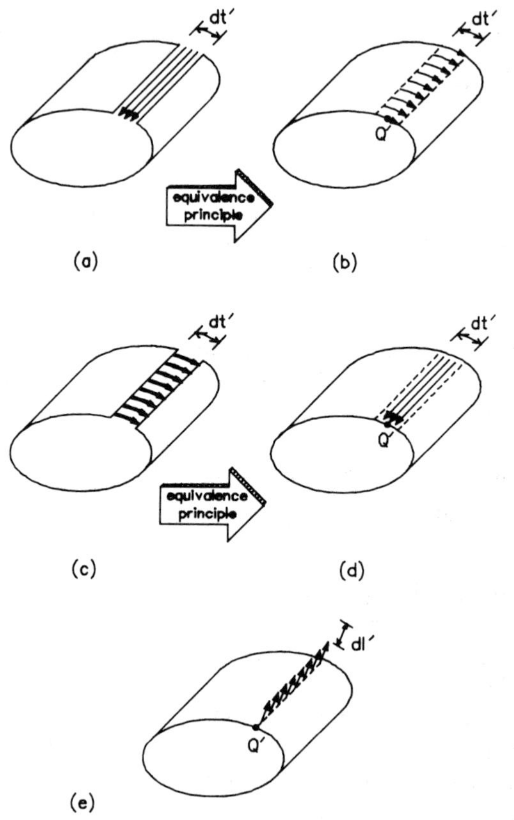

Figure 8.16 Source types for the radiation problem: (a) slot, in a 2D surface, with an axially directed electric field in its aperture; (b) circumferentially directed magnetic current distribution above a closed 2D surface, equivalent to the situation in (a); (c) slot, in a 2D surface, with a circumferentially directed electric field in its aperture; (d) axially directed magnetic current distribution above a closed 2D surface, equivalent to the situation in (c); (e) electric monopole on a 2D surface

We may apply the equivalence principle and replace the aperture by a magnetic current source:

$$M_t(Q')\delta(Q_\tau - Q')\hat{\mathbf{t}}_s(Q') = E_z(Q')\hat{\mathbf{z}} \times \hat{\mathbf{n}}(Q') \tag{8.183}$$

where $\hat{\mathbf{t}}_s(Q')$ is tangent to the surface at Q' and directed counterclockwise. This current element now radiates in the presence of the cylindrical surface with the original aperture removed (that is, shorted out), as in Figure 8.16 (b). We refer to the source (8.183) as a magnetic dipole source and, because it radiates an electric field that is entirely $\hat{\mathbf{z}}$-directed, it corresponds to the TM or soft case.

Similarly, the elemental aperture of Figure 8.16 (c), in which the electric field is

$$E_t(Q')\hat{\mathbf{t}}(Q') \tag{8.184}$$

and is represented by the magnetic current:

$$M_z(Q')\delta(Q_T - Q')\hat{\mathbf{b}} = E_t(Q')\hat{\mathbf{t}}(Q') \times \hat{\mathbf{n}}(Q') \tag{8.185}$$

of Figure 8.16 (d). This $\hat{\mathbf{z}}$-directed elemental magnetic current is the usual magnetic line source already considered in the previous section and elsewhere throughout the text. Because it radiates a magnetic field that is entirely $\hat{\mathbf{z}}$-directed, this current corresponds to the TE or hard case, as we have seen. The general magnetic source can be considered to be

$$\mathbf{M}(Q') = M_z(Q')\delta(Q_T - Q')\hat{\mathbf{z}} + M_t(Q')\delta(Q_T - Q')\hat{\mathbf{t}}_s(Q') \tag{8.186}$$

and its electric field can be expressed as a superposition of independent TM and TE solutions.

Electric Current Sources

Finally, the electric monopole line current (and infinitely long "blade" of electric current normal to the surface) is

$$\mathbf{J}(Q') = J_n(Q')\delta(Q_T - Q')\hat{\mathbf{n}}(Q') \tag{8.187}$$

located at source position Q'.

All three types of source, of course, are infinite in extent along the z-direction, to preserve the 2D nature of the problem. Following [9] and [12], the final UTD solutions for the radiated fields will be written in terms of those for elemental sources. Once this has been done, we will consider what to do about nonelemental sources.

8.3.3 UTD Solution for the Radiation Problem: Observation Point in the Lit Zone

The UTD solution follows for the radiated field at the observation point P_l in the lit zone for the infinitesimal magnetic and electric current source excitations $\mathbf{M}(Q')$ and $\mathbf{J}(Q')$ in expressions (8.186) and (8.187), respectively. We first specify the final expressions and then explain the various terms in them.

For Magnetic Current Source $\mathbf{M}(Q')$

$$\mathbf{E}^l(P_l) = E_n^l(P_l)\,\hat{\mathbf{n}}^l + E_b^l(P_l)\,\hat{\mathbf{b}}^l \tag{8.188}$$

where

$$E_n^l(P_l) = \mathbf{E}^l(P_l)\cdot\hat{\mathbf{n}}^l = C_0\,\mathbf{M}(Q')\cdot\hat{\mathbf{b}}^l\,H^l(\xi^l)\,\frac{e^{-jks^l}}{\sqrt{s^l}} \tag{8.189}$$

$$E_b^l(P_l) = \mathbf{E}^l(P_l)\cdot\hat{\mathbf{b}}^l = C_0\,\mathbf{M}(Q')\cdot\hat{\mathbf{t}}^l(Q')S^l(\xi^l)\,\frac{e^{jks^l}}{\sqrt{s^l}} \tag{8.190}$$

with

$$C_0 = -\sqrt{\frac{k}{8\pi}}\,e^{j\pi/4} \tag{8.191}$$

The polarization of the radiated field, which will be discussed in detail shortly, clearly is either $\hat{\mathbf{n}}^l$ or $\hat{\mathbf{b}}^l$, or a linear combination of these. Solution (8.180) sometimes is written more compactly in the form:

$$\mathbf{E}^l(P_l) = C_0\mathbf{M}(Q')\cdot[\hat{\mathbf{b}}^l\hat{\mathbf{n}}^l H^l(\xi^l) + \hat{\mathbf{t}}(Q')\hat{\mathbf{b}}^l S^l(\xi^l)]\,\frac{e^{-jks^l}}{\sqrt{s^l}} \tag{8.192}$$

For Electric Current Source $\mathbf{J}(Q')$

$$\mathbf{E}^l(P_l) = E_n^l(P_l)\,\hat{\mathbf{n}}^l \tag{8.193}$$

where

$$E_n^l(P_l) = C_0\,Z\,J_n(Q')\,H^l(\xi_l)\,\sin\theta^l\,\frac{e^{-jksl}}{\sqrt{s^l}} \tag{8.194}$$

indicating that it is an $\hat{\mathbf{n}}^l$-polarized field. Alternatively,

$$\mathbf{E}^l(P_l) = C_0\,Z\,\mathbf{J}(Q')\cdot[\hat{\mathbf{n}}(Q')\hat{\mathbf{n}}^l\,H^l(\xi^l)\,\sin\theta^l]\,\frac{e^{-jks^l}}{\sqrt{s^l}} \tag{8.195}$$

Geometrical Parameters and Unit Vectors

The origin and meaning of the unit vectors in the preceding expressions need some explanation. The direct ray from the source point Q' to some observation point P_l in the lit region has the direction vector $\hat{\mathbf{s}}^l$. The angle θ^l is that between $\hat{\mathbf{s}}^l$ and the unit normal $\hat{\mathbf{n}}(Q')$ to the surface at Q'. We therefore have

$$\cos\theta^l = \hat{\mathbf{n}}(Q') \cdot \hat{\mathbf{s}}^l \tag{8.196}$$

Concerning the solution (8.193) and the first term in (8.188), recall that in Section 8.3.2 we observed that the $\hat{\mathbf{z}}$-directed infinitesimal magnetic current source $M_z(Q')$ and the normally directed electric source $\mathbf{J}(Q')$ would generate a magnetic field that is totally $\hat{\mathbf{z}}$-directed. Suppose then that we have this field $H_z(P_l)$. Because the field is of the GO type propagating in the direction $\hat{\mathbf{s}}^l$, the electric field easily is found as

$$\mathbf{E}^l(P_l) = -Z\,\hat{\mathbf{s}}^l \times \hat{\mathbf{z}}H_z(P_l) \tag{8.197}$$

Defining a unit vector:

$$\hat{\mathbf{n}}^l = \hat{\mathbf{z}} \times \hat{\mathbf{s}}^l \tag{8.198}$$

which is shown in Figure 8.17(a), we can write the preceding field as

$$\mathbf{E}^l(P_l) = Z\,H_z(P_l)\,\hat{\mathbf{n}}^l \tag{8.199}$$

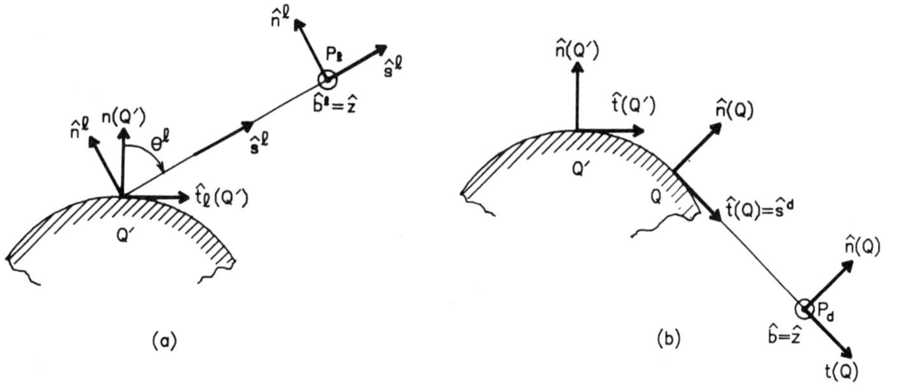

Figure 8.17 (a) Polarization vectors associated with the UTD direct radiation solution in the lit zone (unit vector $\hat{\mathbf{b}}^l = +\hat{\mathbf{z}}$ because $\hat{\mathbf{t}}^l$ has been chosen in the clockwise direction); (b) polarization vectors associated with the UTD diffracted radiation solution (unit vector $\mathbf{b} = +\hat{\mathbf{z}}$ for the clockwise surface ray shown).

The electric field due to $M_z(Q')$, and that radiated by $\mathbf{J}(Q')$ thus is $\hat{\mathbf{n}}^l$-polarized. Remember that $\hat{\mathbf{n}}(Q') \neq \hat{\mathbf{n}}^l$; whereas $\hat{\mathbf{n}}$ is a property of the surface, $\hat{\mathbf{n}}^l$ is a property of the direct ray from Q' to P_l. Note that both $\hat{\mathbf{n}}^l$ and $\hat{\mathbf{s}}^l$ have fixed directions at *any* point on the ray as it propagates from Q' to P_l. The definition (8.198) also implies that

$$\hat{\mathbf{n}}^l \cdot \hat{\mathbf{s}}^l = 0 \tag{8.200}$$

The unit vector $\hat{\mathbf{t}}^l(Q')$ is defined such that

$$\sin\theta^l = \hat{\mathbf{s}}^l \cdot \hat{\mathbf{t}}^l(Q') \tag{8.201}$$

Remember that $\hat{\mathbf{t}}^l(Q')$ is not intended to be associated with any particular one of the two surface rays launched at Q', although in these 2D problems, of course, it always will be equal to the direction of one ray at Q' and tangent to the surface at Q'.

The binormal unit vector $\hat{\mathbf{b}}^l$ is

$$\hat{\mathbf{b}}^l = \hat{\mathbf{t}}^l(Q') \times \hat{\mathbf{n}}(Q') = \hat{\mathbf{s}}^l \times \hat{\mathbf{n}}^l \tag{8.202}$$

If the vector $\hat{\mathbf{t}}^l(Q')$ is chosen to point clockwise so that angle θ^l is measured from the normal $\hat{\mathbf{n}}(Q')$ toward the SSB in the clockwise direction, as in Figure 8.17 (a), then $\hat{\mathbf{b}}^l = \hat{\mathbf{z}}$ and $\hat{\mathbf{t}}^l(Q') = -\hat{\mathbf{t}}_s(Q')$, the latter being that in the source expression (8.183). If the vector $\hat{\mathbf{t}}^l(Q')$ is chosen to point counterclockwise so that angle θ^l is measured from the normal $\hat{\mathbf{n}}(Q')$ toward the SSB in the counterclockwise direction, then $\hat{\mathbf{b}}^l = -\hat{\mathbf{z}}$ and $\hat{\mathbf{t}}^l(Q') = \hat{\mathbf{t}}_s(Q')$. The source $M_t(Q')$ produces a $\hat{\mathbf{z}}$-directed electric field; hence, the polarization of the second term in (8.188).

Fock Parameter Associated with the lit Region

The Fock parameter associated with the lit zone is

$$\xi^l = -m(Q')\cos\theta^l \tag{8.203}$$

always such that

$$\xi^l \leqslant 0 \tag{8.204}$$

and real. In the deep lit region $\xi_l \ll 0$. On the SSB we have $\theta^l = \pi/2$ and $\xi^l = 0$.

As earlier, the curvature parameter

$$m(Q') = \left[\frac{ka_0(Q')}{2}\right]^{1/3} \tag{8.205}$$

with $a_0(Q')$ being the radius of curvature of the surface at the source point Q' in this case.

Special Functions

In (8.188) and (8.194), the intermediate functions $H^l(\xi)$ and $S^l(\xi)$ are defined in terms of the hard (H) and soft (S) Fock radiation functions $g(x)$ and $\tilde{g}(x)$ as

$$H^l(x) = g(x)\,\mathrm{e}^{-jx^3/3} \tag{8.206}$$

$$S^l(x) = \frac{-j}{m(Q')}\tilde{g}(x)\,\mathrm{e}^{-jx^3/3} \tag{8.207}$$

The universal Fock radiation functions $G(x)$ and $\tilde{G}(x)$, defined by expressions (B.37) and (B.38) of Appendix B and computed using routines given in Appendix F, are

$$G(x) = \begin{cases} \mathrm{e}^{-jx^3/3}g(x), & x \leqslant 0 \\ g(x), & x \geqslant 0 \end{cases} \tag{8.208}$$

$$\tilde{G}(x) = \begin{cases} \mathrm{e}^{-jx^3/3}\tilde{g}(x), & x \leqslant 0 \\ \tilde{g}(x), & x \geqslant 0 \end{cases} \tag{8.209}$$

Therefore, because of condition (8.204), (8.206), and (8.207), can be written as

$$H^l(x) = G(x) \tag{8.210}$$

$$S^l(x) = \frac{-j}{m(Q')}\tilde{G}(x) \tag{8.211}$$

8.3.4 UTD Solution for the Radiation Problem: Observation Point in the Shadow Zone

The UTD solution for the radiated field at observation point P_d in the shadow zone follows for both magnetic and electric current source excitation.

For Magnetic Current Source $\mathbf{M}(Q')$

$$\mathbf{E}^d(P_d) = E_n^d(P_d)\hat{\mathbf{n}}(Q) + E_b^d(P_d)\hat{\mathbf{b}} \tag{8.212}$$

where

$$E_n^d(P_d) = C_0\mathbf{M}(Q') \cdot \hat{\mathbf{b}}H(\xi_c)\left[\frac{a_0(Q)}{a_0(Q')}\right]^{1/6} \mathrm{e}^{-jkt}\,\frac{\mathrm{e}^{-jks^d}}{\sqrt{s^d}} \tag{8.213}$$

$$E_b^d(P_d) = C_0\mathbf{M}(Q') \cdot \hat{\mathbf{t}}(Q')S(\xi_c)\left[\frac{a_0(Q)}{a_0(Q')}\right]^{1/6} \mathrm{e}^{-jkt}\,\frac{\mathrm{e}^{-jks^d}}{\sqrt{s^d}} \tag{8.214}$$

These fields clearly are $\hat{\mathbf{n}}(Q)$-polarized and $\hat{\mathbf{b}}$-polarized, as shown in Figure 8.17 (b). Note that both $\hat{\mathbf{n}}(Q)$ and $\hat{\mathbf{s}}^d(Q)$ have fixed directions at *any* point on the ray as it propagates from Q to P_d. Further discussion of the various unit vectors will follow. Note that expression (8.214) may be written in the more compact form in which it often is seen in the literature as

$$\begin{aligned}\mathbf{E}^d(P_d) = {} & C_0\mathbf{M}(Q') \cdot [\hat{\mathbf{b}}\hat{\mathbf{n}}(Q)H(\xi_c) + \hat{\mathbf{t}}(Q')\hat{\mathbf{b}}S(\xi_c)] \\ & \times \left[\frac{a_0(Q)}{a_0(Q')}\right]^{1/6} \mathrm{e}^{-jkt}\,\frac{\mathrm{e}^{-jks^d}}{\sqrt{s^d}}\end{aligned} \tag{8.215}$$

The dyadic factor inside the square brackets may be viewed as transfer functions that describe the launching of the surface ray field at Q', the field's amplitude and phase variation along the path from Q' to Q, and, finally, the field's shedding (diffraction) at the shedding point Q.

For Electric Current Source $\mathbf{J}(Q')$

$$\mathbf{E}^d(P_d) = E_n^d(P_d)\hat{\mathbf{n}}(Q) \tag{8.216}$$

where

$$E_n^d(P_d) = C_0ZJ_n(Q')H(\xi_c)\left[\frac{a_0(Q)}{a_0(Q')}\right]^{1/6} \mathrm{e}^{-jkt}\,\frac{\mathrm{e}^{-jks^d}}{\sqrt{s^d}} \tag{8.217}$$

Alternatively, this result sometimes is presented as

$$\mathbf{E}^d(P_d) = C_0 Z \mathbf{J}(Q') \cdot \hat{\mathbf{n}}(Q')\hat{\mathbf{n}}(Q) H(\xi_c) \left[\frac{a_0(Q)}{a_0(Q')}\right]^{1/6} \mathrm{e}^{-jkt} \frac{\mathrm{e}^{-jks^d}}{\sqrt{s^d}} \tag{8.218}$$

Geometry and Unit Vectors

As before, unit vector $\hat{\mathbf{t}}(Q')$ is the direction of the surface ray launched at Q' (and hence tangential to the surface there). Its direction will be clockwise or counterclockwise, depending on which of the rays launched at Q' is considered. For the ray shown in Figure 8.15, it clearly is clockwise. A similar statement applies to $\hat{\mathbf{t}}(Q)$, the surface ray direction at the shedding point. Unit binormal:

$$\hat{\mathbf{b}} = \hat{\mathbf{t}}(Q') \times \hat{\mathbf{n}}(Q') = \hat{\mathbf{t}}(Q) \times \hat{\mathbf{n}}(Q)$$

was introduced earlier. For the 2D problems considered here, we simply have

$$\hat{\mathbf{b}} = \pm\hat{\mathbf{z}} \tag{8.219}$$

In (8.219), the plus sign applies if the clockwise ray is considered (as for the case illustrated in Figure 8.17 (b)), and the minus sign for the counterclockwise ray. Note that $\hat{\mathbf{t}}^l(Q')$ will be related to $\hat{\mathbf{t}}_s(Q')$ of (8.183) as $\hat{\mathbf{t}}^l(Q') = \mp\hat{\mathbf{t}}_s(Q')$, the minus sign now applying for a clockwise surface ray.

Fock Parameters Associated with the Shadow Zone

In (8.212) and (8.216), the Fock parameter associated with the shadow zone is

$$\xi_c = \int_{\tau(Q')}^{\tau(Q)} \frac{m(\tau)}{a_0(\tau)} d\tau \tag{8.220}$$

and we always have

$$\xi_c \geqslant 0 \tag{8.221}$$

and it is always real, with $\xi_c \gg 0$ in the deep shadow region. The curvature parameter $m(\tau)$ in (8.220) is

$$m(\tau) = \left[\frac{k a_0(\tau)}{2}\right]^{1/3} \tag{8.222}$$

with $a_0(\tau)$ the radius of curvature at some general point identified by the parameter τ. The path length t is given by

$$t = \int_{\tau(Q')}^{\tau(Q)} d\tau \tag{8.223}$$

and is the distance over the surface between the source and shedding points Q' and Q, respectively. Expressions (8.221), (8.222) and (8.223) have the same form as (8.46) to (8.48).

Special Functions

In (8.212) and (8.216) the functions $H(x)$ and $S(x)$ are defined in terms of the hard and soft Fock radiation functions $g(x)$ and $\tilde{g}(x)$ as

$$H(x) = g(x) \tag{8.224}$$

$$S(x) = \frac{-j}{m(Q')} \tilde{g}(x) \tag{8.225}$$

which means that, because of condition (8.221), in terms of the universal Fock radiation functions $G(x)$ and $\tilde{G}(x)$, we have

$$H(x) = G(x) \tag{8.226}$$

$$S(x) = \frac{-j}{m(Q')} \tilde{G}(x) \tag{8.227}$$

The use of the universal Fock functions simplifies computational work. Note that $m(Q')$ in (8.227) is the same as that in (8.205).

Note: The validity of the preceding UTD radiation solutions relies on similar conditions being satisfied as for the scattering problem. The curvature parameter m must be large everywhere on the surface ray path between the launching or source point Q' and shedding or diffraction point Q. The surface curvature must be slowly varying with respect to wavelength, and the observation point must be off the surface.

Finally, note that, although the UTD solutions (8.212) and (8.216) have been called the *shadow zone solutions,* they also are valid in the lit zone for the case of closed bodies that permit rays to creep right around their surface and then shed into the lit zone. In the case of electrically large bodies, such ray contributions usually are very small.

8.3.5 Noninfinitesimal Sources

In many cases, we are able to draw conclusions of practical importance by considering the fields of the infinitesimal sources considered earlier. We here wish to discuss briefly what should be done for sources of finite width or length. One possibility is suggested upon consideration of the form of the fields in the deep lit region. Recall that $\xi_l \ll 0$ in the deep lit region; that is, away from the transition region about the SSB in Figure 8.15. We use the large negative argument form of $G(x)$ and $\tilde{G}(x)$ given in Appendix B, for $\xi_l \ll 0$, to yield the approximations [12, p. 616]:

$$S^l(\xi_l) \approx 2\cos\theta^l \tag{8.228}$$

$$H^l(\xi_l) \approx 2 \tag{8.229}$$

which means that the electric field components in (8.188), due to the magnetic source (8.186), become

$$E_n^l(P_l) = 2\, C_0 M_z(Q')\,\frac{e^{-jks^l}}{\sqrt{s^l}} \tag{8.230}$$

$$E_b^l(P_l) = 2\cos\theta^l C_0\, M_t(Q')\,\frac{e^{-jks^l}}{\sqrt{s^l}} \tag{8.231}$$

whereas that for the electric line monopole (8.187) now is

$$\mathbf{E}^l(P_l)\cdot\hat{\mathbf{b}}^l = 2\sin\theta^l C_0 M_t(Q')\,\frac{e^{-jks^l}}{\sqrt{s^l}} \tag{8.232}$$

These would be the fields of the given sources if they were mounted on a perfectly flat conducting surface. Factor 2 results from the effective doubling of the strength of the sources due to their images in the planar conductor, besides which the radiated fields in the region above the conductor will be the same as those obtained when these sources plus their images radiate in free space. The term $\cos\theta^l$ simply is a pattern factor appropriate to a $\hat{\mathbf{t}}$-directed infinitesimal (Hertzian) magnetic dipole radiating in free space. The $\hat{\mathbf{z}}$-directed field is omnidirectional and therefore has no pattern factor. Because the $\hat{\mathbf{n}}$-directed electric monopole plus its image in the planar conductor form an electric dipole, the $\sin\theta^l$ term is the pattern factor applicable to an infinitesimal (Hertzian) electric dipole radiating in free space [20, p. 233].

Therefore, if we wish to model sources other than the infinitesimal ones in (8.186) and (8.187), we can include in the UTD field expressions multiplicative pattern factors representing the far-zone patterns obtained for the given sources when they are mounted on planar conductors (ground planes).

As an example, for an aperture of width w, over which there is a uniform aperture field (and thus equivalent magnetic current) distribution, the appropriate pattern factor would be that extracted from (2.116) as

$$f(\theta^l) = \frac{\sin[(kw/2)\sin\theta^l]}{(kw/2)\sin\theta^l} \tag{8.233}$$

or for the cosine distribution as

$$f(\theta^l) = \frac{\cos[(kw/2)\sin\theta^l]}{1 - [(kw/2)\sin\theta^l]^2} \tag{8.234}$$

If we wish to model an electric monopole of finite length h, for example, because its pattern when mounted on the planar conducting ground plane is that of a dipole of length $2h$, the pattern factor that could be used [20, p. 244] is

$$f(\theta^l) = \frac{\cos(kh\cos\theta^l) - \cos(kh)}{\sin^2\theta^l} \tag{8.235}$$

Such factors are used in [21] in the study of airborne antenna patterns via the UTD solutions.

Note: Of the utmost importance to remember is that, if such pattern factors indeed are used, their value at $\theta^l = \pi/2$—that is, $f(\theta^l = \pi/2)$—must be included as a multiplicative factor in the appropriate source terms for the shadow region solution as well.

The preceding pattern factor approach obviously is not an exact one. The method that would be correct within the same accuracy as that of the UTD solutions themselves would be one that treated the fields of the infinitesimal sources as Green's functions for sources located on the particular cylindrical surface in question and then integrated over the actual source distribution, the integrand being the Green's function weighted by the specific source distribution. This technique is outlined in more detail in [12]. However, the approach is seldom used in practice, except for the coupling problems discussed in Section 8.4. More often, a given source distribution is quantized into constant amplitude segments of small but finite width or length, and the fields of each segment found from the UTD expressions as if these segments were infinitesimal. The field of the source then is the superposition of the fields due to each of these segments. Most often, the pattern factor method is the one to which we resort.

Remark 7: Continuity at the SSB. As has been done in Section 8.2.4, we can show, through use of the limiting forms of the functions $G(x)$ and $\bar{G}(x)$, that the UTD radiation solutions indeed are continuous on the SSB. Such an exercise must show continuity not only of the complex amplitudes of the fields but of their polarizations as well. Clearly, the $\hat{\mathbf{z}}$-components of the fields are continuous in polarization at the SSB. That the $\hat{\mathbf{n}}^l$-polarized lit zone solutions match the $\hat{\mathbf{n}}(Q)$-polarized shadow zone solutions at the SSB can be seen in Figure 8.18. When angle $\delta \to 0$ unit vectors $\hat{\mathbf{n}}^l$ and $\hat{\mathbf{n}}(Q)$ become the same vector (at the same distance from Q' of course).

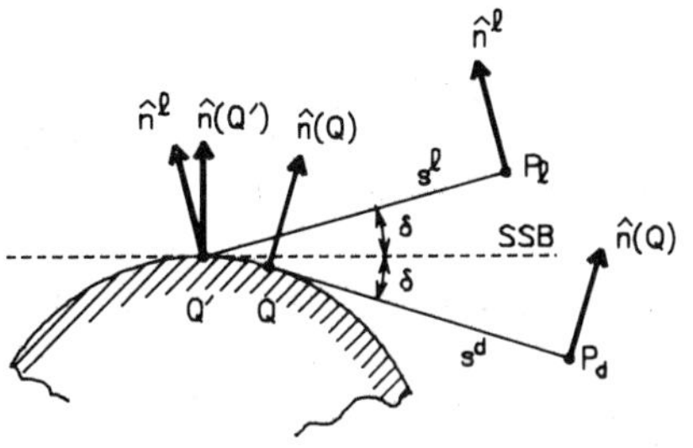

Figure 8.18 Illustration suggesting the continuity of the UTD field solution polarization at the SSB.

8.3.6 Deep Shadow Zone Field Expressions and Their Interpretation

The UTD is designed with applications in mind, and it indeed is possible to simply use the previous solutions without a detailed knowledge of their derivation, even though these are available [9, 12]. However, an examination of limiting cases provides some insight into the significance of the various terms and often can help us check the correctness of a particular application.

Consider, therefore, observation points in the deep shadow region and write

$$H_0(\xi_c) = H(\xi_c)\left[\frac{a_0(Q)}{a_0(Q')}\right]^{1/6} e^{-jkt} \tag{8.236}$$

$$S_0(\xi_c) = S(\xi_c)\left[\frac{a_0(Q)}{a_0(Q')}\right]^{1/6} e^{-jkt} \tag{8.237}$$

Then expression (8.215) can be written

$$\mathbf{E}^d(P_d) = C_0\mathbf{M}(Q') \cdot [\hat{\mathbf{b}}\hat{\mathbf{n}}(Q)H(\xi_c) + \hat{\mathbf{t}}(Q')\hat{\mathbf{b}}S_0(\xi_c)] \frac{e^{-jks^d}}{\sqrt{s^d}} \tag{8.238}$$

If $\xi_c \gg 0$, as in the deep shadow region exterior to the SSB transition region, we can show that [9, 12]

$$H_0 \approx \sum_{n=1}^{N} L_n^h(Q')\, e^{-jkt} \exp\left[-\int_{\tau(Q')}^{\tau(Q)} \alpha_n^h(\tau)d\tau\right] D_n^h(Q) \tag{8.239}$$

$$S_0 \approx \sum_{n=1}^{N} L_n^s(Q')\, e^{-jkt} \exp\left[-\int_{\tau(Q')}^{\tau(Q)} \alpha_n^s(\tau)d\tau\right] D_n^s(Q) \tag{8.240}$$

In (8.239) and (8.240), quantities $L_n^h(Q')$ and $L_n^s(Q')$ are known as the *surface ray modal launching coefficients* at Q', and $D_n^h(Q)$ and $D_n^s(Q)$ are the diffraction coefficients for the nth mode at Q, as for (8.24).

The attenuation constants α_n^s and α_n^h also are the same as those in (8.24). The launching coefficients, introduced by Kouyoumjian and Pathak [9] in 1974, relate source distributions at Q' to the surface ray field launched there. Because, as noted earlier, the UTD surface ray field, in general, is not a true field on the surface, describing the launching or initial value of this boundary layer field by the modal launching coefficient is necessary, as was argued for the modal surface diffraction coefficient in (8.15) in the scattering problem. Recall that, in the scattering solutions, for observation in the deep shadow zone, the launching coefficients are replaced by the modal surface diffraction coefficients at Q', which relate the incident field there to the initial value of the boundary layer field. The launching coefficients, in fact, are approximately related to the diffraction coefficients as [9, p. 1443]

$$L_n^h(Q') \approx \sqrt{2\pi k}\, \frac{e^{-j\pi/2}}{m^2(Q')} Ai'(-q_n) D_n^h(Q') \tag{8.241}$$

$$L_n^s(Q') \approx \sqrt{2\pi k}\, \frac{e^{j\pi/2}}{m(Q')} Ai(-q_n') D_n^s(Q') \tag{8.242}$$

with $Ai'(-q_n)$ and $Ai(-q_n')$ as defined earlier. We would expect the attenuation and phase shift along the surface boundary layer to be the same in the radiation case as in the scattering problem; this indeed is the case, in that the exponential terms in (8.24) and (8.239) or (8.240) are identical.

In applications of the UTD, the forms (8.239) and (8.240) should not be used. They have been mentioned here simply to provide some insight into the problem. Alternative large argument forms for $G(x)$ and $\tilde{G}(x)$ are given in Appendix B, and are those used in the routines $G(X)$ and $GB(X)$ of Appendix F.

Remark 8: The radiation problem is directly related by the reciprocity theorem of electromagnetics to the problem of calculating the fields induced *on* the convex surface by a source located off the surface [15]. This is the reason for the latter case not being considered as a special class of problem in Section 8.1.6.

Remark 9: The radiation patterns of a 2D line source are the same as those of a 3D point source on an infinitely long cylindrical surface in the plane transverse to the axis of the cylinder that contains the 3D point source. Thus, the UTD 2D solutions presented in this chapter can be used to analyze the radiation patterns of finite slots, for example, in the plane specified earlier.

Remark 10: Edge-Excited Surface Rays. If we have a line source in the presence of a curved surface containing an edge, as illustrated in Figure 8.19, then the edge-diffracted field acts like a source term in the radiation solutions that have been described in this section (or indeed for the coupling solutions to be discussed next); hence, we have the term *edge-excited surface rays*. Expressions relating the effective source strengths to the incident fields at the edge and the edge-diffraction coefficients are given in [15].

Example 8.4: Radiation from a magnetic line source $M_z(Q')$ on a conducting elliptical cylinder

Referring to Figure 8.20, the basic geometrical information for the elliptical cylinder already has been considered in Example 8.3; we will use many of the results obtained there. All distances used here are shown in Figure 8.20 and need no further explanation.

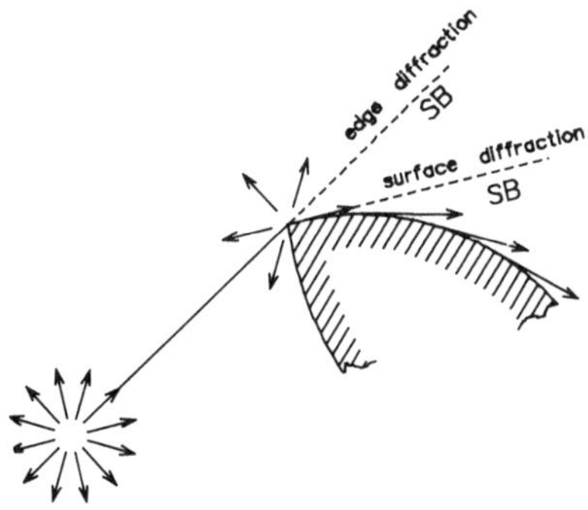

Figure 8.19 Edge-excited surface rays.

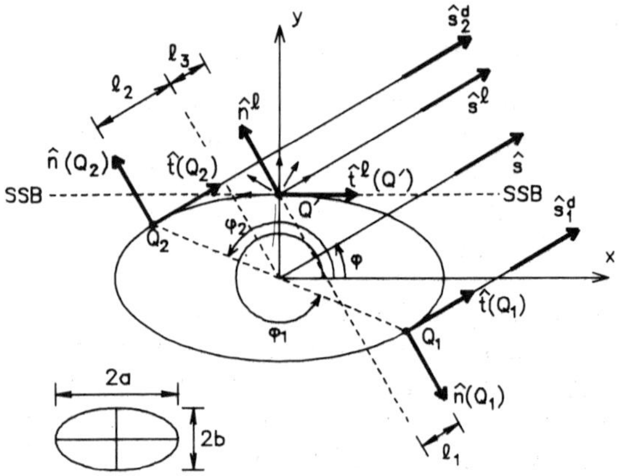

Figure 8.20 Magnetic line source radiating from an elliptic cylinder.

Geometry and UTD Parameters

The observation point is in the far-zone and will be specified as (s, ϕ) both in the lit and shadow regions. There are two surface diffracted rays. Ray 1 travels counterclockwise over the surface and sheds at Q_1; ray 2 travels clockwise and sheds at Q_2. Subscripts 1 and 2 will be used with unit vectors to indicate their association with ray 1 or 2, respectively. In the lit region $(0 \leqslant \phi \leqslant \pi)$, there are both direct rays from the source and surface-diffracted rays. In the shadow zone, the surface-diffracted rays are the only contributors. The observation direction is specified by angle ϕ, the unit vector in this direction being

$$\hat{\mathbf{s}} = \cos\phi \; \hat{\mathbf{x}} + \sin\phi \; \hat{\mathbf{y}} \tag{8.243}$$

Because the observation is in the far-zone:

$$\hat{\mathbf{s}}_1^d = \hat{\mathbf{s}}_2^d = \hat{\mathbf{s}}^l = \hat{\mathbf{s}} \tag{8.244}$$

and angle θ^l is related to ϕ as

$$\theta^l = \pi/2 - \phi \tag{8.245}$$

As far as the various indicated distances are concerned,

$$s_1^d = s - l_1 - l_3 \tag{8.246}$$

$$s_2^d = s + l_2 \tag{8.247}$$

$$s^l = s - l_3 \tag{8.248}$$

Because $s \gg l_{1,2,3}$, we can write

$$1/\sqrt{s_1^d} \approx 1/\sqrt{s_2^d} \approx 1/\sqrt{s^l} \approx 1/\sqrt{s} \tag{8.249}$$

Also, because by definition $\hat{\mathbf{n}}^l \cdot \hat{\mathbf{s}}^l = 0$, from (8.240) we have

$$\hat{\mathbf{n}}^l = -\sin\phi\hat{\mathbf{x}} + \cos\phi\hat{\mathbf{y}} \tag{8.250}$$

Unit vector $\hat{\mathbf{t}}^l(Q')$ has been chosen to point in the clockwise direction, and so

$$\hat{\mathbf{t}}^l(Q') = \hat{\mathbf{x}} \text{ and } \hat{\mathbf{b}}^l = \hat{\mathbf{z}} \tag{8.251}$$

Ray 1 proceeds counterclockwise and thus we can immediately write

$$\hat{\mathbf{t}}_1(Q') = -\hat{\mathbf{x}} \tag{8.252}$$

$$\hat{\mathbf{b}}_1 = \hat{\mathbf{t}}_1(Q') \times \hat{\mathbf{n}}(Q') = -\hat{\mathbf{x}} \times \hat{\mathbf{y}} = -\hat{\mathbf{z}} \tag{8.253}$$

Ray 2 travels in a clockwise direction over the surface, and so

$$\hat{\mathbf{t}}_2(Q') = \hat{\mathbf{x}} \tag{8.254}$$

$$\hat{\mathbf{b}}_2 = \hat{\mathbf{t}}_2(Q') \times \hat{\mathbf{n}}(Q') = \hat{\mathbf{x}} \times \hat{\mathbf{y}} = \hat{\mathbf{z}} \tag{8.255}$$

We will use the information in Example 8.3 to locate the shedding points $Q_1(x_1, y_1)$ and $Q_2(x_2, y_2)$:

$$x_1 = \frac{a^2 \sin\phi}{\sqrt{a^2 \sin^2\phi + b^2 \cos^2\phi}} \tag{8.256}$$

$$y_1 = \frac{-b^2 \cos\phi}{\sqrt{a^2 \sin^2\phi + b^2 \cos^2\phi}} \tag{8.257}$$

$$x_2 = -x_1 \tag{8.258}$$

$$y_2 = -y_1 \tag{8.259}$$

For $j = 1, 2$, we have

$$\gamma_j = \tan^{-1}(y_j/x_j) \tag{8.260}$$

$$\phi_j = \tan^{-1}\left[\frac{b}{a}\tan\gamma_j\right] \tag{8.261}$$

in terms of which we have distances:

$$l_3 = b\cos(\pi/2 - \phi) = b\sin\phi \tag{8.262}$$

$$l_1 = \sqrt{x_1^2 + y_1^2}\cos(\phi + 2\pi - \phi_1) - l_3 \tag{8.263}$$

$$l_2 = \sqrt{x_2^2 + y_2^2}\sin(\phi_2 - \phi - \pi/2) \tag{8.264}$$

The units normal to the surface at Q_1 and Q_2 are parallel to $\hat{\mathbf{n}}^l$ (but only in this case because we are dealing with observation points in the far-zone) and easily are found by inspection to be

$$\hat{\mathbf{n}}(Q_1) = \sin\phi\hat{\mathbf{x}} - \cos\phi\hat{\mathbf{y}} \tag{8.265}$$

$$\hat{\mathbf{n}}(Q_2) = -\sin\phi\hat{\mathbf{x}} + \cos\phi\hat{\mathbf{y}} \tag{8.266}$$

For $j = 1, 2$, ray j travels from Q' (at which parameter $\gamma = \pi/2$) to Q_j and therefore

$$t_j = \pm\int_{\pi/2}^{\gamma_j}\sqrt{a^2\sin^2\gamma + b^2\cos^2\gamma}\,d\gamma \tag{8.267}$$

the plus sign applying to $j = 1$ and the minus when $j = 2$. Note that $t_1 \neq t_2$, in general.

Furthermore, the radius of curvature at any point γ on the elliptical surface is

$$a_0(\gamma) = \frac{(a^2\sin^2\gamma + b^2\cos^2\gamma)^{3/2}}{ab} \tag{8.268}$$

At Q', we have $\gamma = \pi/2$, and so at Q' radius of curvature $a_0(\pi/2) = a^2/b$. Thus, the value of the curvature parameter at Q' becomes

$$m(Q') = \left[\frac{ka^2}{2b}\right]^{1/3} \tag{8.269}$$

Therefore, the lit region Fock parameter is

$$\xi^l = -\left[\frac{ka^2}{2b}\right]^{1/3} \sin\phi \tag{8.270}$$

Because in the lit region we have $0 \leqslant \phi \leqslant \pi$, and $\sin\phi \geqslant 0$, $\xi^l \leqslant 0$ as required. Finally, the Fock parameter for the shadow region, after altering equation (8.133) in Example 8.3, is

$$\xi_{cj} = \left[\frac{ka^2b^2}{2}\right]^{1/3} \int_{\pi/2}^{\gamma_j} \frac{d\gamma}{\sqrt{a^2\sin^2\gamma + b^2\cos^2\gamma}} \tag{8.271}$$

Note that $\xi_{c1} \neq \xi_{c2}$, in general. We can evaluate the radius of curvature at shedding points Q_j using (8.268) and (8.260). For notational convenience, we let

$$\beta_j = \left[\frac{a_0(\gamma_j)}{a_0(\pi/2)}\right]^{1/6} = \left[\frac{ba_0(\gamma_j)}{a^2}\right]^{1/6} \tag{8.272}$$

At this stage we have expressions for all the UTD parameters in terms of the observation angle ϕ.

UTD Solutions for the Total Field $\mathbf{E}(s, \phi)$

We consider the direct field in the lit region first. The magnetic current $\mathbf{M}(Q')$ is entirely $\hat{\mathbf{z}}$-directed. Because $\hat{\mathbf{t}}^l(Q') = \hat{\mathbf{x}}$ we see that $\mathbf{M}(Q') \cdot \hat{\mathbf{t}}^l(Q') = 0$ and thus from (8.190) $E_b^l(s, \phi) = 0$. However, with $\hat{\mathbf{b}}^l = \hat{\mathbf{z}}$, expression (8.189) gives

$$E_n^l(s, \phi) = C_0 M_z(Q') H^l(\xi^l) \frac{e^{-jks^l}}{\sqrt{s^l}}$$

Substitution of (8.248) and (8.249) reduces this to

$$E_n^l(s, \phi) = C_0 M_z(Q') H^l(\xi^l)\, e^{jkl_3} \frac{e^{-jks}}{\sqrt{s}} \tag{8.273}$$

Recalling from (8.188) that this field is $\hat{\mathbf{n}}^l$ – polarized, we use (8.250) to write

$$\mathbf{E}^l(s, \phi) = \begin{cases} -E_n^l(s, \phi)\sin\phi\hat{\mathbf{x}} + E_n^l(s, \phi)\cos\phi\hat{\mathbf{y}}, & 0 \leqslant \phi \leqslant \pi \\ 0, & \text{otherwise} \end{cases} \tag{8.274}$$

We next consider the surface diffracted fields at *any* point (s, ϕ) in the lit or shadow regions. Because of (8.252) and (8.254), the term $\mathbf{M}(Q') \cdot \hat{\mathbf{t}}(Q') = 0$ in (8.214), and hence $E_n^d(s, \phi) = 0$ for both the surface diffracted rays.

For ray 1, equation (8.253) gives $\mathbf{M}(Q') \cdot \hat{\mathbf{b}}_1 = -M_z(Q')$ in (8.213). Thus, using (8.249) and (8.246), expression (8.213) for ray 1 becomes

$$E_{n1}^d(s, \phi) = -C_0 M_z(Q') H(\xi_{c1}) \beta_1 \, e^{-jkt_1} \, e^{jk(l_1+l_3)} \frac{e^{-jks}}{\sqrt{s}} \tag{8.275}$$

This field is $\hat{\mathbf{n}}(Q_1)$-polarized, so that by using (8.265) we may write for $0 \leqslant \phi < 2\pi$

$$\mathbf{E}_1^d(s, \phi) = E_{n1}^d(s, \phi) \sin\phi \hat{\mathbf{x}} - E_{n1}^d(s, \phi) \cos\phi \hat{\mathbf{y}} \tag{8.276}$$

The advantage of expressing each of the field contributions in terms of their $\hat{\mathbf{x}}$ and $\hat{\mathbf{y}}$ components is that they all are easily summed to give the final resultant field.

For ray 2, equation (8.255) gives $\mathbf{M}(Q') \cdot \hat{\mathbf{b}}_2 = M_z(Q')$ in (8.213). Furthermore, (8.249) and (8.247) substituted into (8.213) give

$$E_{n2}^d(s, \phi) = C_0 M_z(Q') H(\xi_{c2}) \beta_2 \, e^{-jkt_2} \, e^{-jkl_2} \frac{e^{-jks}}{\sqrt{s}} \tag{8.277}$$

This field is $\hat{\mathbf{n}}(Q_2)$-polarized, which, in general, is not the same polarization as that of E_{n1}^d, and using (8.263) we may write for $0 \leqslant \phi < 2\pi$

$$\mathbf{E}_2^d(s, \phi) = -E_{n2}^d(s, \phi) \sin\phi \hat{\mathbf{x}} + E_{n2}^d(s, \phi) \cos\phi \hat{\mathbf{y}} \tag{8.278}$$

Thus, the final expression for $\mathbf{E}(s, \phi)$ is

$$\mathbf{E}(s, \phi) = \begin{bmatrix} \mathbf{E}^l(s, \phi) + \mathbf{E}_1^d(s, \phi) + \mathbf{E}_2^d(s, \phi) & 0 \leqslant \phi \leqslant \pi \\ \mathbf{E}_1^d(s, \phi) + \mathbf{E}_2^d(s, \phi) & \pi < \phi < 2\pi \end{bmatrix} \tag{8.279}$$

Alternatively, collecting terms in (8.274), (8.276), and (8.278), we find that (8.279) becomes

$$\mathbf{E}(s, \phi) = E_x(s, \phi)\hat{\mathbf{x}} + E_y(s, \phi)\hat{\mathbf{y}} \tag{8.280}$$

where

$$E_x(s, \phi) = \begin{cases} [-E_n^l(s, \phi) + E_{n1}^d(s, \phi) - E_{n2}^d(s, \phi)] \sin\phi, & 0 \leqslant \phi \leqslant \pi \\ [E_{n1}^d(s, \phi) - E_{n2}^d(s, \phi)] \sin\phi, & \pi < \phi < 2\pi \end{cases} \tag{8.281}$$

$$E_y(s,\ \phi) = \begin{cases} [E_n^l(s,\ \phi) - E_{n1}^d(s,\ \phi) + E_{n2}^d(s,\phi)]\ \cos\phi, & 0 \leqslant \phi \leqslant \pi \\ [-E_{n1}^d(s,\ \phi) + E_{n2}^d(s,\ \phi)]\ \cos\phi, & \pi < \phi < 2\pi \end{cases} \tag{8.282}$$

from which the $\hat{\phi}$-component of the far-zone field can be found as

$$E_\phi(s,\ \phi) = -E_x(s,\ \phi)\ \sin\phi + E_y(s,\ \phi)\ \cos\phi \tag{8.283}$$

Inclusion of a Pattern Factor

Suppose that instead of a magnetic line source at Q', we had a slot of finite width w, with w very much less than the dimensions of the cylinder. The effect of the slot then could be included by including a pattern factor such as described by (8.233); for example,

$$f(\theta^l) = \frac{\sin[(kw/2)\ \sin\theta^l]}{(kw/2)\ \sin\theta^l}$$

If we substitute (8.245), this can be written in terms of observation angle ϕ as

$$f(\phi) = \frac{\sin[(kw/2)\ \cos\phi]}{(kw/2)\ \cos\phi} \tag{8.284}$$

In (8.273), we replace $E_n^l(s,\ \phi)$ by

$$E_n^l(s,\ \phi)\frac{\sin[(kw/2)\ \cos\phi]}{(kw/2)\ \cos\phi} \tag{8.285}$$

and because

$$f(\theta^l = \pm\pi/2) = \frac{\sin(kw/2)}{(kw/2)} \tag{8.286}$$

in (8.275) and (8.277),

$$E_{n1}^d(s,\ \phi) \rightarrow E_{n1}^d(s,\ \phi)\frac{\sin(kw/2)}{(kw/2)} \tag{8.287}$$

$$E_{n2}^d(s,\ \phi) \rightarrow E_{n2}^d(s,\ \phi)\frac{\sin(kw/2)}{(kw/2)} \tag{8.288}$$

where the arrow means "replace with." Note that $f(\theta^l)$ need not be a symmetrical function; it could be "scanned" to one side, for instance. Then, in the preceding considerations, we would use $f(\theta^l = -\pi/2)$ for ray 1 and $f(\theta^l = \pi/2)$ for ray 2.

Figure 8.21 shows the computed solution (without a pattern factor) as a function of ϕ for the case $a = 3\lambda$ and $b = \lambda$. The common factor $e^{-jks}/\sqrt{s}$ has been factored out of each expression. No multiply encircling rays have been included; as is clear from the results given, these rays will be extremely small.

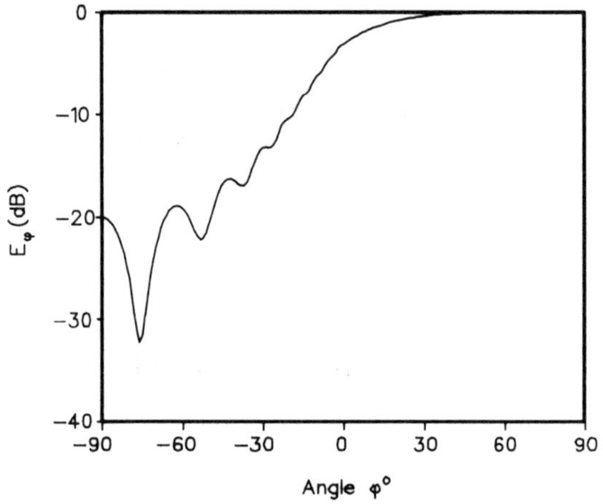

Figure 8.21 UTD solution for the radiation problem shown in Figure 8.20, with $a = 3\lambda$ and $b = 2\lambda$.

Example 8.5: Near-zone fields of an electric monopole $J_n(Q')$ on a circular cylinder

The purpose of this example is simply to work through an exercise for which the polarization vectors are of a more general nature than those obtained in Example 8.4 as a result of far-zone observation. Numerical results will not be shown. This is left to the reader as an exercise in the use of the computational routines provided in Appendix F for the Fock radiation functions.

The problem involves an electric monopole located at $\phi = \phi'$ on a circular conducting cylinder of radius a, as shown in Figure 8.22. The observation point (s, ϕ) is in the near zone. Two singly encircling surface-diffracted rays reach the observation point, for all $0 \leq \phi \leq 2\pi$. In the lit zone (which we will delineate later) a direct ray from the source must be added to these contributions.

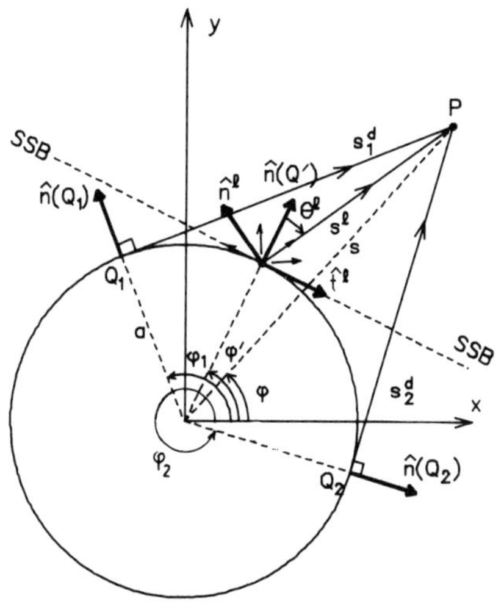

Figure 8.22 Electric current monopole on a circular cylinder of radius a.

Direct Ray Contribution

With $\hat{\mathbf{t}}^l(Q')$ selected to be directed clockwise, as shown in Figure 8.22, we thus have

$$\hat{\mathbf{b}}^l = \hat{\mathbf{z}} \tag{8.289}$$

From equation (3.90), we can write the unit normal to the surface at Q' as

$$\hat{\mathbf{n}}(Q') = \cos\phi'\hat{\mathbf{x}} + \sin\phi'\hat{\mathbf{y}} \tag{8.290}$$

Because $\hat{\mathbf{t}}' \cdot \hat{\mathbf{n}}(Q') = 0$, we then have

$$\hat{\mathbf{t}}^l(Q') = \sin\phi'\hat{\mathbf{x}} - \cos\phi'\hat{\mathbf{y}} \tag{8.291}$$

Denote the source and observation point rectangular coordinates by (x', y') and (x_0, y_0), respectively. Then,

$$(x', y') = (a\cos\phi', a\sin\phi') \tag{8.292}$$

$$(x_0, y_0) = (s\cos\phi, s\sin\phi) \tag{8.293}$$

and, hence,

$$s^l = \sqrt{(x_0 - x')^2 + (y_0 - y')^2} \tag{8.294}$$

and therefore the direct ray direction is

$$\hat{\mathbf{s}}^l = \frac{(x_0 - x')\hat{\mathbf{x}} + (y_0 - y')\hat{\mathbf{y}}}{s^l} \tag{8.295}$$

Thus,

$$\hat{\mathbf{s}}^l \cdot \hat{\mathbf{t}}^l(Q') = \frac{(x_0 - x')\sin\phi' - (y_0 - y')\cos\phi'}{s^l} \tag{8.296}$$

and from (8.201) this yields θ^l for a specified observation point as

$$\theta^l = \sin^{-1}[\hat{\mathbf{s}}^l \cdot \hat{\mathbf{t}}^l(Q')] \tag{8.297}$$

The lit region is delineated by those values of s and ϕ for which $-\pi/2 \leqslant \theta^l \leqslant \pi/2$. For the circular cylinder, the curvature parameter reduces to

$$m = (ka/2)^{1/3} \tag{8.298}$$

at all points on the surface. The Fock parameter for the lit region is

$$\xi^l = -(ka/2)^{1/3}\cos\theta^l \tag{8.299}$$

the value of θ^l being obtained from (8.297). The final step in obtaining the UTD parameters for the direct ray is that of finding the polarization vector $\hat{\mathbf{n}}^l$ of these fields. Substitution of (8.289) and (8.295) into (8.202) yields

$$\hat{\mathbf{n}}^l = -\frac{(y_0 - y')}{s^l}\hat{\mathbf{x}} + \frac{(x_0 - x')}{s^l}\hat{\mathbf{y}} \tag{8.300}$$

Note that $\hat{\mathbf{n}}^l \neq \hat{\mathbf{n}}(Q')$. The direct ray field at (s, ϕ) thus is given from (8.193) and (8.194) as

$$\mathbf{E}^l(s, \phi) = -\frac{(y_0 - y')}{s^l}E_n^l(s, \phi)\hat{\mathbf{x}} + \frac{(x_0 - x')}{s^l}E_n^l(s, \phi)\hat{\mathbf{y}} \tag{8.301}$$

with

$$E_n^l(s, \phi) = C_0 Z J_n(Q')H^l(\xi_l) \sin\theta^l \frac{e^{-jks^l}}{\sqrt{s^l}} \tag{8.302}$$

Surface-Diffracted Ray Contributions

Let us assume that ray 1 travels clockwise and sheds at Q_1, and ray 2 moves counterclockwise and sheds at Q_2. With the subscripts indicating the ray, we then have

$$\hat{\mathbf{b}}_1 = \hat{\mathbf{z}} \tag{8.303}$$

$$\hat{\mathbf{b}}_2 = -\hat{\mathbf{z}} \tag{8.304}$$

Like s^l, the distances s_1^d and s_2^d are finite. Note the right angles indicated in the vicinity of Q_1 and Q_2. These immediately allow us to write

$$s_1^d = \sqrt{s^2 - a^2} \tag{8.305}$$

which is known once the observation point has been specified, and from which we have

$$\tan(\phi_1 - \phi) = s_1^d/a \tag{8.306}$$

and therefore the shedding point Q_1 is located by

$$\phi_1 = \phi + \tan^{-1}(s_1^d/a) \tag{8.307}$$

Similarly,

$$s_2^d = \sqrt{s^2 - a^2} = s_1^d \tag{8.308}$$

and the shedding point Q_2 is at

$$\phi_2 = \phi - \tan^{-1}(s_2^d/a) \tag{8.309}$$

The distances traveled by ray 1 over the surface is

$$t_1 = a\,(2\pi - \phi_1 + \phi') \tag{8.310}$$

and that by ray 2 is

$$t_2 = a(2\pi + \phi_2 - \phi') \tag{8.311}$$

The Fock parameters ξ_c for the two rays are

$$\xi_{c1} = (k/2)^{1/3}\, a^{-2/3} t_1 \tag{8.312}$$

$$\xi_{c2} = (k/2)^{1/3}\, a^{-2/3} t_2 \tag{8.313}$$

Because the radius of curvature is simply a at all points on the surface, the factor:

$$\left[\frac{a_0(Q)}{a_0(Q')}\right]^{1/6} = 1 \tag{8.314}$$

We must still determine expressions for the diffracted field polarization vectors $\hat{\mathbf{n}}(Q_1)$ and $\hat{\mathbf{n}}(Q_2)$. Because these are the normals to the surface at the shedding points they, too, may be determined from (3.90) as

$$\hat{\mathbf{n}}(Q_1) = \cos\phi_1\hat{\mathbf{x}} + \sin\phi_1\hat{\mathbf{y}} \tag{8.315}$$

$$\hat{\mathbf{n}}(Q_2) = \cos\phi_2\hat{\mathbf{x}} + \sin\phi_2\hat{\mathbf{y}} \tag{8.316}$$

The surface diffracted field contribution of ray 1 at (s, ϕ) then is obtained using (8.216), (8.217), and (8.315) to be

$$\mathbf{E}_1^d(s, \phi) = E_{n1}^d(s, \phi) \cos\phi_1\hat{\mathbf{x}} + E_{n1}^d(s, \phi) \sin\phi_1\hat{\mathbf{y}} \tag{8.317}$$

with

$$E_{n1}^d(s, \phi) = C_0 Z J_n(Q') H(\xi_{c1})\, e^{-jkt_1} \frac{e^{-jks^{d_1}}}{\sqrt{s^{d_1}}} \tag{8.318}$$

Similarly, the contribution due to ray 2 is

$$E_2^d(s, \phi) = E_{n2}^d(s, \phi) \cos\phi_2\, \hat{\mathbf{x}} + E_{n2}^d(s, \phi) \sin\phi_2\hat{\mathbf{y}} \tag{8.319}$$

with

$$E_{n2}^d(s, \phi) = C_0 Z J_n(Q') H(\xi_{c2})\, e^{-jkt_2} \frac{e^{-jks^{d_2}}}{\sqrt{s^{d_2}}} \tag{8.320}$$

The total field is the sum of (8.319) and (8.317) if the observation point is in the shadow zone; and it is the sum of these two terms plus that in (8.301) if the observer is in the lit region.

8.4 THE TWO-DIMENSIONAL CONVEX CONDUCTING SURFACE COUPLING PROBLEM

8.4.1 Detailed Geometry for the Coupling Problem

We finally have come to the third class of surface diffraction problems: coupling between sources mounted on the same convex conducting surface. Both the source at Q' and observation (or coupling) point Q are located on the surface, as shown in Figure 8.23. The surface ray path satisfies the generalized Fermat principle and hence is a geodesic on the surface between Q' and Q. For 2D problems, the path thus simply follows the cross-sectional contour, as for the other two problem classes in Sections 8.2 and 8.3.

8.4.2 Preliminaries

We need to use two further special functions $u(x)$ and $v(x)$, known as the soft and hard Fock coupling functions, respectively. As with the other functions, the definition and discussion of $u(v)$ and $v(x)$ have been relegated to Appendix B, with routines *FU(X, UU)* and *FV(X, VV)* given in Appendix F for their numerical evaluation. In the UTD expressions that follow, the following functions will be used.

$$F_s(x) = \sqrt{\frac{jk}{2}}\, 2\, f^{-1}\, e^{-j\pi/4}\, x^{-1/2}\, v(x) \tag{8.321}$$

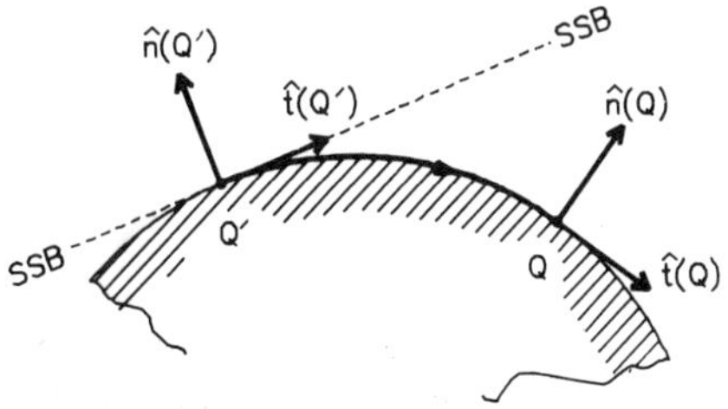

Figure 8.23 Detailed geometry for the coupling problem.

applies to the TE coupling configuration, and

$$G_s(x) = -\sqrt{\frac{jk}{2}}\, f^{-3}\, e^{-j3\pi/4}\, x^{-3/2}\, u(x) \tag{8.322}$$

is applicable to the TM situation, where

$$f = m(Q')\left[\frac{a_0(Q)}{a_0(Q')}\right]^{1/6} \tag{8.323}$$

will be used. Note that f in (8.321) depends on the radius of curvature of the surface at both Q' and Q. The curvature parameter $m(Q')$ is that defined in (8.205). All the unit vectors that are used have already been defined in Section 8.3.

The UTD solutions that follow are those of Pathak and Wang [13], specialized to the 2D situation [22].

8.4.3 UTD Coupling Solution for Magnetic Current Sources

The source is assumed to be the infinitesimal magnetic current of (8.186) given by

$$\mathbf{M}(Q') = M_z(Q')\delta(Q_\tau - Q')\hat{\mathbf{z}} + M_t(Q')\delta(Q_\tau - Q')\hat{\mathbf{t}}_s \tag{8.324}$$

Boundary conditions forbid a nonzero normal component of a magnetic field at a conducting surface, so only a tangential magnetic field exists at Q on the surface. By substituting (8.324) into (8.186) this field is given by [13]

$$\mathbf{H}(Q) = H_b(Q)\hat{\mathbf{b}} + H_t(Q)\hat{\mathbf{t}}(Q) \tag{8.325}$$

with

$$H_b(Q) = C_0 Y \mathbf{M} \cdot \hat{\mathbf{b}}\, F_S(\xi_c)\, e^{-jkt} \tag{8.326}$$

$$H_t(Q) = C_0 Y\, \mathbf{M} \cdot \hat{\mathbf{t}}(Q')\, G_S(\xi_c)\, e^{-jkt} \tag{8.327}$$

The tangential magnetic field $H_b(Q)$ is $\hat{\mathbf{b}}$-polarized whereas $H_t(Q)$ is $\hat{\mathbf{t}}(Q)$-polarized. In these solutions t denotes the surface ray distance from Q' to Q, as given by (8.223), and the Fock parameter ξ_c is identical to that in (8.220). Note that $\xi_c \geqslant 0$, is a real quantity, and in the deep shadow region (that is, when Q is far from Q') we have $\xi_c \gg 0$. The constant C_0 is given in (8.191).

Alternatively, we can write the result in dyadic notation as

$$\mathbf{H}(Q) = C_0 Y\mathbf{M} \cdot [\hat{\mathbf{b}}(Q')\hat{\mathbf{b}}(Q)F_S(\xi_c) + \hat{\mathbf{t}}(Q')\hat{\mathbf{t}}(Q)G_S(\xi_c)]\, e^{-jkt} \tag{8.328}$$

8.4.4 UTD Coupling Solution for Electric Current Sources

If instead an electric line monopole current source:

$$\mathbf{J}(Q') = J_n(Q')\delta(Q_\tau - Q')\hat{\mathbf{n}}(Q') \tag{8.329}$$

as defined in (8.187) is located at source position Q', then the UTD solution for the electric field at Q is [13]

$$\mathbf{E}(Q) = E_n(Q)\hat{\mathbf{n}}(Q) \tag{8.330}$$

where

$$E_n(Q) = C_0 Z\mathbf{J}(Q') \cdot \hat{\mathbf{n}}(Q')F_S(\xi_c)\, e^{-jkt} \tag{8.331}$$

or we can write it as

$$\mathbf{E}(Q) = C_0 Z\mathbf{J}(Q') \cdot [\hat{\mathbf{n}}(Q')\hat{\mathbf{n}}(Q)]F_S(Q)\, e^{-jkt} \tag{8.332}$$

Note that the field in (8.330) is $\hat{\mathbf{n}}(Q)$-polarized. The field must be entirely normal to the surface at Q, because there cannot be a nonzero component of electric field tangential to a conducting surface.

8.4.5 Special Geometries

Circular Cylinder

Suppose the cylindrical surface has a circular cross section of radius a. Then the curvature parameter simplifies to

$$m = (ka/2)^{1/3} \tag{8.333}$$

and thus ξ_c becomes

$$\xi_c = (k/2)^{1/3}\, a^{-2/3}\, t \tag{8.334}$$

with t the distance along the surface from Q' to Q. The term f in (8.323) is

$$f = (ka/2)^{1/3} \tag{8.335}$$

and, therefore,

$$f^{-1} = (2/ka)^{1/3} \tag{8.336}$$

$$f^{-3} = (2/ka) \tag{8.337}$$

Also, we then have

$$\sqrt{\frac{jk}{2}}\, 2\, f^{-1}\, e^{-j\pi/4}\, x^{-1/2} = \frac{2}{\sqrt{t}} \tag{8.338}$$

which implies that

$$F_s(\xi_c) = 2\, t^{-1/2}\, v(\xi_c) \tag{8.339}$$

In a similar fashion, we find from (8.322), (8.334), and (8.336) that

$$G_s(\xi_c) = \frac{2j}{k}\, t^{-3/2}\, v(\xi_c) \tag{8.340}$$

Thus, we can write

$$H_b(Q) = 2\, C_0\, Y\, \mathbf{M}(Q') \cdot \hat{\mathbf{b}} \left[\frac{v(\xi_c)\, e^{-jkt}}{t^{1/2}}\right] \tag{8.341}$$

$$H_t(Q) = \frac{2\, j\, C_0\, Y\mathbf{M}(Q') \cdot \hat{\mathbf{t}}(Q')}{k} \left[\frac{u(\xi_c)\, e^{-jkt}}{t^{3/2}}\right] \tag{8.342}$$

$$E_n(Q) = 2\, C_0 Z\, J_n(Q') \left[\frac{v(\xi_c)\, e^{-jkt}}{t^{1/2}}\right] \tag{8.343}$$

Planar Conducting Surface

This surface can be considered to be the circularly cylindrical problem for which $a \to \infty$. From (8.334) we note that as $a \to \infty$ the Fock parameter $\xi_c \to 0$. As is clear from the small argument forms of $u(x)$ and $v(x)$ given in equations (B.47) and (B.48), respectively, of Appendix B, $v(0) = u(0) = 1$. Therefore, for $\xi_c = 0$, we have from (8.340):

$$H_b(Q) = 2\, C_0\, Y\, M_Z(Q')\, \frac{e^{-jkt}}{\sqrt{t}} \tag{8.344}$$

which, as expected, is the field of a magnetic line source placed on a flat conducting ground plane and observed on the ground plane itself, at a distance t away from the source. Comparing this with expression (2.91), we see that the factor 2 results from the imaging of the source causing an effective doubling of the source strength and the resultant field in the region above the conducting surface being equivalent to that of a line source of strength $2M_z(Q')$ radiating in free space.

8.4.6 A Form of the Coupling Solution in the Deep Shadow Region and Its Interpretation

When the coupling point Q is far from the source point Q'—that is, when Q is in the deep shadow region outside the transition region surrounding the SSB—surface ray modal representations for $F_s(\xi_c)$ and $G_s(\xi_c)$ are [9, p. 1443]

$$F_S\, e^{-jkt} \approx \sum_{n=1}^{N} L_n^h(Q')\, e^{-jkt} \exp\left[\int_{\tau(Q')}^{\tau(Q)} \alpha_n^h(\tau)\, d\tau\right] A_n^h(Q) \tag{8.345}$$

$$G_S\, e^{-jkt} \approx \sum_{n=1}^{N} L_n^s(Q')\, e^{-jkt} \exp\left[\int_{\tau(Q')}^{\tau(Q)} \alpha_n^s(\tau)\, d\tau\right] A_n^s(Q) \tag{8.346}$$

$L_n^h(Q')$ and $L_n^s(Q')$ are the surface-ray modal launching coefficients at Q' mentioned in Section 8.3.6 in connection with the radiation problem. If we remember that these relate the source at Q' to the fictitious surface-ray fields there, they ought to appear in the coupling solutions as well, because that part of the problem concerned with matters in the vicinity of the source indeed is the same for the radiation and coupling problems. Also, the exponential terms in (8.345) and (8.346), which describe that attenuation and phase shift along the surface boundary, are identical to the corresponding ones for the scattering and radiation problems, as we may expect.

Once the surface ray modes have been related to a source or an incident field at Q', there is no difference between the three classes of problem until we "reach" the point Q. In the scattering and radiation problems the surface-ray field had to be related to a field that is shed at Q; hence, they both utilized the modal diffraction coefficients $D_n^h(Q)$ and $D_n^s(Q)$ there. For the coupling problem, we wish to relate the fictitious surface ray modal fields to an observable physical field at Q, on the surface itself. Rather than the diffraction coefficients we must use *attachment coefficients* [9] $A_n^h(Q)$ and $A_n^s(Q)$. By reciprocity these can be shown to be related to the launching coefficients as [9]

$$L_n^h(Q) = A_n^h(Q) \tag{8.347}$$

$$L_n^s(Q') = -A_n^s(Q) \tag{8.348}$$

These results have been included simply for the satisfying correspondence they show between the three different problem classes considered in this chapter. When applying the UTD to particular problems and performing numerical computations we use the solutions given in Sections 8.4.3 and 8.4.4 in terms of the Fock radiation functions and the computer routines provided in Appendix F for their numerical evaluation.

Example 8.6 Coupling between slots on a conducting circular cylinder

Mutual Admittance

The mutual admittance between any two slots has been given by Richmond as [23]

$$Y_{12} = Y_{21} = \frac{1}{V_1 V_2} \iint_{S_2} \mathbf{M}_2(Q) \cdot \mathbf{H}(Q', Q)\, dS \tag{8.349}$$

where

(a) S_2 is the surface of slot 2.
(b) $\mathbf{M}_2(Q)$ is the equivalent magnetic current distribution representing slot 2, at points Q on S_2.
(c) $\mathbf{H}(Q', Q)$ is the magnetic field at points Q on S_2, due to source $\mathbf{M}_1(Q')$ at points Q' over slot 1 surface S_1, but with slot 2 shorted out.
(d) V_1 and V_2 are network model terminal voltages, taken to be the amplitudes of the aperture fields in the slots and thus of their respective equivalent magnetic currents.

We next apply (8.349) to determine the mutual admittance between two slots on a conducting cylinder.

Defining Geometry

Consider a circular cylinder of radius a, at whose surface are two radiating slots, each of width w and each having an azimuthally directed aperture field distribution of amplitudes V_1 and V_2, respectively. As illustrated in Figure 8.24, these slots are centered at $\phi = \phi_1$ and $\phi = \phi_2$, respectively. We replace the slots by their equivalent magnetic current distributions M_{z1} and M_{z2} on the (now closed) conducting surface. Each such distribution thus subtends an angle w/a.

We specify positions on slot 1 by angular variable ϕ', and those on slot 2 by ϕ. The slot field distributions are assumed to be cosinusoidal and so the equivalent magnetic current distributions are

$$M_{z1}(\phi') = V_1 \cos(\pi a\phi'/w), \quad \phi_1 - (2w/a) \leq \phi' \leq \phi_1 + (2w/a) \tag{8.350}$$

$$M_{z2}(\phi) = V_2 \cos(\pi a\phi/w), \quad \phi_2 - (2w/a) \leq \phi \leq \phi_2 + (2w/a) \tag{8.351}$$

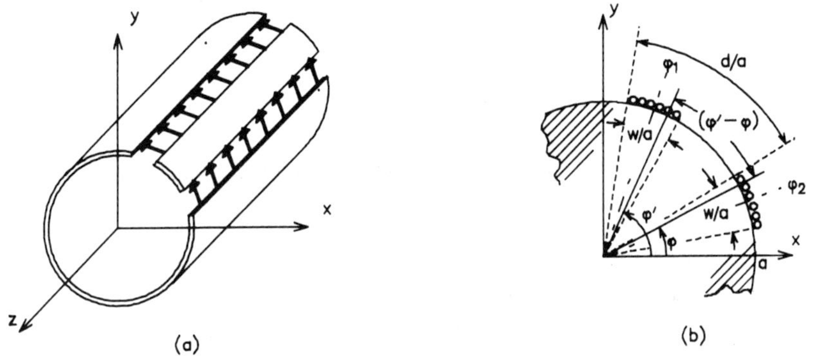

Figure 8.24 (a) Slots in a circular cylinder; (b) equivalent magnetic current formulation for the geometry in (a).

Surface Field

We are considering coupling via a ray that travels clockwise from slot 1 to slot 2, and so in (8.341) we have $\hat{\mathbf{b}} = \hat{\mathbf{z}}$ and thus $\mathbf{M}(Q') \cdot \hat{\mathbf{b}} = M_{z1}(\phi')$. The magnetic field at some point ϕ due to an infinitesimal *line source* positioned at ϕ' is given directly from (8.341) to be, for example,

$$g(\phi', \phi) = \frac{2C_0 Y}{\sqrt{t}} \, v(\xi_c) \, e^{-jkt} \tag{8.352}$$

where, from (8.334)

$$\xi_c = \left[\frac{k}{2a^2}\right]^{1/3} t \tag{8.353}$$

$$t = a(\phi' - \phi) \tag{8.354}$$

The field $H_z(\phi)$ at any point ϕ on the surface is given by equation (8.352) integrated over the source distribution $M_{z1}(\phi')$. In other words, we view $g(\phi', \phi)$ in (8.352) as the asymptotic Green's function for the problem at hand [24]. Thus,

$$H_z(\phi) = \int_{\phi_1-(2w/a)}^{\phi_1+(2w/a)} M_{z1}(\phi')g(\phi', \phi)ad\phi' \tag{8.355}$$

$$= a\int_{\phi_1-(2w/a)}^{\phi_1+(2w/a)} \cos(\pi a\phi'/w)g(\phi', \phi)\, d\phi' \tag{8.356}$$

Using the definition of Y_{21} in (8.349) along with (8.355) and (8.351), we obtain

$$Y_{21} = \int_{\phi_2-(2w/a)}^{\phi_2+(2w/a)} M_{z2}(\phi)H_z(\phi)ad\phi \tag{8.357}$$

Sustitution of (8.356) reduces this to

$$Y_{21} = 2a^2\int_{\phi_2-(2w/a)}^{\phi_2+(2w/a)}\int_{\phi_1-(2w/a)}^{\phi_1+(2w/a)} \cos(\pi a\phi/w)\cos(\pi a\phi'/w)g(\phi', \phi)d\phi' d\phi \tag{8.358}$$

Quantitative results for Y_{21} as a function of w, and the separation t, can be obtained by numerical evaluation of (8.358).

Remark 11: Discussions of the determination of mutual coupling between slots on cylinders and cones, using asymptotic techniques (albeit not necessarily using the UTD solutions) can be found in the journal articles [24–27] and the book chapter by Borgiotti [28].

Remark 12: The UTD solutions also can be used in determining self-impedance or self-admittance of slots and monopoles on curved surfaces. Special care must be taken in the evaluation of the resulting integrals because of troublesome singularities that arise. The reader is referred to the paper by Lee and Mittra [29] for a discussion of such matters.

8.5 BIBLIOGRAPHIC REMARKS

The circularly cylindrical canonical problem geometry discussed at the beginning of this chapter clearly is not sufficient for deriving curved-surface–diffraction coefficients for 3D geometries. The only two canonical geometries for which asymptotic solutions can be manipulated to yield a ray description are the circular cylinder and the sphere. These are used in the following references in the derivation of 3D UTD solutions for the three classes of problem:

(a) References [11, 14, 15] for the scattering case.

(b) References [9, 12, 15] for the radiation problem.
(c) References [13, 15, 22] for the coupling arrangement.

The UTD solution for the reflected field of 3D problems simply is the 3D reflection formulation discussed in Chapter 3, but with R_s and R_h the UTD reflection coefficients of expression (8.36) and applying to convex surfaces.

Fortunately, the 2D considerations contain most of what can be considered the essential features of curved-surface diffraction, and the move to 3D cases is *relatively* painless once the 2D cases have been understood. However, we would expect the purely geometrical aspects (e.g., locating attachment or shedding points and the like, determining the location of appropriate geodesics between the points Q' and Q) to become more complicated; and this indeed is so.

The UTD solutions for the 2D case have been given in a form that relates easily to the 3D results. Indeed, the latter are obtained as a special case of the former. A number of additional or altered terms appear in the 3D solutions that are not in the 2D ones. For instance, in 3D problems, as the ray travels across the surface the binormal vector related to the ray field polarization does not necessarily remain pointed in a constant direction (it is not always $\pm\hat{\mathbf{z}}$). This because of a property of 3D surfaces known as *torsion.* The notation $\hat{\mathbf{b}}(Q)$ and $\hat{\mathbf{b}}(Q')$ then is used to identify precisely where the $\hat{\mathbf{b}}$ vector is being determined. This must be specified in any UTD expressions employing the $\hat{\mathbf{b}}$ vector. Generally speaking, when any binormal is dotted with a source distribution it will be $\hat{\mathbf{b}}(Q')$. When it describes the polarization of a surface diffracted ray, it is $\hat{\mathbf{b}}(Q)$. The reader is referred to the references listed earlier for precise details of the 3D forms.

Here, we discussed 2D problems of diffraction by smooth *convex* bodies. Situations applicable to concave conducting surfaces have been discussed by Ishihara and Felsen [30–32]. Note that the concave problem is complicated by the appearance of multiply reflected fields, and to cast the phenomena solely in ray-optical terms is not always possible.

PROBLEMS

8.1 Repeat the computations of Examples 8.1 and 8.2 but for the case of a magnetic line source parallel to the cylinder.

8.2 In Section 8.2.4, the continuity at the SSB of the UTD solutions for the scattering problem was examined. Devise a similar examination for the UTD radiation solution given in Section 8.3.

8.3 Formulate and code the problem geometry of Example 8.4, but for an arbitrary line source position, to study the effect of the magnetic line source location on the radiation patterns. Include the option of a pattern factor for the line source.

8.4 Develop a computer program to obtain numerical results for Example 8.5.

8.5 Derive expressions for the near-zone fields of an electric monopole on an elliptic cylinder.

8.6 Write a computer program to perform numerical calculations for the mutual admittance in Example 8.6.

8.7 Consider the *E*-plane horn in Figure 5.15, where straight flanges were added to improve the FBR. Replace the straight flanges with circular rolled edges of radius *a* (similar to those added to the parabolic reflector in Figure 4.39). Determine the value of radius *a* required to obtain an FBR of 60 dB. Compare the size of this horn with that in Figure 5.15 which gives the same FBR.

REFERENCES

[1] J.B. Keller, "Diffraction by a Convex Cylinder," *IRE Trans. Antennas and Propagation,* Vol. AP-24, 1956, pp. 312–321.

[2] D.S. Jones, *Methods in Electromagnetic Wave Propagation,* Oxford University Press, London, 1979.

[3] R.G. Kouyoumjian, "The Geometrical Theory of Diffraction and Its Applications" in *Numerical and Asymptotic Techniques in Electromagnetics,* R. Mittra (Ed.), Springer-Verlag, New York, 1975.

[4] P.L. Christiansen, "Canonical Problems and Diffraction Coefficients," in *Theoretical Methods for Determining the Interaction of Electromagnetic Waves with Structures,* J.K. Skwirzynski (Ed.), Sijthoff and Noordhoff, Amsterdam, 1981, pp. 455–475.

[5] M. Levy, and J.B. Keller, "Diffraction by a Smooth Object," *Comm. Pure Appl. Math.,* Vol. 12, 1959, pp. 159–209.

[6] R.G. Kouyoumjian, "Asymptotic High-Frequency Methods," *Proc. IEEE,* Vol. 53, No. 8, August 1965, pp. 864–876.

[7] W. Franz and K. Depperman, "Theorie der Beugang am Zylinder unter Berucksichtigung der Kreichwelle," *Am. Physik,* Vol. 10, June 1952, pp. 361–373.

[8] W. Franz and K. Depperman, "Theorie der Beugang der Kugel unter Berucksichtigung der Kreichwelle," *Am. Physik,* Vol. 14, June 1954, pp. 253–264.

[9] P.H. Pathak and R.G. Kouyoumjian, "An Analysis of the Radiation from Apertures in Curved Surfaces by the Geometrical Theory of Diffraction," *Proc. IEEE,* November 1974, pp. 1438–1447.

[10] H. Bach, "Introduction to GTD Applications," in *Theoretical Methods for Determining the Interaction of Electromagnetic Waves with Structures,* J.K. Skwirzynski (Ed.), Sijthoff and Noordhoff, Amsterdam, 1981, pp. 563–594.

[11] P.H. Pathak, W.D. Burnside, and R.J. Marhefka, "A Uniform GTD Analysis of the Diffraction of Electromagnetic Waves by a Smooth Convex Surface," *IEEE Trans. Antennas and Propagation,* Vol. AP-28, No. 5, September 1980, pp. 631–642.

[12] P.H. Pathak, N. Wang, W.D. Burnside, and R.G. Kouyoumjian, "A Uniform GTD Solution for the Radiation from Sources on a Convex Surface," *IEEE Trans. Antennas and Propagation,* Vol. AP-29, No. 4, July 1981, pp. 609-621.

[13] P.H. Pathak and N. Wang, "Ray Analysis of Mutual Coupling between Antennas on a Convex Surface," *IEEE Trans. Antennas and Propagation,* Vol. AP-29, No. 6, Nov. 1981, pp. 911–922.

[14] P.H. Pathak, "An Asymptotic Analysis of the Scattering of Plane Waves by a Smooth Convex Surface," *Radio Science,* Vol. 14, No. 3, May–June 1979, pp. 419–435.
[15] R.G. Kouyoumjian, P.H. Pathak, and W.D. Burnside, "A Uniform GTD for the Diffraction by Edges, Vertices and Convex Surfaces," in *Theoretical Methods for Determining the Interaction of Electromagnetic Waves with Structures,* J.K. Skwirzynski, (Ed.) Sijthoff and Noordhoff, Amsterdam, 1981, pp. 497–561.
[16] W.D. Burnside, R.J. Markefka, and C.L. Yu, "Roll-Plane Analysis of On-Aircraft Antennas," *IEEE Trans. Antennas and Propagation,* Vol. AP-21, No. 6, November 1973, pp. 780–786.
[17] W.D. Burnside, M.C. Gilreath, R.J. Markefka, and C.L. Yu, "A Study of KC-135 Aircraft Antenna Patterns," *IEEE Trans. Antennas and Propagation,* Vol. AP-23, No. 3, May 1975, pp. 309–316.
[18] C.L. Yu, W.D. Burnside, and M.C. Gilreath, "Volumetric Pattern Analysis of Airborne Antennas," *IEEE Trans. Antennas and Propagation,* Vol. AP-26, No. 5, September 1978, pp. 636–641.
[19] W.D. Burnside, "Analysis of On-Aircraft Antenna Patterns," Ph.D. dissertation, Ohio State University, 1972, pp. 112–137.
[20] J.A. Kong, *Electromagnetic Wave Theory,* John Wiley and Sons, New York, 1986.
[21] W.D. Burnside, N. Wang, and E.L. Pelton, "Near-Field Pattern Analysis of Airborne Antennas," *IEEE Trans. Antennas and Propagation,* Vol. AP-28, May 1980, pp. 318–327.
[22] "Notes for the Short Course on the Modern Geometrical Theory of Diffraction," Ohio State University.
[23] J.H. Richmond, "A Reaction Theorem and Its Application to Antenna Impedance Calculation," *IRE Trans. Antennas and Propagation,* Vol. AP-9, No. 6, November 1961, pp. 515–520.
[24] S.W. Lee, "A review of GTD calculations of Mutual Admittance of Slot Conformal Array, Electromagnetics, Vol. 2, No. 2, April–June 1982 pp. 85–127.
[25] S.W. Lee and S. Safavi-Naini, "Approximate Asymptotic Solution of Surface Field Due to a Magnetic Dipole on a Cylinder," *IEEE Trans. Antennas and Propagation,* Vol. AP-26, 1978, pp. 593–598.
[26] S.W. Lee, "Mutual Admittance of Slots on a Cone: Solution by Ray Technique," *IEEE Trans. Antennas and Propagation,* Vol. AP-26, November 1978, pp. 768–773.
[27] R.M. Jha, V. Sudhakar, and N. Balakrishnan, "Ray Analysis of Mutual Coupling between Antennas on a General Paraboloid of Revolution (GPOR)," *Electronics Lett.,* Vol. 23, No. 11, May 1987, pp. 583–584.
[28] G.V. Borgiotti, "Conformal Arrays," in *The Handbook of Antenna Design,* A.W. Rudge, K. Milne, A.D. Olver, and P. Knight (Eds.), Peter Peregrinus Ltd., 1986.
[29] S.W. Lee and R. Mittra, "GTD Solution of Slot Admittance on a Cone or Cylinder," *Proc. IEE,* Vol. 126, No. 6, June 1979, pp. 487–492.
[30] T. Ishihara and L.B. Felsen, "High Frequency Fields Excited by a Line Source Located on a Perfectly Conducting Concave Cylindrical Surface," *IEEE Trans. Antennas and Propagation,* Vol. AP-26, 1979, pp. 757–767.
[31] T. Ishihara and L.B. Felsen, "High Frequency Surface Fields Excited by a Point Source on a Concave Perfectly Conducting Cylindrical Boundary," *Radio Science,* Vol. 14, March–April 1979, pp. 205–216.
[32] T. Ishihara and L.B. Felsen, "High Frequency Propagation at Long Ranges Near a Concave Boundary," *Radio Science,* Vol. 23, No. 6, November–December 1988, pp. 997–1012.

Appendix A
Unit Vectors

Unit vectors are a very elegant and mathematically convenient way of indicating direction and polarization. Depending on the nature of the problem, it may be easier to formulate it in a Cartesian, spherical, or cylindrical coordinate system.

A.1 CARTESIAN COORDINATE SYSTEM

The three perpendicular unit vectors $\hat{\mathbf{x}}$, $\hat{\mathbf{y}}$, and $\hat{\mathbf{z}}$ are shown in Figure A.1. Note that these vectors are constants; that is, their directions do not depend on their spatial positions, as is the case with unit vectors in spherical and cylindrical coordinate systems.

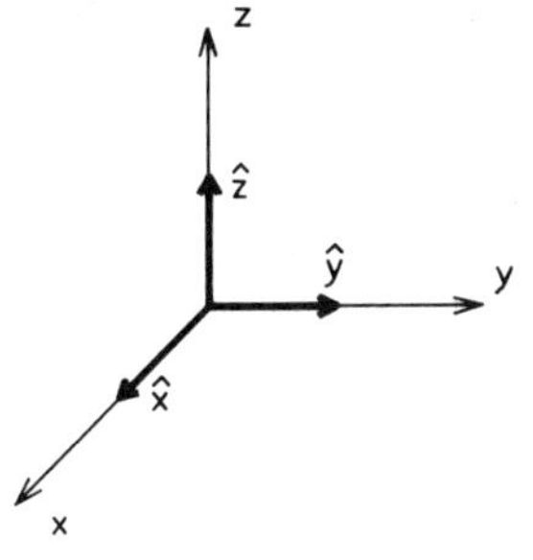

Figure A.1 Cartesian coordinate system.

A.2 SPHERICAL COORDINATE SYSTEM

The three perpendicular unit vectors, $\hat{\mathbf{r}}$, $\hat{\theta}$, and $\hat{\Phi}$, shown in Figure A.2, are given by

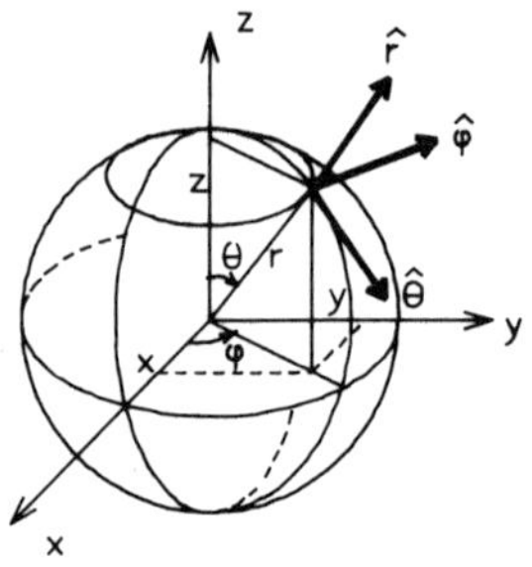

Figure A.2 Spherical coordinate system.

$$\hat{\mathbf{r}} = \sin\theta \cos\Phi \, \hat{\mathbf{x}} + \sin\theta \sin\Phi \, \hat{\mathbf{y}} + \cos\theta \, \hat{\mathbf{z}} \tag{A.1}$$

$$\hat{\theta} = \cos\theta \cos\Phi \, \hat{\mathbf{x}} + \cos\theta \sin\Phi \, \hat{\mathbf{y}} - \sin\theta \, \hat{\mathbf{z}} \tag{A.2}$$

$$\hat{\Phi} = - \sin\Phi \, \hat{\mathbf{x}} + \cos\Phi \, \hat{\mathbf{y}} \tag{A.3}$$

Note that these three unit vectors are not constants, because they depend on the angles θ and Φ.

A.3 CYLINDRICAL COORDINATE SYSTEM

The three perpendicular unit vectors, $\hat{\rho}$, $\hat{\Phi}$, and $\hat{\mathbf{z}}$, shown in Figure A.3, are given by

$$\hat{\rho} = \cos\Phi \, \hat{\mathbf{x}} + \sin\Phi \, \hat{\mathbf{y}} \tag{A.4}$$

$$\hat{\Phi} = -\sin\Phi \, \hat{\mathbf{x}} + \cos\Phi \, \hat{\mathbf{y}} \tag{A.5}$$

$$\hat{\mathbf{z}} = \hat{\mathbf{z}} \tag{A.6}$$

Note that $\hat{\mathbf{z}}$ is a constant, whereas $\hat{\rho}$ and $\hat{\Phi}$ are not.

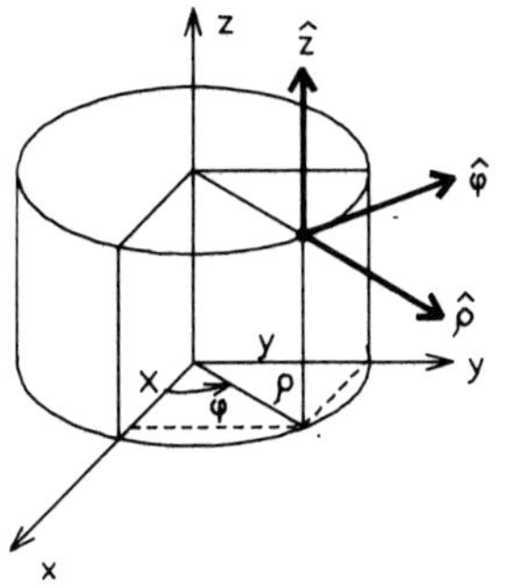

Figure A.3 Cylindrical coordinate system.

Appendix B
Special Functions for the Uniform Geometrical Theory of Diffraction

"A man should keep his little brain attic stocked with all the furniture that he is likely to use, and the rest he can put away in the lumber-room of his library, where he can get it if he wants it."

Sir Arthur Conan Doyle,
Five Orange Pips

B.1 INTRODUCTION

A mathematical function $f(x)$ may be called *special* when it arises so frequently or naturally, and is used so often, that it is worthwhile spending some time to study of its properties. This may include a study of the behavior of the function—perhaps, a simplification in its mathematical form or expression—when the argument x is either very small or very large. These are known as the *small argument* and *large argument* (or *asymptotic*) *forms* of the particular function, respectively. Alternatively, the efficient numerical computation of values of the function for a given argument often is the matter of concern.

A number of special functions are used or mentioned in this text: the Fresnel integrals; the transition function; the Bessel, Hankel, and Airy functions; and the various Fock functions. These functions need be no more intimidating than the trigonometric functions, for example, because the $\sin(x)$ function really is no less complicated than the Fock function $p^*(x)$! If someone were to request the value of $\sin(2.135)$, we would be in no better a position than if the value of $p^*(2.135)$ had been required. We may feel more comfortable with the former question only because we have spent many years studying and evaluating trigonometric functions, and thus know its value at special values of its argument, its bounds, and so on.

Yet, to actually determine the numerical value of sin(2.135) we have to use some expansion in terms of elementary arithmetic operations, just as could be done for $p^*(x)$, for example. Admittedly, most handheld calculators will evaluate trigonometric functions automatically, and all computers have intrinsic trigonometric functions that may be used at will. However, once *any* special function has been programmed reliably, it is no more difficult to calculate with than the "simpler" functions.

This appendix provides formal definitions of the special functions used in this text. Their large and small argument behavior is considered, interpolatory formulas enabling the numerical evaluation of such functions for intermediate values of the argument are provided, and a graph is plotted for each function to show the form of the function at a glance.

A complete treatise on these functions is not intended. Only those aspects relevant to the topics in this book are discussed. In UTD applications the arguments of the special functions considered always are real, usually arising from some combination of geometrical properties associated with a specific problem. For this reason, we are concerned here only with definitions of the functions convergent for *real arguments* x. The functions themselves however will be complex in most cases. FORTRAN subroutines for evaluating these functions are provided in Appendix F.

B.2 THE FRESNEL INTEGRALS AND TRANSITION FUNCTION

Special functions of importance in the study of optical phenomena since the early quantitative treatments of the subject are the Fresnel cosine and sine integrals defined by

$$C(x) = \int_0^x \cos\frac{\pi}{2}u^2\,du \tag{B.1}$$

$$S(x) = \int_0^x \sin\frac{\pi}{2}u^2\,du \tag{B.2}$$

and named in honor of the French physicist Augustin Fresnel (1788–1827). The complex *Fresnel integral* is defined in terms of $C(x)$ and $S(x)$ as

$$C(x) - jS(x) = \int_0^x e^{-j(\pi/2)u^2}\,du$$

The special function used in both the UTD edge and curved-surface diffraction coefficient expressions is the *transition function* [1] defined by

$$F(x) = 2j\sqrt{x}\, e^{jx} \int_{\sqrt{x}}^{\infty} e^{-ju^2}\, du \tag{B.3}$$

Note that definition (B.3) is intended for $x > 0$. For $x < 0$ the transition function is given by $F^*(|x|)$, where the asterisk indicates complex conjugation. The transition function is always complex. Figure B.1 gives a plot of the transition function $F(x)$ as would be obtained using the FORTRAN routine *FFCT(XF)* of Appendix F. This routine automatically performs the conjugation operation when $x < 0$. It uses the following limiting forms.

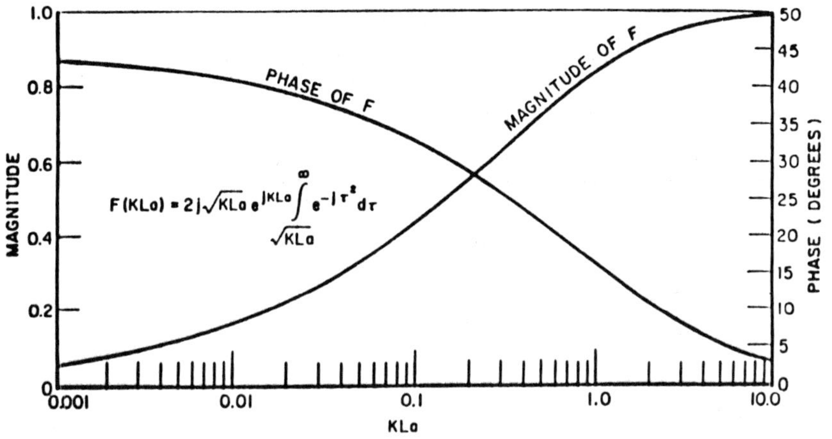

Figure B.1 Plot of the transition function $F(x)$ (after [1], reproduced with permission).

(a) For x small (which pertains when observation points are close to shadow boundaries):

$$F(x) \approx [\sqrt{\pi x} - 2x\, e^{j\pi/4} - (2/3)x^2\, e^{-j\pi/4}]\, e^{j\pi/4}\, e^{jx} \tag{B.4}$$

from which we can easily see that

$$F(0) = 0 \tag{B.5}$$

(b) For x large,

$$F(x) \approx 1 + \frac{j}{2x} - \frac{3}{4x^2} - \frac{j15}{8x^3} + \frac{75}{16x^4} \tag{B.6}$$

Clearly, for $x \gg 1$ (which is the case when observation points are far from shadow boundaries):

$$F(x) \approx 1 \tag{B.7}$$

In the interval $0.3 \leqslant x \leqslant 5.5$ the interpolation scheme used for numerical computation is

$$F(x) \approx F(x_i) + \frac{[F(x_{i+1}) - F(x_i)]}{(x_{i+1} - x_i)}(x - x_i) \tag{B.8}$$

with x_i, $F(x_i)$, and $[F(x_{i+1}) - F(x_i)]/(x_{i+1} - x_i)$ tabulated in Table B.1.

Table B.1

Numerical Data for the Interpolation Formula for the Transition Function $F(x)$ in the Range $0.3 \leqslant x \leqslant 5.5$ (data from [23])

x_i	$F(x_i)$ Real	$F(x_i)$ Imaginary	$[F(x_{i+1}) - F(x_i)]/(x_{i+1} - x_i)$ Real	$[F(x_{i+1}) - F(x_i)]/(x_{i+1} - x_i)$ Imaginary
0.3	0.0	0.0	0.5729	0.2677
0.5	0.5195	0.0025	0.6768	0.2682
0.7	0.3355	−0.0665	0.7439	0.2549
1.0	0.2187	−0.0757	−0.8095	0.2322
1.5	0.1270	−0.680	0.8730	0.1982
2.3	0.0638	−0.0506	0.9240	0.1577
4.0	0.0246	−0.0296	0.9658	0.1073
5.5	0.0093	−0.0163	0.9797	0.0828

B.3 BESSEL AND HANKEL FUNCTIONS

Apart from the transcendental functions, the Bessel and Hankel functions, named after the German mathematicians Friedrich Wilhelm Bessel (1784–1846) and Hermann Hankel (1839–1873), probably are the most comprehensively studied and widely used special functions of mathematical physics. As such they are well documented throughout the literature. Possibly the most detailed treatise is by Watson [2]. In engineering use, the information and tabulations in [3] usually are more than sufficient. In fact, the essential results required for most electromagnetic theory applications are available in the now standard texts, for example, by Harrington [4], Jones [5], or Felsen and Marcuvitz [6].

The *Bessel function* of the first kind and order ν (which is real, but not necessarily an integer) is denoted $J_\nu(x)$, and that of the second kind by $Y_\nu(x)$. These functions are defined (though not necessarily computed) by

$$J_\nu(x) = \sum_{m=0}^{\infty} (-1)^m \frac{(x/2)^{\nu+2m}}{m!(\nu + m)!} \tag{B.9}$$

$$Y_\nu(x) = \frac{J_\nu(x) \cos\nu\pi - J_{-\nu}(x)}{\sin\nu\pi} \tag{B.10}$$

Although the argument x need not be real, in most electromagnetic scattering applications it is. From these definitions, we thus have $J_\nu(0) = 0$, except for $\nu = 0$, for which $J_0(0) = 1$, and $Y_\nu(0) \to -\infty$.

The *Hankel functions* of order ν, of the first and second kind, can be defined as a linear combination of Bessel functions as

$$H_\nu^{(1)}(x) = J_\nu(x) + jY_\nu(x) \tag{B.11}$$

$$H_\nu^{(2)}(x) = J_\nu(x) - jY_\nu(x) \tag{B.12}$$

respectively. For real x, both $J_\nu(x)$ and $Y_\nu(x)$ are real, whereas the Hankel functions are complex.

Integral expressions for the Bessel and Hankel functions also are available [6, 7] and are the forms used to obtain large argument expressions for the functions. The validity of a particular large argument form depends on both the size of argument x *and* the order ν of the function. Large argument expressions for the Hankel functions are

$$H_\nu^{(1)}(x) = \sqrt{\frac{2}{\pi x}}\, e^{jx}\, e^{-j2\nu\pi/4} \tag{B.13}$$

$$H_\nu^{(2)}(x) = \sqrt{\frac{2}{\pi x}}\, e^{-jx}\, e^{j2\nu\pi/4} \tag{B.14}$$

but are only valid for those orders $\nu \ll x$. If this condition is not satisfied, alternative large argument forms must be used [5]. These have not been utilized at any point in the text and thus are not considered here.

Computer routines for the generation of Bessel and Hankel functions, at least when ν is an integer, are included in most mathematical software libraries [8, 9]. A good discussion on the numerical evaluation of Bessel functions (including sound warnings on the careful use of recurrence relations), along with computer subroutines, is given in [10].

Note that accurate computation of Bessel, and hence Hankel, functions of noninteger order (other than half an odd integer order) is more difficult.

B.4 THE AIRY FUNCTIONS

The Airy function $Ai(\tau)$, named after the British astronomer Sir George Biddell Airy (1801–1892), may be defined for complex τ. However, for the case of a real argument x, the only situation for which numerical data is listed in this text, $Ai(x)$ can be defined by the integral representation (with integration along the real axis):

$$Ai(x) = \frac{1}{2\pi}\int_{-\infty}^{\infty} e^{-j(t^3/3 + xt)}\, dt \tag{B.15}$$

Alternatively, because the imaginary part of (B.15) always will be zero, and the real part of the integrand is an even function, we may write the Airy function for real arguments as [12]

$$Ai(x) = \frac{1}{\pi}\int_{0}^{\infty} \cos(t^3/3 + xt)\, dt \tag{B.16}$$

Thus, for the real values of x of interest here, $Ai(x)$, as well as its derivative, $Ai'(x) = d/dx\{Ai(x)\}$, itself is a real function. Numerical data for the *real argument form* (B.16) was comprehensively tabulated by Miller [11], and thus sometimes is referred to as the *Miller-type Airy function* [12]. However, because of the part it plays in the surface diffraction results first written in GTD format by Keller [13, 14], it also has been called [15] the *Keller-type Airy function.*

The first five zeros of $Ai(x)$, and those of its derivative $Ai'(x)$, are given in Table B.2. These were obtained from the results of Miller [11], which are reproduced in Abramowitz and Stegun [3, p. 478].

Table B.2
The First Five Zeros of $Ai(x)$ and $Ai'(x)$: $Ai(-q_n) = 0$ and $Ai'(-q_n') = 0$

n	q_n	q_n'	$Ai'(-q_n)$	$Ai(-q_n)$
1	2.33811	1.01879	0.70121	0.53566
2	4.08795	3.24820	−0.80311	−0.41902
3	5.52055	4.82009	0.86520	0.38041
4	6.78670	6.16330	−0.91085	−0.35791
5	7.94413	7.37217	0.94734	0.34230

There are three other Airy functions, referred to in the electromagnetics literature [15–18] as the *Fock-type Airy functions,* and denoted $W_1(\tau)$, $W_2(\tau)$, and $V(\tau)$, where τ, in general, is complex. Indeed, the first few complex roots of these functions are given in Table B.3. The Fock-type Airy function $W_1(\tau)$ is defined by the following contour integral:

$$W_1(\tau) = \frac{1}{\sqrt{\pi}} \int_{C_1} e^{-z^3/3} e^{\tau z} \, dz \tag{B.17}$$

with contour C_1 in the complex z plane shown in Figure B.2. The function $W_2(\tau)$ is [15, 17, 19]

$$W_2(\tau) = \frac{1}{\sqrt{\pi}} \int_{C_2} e^{-z^3/3} e^{\tau z} \, dz \tag{B.18}$$

with contour C_2 also shown in Figure B.2. Expressions (B.17) and (B.18) usually are written in the following forms:

$$W_1(\tau) = \frac{1}{\sqrt{\pi}} \int_{\infty e^{-j2\pi/3}}^{\infty} e^{-z^3/3} e^{\tau z} \, dz \tag{B.19}$$

$$W_2(\tau) = \frac{1}{\sqrt{\pi}} \int_{\infty e^{j2\pi/3}}^{\infty} e^{-z^3/3} e^{\tau z} \, dz \tag{B.20}$$

Table B.3

The First Ten Zeros of $W_2(\tau)$ and $W_2'(\tau)$: $W_2(\tau_n') = 0$ and $W_2'(\tau_n) = 0$, where $\tau_n = |\tau_n| \, e(-j\pi/3)$ and $\tau_n' = |\tau_n'| \, e(-j\pi/3)$

n	$\lvert\tau_n\rvert$	$\lvert\tau_n'\rvert$
1	2.33811	1.01879
2	4.08795	3.24819
3	5.52056	4.82010
4	6.78661	6.16331
5	7.94413	7.37218
6	9.02265	8.48849
7	10.0402	9.53545
8	11.0085	10.5277
9	11.9300	11.4751
10	12.8288	12.3848

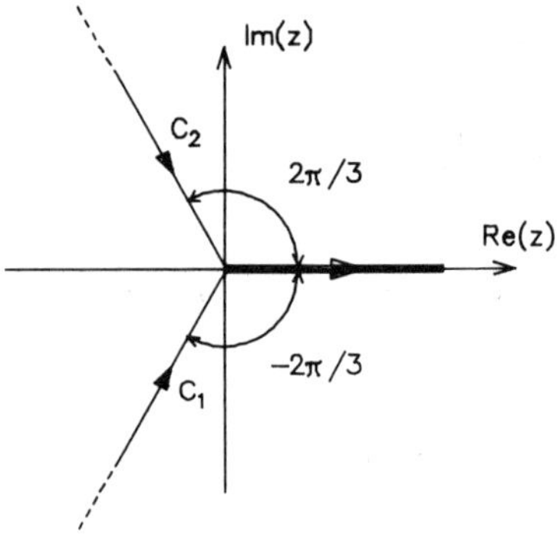

Figure B.2 Contours C_1 and C_2, used in the definition of various special functions: contour C_1 goes from ∞ to 0 along the line arg $z = -2\pi/3$ and from 0 to ∞ along the real line; contour C_2 goes from ∞ to 0 along the line arg $z = +2\pi/3$ and from 0 to ∞ along the real line.

The Fock-type Airy function $V(\tau)$ is related to $W_1(\tau)$ and $W_2(\tau)$ by

$$V(\tau) = \frac{W_1(\tau) - W_2(\tau)}{2j} \tag{B.21}$$

The function $V(\tau)$ can be related to $Ai(\tau)$ as

$$V(\tau) = \sqrt{\pi}\, Ai(\tau) \tag{B.22}$$

if we recognize that $W_1(\tau) = W_2^*(\tau)$.

In the UTD, the various Airy functions are embedded in the definition of other special functions. The routines used in Appendix F to compute special functions numerically do so directly. In applications of the UTD we therefore seldom need to actually compute the Airy functions separately. Their definition has been given both to enhance an understanding of the basis of the UTD and because the detailed derivation of the UTD formulations given in journal articles make use of these functions. Because of the difference in time conventions, $W_1(W_2)$ defined earlier is equal to $W_2(W_1)$ originally defined by Fock [20] and are those used in the engineering literature. Information on the efficient numerical evaluation of certain Airy functions can be found in [21].

B.5 THE FOCK SCATTERING FUNCTIONS

The special function $\hat{P}_{s,h}(x)$ can be written in terms of the functions $p^*(x)$ and $q^*(x)$ [15, 16, 19, 22] as

$$\hat{P}_s(x) = p^*(x)\, e^{-j\pi/4} - \frac{e^{-j\pi/4}}{2x\sqrt{\pi}} \tag{B.23}$$

$$\hat{P}_h(x) = q^*(x)\, e^{-j\pi/4} - \frac{e^{-j\pi/4}}{2x\sqrt{\pi}} \tag{B.24}$$

The asterisk in (B.23) and (B.24) does not imply that complex conjugation is to be performed, but simply indicates that these *already* are the complex conjugates of functions $p(x)$ and $q(x)$, which are applicable to the same problems as those in question in this text, but when $e^{-j\omega t}$ time-dependence is used instead of the $e^{j\omega t}$ dependence assumed in this and most other engineering texts.

There is no complete unanimity in the literature on what exactly to call certain of the functions to be defined in this section. We have selected what we hope by now are the most common means of identification of such functions. Therefore, $p^*(x)$ and $q^*(x)$ we will call the *Fock scattering functions* (to distinguish them from several other Fock functions defined in later sections of this appendix). It indeed is proper that many of the functions used in surface diffraction treatments are named after V.A. Fock [20], as most of the available formulations are based on his early work. The $\hat{P}_{s,h}(x)$ are termed the Pekeris caret functions [16, 22], in reference to the caret above P.

Sometimes [17] authors call p^* and q^* themselves the Pekeris functions, and $\hat{P}_{s,h}$ the Fock-type reflection function. Oftentimes, p^* and q^* simply are termed *the* Fock functions. Nonetheless, they are defined formally as

$$p^*(x) = \frac{1}{\sqrt{\pi}} \int_{-\infty}^{\infty} \frac{V(\tau)\, e^{-jx\tau}}{W_2(\tau)}\, d\tau \tag{B.25}$$

$$q^*(x) = \frac{1}{\sqrt{\pi}} \int_{-\infty}^{\infty} \frac{V'(\tau)\, e^{-jx\tau}}{W_2'(\tau)}\, d\tau \tag{B.26}$$

so that we often find a definition for $\hat{P}_{s,h}(x)$ written as

$$\hat{\mathbf{P}}_{s,h}(x) = \frac{1}{\sqrt{\pi}} \int_{-\infty}^{\infty} \frac{\tilde{Q}\, V(\tau)\, e^{-jx\tau}}{\tilde{Q}\, W_2(\tau)}\, d\tau \tag{B.27}$$

with

$$\tilde{Q} = \begin{cases} 1, & \text{for the soft case} \\ d/d\tau, & \text{for the hard case} \end{cases}$$

Although the integration in (B.25), (B.26), or (B.27) is along the real axis, and argument x is assumed real as well, the function $p^*(x)$, $q^*(x)$, and $\hat{\mathbf{P}}_{s,h}(x)$ are complex. Functions $p^*(x)$ and $q^*(x)$ can be computed using the routines *PFUN(X)* and *QFUN(X)*, respectively, given in Appendix F. Graphical representations of $p^*(x)$ and $q^*(x)$, obtained using these routines, are shown in Figures B.3 and B.4.

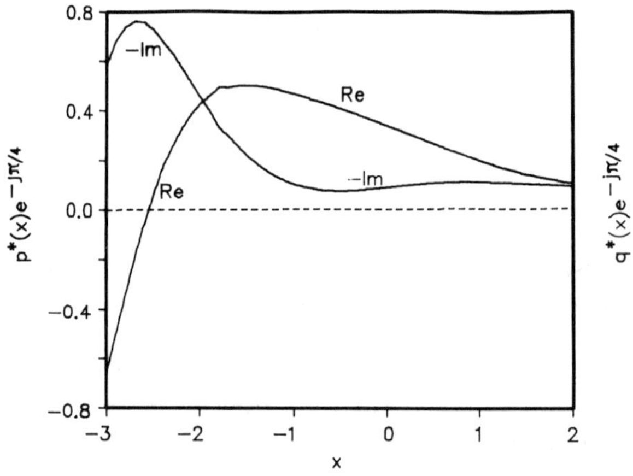

Figure B.3 Plot of the soft Fock scattering function $p^*(x)$ multiplied by the factor e$(-j\pi/4)$.

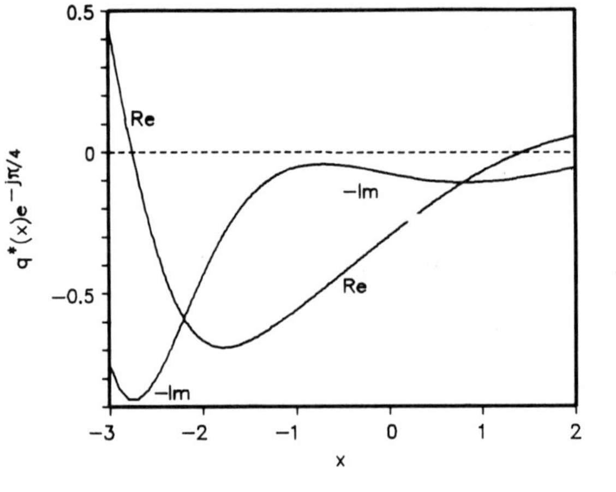

Figure B.4 Plot of the hard Fock scattering function $q^*(x)$ multiplied by the factor e$(-j\pi/4)$.

Remark: Note that the $q^*(x)$ function, applicable to the *hard* case (for which the electric field is normal to the smooth surface and for which the boundary condition applied in mathematical derivations is that of the normal *derivative* of the magnetic field being zero at the surface), involves the *derivative* of the function $W_2(\tau)$. On the other hand, the $p^*(x)$ function, applicable to the *soft* case (for which the electric field is tangential to the smooth surface and for which the boundary condition used in mathematical derivations is that of zero tangential electric field at the surface), involves the function $W_2(\tau)$ directly and *not* its derivative. We find that this is so for the special functions of the radiation case as well, as can be seen in Section B.6 of this appendix.

Limiting Forms

(a) When $x > 2$ (which is so when observation points are in the deep shadow region) we can see that [15]

$$p^*(x) = \frac{1}{2x\sqrt{\pi}} - \frac{e^{-j\pi/6}}{2\sqrt{\pi}} \sum_{n=1}^{5} \frac{e^{xq_n e^{-j5\pi/6}}}{[Ai'(-q_n)]^2} \tag{B.28}$$

$$q^*(x) = \frac{1}{2x\sqrt{\pi}} - \frac{e^{j\pi/6}}{2\sqrt{\pi}} \sum_{n=1}^{5} \frac{e^{xq'_n e^{-j5\pi/6}}}{q'_n[Ai(-q'_n)]^2} \tag{B.29}$$

where $Ai(-q_n) = 0$ and $Ai'(-q'_n) = 0$, these zeros being given in Table B.2. In this series, no more than the first five terms are required if the curvature of the diffracting surfaces are large enough electrically so that a UTD analysis in fact is valid.

(b) When $x < -3.0$ (that is, in the deep lit region):

$$p^*(x) \approx \frac{1}{2x\sqrt{\pi}} + \frac{1}{2}\sqrt{|x|}\left[1 + \frac{j2}{x^3}\right] e^{jx^3/12}\, e^{j\pi/4} \tag{B.30}$$

$$q^*(x) \approx \frac{1}{2x\sqrt{\pi}} - \frac{1}{2}\sqrt{|x|}\left[1 - \frac{j2}{x^3}\right] e^{jx^3/12}\, e^{j\pi/4} \tag{B.31}$$

(c) Examining the Pekeris caret function directly, from (B.23), (B.24), (B.28), and (B.29), we have, for $x \gg 0$ (that is, in the deep shadow region):

$$\hat{P}_s(x) = -\frac{e^{-j\pi/4}}{\sqrt{\pi}} \sum_{n=1}^{5} \frac{e^{j\pi/6}\, e^{xq_n e^{-j5\pi/6}}}{2[Ai'(-q_n)]^2} \tag{B.32}$$

$$\hat{P}_h(x) = -\frac{e^{j\pi/4}}{\sqrt{\pi}} \sum_{n=1}^{5} \frac{e^{j\pi/6} e^{xq'_n e^{-j5\pi/6}}}{2q'_n[Ai(-q'_n)]^2} \tag{B.33}$$

When $x \ll 0$ (in the deep lit region), the asymptotic form of the Pekeris caret function given by

$$\hat{P}_{s,h}(x) \approx \pm \sqrt{\frac{-x}{4}}\, e^{jx^3/12} \tag{B.34}$$

can be shown to apply [15, 16]. Note that, at $x = 0$, we have $\hat{P}_{s,h} = 0$.

Interpolatory Scheme

In the range $-3.0 < x < 2.0$, the functions $p^*(x)$ and $q^*(x)$ can be computed using a linear interpolation of tabulated values as was done for $F(x)$ in (B.8). This information is embedded in the function subroutines *PFUN(X)* and *QFUN(X)* listed in Appendix F and will not be listed here.

B.6 THE FOCK RADIATION FUNCTIONS

After [19], we define the following functions:

$$g(x) = \frac{1}{\sqrt{\pi}} \int_{\infty e^{-j2\pi/3}}^{\infty} \frac{e^{-jx\tau}}{W_2'(\tau)}\, d\tau \tag{B.35}$$

$$\tilde{g}(x) = \frac{1}{\sqrt{\pi}} \int_{\infty e^{-j2\pi/3}}^{\infty} \frac{e^{-jx\tau}}{W_2(\tau)}\, d\tau \tag{B.36}$$

where the contour of integration is C_1 of Figure B.2. The special functions $g(x)$ and $\tilde{g}(x)$ are known as the *hard* and *soft Fock functions,* respectively. We refer to them in this text as the *hard* and *soft Fock radiation functions,* to distinguish them from the Fock scattering functions described in the previous section. Although x is real, the functions $g(x)$ and $\tilde{g}(x)$ are complex.

Next we define two further functions $G(x)$ and $\tilde{G}(x)$, which we refer to as the *universal Fock radiation functions,* simply related to $g(x)$ and $\tilde{g}(x)$ as

$$G(x) = \begin{cases} e^{-jx^3/3} g(x), & x \leq 0 \\ g(x), & x \geq 0 \end{cases} \tag{B.37}$$

$$\tilde{G}(x) = \begin{cases} e^{-jx^3/3} \tilde{g}(x), & x \leq 0 \\ \tilde{g}(x), & x \geq 0 \end{cases} \tag{B.38}$$

Function subroutines for the numerical evaluation of $G(x)$ and $\hat{G}(x)$ are provided in Appendix F, named *G(X)* and *GB(X)*, respectively. Graphs of the functions, obtained using these function subroutines, are shown in Figures B.5 and B.6, respectively. The results given here are for the $e^{j\omega t}$ time-dependence assumed in this text. The reader should be aware that some authors would assume an $e^{-j\omega t}$ dependence.

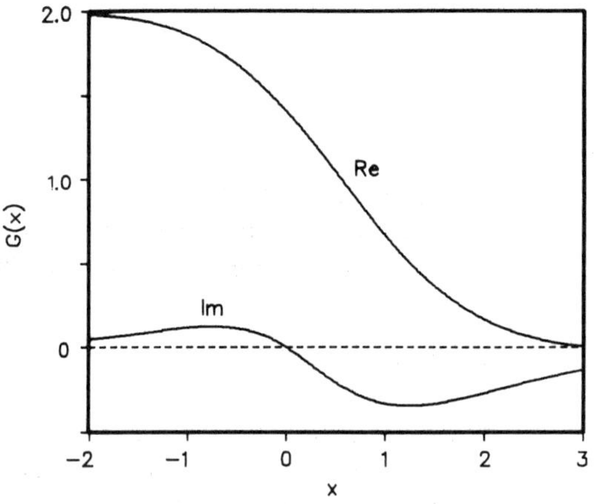

Figure B.5 Plot of the hard universal Fock radiation function $G(x)$.

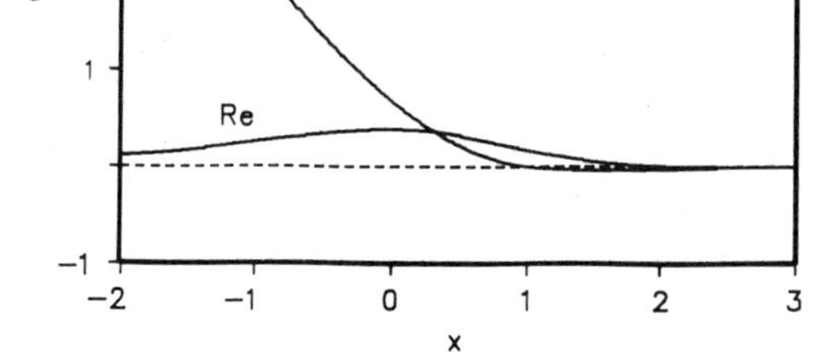

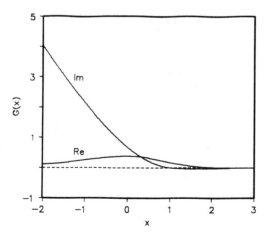

Figure B.6 Plot of the soft universal Fock radiation function $\tilde{G}\,(x)$.

Limiting Forms [23]

(a) $\tilde{G}(x) \approx 0$, for $x > 4$ (B.39)

(b) $\tilde{G}(x) \approx -jx\,(2 + j/2x^3)$, for $x < -4$ (B.40)

(c) $G(x) \approx 1.8325\, e^{(-0.8823 - j0.5094)x}$, for $x > 4$ (B.41)

(d) $G(x) \approx 2 - j/2x^3$, for $x < -2.5$ (B.42)

Interpolatory Scheme

In the argument range not covered by the preceding limiting forms, the functions $G(x)$ and $\tilde{G}(x)$ can be computed by using a linear interpolation of tabulated values, as has been done for $F(x)$ in (B.8). This information is embedded in the function subroutines *G(X)* and *GB(X)* listed in Appendix F and will not be given here.

B.7 THE FOCK COUPLING FUNCTIONS

We need to introduce two further special functions $u(x)$ and $v(x)$: the soft and hard Fock coupling functions, sometimes referred to as the generalized Fock functions. The function $v(x)$ should not be confused with the Fock-type Airy function defined by expressions (B.21) or (B.22). Functions $u(x)$ and $v(x)$ are defined (once more, only real arguments x being of interest) as [24]

$$u(x) = \frac{x^{3/2}\, e^{j3\pi/4}}{\sqrt{\pi}} \int_{\infty e^{-j2\pi/3}}^{\infty} \frac{W_2'(\tau)\, e^{-jx\tau}}{W_2(\tau)}\, d\tau \tag{B.43}$$

$$v(x) = \frac{x^{1/2}\, e^{j\pi/4}}{2\sqrt{\pi}} \int_{\infty e^{-j2\pi/3}}^{\infty} \frac{W_2(\tau)\, e^{-jx\tau}}{W_2'(\tau)}\, d\tau \tag{B.44}$$

Although x is real, $u(x)$ and $v(x)$ are complex. Note that the contour of integration in (B.43) and (B.44) is the C_1 shown in Figure B.2. The functions $u(x)$ and $v(x)$ are computed by subroutines *FU* and *FV*, respectively, of Appendix F. They are shown in Figures B.7 and B.8.

Limiting Cases

(a) For $x > 0.6$, which arises in practice when the coupling point is in the deep shadow region, rapidly converging residue series for the functions $v(x)$ and $u(x)$ are [22]

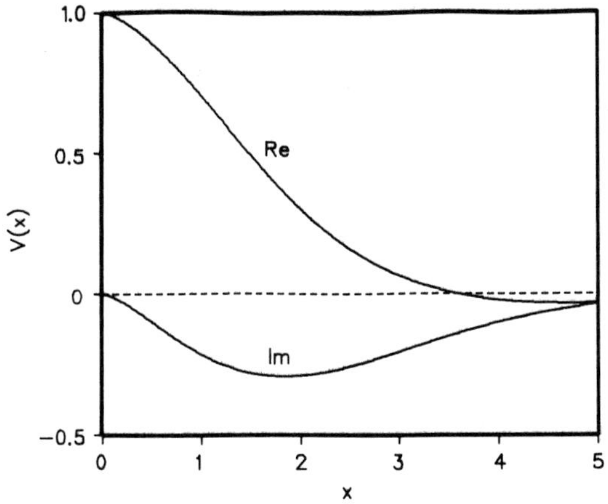

Figure B.7 Plot of the hard Fock coupling function $v(x)$.

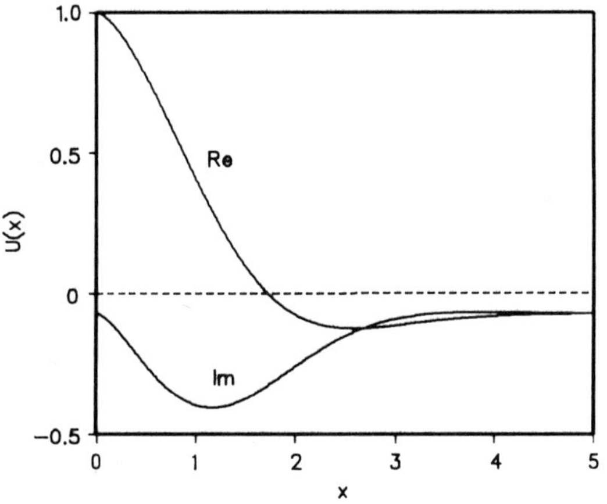

Figure B.8 Plot of the soft Fock coupling function $u(x)$.

$$u(x) \approx e^{j\pi/4}\, 2\sqrt{\pi}\, x^{3/2} \sum_{n=1}^{10} e^{-jx\tau_n} \tag{B.45}$$

$$v(x) \approx e^{-j\pi/4}\, \sqrt{\pi}\, x^{1/2} \sum_{n=1}^{10} (\tau_n')^{-1}\, e^{-jx\tau_n'} \tag{B.46}$$

If we write $\tau_n = |\tau_n|\, e^{-j\pi/3}$ and $\tau'_n = |\tau'_n|\, e^{-j\pi/3}$, the first ten terms (although four usually are sufficient) in (B.45) and (B.46) can be calculated by using the tabulated τ_n and τ'_n values in Table B.3. Use of these series, of course, assumes that the geometry is large enough for a UTD treatment to be valid in the first place.

(b) Conversely, when $x < 0.6$, the small argument expansions [24]:

$$u(x) \approx 1 - \frac{\sqrt{\pi}}{2}\, e^{j\pi/4}\, x^{3/2} + \frac{j5}{12}\, x^3 + \frac{5\sqrt{\pi}}{64}\, e^{-j\pi/4}\, x^{9/2} \tag{B.47}$$

$$v(x) \approx 1 - \frac{\sqrt{\pi}}{4}\, e^{j\pi/4}\, x^{3/2} + \frac{j7}{60}\, x^3 + \frac{7\sqrt{\pi}}{512}\, x^{9/2} \tag{B.48}$$

suffice.

B.8 CONCLUDING REMARKS

The special functions discussed here and that can be evaluated numerically by using the subroutines given in Appendix F are the only such functions required in the application of the UTD described in this text. For the 3D extension of the UTD curved-surface–diffraction analysis, the coupling case [25] may require the numerical evaluation of the derivatives $u'(x)$ and $v'(x)$, as well as an additional function $v_1(x)$. Both small and large argument forms for these are given by Lee and Safavi-Naini [24].

REFERENCES

[1] R.G. Kouyoumjian and P.H. Pathak, "A Uniform Geometrical Theory of Diffraction for an Edge in a Perfectly Conducting Surface," *Proc. IEEE,* Vol. 62, November 1974, pp. 1448–1461.

[2] G.N. Watson, *A Treatise on the Theory Of Bessel Functions,* Cambridge University Press, Cambridge, England, 1958.

[3] M. Abramowitz and I.A. Stegun, *Handbook of Mathematical Functions,* Dover, New York, 1972.

[4] R.F. Harrington, *Time-Harmonic Electromagnetic Fields,* McGraw-Hill, New York, 1961.

[5] D.S. Jones, *The Theory of Electromagnetism,* Oxford University Press, London, 1964.

[6] L.B. Felsen and N. Marcuvitz, *Radiation and Scattering of Waves,* Prentice-Hall, Englewood Cliffs, NJ, 1973.

[7] D.S. Jones, *Acoustic and Electromagnetic Waves,* Oxford University Press, London, 1986.

[8] Numerical Algorithms Group (NAG), Mathematical Software Library, Oxfordshire, England, United Kingdom.

[9] IMSL Inc., Library of Mathematical Software, Houston.

[10] W.H. Press, B.P. Flannery, S.A. Teukolsky, and W.T. Vetterling, *Numerical Recipes,* Cambridge University Press, Cambridge, 1986.

[11] J.C.P. Miller, *The Airy Integral,* Cambridge University Press, Cambridge, 1946.
[12] P.H. Pathak and R.G. Kouyoumjian, "An Analysis of the Radiation from Apertures in Curved Surfaces by the Geometrical Theory of Diffraction," *Proc. IEEE,* Vol. 62, November 1974, pp. 1438–1447.
[13] J.B. Keller, "Diffraction by a Convex Cylinder," *IRE Trans. Antennas and Propagation,* Vol. AP-24, 1956, pp. 312–321.
[14] Levy, M., and J.B. Keller, "Diffraction by a Smooth Object," *Comm. Pure Appl. Math.,* Vol. 12, 1959, pp. 159–209.
[15] P.H. Pathak, W.D. Burnside, and R.J. Marhefka, "A Uniform GTD Analysis of the Diffraction of Electromagnetic Waves by a Smooth Convex Surface," *IEEE Trans. Antennas and Propagation,* Vol. AP-28, No. 5, September 1980, pp. 631–642.
[16] P.H. Pathak, "An Asymptotic Analysis of the Scattering of Plane Waves by a Smooth Convex Surface," *Radio Science,* Vol. 14, No. 3, May–June 1979, pp. 419–435.
[17] R.G. Kouyoumjian, P.H. Pathak, and W.D. Burnside, "A Uniform GTD for the Diffraction by Edges, Vertices and Convex Surfaces" in *Theoretical Methods for Determining the Interaction of Electromagnetic Waves with Structures,* J.K. Skwirzynski (Ed.), Sijthoff and Noordhoff, Amsterdam, 1981, pp. 497–561.
[18] J.R. Wait, *Introduction to Antennas and Propagation,* Peter Peregrinus, London, 1986, pp. 245–254.
[19] P.H. Pathak, N. Wang, W.D. Burnside, and R.G. Kouyoumjian, "A Uniform GTD Solution for the Radiation from Sources on a Convex Surface," *IEEE Trans. Antennas and Propagation,* Vol. AP-29, No. 4, July 1981, pp. 609–621.
[20] V.A. Fock, *Electromagnetic Diffraction and Propagation Problems,* Pergamon Press, London, 1965.
[21] T. Cwik, "On the Efficient Calculation of the Incomplete Airy Function with Application to Edge Diffraction," *Radio Science,* Vol. 23, No. 6, November–December 1988, pp. 1133–1140.
[22] J.J. Bowman, T.B.A. Senior, and P.L.E. Uslenghi, *Electromagnetic and Acoustic Scattering by Simple Shapes,* North Holland, Doderecht, Holland, pp. 380–395.
[23] W.D. Burnside, R.J. Marhefka, and N. Wang, "Computer Programs, Subroutines and Functions for the Short Course on the Modern Geometrical Theory of Diffraction," in "Notes for the Short Course on the Modern Geometrical Theory of Diffraction," Ohio State University.
[24] S.W. Lee and S. Safavi-Naini, "Approximate Asymptotic Solution of Surface Field Due to a Magnetic Dipole on a Cylinder," *IEEE Trans. Antennas and Propagation,* Vol. AP-26, No. 4, July 1978, pp. 593–597.
[25] P.H. Pathak and N. Wang, "Ray Analysis of Mutual Coupling between Antennas on a Convex Surface," *IEEE Trans. Antennas and Propagation,* Vol. AP-29, No. 6, November 1981, pp. 911–922.

Appendix C
Differential Geometry

A working knowledge of differential geometry is required to use the UTD. In this appendix, some elementary aspects of differential geometry dealing with curves and surfaces will be considered, specifically the parametric description of curves and surfaces, the calculation of unit vectors tangential and normal to curves and surfaces, and the calculation of the radii of curvature. Many excellent texts are available that deal with the subject in great detail, some of which are listed at the back of this appendix.

C.1 CURVES

There are many ways in which curves can be described mathematically. In the two-dimensional x-y plane, a curve can be expressed in explicit form as $y = f(x)$ and in implicit form as $f(x, y) = 0$. These forms are sufficient for describing surfaces in two dimensions. Three-dimensional edges, however, are not necessarily contained in a single plane, but can curve arbitrarily through space. A parametric description of curves rather than an implicit or explicit formulation is found to be much more useful, in general, and will be used in this text.

A curve can be expressed in parametric form as

$$\mathbf{r}(\gamma) = x(\gamma)\hat{\mathbf{x}} + y(\gamma)\hat{\mathbf{y}} + z(\gamma)\hat{\mathbf{z}} \tag{C.1}$$

where $\mathbf{r}$ is a vector indicating the spatial position of a point on the curve for a particular value of the parameter γ, with $(\gamma_{min} \leqslant \gamma \leqslant \gamma_{max})$. The functions $x(\gamma)$, $y(\gamma)$, and $z(\gamma)$ indicate the spatial displacement in the $\hat{\mathbf{x}}$, $\hat{\mathbf{y}}$, and $\hat{\mathbf{z}}$ directions, respectively.

Having expressed the curve in parametric form as in (C.1), the vector that is tangential to the curve is given by

$$\mathbf{t} = \frac{d\mathbf{r}}{d\gamma} \tag{C.2}$$

where

$$\mathbf{r}' = x'(\gamma)\,\hat{\mathbf{x}} + y'(\gamma)\,\hat{\mathbf{y}} + z'(\gamma)\,\hat{\mathbf{z}} \tag{C.3}$$

with the prime denoting the derivative with respect to the parameter γ. The unit vector tangential to the curve thus is given by

$$\hat{\mathbf{t}} = \frac{\mathbf{t}}{|\mathbf{t}|} \tag{C.4}$$

A curve with $|\mathbf{t}(\gamma)| = 1$ for all γ, is known as a *unit-speed curve.* In general, γ can be any parameter; however, when the arc length (s) is used as the parameter [1], it can be shown that

$$\hat{\mathbf{t}} = \frac{d\mathbf{r}}{ds} \tag{C.5}$$

Having found $\hat{\mathbf{t}}$, the next step is to find a unit vector perpendicular to the curve; that is, a vector ($\hat{\mathbf{n}}$) that satisfies the relationship:

$$\hat{\mathbf{t}} \cdot \hat{\mathbf{n}} = 0 \tag{C.6}$$

In fact, there are an infinite number of such vectors, as can been seen in Figure 4.6, for example.

The particular unit vector perpendicular to the curve given by

$$\hat{\mathbf{n}}_p = \frac{\hat{\mathbf{t}}'}{|\hat{\mathbf{t}}'|} \tag{C.7}$$

is known as the *principal normal vector.* Note that $\hat{\mathbf{t}}'$ is the derivative of $\hat{\mathbf{t}}$ with respect to γ, and not necessarily a unit vector. The unit vector $\hat{\mathbf{n}}_e$, referred to as the *unit vector normal to the edge* in Chapter 6, is related to $\hat{\mathbf{n}}_p$ by

$$\hat{\mathbf{n}}_e = -\hat{\mathbf{n}}_p \tag{C.8}$$

Note that

$$\hat{\mathbf{t}}' = \frac{d\hat{\mathbf{t}}}{d\gamma} = \frac{d\hat{\mathbf{t}}}{ds}\frac{ds}{d\gamma} \tag{C.9}$$

where

$$\frac{ds}{d\gamma} = |\mathbf{t}| \tag{C.10}$$

The plane containing $\hat{\mathbf{t}}$ and $\hat{\mathbf{n}}_p$, known as the *osculating plane,* is shown in Figure C.1. The radius of curvature (a_e) of the curve at Q is given by

$$a_e = \frac{|\mathbf{r}'|^3}{|\mathbf{r}' \times \mathbf{r}''|^2} \tag{C.11}$$

or

$$a_e = \frac{|\mathbf{t}|}{|\hat{\mathbf{t}}'|} \tag{C.12}$$

By convention, $a_e > 0$. The direction of the curve is indicated by $\hat{\mathbf{n}}_p$. The *curvature* (κ_e) is given by

$$\kappa_e = \frac{1}{a_e} \tag{C.13}$$

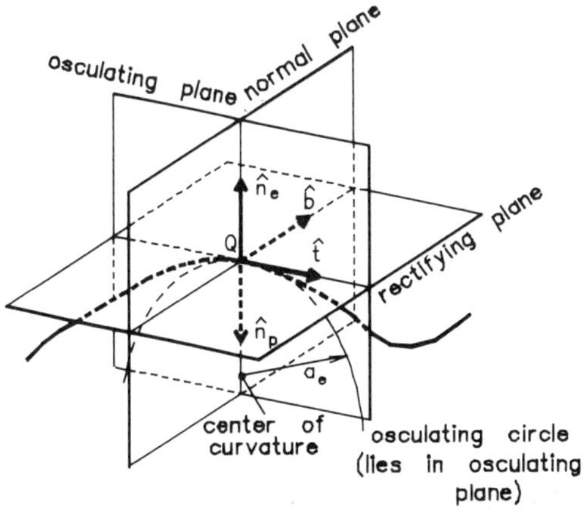

Figure C.1 Coordinate frame showing osculating plane and the unit vectors $\hat{\mathbf{t}}$, $\hat{\mathbf{n}}_p$ $\hat{\mathbf{n}}_e$, and $\hat{\mathbf{b}}$.

and the *curvature vector* by

$$\mathbf{K} = \frac{\hat{\mathbf{t}}'}{|\mathbf{t}|} = \frac{d\hat{\mathbf{t}}}{ds} \tag{C.14}$$

The curvature, and hence the radius of curvature, is related to $\hat{\mathbf{t}}$ and $\hat{\mathbf{n}}_p$ by

$$\frac{d\hat{\mathbf{t}}}{ds} = \kappa\hat{\mathbf{n}}_p = \mathbf{K} \tag{C.15}$$

The so-called osculating plane contains the vectors $\hat{\mathbf{t}}$ and $\hat{\mathbf{n}}_p$. The center of the corresponding osculating circle (having the same radius of curvature as the curve at Q) lies in the osculating plane in the direction of $\hat{\mathbf{n}}_p$.

The *binormal vector* is given by

$$\hat{\mathbf{b}} = \hat{\mathbf{t}} \times \hat{\mathbf{n}}_p \tag{C.16}$$

If $\hat{\mathbf{b}}$ is a constant, the curve is contained in a plane. If the curve is known to lie in a plane, the calculation of $\hat{\mathbf{n}}_p$ can be simplified considerably by applying (C.6). Hence, if

$$\hat{\mathbf{t}} = (t_1, t_2) \tag{C.17}$$

then

$$\hat{\mathbf{n}} = (\pm t_2, \mp t_1) \tag{C.18}$$

where $\hat{\mathbf{n}}$ is $\hat{\mathbf{n}}_p$ or $\hat{\mathbf{n}}_e$, depending on the sign. The *torsion* (τ) of the curve is given by

$$\frac{d\hat{\mathbf{b}}}{ds} = -\tau\hat{\mathbf{n}}_p \tag{C.19}$$

Example C.1: Ellipse

Consider the ellipse with a semiminor axis of length c and a semimajor axis of length d, as shown in Figure C.2. Note that the ellipse is contained in the x-y plane. In implicit form, the ellipse is given by

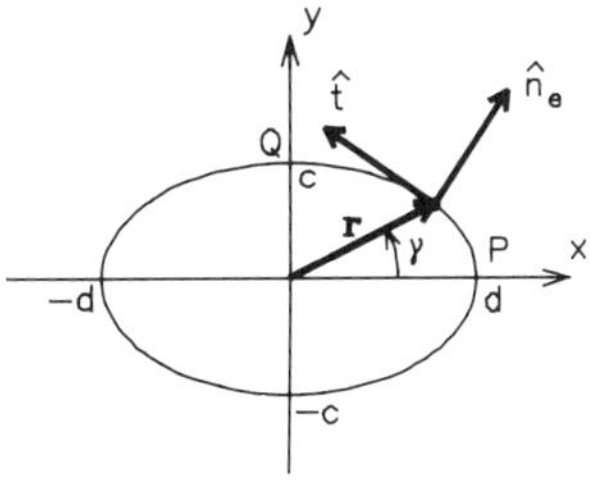

Figure C.2 Ellipse.

$$\frac{x^2}{d^2} + \frac{y^2}{c^2} = 1 \tag{C.20}$$

In parametric form, the ellipse can be described by

$$\mathbf{r}(\gamma) = d\cos\gamma\,\hat{\mathbf{x}} + c\sin\gamma\,\hat{\mathbf{y}}, \quad 0 \leq \gamma \leq 2\pi \tag{C.21}$$

Note that γ is a parametric angle and not a geometric angle. The tangent of the geometric angle (ϕ) is given by

$$\tan\phi = \frac{y}{x} \tag{C.22}$$

Substituting $y(\gamma)$ and $x(\gamma)$ as given in (C.21) into (C.22), we find that

$$\tan\phi = \frac{c}{d}\tan\gamma \tag{C.23}$$

Applying (C.2) to (C.21), it follows that

$$\mathbf{t} = -d\sin\gamma\,\hat{\mathbf{x}} + c\cos\gamma\,\hat{\mathbf{y}} \tag{C.24}$$

so that

$$|\mathbf{t}| = \sqrt{d^2\sin^2\gamma + c^2\cos^2\gamma} \tag{C.25}$$

The unit vector tangential to the curve thus is given by

$$\hat{\mathbf{t}} = -\frac{d\sin\gamma}{\sqrt{d^2\sin^2\gamma + c^2\cos^2\gamma}}\hat{\mathbf{x}} + \frac{c\cos\gamma}{\sqrt{d^2\sin^2\gamma + c^2\cos^2\gamma}}\hat{\mathbf{y}} \tag{C.26}$$

The radius of curvature is found from (C.12) and $\hat{\mathbf{n}}_p$ from (C.7). Because $|\mathbf{t}|$ already has been calculated, it remains to find $|\hat{\mathbf{t}}'|$. Differentiating (C.26) with respect to γ, we find that

$$\hat{\mathbf{t}}' = -\frac{c^2 d \cos\gamma}{(d^2 \sin^2\gamma + c^2 \cos^2\gamma)^{3/2}}\hat{\mathbf{x}} - \frac{c d^2 \sin\gamma}{(d^2 \sin^2\gamma + c^2 \cos^2\gamma)^{3/2}}\hat{\mathbf{y}} \qquad \text{(C.27)}$$

$$|\hat{\mathbf{t}}'| = \frac{cd}{d^2 \sin^2\gamma + c^2 \cos^2\gamma} \qquad \text{(C.28)}$$

Substituting (C.27) and (C.28) into (C.8), it follows that

$$\hat{\mathbf{n}}_p = -\frac{c \cos\gamma}{\sqrt{d^2 \sin^2\gamma + c^2 \cos^2\gamma}}\hat{\mathbf{x}} - \frac{d \sin\gamma}{\sqrt{d^2 \sin^2\gamma + c^2 \cos^2\gamma}}\hat{\mathbf{y}} \qquad \text{(C.29)}$$

Because the ellipse is contained in the x-y plane, $\hat{\mathbf{n}}_p$ could have been determined without having calculated $\hat{\mathbf{t}}'$, by using (C.17) and (C.18). Substituting (C.25) and (C.28) into (C.12), we find that the radius of curvature is given by

$$a_e(\gamma) = \frac{(d^2 \sin^2\gamma + c^2 \cos^2\gamma)^{3/2}}{cd} \qquad \text{(C.30)}$$

In particular, we find that at P

$$a_e(\gamma = 0) = \frac{c^2}{d} \qquad \text{(C.31)}$$

and at Q

$$a_e(\gamma = \pi/2) = \frac{d^2}{c} \qquad \text{(C.32)}$$

An inspection of Figure C.2 reveals that $a_e(P) < a_e(Q)$. This is borne out by (C.31) and (C.32), as $d > c$.

C.2 SURFACES

C.2.1 Unit Vector Normal to a Surface

A surface can be expressed in parametric form as

$$\mathbf{r}(u, v) = x(u, v)\,\hat{\mathbf{x}} + y(u, v)\,\hat{\mathbf{y}} + z(u, v)\,\hat{\mathbf{z}} \tag{C.33}$$

with ($u_{min} \leqslant u \leqslant u_{max}$) and ($v_{min} \leqslant v \leqslant v_{max}$). The partial derivatives of $\mathbf{r}$ with respect to u and v are given by

$$\mathbf{r}_u = \frac{\partial x(u, v)}{\partial u}\,\hat{\mathbf{x}} + \frac{\partial y(u, v)}{\partial u}\,\hat{\mathbf{y}} + \frac{\partial z(u, v)}{\partial u}\,\hat{\mathbf{z}} \tag{C.34}$$

$$\mathbf{r}_v = \frac{\partial x(u, v)}{\partial v}\,\hat{\mathbf{x}} + \frac{\partial y(u, v)}{\partial v}\,\hat{\mathbf{y}} + \frac{\partial z(u, v)}{\partial v}\,\hat{\mathbf{z}} \tag{C.35}$$

The vectors $\mathbf{r}_u$ and $\mathbf{r}_v$ are tangential to the surface [1], as shown in Figure C.3, so that the unit vector normal to the surface is given by

$$\hat{\mathbf{n}}(u, v) = \pm \frac{\mathbf{r}_u \times \mathbf{r}_v}{|\mathbf{r}_u \times \mathbf{r}_v|} \tag{C.36}$$

The plus or minus sign depends on the orientation of the vector required; that is, outward or inward.

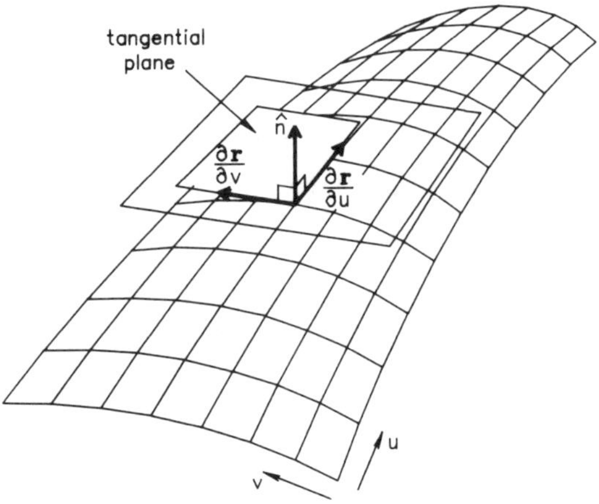

Figure C.3 Tangential and normal vectors to a surface.

Example C.2: Paraboloid

The paraboloid with focal length f shown in Figure C.4 can be represented in parametric form as

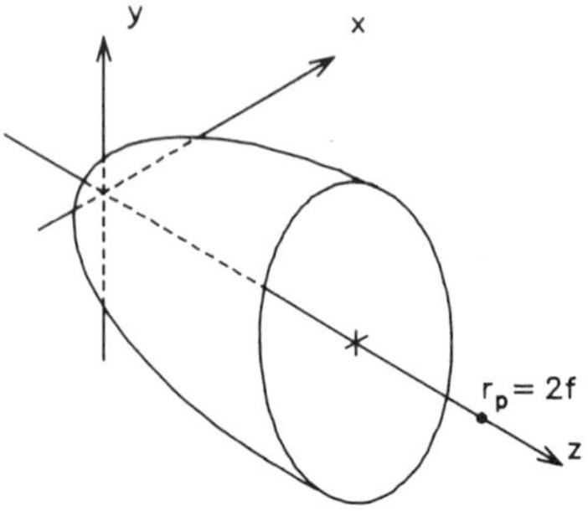

Figure C.4 Paraboloid.

$$\mathbf{r}(u, v) = u\hat{\mathbf{x}} + v\hat{\mathbf{y}} + \frac{u^2 + v^2}{4f}\hat{\mathbf{z}} \tag{C.37}$$

where u and v are the displacements in the x and y directions, respectively. Applying (C.34) and (C.35) to (C.37), we find that

$$\mathbf{r}_u = \hat{\mathbf{x}} + \frac{u}{2f}\hat{\mathbf{z}} \tag{C.38}$$

$$\mathbf{r}_v = \hat{\mathbf{y}} + \frac{v}{2f}\hat{\mathbf{z}} \tag{C.39}$$

Substituting (C.38) and (C.39) into (C.36), the unit vector normal to the surface is thus given by

$$\hat{\mathbf{n}}(u, v) = -\frac{u\hat{\mathbf{x}}}{\sqrt{u^2 + v^2 + 4f^2}} - \frac{v\hat{\mathbf{y}}}{\sqrt{u^2 + v^2 + 4f^2}} + \frac{2f\,\hat{\mathbf{z}}}{\sqrt{u^2 + v^2 + 4f^2}} \tag{C.40}$$

The unit vector $\hat{\mathbf{n}}$ given in (C.40) is attached to the inside of the parabolical shell and points toward $\mathbf{r}_p = 2f\,\hat{\mathbf{z}}$. The outward directed normal vector is given by $-\hat{\mathbf{n}}$ in this case.

C.2.2 Radius of Curvature of a Surface

Consider now the surface shown in Figure C.5. Let there be a curve C through a point Q on the surface as shown. We can see [2] that the radius of curvature of C on the surface in the direction of the unit vector $\hat{\mathbf{t}}$ is given by

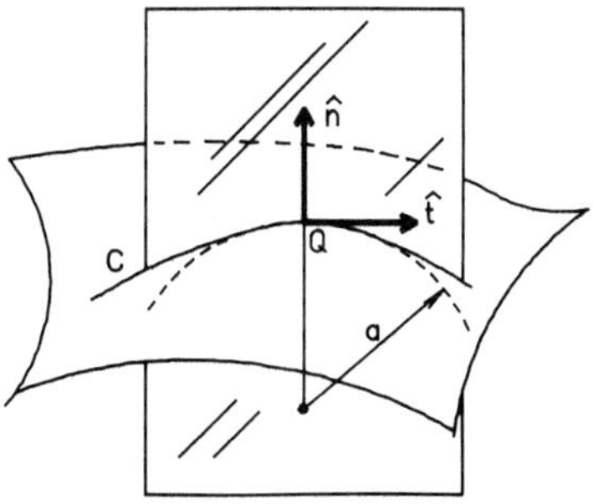

Figure C.5 Radius of curvature of the curve (*C*) on a surface.

$$a = \frac{\mathrm{I}}{\mathrm{II}} \tag{C.41}$$

where I is known as the *first fundamental form* and II is the *second fundamental form*. The first fundamental form is the square of the differential arc length along a curve on the surface [2], given by

$$\mathrm{I} = d\mathbf{r} \cdot d\mathbf{r} = ds^2 \tag{C.42}$$

or

$$\mathrm{I} = Edu^2 + Fdu\,dv + Gdv^2 \tag{C.43}$$

where

$$E = \mathbf{r}_u \cdot \mathbf{r}_u \tag{C.44}$$

$$F = \mathbf{r}_u \cdot \mathbf{r}_v \tag{C.45}$$

$$G = \mathbf{r}_v \cdot \mathbf{r}_v \tag{C.46}$$

We can see that the second fundamental form is twice the deviation of the surface at $(u + du, v + dv)$ from the tangent plane at (u, v) [2], given by

$$\mathrm{II} = -d\mathbf{r} \cdot d\hat{\mathbf{n}} \tag{C.47}$$

or

$$\mathrm{II} = e du^2 + f du dv + g dv^2 \tag{C.48}$$

where

$$e = \mathbf{r}_{uu} \cdot \hat{\mathbf{n}} \tag{C.49}$$

$$f = \mathbf{r}_{uv} \cdot \hat{\mathbf{n}} \tag{C.50}$$

$$g = \mathbf{r}_{vv} \cdot \hat{\mathbf{n}} \tag{C.51}$$

The function $\mathbf{r}_{uu}$ is the second partial derivative of $\mathbf{r}$ with respect to u, and similarly for $\mathbf{r}_{uv}$ and $\mathbf{r}_{vv}$. If $\hat{\mathbf{t}}$ is expressed as

$$\hat{\mathbf{t}} = t_1 \mathbf{r}_u + t_2 \mathbf{r}_v \tag{C.52}$$

then I and II also can be written as

$$\mathrm{I} = E t_1^2 + F t_1 t_2 + G t_2^2 \tag{C.53}$$

$$\mathrm{II} = e t_1^2 + f t_1 t_2 + g t_2^2 \tag{C.54}$$

Alternatively, the curvature can be found from [3]

$$\kappa = \mathbf{K} \cdot \hat{\mathbf{n}} \tag{C.55}$$

where $\mathbf{K}$ is given in (C.14). Careful attention should be paid to the sign of κ. If $\kappa < 0$, the surface curves away from $\hat{\mathbf{n}}$.

As $\hat{\mathbf{t}}$ rotates in the plane perpendicular to $\hat{\mathbf{n}}$ at Q, the radius of curvature of the surface in the direction of $\hat{\mathbf{t}}$ generally will change. We can see [2] that a pair of orthogonal directions $\hat{\mathbf{U}}_1$ and $\hat{\mathbf{U}}_2$ (i.e., special cases of $\hat{\mathbf{t}}$) exist at Q, for which the radius of curvature assumes minimum and maximum values. These two directions are known as the *principal directions* and the corresponding radii of curvature as the *principal radii of curvature*. The *principal curvatures* are related to the principal radii of curvature by

$$\kappa_1 = \frac{1}{a_1} \tag{C.56}$$

$$\kappa_2 = \frac{1}{a_2} \tag{C.57}$$

The principal curvatures are the roots of the quadratic equation [2]

$$\kappa^2 - 2\,\kappa_m\kappa + \kappa_g = 0 \tag{C.58}$$

where κ_m and κ_g are known as the *mean curvature* and *Gaussian curvature*, respectively, given by

$$\kappa_m = \frac{Eg - 2Ff + Ge}{2(EG - F^2)} \tag{C.59}$$

$$\kappa_g = \frac{eg - f^2}{EG - F^2} \tag{C.60}$$

Equation (C.58) has two real roots, which are given by

$$\kappa_{1,2} = \kappa_m \pm \sqrt{\kappa_m^2 - \kappa_g} \tag{C.61}$$

If $\kappa_1 = \kappa_2$, the curvature at Q is independent of the direction of $\hat{\mathbf{t}}$, and Q is called *umbilic*. It is interesting to note that

$$\kappa_m = \frac{\kappa_1 + \kappa_2}{2} \tag{C.62}$$

$$\kappa_g = \kappa_1\,\kappa_2 \tag{C.63}$$

The unit vectors in the principal directions are given by [2]

$$\hat{\mathbf{U}}_1 = \frac{\mathbf{r}_u + \alpha\mathbf{r}_v}{\xi_1} \tag{C.64}$$

$$\hat{\mathbf{U}}_2 = \frac{\beta\mathbf{r}_u + \mathbf{r}_v}{\xi_2} \tag{C.65}$$

where

$$\alpha = \frac{e - \kappa_1 E}{\kappa_1 F - f} = \frac{f - \kappa_1 F}{\kappa_1 G - g} \tag{C.66}$$

$$\beta = \frac{f - \kappa_2 F}{\kappa_2 E - e} = \frac{g - \kappa_2 G}{\kappa_2 F - f} \tag{C.67}$$

$$\xi_1 = \sqrt{E + 2\alpha F + \alpha^2 G} \tag{C.68}$$

$$\xi_2 = \sqrt{\beta^2 E + 2\beta F + G} \tag{C.69}$$

Example C.3: Spheroid

The spheroid in Figure C.6 can be described by

$$\mathbf{r}(u, v) = c \cos v \sin u\hat{\mathbf{x}} + c \sin v \sin u\hat{\mathbf{y}} + d \cos u\hat{\mathbf{z}} \tag{C.70}$$

with $(0 \leqslant v \leqslant 2\pi)$ and $(0 \leqslant u \leqslant \pi)$. The spheroid is generated by rotating the ellipse in the y-z plane around the z-axis. We now calculate the unit vector normal to the surface as well as the radii of curvature, specifically at a point $Q(u = \pi/2, v = \pi/2)$. The first and second partial derivatives of $\mathbf{r}$ are given by

$$\mathbf{r}_u = c \cos v \cos u\hat{\mathbf{x}} + c \sin v \cos u\hat{\mathbf{y}} - d \sin u\hat{\mathbf{z}} \tag{C.71}$$

$$\mathbf{r}_{uu} = -c \cos v \sin u\hat{\mathbf{x}} - c \sin v \sin u\hat{\mathbf{y}} - d \cos u\hat{\mathbf{z}} \tag{C.72}$$

$$\mathbf{r}_{uv} = -c \sin v \cos u\hat{\mathbf{x}} + c \cos v \cos u\hat{\mathbf{y}} \tag{C.73}$$

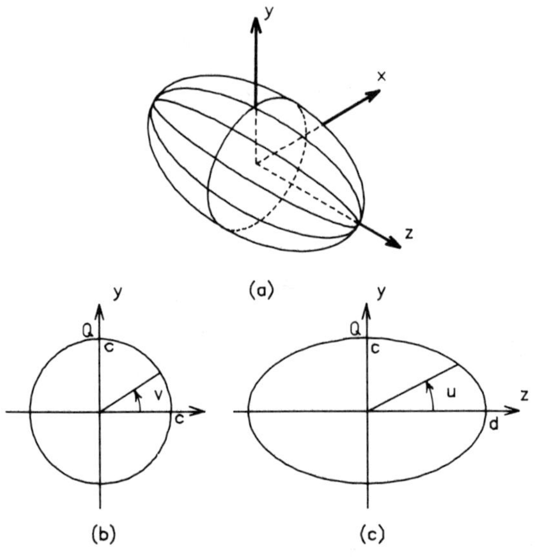

Figure C.6 Spheroid: (a) oblique view; (b) x-y plane; (c) y-z plane.

$$\mathbf{r}_v = -c\ \sin v\ \sin u\hat{\mathbf{x}} + c\ \cos v\ \sin u\hat{\mathbf{y}} \tag{C.74}$$

$$\mathbf{r}_{vv} = -c\ \cos v\ \sin u\hat{\mathbf{x}} - c\ \sin v\ \sin u\hat{\mathbf{y}} \tag{C.75}$$

Substituting (C.71) and (C.74) into (C.36), we find that the unit vector normal to the spheroid is given by

$$\hat{\mathbf{n}} = \frac{d\ \cos v\ \sin u}{\sqrt{c^2\ \cos^2 u + d^2\ \sin^2 u}}\hat{\mathbf{x}} + \frac{d\ \sin v\ \sin u}{\sqrt{c^2\ \cos^2 u + d^2\ \sin^2 u}}\hat{\mathbf{y}} + \frac{c\ \cos u}{\sqrt{c^2\ \cos^2 u + d^2\ \sin^2 u}}\hat{\mathbf{z}} \tag{C.76}$$

Evaluating (C.44) to (C.46) and (C.49) to (C.51), we find that the components of the first and second fundamental forms are given by

$$E = c^2\ \cos^2 u + d^2\ \sin^2 u \tag{C.77}$$

$$F = 0 \tag{C.78}$$

$$G = c^2\ \sin^2 u \tag{C.79}$$

$$e = \frac{-\ c\ d}{\sqrt{c^2\ \cos^2 u + d^2\ \sin^2 u}} \tag{C.80}$$

$$f = 0 \tag{C.81}$$

$$g = \frac{-\ ab\ \sin^2 u}{\sqrt{c^2\ \cos^2 u + d^2\ \sin^2 u}} \tag{C.82}$$

In the x-y plane, $u = \pi/2$, and

$$\hat{\mathbf{t}} = -\ \sin v\ \hat{\mathbf{x}} + \cos v\ \hat{\mathbf{y}} \tag{C.83}$$

Comparing (C.83) with (C.52), we find that

$$t_1 = 0 \tag{C.84}$$

$$t_2 = \frac{1}{c} \tag{C.85}$$

The radius of curvature of the surface in the *x*-*y* plane is found by substituting (C.77) into (C.82), (C.84) and (C.85) into (C.41); that is,

$$a = \frac{G}{g} = -c \tag{C.86}$$

The negative sign in (C.86) signifies that the surface curves away from $\hat{\mathbf{n}}$. Recall from (C.11) and (C.12) that the radius of curvature of a curve (a_e) was assumed to be positive. We therefore should be very careful in interpreting the sign of the radius of curvature.

The radius of curvature at Q in the *y*-*z* plane is found by considering

$$\hat{\mathbf{t}} = -\hat{\mathbf{z}} \tag{C.87}$$

so that

$$t_1 = \frac{1}{b} \tag{C.88}$$

$$t_2 = 0 \tag{C.89}$$

and, hence,

$$a = \frac{E}{e} = -\frac{d^2}{c} \tag{C.90}$$

From (C.59) and (C.60), the mean and Gaussian curvatures at Q are found to be

$$\kappa_m = -\frac{c^2 + d^2}{2cd^2} \tag{C.91}$$

$$\kappa_g = \frac{1}{d^2} \tag{C.92}$$

Substituting (C.91) and (C.92) into (C.61), we find that the principal curvatures at Q are

$$\kappa_1 = -\frac{1}{c} \tag{C.93}$$

$$\kappa_2 = -\frac{c}{d^2} \tag{C.94}$$

with the corresponding principal directions being

$$\hat{\mathbf{U}}_1 = -\hat{\mathbf{z}} \tag{C.95}$$

$$\hat{\mathbf{U}}_2 = -\sin v\,\hat{\mathbf{x}} + \cos v\,\hat{\mathbf{y}} \tag{C.96}$$

Alternatively, we can use (C.55) to calculate the radius of curvature of the surface in the y-z plane, for example. Following (C.27), we know that

$$\hat{\mathbf{t}}' = -\frac{c\,d^2 \sin\gamma\,\hat{\mathbf{y}}}{(c^2 \cos^2\gamma + d^2 \sin^2\gamma)^{3/2}} - \frac{c^2\,d \cos\gamma\,\hat{\mathbf{z}}}{(c^2 \cos^2\gamma + d^2 \sin^2\gamma)^{3/2}} \tag{C.97}$$

From (C.14), we thus have

$$\mathbf{K} = -\frac{c\,d^2 \sin\gamma}{(c^2 \cos^2\gamma + d^2 \sin^2\gamma)^2}\,\hat{\mathbf{y}} - \frac{c^2\,d \cos\gamma}{(c^2 \cos^2\gamma + d^2 \sin^2\gamma)^2}\,\hat{\mathbf{z}} \tag{C.98}$$

The radius of curvature in the y-z plane thus is given by (C.55) as

$$a(u) = -\frac{(c^2 \cos^2 u + d^2 \sin^2 u)^{3/2}}{c\,d} \tag{C.99}$$

Equation (C.99) is similar to (C.30) except for the sign. The radius of curvature of a curve (as in (C.30)) was defined to be positive, whereas the negative sign in (C.99) indicates that the surface curves away from $\hat{\mathbf{n}}$.

REFERENCES

[1] I.D. Faux and M.J. Pratt, *Computational Geometry for Design and Manufacture,* Ellis Horwood, 1979.

[2] S.W. Lee, *Differential Geometry for GTD Applications,* Electromagnetics Laboratory Report No. 77–21, University of Illinois at Urbana-Champaign, 1977.

[3] M.M. Lipschutz, *Differential Geometry,* Schaum's Outline Series, McGraw-Hill, New York, 1969.

[4] J.D. Lawrence, *A Catalog of Special Plane Curves,* Dover, New York, 1972.

[5] D.M.Y. Somerville, *Analytic Geometry of Three Dimensions,* Cambridge University Press, Cambridge, 1934.

[6] C.E. Weatherburn, *Differential Geometry of Three Dimensions,* Cambridge University Press, Cambridge, 1961.

[7] T.J. Willmore, *Differential Geometry,* Oxford University Press, London, 1958.
[8] E.P. Lane, *Metric Differential Geometry of Curves and Surfaces,* University of Chicago Press, Chicago, 1940.

Appendix D
The Method of Stationary Phase

"This case grows upon me, Watson . . . . There are decidedly some points of interest in connection with it. In this early stage I want you to realize these geographical features, which may have a good deal to do with our investigation."

Sir Arthur Conan Doyle
The Return of Sherlock Holmes

D.1 INTRODUCTION

Perusal of the electromagnetics literature on any high-frequency theories of diffraction soon reveals terminology such as the *method of stationary phase,* the *Watson transformation,* and the *method of steepest descent.* These represent only one aspect of the vast branch of mathematics known as *asymptotic techniques* and are concerned with obtaining, for certain types of integrals, approximate expressions that become increasingly accurate as some parameter becomes "asymptotically" large.

In line with the stated aims of an introductory text, we have not included a detailed derivation of any of the diffraction coefficients used. Reference however has been made to the *method of stationary phase* to illustrate the relationship of high-frequency techniques to fundamental electromagnetic theory. Therefore, in Section D.2 of this appendix, a brief introduction to this method is presented.

Having grasped the material on the application and underlying concepts of the UTD, the reader may wish to proceed with a study of the derivations of the UTD diffraction coefficients. This is best done through journal articles, many of which are referenced in the text. However, the authors of such articles, being primarily concerned with new developments, justifiably assume the reader has the required mathematical background. In Section D.3 a suggested reading list for obtaining this is given.

D.2 THE METHOD OF STATIONARY PHASE

Perhaps the most easily understood asymptotic technique is that applicable to the evaluation, for $\kappa \gg 1$, of an integral of the form:

$$I(k) = \int_{\xi_2}^{\xi_1} F(\xi)\, e^{j\kappa\Psi(\xi)}\, d\xi \tag{D.1}$$

where the limits ξ_1 and ξ_2 are real constants, $F(\xi)$ is a complex function that is slowly varying and well behaved for values of its argument in the range of integration, and $\Psi(\xi)$ is a real function of a real variable ξ. Quantity κ is real, positive, and large; it is called the *largeness parameter.*

Because $\kappa \gg 1$ is part of the purely imaginary argument of an exponential function, the integrand in (D.1) will fluctuate rapidly with ξ from positive to negative values over the integration interval. The method of stationary phase makes use of the self-cancelling oscillation of the exponential factor of this integrand to allow its contribution to be neglected everywhere except near one or more critical points. This is illustrated graphically for a particular example in Section 3.3.11. In the integral (D.1) the critical points are the *stationary points* ξ_s at which the derivative of the phase with respect to ξ is zero (that is, $\Psi'(\xi_s) = 0$) and at the integration limits or *end points.*

Consider first just the contribution (call it I_{sp}) from the small interval $\xi_s - \delta \leqslant \xi \leqslant \xi_s + \delta$ about the stationary point ξ_s, with $\delta \ll 1$ but positive. Following [1], if $F(\xi)$ is a smoothly varying continuous function, then throughout this interval, we have

$$F(\xi) \approx F(\xi_s) \tag{D.2}$$

The function $\Psi(\xi)$ in the same interval can be expanded to $O(\delta^3)$ in a Taylor series about ξ_s as,

$$\Psi(\xi) \approx \Psi(\xi_s) + \frac{\Psi''(\xi_s)}{2}(\xi - \xi_s)^2 \tag{D.3}$$

as $\Psi'(\xi_s) = 0$, and on the assumption that $\Psi''(\xi_s) \neq 0$. Substitution of (D.3) into (D.1) yields

$$I(\kappa) \approx F(\xi_s)\, e^{j\kappa\Psi(\xi_s)} \int_{\xi_s-\delta}^{\xi_s+\delta} e^{j(\kappa/2)\Psi''(\xi_s)(\xi-\xi_s)^2}\, d\xi \tag{D.4}$$

If we let

$$\tau = \sqrt{(\kappa/2)\Psi''(\xi_s)}\,(\xi - \xi_s) \tag{D.5}$$

with τ at the limits of the integral being $\tau_0 = \sqrt{(\kappa/2)\Psi''(\xi_s)}\,\delta$, then (D.4) transforms to

$$I(\kappa) \approx F(\xi_s)\,e^{j\kappa\Psi(\xi_s)}\sqrt{\frac{2}{\kappa\Psi''(\xi_s)}}\int_{-\tau_0}^{\tau_0} e^{j\xi^2}\,d\xi \tag{D.6}$$

We can extend, without significant error [1], the limits of the integral in (D.6) to $\pm\infty$. Use of the result:

$$\int_{-\infty}^{\infty} e^{j\xi^2}\,d\xi = \sqrt{\pi}\,e^{j\pi/4} \tag{D.7}$$

reduces (D.6) to

$$I(\kappa) \approx F(\xi_s)\sqrt{\frac{2\pi}{\kappa|\Psi''(\xi_s)|}}\,e^{j\kappa\Psi(\xi_s)}\,e^{\pm j\pi/4} \tag{D.8}$$

the plus sign is chosen when $\Psi''(\xi_s) > 0$, and the minus sign when $\Psi''(\xi_s) < 0$.

Additional analysis [2, 3] is required to determine the end-point contributions. The final result is as follows. For the integral (D.1), the stationary phase asymptotic approximation is given by the following result, the integral being denoted $I(\kappa)$ to indicate that κ is the largeness parameter:

$$I(\kappa) \sim I_{sp} + I_{\xi_2} - I_{\xi_1} \tag{D.9}$$

where we have

$$I_{sp} = F(\xi_s)\sqrt{\frac{2\pi}{\kappa|\Psi''(\xi_s)|}}\,e^{j[\kappa\Psi(\xi_s) + (\pi/4)\mathrm{sgn}(\Psi''(\xi_s))]} \tag{D.10}$$

$$I_{\xi_1} = \frac{F(\xi_1)}{\kappa\Psi'(\xi_1)}\,e^{j[\kappa\Psi(\xi_1) - \pi/2]} \tag{D.11}$$

$$I_{\xi_2} = \frac{F(\xi_2)}{\kappa\Psi'(\xi_2)}\,e^{j[\kappa\Psi(\xi_2) - \pi/2]} \tag{D.12}$$

where the primes denote differentiation with respect to ξ, and $\Psi'(\xi_s) = 0$; that is, ξ_s is the stationary point. Note that the contribution from the stationary phase point is on order $1/\sqrt{\kappa}$, whereas the end-point contributions are on order $1/\kappa$.

Although we have referred to *the* stationary point, note that there may be more than one stationary point in the interval $a \leqslant \xi \leqslant b$. For each such point, a contribution to $I(\kappa)$ of the form of the first term in (D.9) must be added to the expression in (D.9).

The approximation (D.9) of course is subject to a number of conditions:

(a) $\Psi(\xi)$ is a continuous function with continuous derivatives for $\xi_1 \leqslant \xi \leqslant \xi_2$.
(b) $\Psi(\xi_s) \neq 0$.
(c) $F(\xi)$ is slowly varying and does not have any singularities in the range $\xi_1 \leqslant \xi \leqslant \xi_2$.
(d) The stationary point ξ_s is not close to either of the end points ξ_1 or ξ_2.

Jones [4, pp. 715–716] discussed what should be done if condition (b) is not met, the asymptotic expression for I_{sp} then involving Airy functions of the type discussed in Appendix B. The following also should be noted:

(a) If the interval $\xi_1 \leqslant \xi \leqslant \xi_2$ contains several stationary points, I_{sp} is a sum comprising terms representative of each ξ_s.
(b) When the stationary point coincides with either of the end points, the corresponding end-point contribution in the asymptotic expansion (D.9) is omitted and we take one-half times the stationary point contribution for that particular stationary point.
(c) When the interval $\xi_1 \leqslant \xi \leqslant \xi_2$ lacks a stationary point the integral $I(\kappa)$ is approximated by the end-point contributions only.
(d) When an end point moves to infinity, the corresponding end-point contribution is omitted.
(e) When the stationary point ξ_s approaches one of the end points but does not coincide with it, the asymptotic approximation (D.9) for $I(\kappa)$ no longer holds.

Complete derivations of the stationary phase results presented above may be found in the texts by Born and Wolf [1], Jones [4], and Felsen and Marcuvitz [2]. These also consider stationary phase methods applicable to functions of two variables. In such cases the stationary points are sometimes called *saddle points.*

Example D.1: Stationary phase evaluation of the physical optics integral in Section 3.3.11

Consider as an example the integral:

$$g(\phi) = \int_{-\pi/2}^{\pi/2} \cos\tau \, e^{jka\cos\tau} \, e^{jka\cos(\phi-\tau)} \, d\tau \qquad \text{(D.13)}$$

This clearly is an integral of the form (D.1) with $\xi = \tau$, $\xi_2 = \pi/2$, $\xi_1 = -\pi/2$, largeness parameter $\kappa = ka$, and

$$\Psi(\tau) = [\cos\tau + \cos(\phi - \tau)] \tag{D.14}$$

$$F(\tau) = \cos\tau \tag{D.15}$$

The derivative of the phase term $\Psi(\tau)$ with respect to τ is

$$\Psi'(\tau) = -[\sin\tau - \sin(\phi - \tau)] \tag{D.16}$$

and its second derivative is

$$\Psi''(\tau) = -[\cos\tau + \cos(\phi - \tau)] \tag{D.17}$$

The stationary phase point τ is given by the solution of $\Psi'(\tau_s) = 0$ to be $\tau_s = \phi/2$. Therefore, we have

$$\begin{aligned} \Psi(\tau_s) &= 2\cos(\phi/2) \\ \Psi''(\tau_s) &= -2\cos(\phi/2) \\ F(\tau_s) &= \cos(\phi/2) \\ \operatorname{sgn}[\Psi''(\tau_s)] &= -1 \end{aligned}$$

At the end-points $\pm\pi/2$, we have

$$\begin{aligned} \Psi'(\pi/2) &= -(\cos\phi + 1) \\ \Psi'(-\pi/2) &= (\cos\phi + 1) \\ F(\pm\pi/2) &= \sqrt{2} \\ \Psi(\pm\pi/2) &= \pm\sin\phi \end{aligned}$$

Assembling these results according to (D.9) we obtain,

$$g(\phi) \sim \sqrt{\frac{\pi\cos(\phi/2)}{ka}}\, e^{j2ka\cos(\phi/2)}\, e^{-j\pi/4} - \frac{\sqrt{2}}{ka(\cos\phi + 1)}\, e^{j(ka\sin\phi - \pi/2} - \frac{\sqrt{2}}{ka(\cos\phi + 1)}\, e^{-j(ka\sin\phi + \pi/2)} \tag{D.18}$$

Expression (D.18) is *an* asymptotic approximation to the integral (D.13), which becomes increasingly accurate as the largeness parameter ka increases in size. We emphasize the word *an* because asymptotic approximations are not unique. Any chosen approximation is simply one of many possible approximations. Some asymptotic approximations for a specific integral of course may be better than others. The mathematical result (D.18) is used in Section 3.3.11 of the text.

Asymptotic expansions are unlike the expansions with which most engineers are familiar (e.g., Taylor series or Fourier series). When we speak of the convergence of an asymptotic expansion we do not mean the same thing as when considering eigenfunction expansions, for example. Convergence of an asymptotic expansion means only that it approximates some quantity with increasing accuracy as some largeness parameter increases in size.

D.3 BIBLIOGRAPHICAL REMARKS

A good start to a study of additional asymptotic techniques used in derivations of the UTD formulations is the broad overview provided by Kouyoumjian [5]. It will be clear from [5] that there are two mathematical techniques specifically that need to be studied: the Watson transformation and the method of steepest descent. Although the Watson transformation need not be viewed as an asymptotic technique, its use in electromagnetics usually is one of facilitating the application of some asymptotic method (especially the method of steepest descent).

An understanding of these approaches relies on a knowledge of the behavior of functions of a complex variable. Many texts are available on complex analysis; the predicament usually is to select the most suitable one. Highly readable, and possibly the most relevant, sources of this information are by Wyld [6] and Carrier, Krook, and Pearson [7]. Reference [7] is particularly useful because it begins with the fundamentals of complex analysis and proceeds to an easily understood discussion of asymptotic methods, including the method of stationary phase and both the Watson transformation and the method of steepest descent.

The list of essential concepts from the theory of functions of a complex variable consists of the following. A complex function is analytic over some region of the complex plane if it has a unique derivative at points in this region. The functions of interest in electromagnetics usually are analytic over specific regions of interest except at certain points (*singularities*) or along particular curves (*branch cuts*). A complex function defined over the complex plane is a function of two variables, and thus integrals of such functions involve two variables in general. To specify the end-points of the integration therefore is not sufficient, but also the particular path or contour to be followed, as can be seen from the special functions discussed in Appendix B. The peculiar analytic properties of complex functions result in the requirement that these functions must satisfy the so-called Cauchy-Riemann equations. Cauchy's theorem and Cauchy's integral formula follow. Then, the integral of a complex function along any closed contour can be expressed purely as a sum of weighted terms (called *residues*) associated with the behavior of the function at all singularities within the contour. The Watson transformation allows, for example, a slowly convergent eigenfunction expansion to be converted into a contour integral. This contour integral then can be evaluated as a more rapidly convergent residue series or using the method of steepest descent.

The method of steepest descent consists in deforming the contour of integration in the complex plane in such a way that the major contribution to the integral arises from a small portion of the new path of integration, this major contribution becoming more and more dominant as some largeness parameter grows. The technique thus differs from the method of stationary phase in that the self-cancelling oscillation is not used, but rather the decay of an exponential factor in an integrand. The application of the method usually requires that some specified contour be deformed to a steepest descent path. If this is done, the crossing of any points or regions at which the integrand is not analytic must be taken into account with the utmost care [2, 8, 9].

The method of steepest descent often is called the *saddle point method*. In those cases where the paths of steepest descent involve a saddle point, this may be acceptable. However, examples can be given [7, p. 261] where the steepest descent paths do not pass through a saddle point. Furthermore, we equally well might call the method of stationary phase, in the case of two variables, the *saddle point method*. The terms *method of steepest descent* and *method of stationary stationary phase* therefore are to be preferred, because they eliminate possible ambiguities and are more descriptive of the methods themselves.

An excellent treatment of the asymptotic evaluation of integrals specifically intended for electromagnetics is given in Felsen and Marcuvitz [2], and for integrals in general by Bleistein and Handelsman [8]. Both books are at the relatively advanced level, but very complete.

An introductory treatment of asymptotic methods in general (that is, not only the evaluation of integrals) is given by Bender and Orszag [10]. The works by Copson [9] and Erdelyi [3] are classics. Applications of the stationary phase method directly to radiation problems (not necessarily in a GTD or UTD context) appear in [11] and [12].

REFERENCES

[1] M. Born and E. Wolf, *Principles of Optics,* Pergamon Press, London, 1975, pp. 752–754.

[2] L.B. Felsen and N. Marcuvitz, *Radiation and Scattering of Waves,* Prentice-Hall, Englewood Cliffs, NJ, 1973.

[3] A. Erdelyi, *Asymptotic Expansions,* Dover, New York, 1956.

[4] D.S. Jones, *Acoustic and Electromagnetic Waves,* Oxford University Press, London, 1986.

[5] R.G. Kouyoumjian, "Asymptotic High-Frequency Methods," *Proc. IEEE,* August 1965, pp. 864–876.

[6] H.W. Wyld, *Mathematical Methods for Physics,* W.A. Benjamin, London, 1976.

[7] G.F. Carrier, M. Krook, and C.E. Pearson, *Functions of a Complex Variable,* McGraw-Hill, New York, 1966.

[8] N. Bleistein and R.A. Handelsman, *Asymptotic Expansion of Integrals,* Dover, New York, 1986.
[9] E.T. Copson, *Asymptotic Expansions,* Cambridge University Press, Cambridge, 1965.
[10] C.M. Bender and S.A. Orszag, *Advanced Mathematical Methods for Scientists and Engineers,* McGraw-Hill, New York, 1972.
[11] C.A. Siller, "Evaluation of the Radiation Integral in Terms of End-Point Contributions," *IEEE Trans. Antennas and Propagation,* Vol. AP-23, September 1975, pp. 743–745.
[12] P. Kildal, "Asymptotic Approximations of Radiation Integrals: Endpoint and Double Endpoint Diffraction," *Radio Science,* Vol. 19, No. 3, 1984, pp. 805–811.

Appendix E
Additional References

GENERAL

[1] V.A. Borovikov and B.Y. Kinber, *Geometrical Theory of Diffraction* [in Russian], Svyaz, Moscow, 1978.

[2] R.G. Kouyoumjian and W.D. Burnside, "The Diffraction by a Cylinder-Tipped Half-Plane," *IEEE Trans. Antennas and Propagation,* Vol. AP-18, No. 3, May 1970, pp. 424–426.

[3] A.Q. Howard, Jr., "A Comparison of Mode Match, Geometrical Theory of Diffraction, and Kirchoff Radiation," *IEEE Trans. Antennas and Propagation,* Vol. AP-21, No. 1, January 1973, pp. 100–102.

[4] S.W. Lee, Y. Rahmat-Samii, and R.C. Menendez, "GTD, Ray Field and Comments on Two Papers," *IEEE Trans. Antennas and Propagation,* Vol. AP-26, No. 2, March 1978, pp. 352–354.

[5] G.A. Deschamps, J. Boersma, and S.W. Lee, "Three-Dimensional half-plane diffraction; exact solution and Testing of Uniform Theories," *IEEE Trans. Antennas and Propagation,* Vol. AP-32, No. 3, March 1984, pp. 264–271.

[6] E.F. Knott, "A Progression of High-Frequency RCS Prediction Techniques," *Proc. IEEE,* Vol. 72, No. 2, February 1985, pp. 252–264.

[7] P. Petre and L. Zombory, "Symbolic Code Approach to GTD Ray Tracing," *IEEE Trans. Antennas and Propagation,* Vol. 36, No. 10, October 1988, pp. 1490–1491.

REFLECTOR ANTENNAS

[8] L.L. Tsai, D.R. Wilton, M.G. Harrison, and E.H. Wright, "A Comparison of Geometrical Theory of Diffraction and Integral Equation Formulation for Analysis of Reflector Antennas," *IEEE Trans. Antennas and Propagation,* Vol. AP-20, No. 6, November 1972, pp. 705–712.

[9] G.L. James and V. Kerdemelidis, "Reflector Antenna Radiation Pattern Analysis by Equivalent Currents," *IEEE Trans. Antennas and Propagation,* Vol. AP-21, No. 1, January 1973, pp. 19–24. See also Kerdemelidis' comments and authors' reply in *IEEE Trans. Antennas and Propagation,*Vol. AP-21, No. 7, September 1973, p. 756.

[10] W.V.T. Rusch and O. Sorensen, "The Geometrical Theory of Diffraction of Axially Symmetric Reflectors," *IEEE Trans. Antennas and Propagation,* Vol. AP-23, No. 3, May 1975, pp. 414–419.

[11] C.A. Mentzer and L. Peters, Jr., "A GTD Analysis of the Far-out Sidelobes of Cassegrain Antennas," *IEEE Trans. Antennas and Propagation,* Vol. AP-23, No. 6, September 1975, pp. 702–709.

[12] S.W. Lee, P. Cramer, Jr., K. Woo, and R.C. Menendez, "Diffraction by an Arbitrary Subreflector GTD Solution," *IEEE Trans. Antennas and Propagation,* Vol. AP-27, No. 3, May 1979, pp. 305–316.

[13] G. Morris, "Coupling between Closely Spaced Back-to-Back Paraboloidal Antennas," *IEEE Trans. Antennas and Propagation,* Vol. AP-28, No. 1, January 1980, pp. 60–64.

[14] O.M. Bucci and G. Franceschetti, "Rim Loaded Reflector Antennas," *IEEE Trans. Antennas and Propagation,* Vol. AP-28, No. 3, May 1980, pp. 297–305.

[15] M.S. Narasimhan, P. Ramanujam, and K. Raghavan, "GTD Analysis of Near-Field and Far-Field Patterns of a Parabolic Subreflector Illuminated by a Plane Wave," *IEEE Trans. Antennas and Propagation,* Vol. AP-29, No. 4, July 1981, pp. 654–660.

[16] M.S. Narasimhan, P. Ramanujam, and K. Raghavan, "GTD Analysis of the Radiation Patterns of a Shaped Subreflector," *IEEE Trans. Antennas and Propagation,* Vol. AP-29, No. 5, September 1981, pp. 792–795.

[17] M.S. Narasimhan, P. Ramanujam, and K. Raghavan, "GTD Analysis of a Hyperboloidal Subreflector with Conical Flange Attachment," *IEEE Trans. Antennas and Propagation,* Vol. AP-29, No. 6, November 1981, pp. 865–871.

[18] M.S. Narasimhan and K.M. Prasad, "GTD Analysis of the Near-Field Patterns by a prime-focus symmetric paraboloidal Reflector Antenna," *IEEE Trans. Antennas and Propagation,* Vol. AP-29, No. 6, November 1981, pp. 959–961.

[19] S.H. Lee and R.C. Rudduck, "Aperture Integration and GTD Techniques Used in the NEC Reflector Antenna Code," *IEEE Trans. Antennas and Propagation,* Vol. AP-33, No. 2, February 1982, pp. 189–194.

[20] M.S. Narasimhan, P. Ramanujam, and K. Raghavan, "GTD Analysis of Radiation Patterns of a Prime Focus Paraboloid with Shroud," *IEEE Trans. Antennas and Propagation,* Vol. AP-31, No. 5, September 1983, pp. 792–794.

[21] P.S. Kildal, "The Effects of Subreflector Diffraction on the Aperture Efficiency of a Conventional Cassegrain Antenna—An Analytical Approach," *IEEE Trans. Antennas and Propagation,* Vol. AP-31, No. 6, November 1983, pp. 903–909.

[22] G. Di Massa and C. Savaresse, "Parabolic and Hyperbolic Rim Loaded Antennas Pattern Computation Using the Geometrical Theory of Diffraction," *IEEE Trans. Antennas and Propagation,* Vol. AP-32, No. 6, June 1984, pp. 656–657.

[23] S.W. Lee, P. Cramer, Jr., K. Woo, and Y. Rahmat-Samii, "Diffraction by an Arbitrary Subreflector: GTD Solution," *IEEE Trans. Antennas and Propagation,* Vol. AP-27, No. 2, March 1979, pp. 305–316. See also S.W. Lee and P.T. Lam, "Correction to Diffraction by an Arbitrary Subreflector: GTD Solution," *IEEE Trans. Antennas and Propagation,* Vol. AP-34, No. 2, March 1986, p. 272.

[24] Y. Rahmat-Samii, "Subreflector Extension for Improved Efficiencies in Cassegrain Antennas—GTD/PO Analysis," *IEEE Trans. Antennas and Propagation,* Vol. AP-34, No. 10, October 1986, pp. 1266–1268.

[25] E. Yazgan and M. Safak, "Comparison of UTD and UAT in Axially Symmetric Reflectors," *IEEE Trans. Antennas and Propagation,* Vol. AP-35, No. 1, January 1987, pp. 113–115.

[26] Y.C. Chang, "Computer Techniques for Diffraction," in *Reflector and Lens Antennas,* C.J. Sletten (Ed.), Artech House, Norwood, MA, 1988.

AIRCRAFT ANTENNAS

[27] W.D. Burnside, R.J. Marhefka, and C.L. Yu, "Roll-Plane Analysis of On-Aircraft Anten-

nas," *IEEE Trans. Antennas and Propagation,* Vol. AP-21, No. 6, November 1973, pp. 780–786.

[28] W.D. Burnside, M.C. Gilreath, R.J. Marhefka, and C.L. Yu, "A Study of KC-135 Aircraft Antenna Patterns," *IEEE Trans. Antennas and Propagation,* Vol. AP-23, May 1975, pp. 309–316.

[29] C.L. Yu, W.D. Burnside, and M.C. Gilreath, "Volumetric Pattern Analysis of Airborne Antennas," *IEEE Trans. Antennas and Propagation,* Vol. AP-26, September 1978, pp. 636–641.

[30] W.D. Burnside, N. Wang, and E.L. Pelton, "Near-Field Pattern Analysis of Airborne Antennas," *IEEE Trans. Antennas and Propagation,* Vol. AP-28, No. 3, May 1980, pp. 318–327.

OTHER ANTENNAS

[31] D.L. Hutchins and R.G. Kouyoumjian, "Calculation of the Field of a Baffled Array by the Geometrical Theory of Diffraction," *J. Acoustic Soc. Am.,* Vol. 45, No. 2, February 1969, pp. 485–495.

[32] C.A. Balanis and L. Peters, Jr., "Equatorial Plane Pattern of an Axial-TEM Slot on a Finite Size Ground Plane," *IEEE Trans. Antennas and Propagation,* Vol. AP-17, No. 3, May 1969, pp. 351–353.

[33] C.A. Ryan, Jr., "Analysis of Antennas on Finite Circular Cylinders with Conical or Disk End Caps," *IEEE Trans. Antennas and Propagation,* Vol. AP-20, No. 4, July 1972, pp. 474–476.

[34] S.W. Lee, "Mutual Admittance of Slots on a Cone: Solution by Ray Technique," *IEEE Trans. Antennas and Propagation,* Vol. AP-26, No. 6, November 1978, pp. 768–773.

[35] G. Mazzarella and G. Panariello, "Evaluation of Edge Effects in Slot Arrays Using the Geometrical Theory of Diffraction," *IEEE Trans. Antennas and Propagation,* Vol. 37, No. 3, March 1989, pp. 392–394.

APERTURES

[36] C.A. Balanis and L. Peters, Jr., "Analysis of Aperture Radiation from an Axially Slotted Circular Conducting Cylinder Using Geometrical Theory of Diffraction," *IEEE Trans. Antennas and Propagation,* Vol. AP-17, No. 1, January 1969, pp. 93–97.

[37] C.A. Balanis and L. Peters, Jr., "Aperture Radiation from an Axially Slotted Elliptical Conducting Cylinder Using Geometrical Theory of Diffraction," *IEEE Trans. Antennas and Propagation,* Vol. AP-17, No. 4, July 1969, pp. 507–513.

[38] C.E. Ryan, Jr., "Analysis of Antennas on Finite Circular Cylinders with Conical or Disk End Caps," *IEEE Trans. Antennas and Propagation,* Vol. AP-20, No. 4, July 1972, pp. 474–476.

[39] C.R. Cockrell and P.H. Pathak, "Diffraction Theory Techniques applied to Aperture Antennas on Finite Circular and Square Ground Planes," *IEEE Trans. Antennas and Propagation,* Vol. AP-22, No. 3, May 1974, pp. 443–448.

[40] M.C. Bailey, "Mutual Coupling between Circular Waveguide-fed Apertures in a Rectangular Ground Plane," *IEEE Trans. Antennas and Propagation,* Vol. AP-22, July 1974, pp. 597–599.

[41] P.H. Pathak and R.G. Kouyoumjian, "An Analysis of the Radiation from Apertures in Curved Surfaces by the Geometrical Theory of Diffraction," *Proc. IEEE,* Vol. 62, No. 11, November 1974, pp. 1438–1447.

[42] S.W. Lee, "Mutual Admittance of Slots on a Cone: Solution by Ray Technique," *IEEE Trans. Antennas and Propagation,* Vol. AP-26, November 1978, pp. 768–773.

RCS AND SCATTERING:

[43] R.A. Ross, "Radar Cross Section of Rectangular Flat Plates as a Function of Aspect Angle," *IEEE Trans. Antennas and Propagation,* Vol. AP-14, No. 3, May 1966, pp. 329–335.

[44] J. Boersma, "Electromagnetic Diffraction by a Unidirectionally Conducting Circular Disk," *Siam J. Appl. Math.,* Vol. 14, November 1966, pp. 1471–1495.

[45] P.E. Mast and R.G. Kouyoumjian, "On the GTD Scattering by Polygonal Cylinder," *IEEE Trans. Antennas and Propagation,* Vol. AP-24, No. 1, January 1976, pp. 94–95.

[46] N. Wang, "Self Consistent GTD Formulation for Conducting Cylinders with Arbitrary Convex Cross Section," *IEEE Trans. Antennas and Propagation,* Vol. AP-24, No. 4, July 1976, pp. 463–468.

[47] N.T. Alexander and R.W. Larson, "Phase Analysis of Cone Scattering near Base-on," *IEEE Trans. Antennas and Propagation,* Vol. AP-26, No. 3, July 1976, pp. 541–543.

[48] M.A. Plonus et al., "Radar Cross Section of Curved Plates Using Geometrical and Physical Diffraction Techniques," *IEEE Trans. Antennas and Propagation,* Vol. AP-26, No. 3, May 1978, pp. 488–493.

[49] T.C.K. Rao and M.A.K. Hamid, "GTD Analysis of Scattering from a Dielectric-Coated Cylinder," *Microwave Optics and Antennas,* Vol. 127, June 1980, pp. 143–153.

[50] A. Michaeli, M. Kaye, and A. Geva, "High-Frequency Backscatter from a Finned Cylinder—Comparison of UTD Results with Experiments," *IEEE Trans. Antennas and Propagation,* Vol. AP-32, No. 4, April 1984, pp. 422–425.

[51] T. Griesser and C.A. Balanis, "Dihedral Corner Reflector Backscatter Using Higher-Order Reflections and Diffractions," *IEEE Trans. Antennas and Propagation,* Vol. AP-35, No. 11, November 1987, pp. 1235–1247.

[52] D.P. Marsland, C.A. Balanis, and S. Brumley, "Higher Order Diffractions from a Circular Disk," *IEEE Trans. Antennas and Propagation,* Vol. AP-35, No. 12, December 1987, pp. 1436–1444.

[53] A. Michaeli, "A Uniform GTD Solution for the Far-Field Scattering by Polygonal Cylinders and Strips," *IEEE Trans. Antennas and Propagation,* Vol. AP-35, No. 8, August 1987, pp. 983–984. See also author's correction in *IEEE Trans. Antennas and Propagation,* Vol. 36, No. 10, October 1988, pp. 1498.

[54] A.H. Serbest, A. Buyukaksoy, and G. Uzgoren, "Diffraction of High-Frequency Electromagnetic Waves by Curved Strips," *IEEE Trans. Antennas and Propagation,* Vol. 37, No. 5, May 1989, pp. 529–600.

[55] H.H. Syed and J.L. Volakis, "Multiple diffractions among polygonal impedance cylinders," *IEEE Trans. Antennas Propagation,* Vol. 37, No. 5, May 1989, pp. 664–672.

TERRAIN SCATTERING:

[56] K.C. Chamberlain and R.J. Luebbers, "An Evaluation of Longley-Rice and GTD Propagation Models," *IEEE Trans. Antennas and Propagation,* Vol. AP-30, No. 6, November 1982, pp. 1093–1098.

[57] R.J. Luebbers, "Finite Conductivity Uniform GTD versus Knife Edge Diffraction in Prediction of Propagation Path Loss," *IEEE Trans. Antennas and Propagation,* Vol. AP-32, No. 1, January 1984, pp. 70–76.

[58] R.J. Luebbers, "Propagation Prediction for Hilly Terrain Using GTD Wedge Diffraction," *IEEE Trans. Antennas and Propagation,* Vol. AP-32, No. 9, September 1984, pp. 951–955.

[59] A.R. Lopez, "Application of Wedge Diffraction Theory to Estimating Power Density at Airport Humped Runways", *IEEE Trans. Antennas and Propagation,* Vol. AP-35, No. 6, June 1987, pp. 708–714.

[60] R.J. Luebbers, "Comparison of Lossy Wedge Diffraction Coefficients with Application to Mixed Path Propagation Loss Prediction," *IEEE Trans. Antennas and Propagation,* Vol. 36, No. 7, July 1988, pp. 1031–1033.

[61] R.J. Luebbers, W.A. Foose, and G. Reyner, "Comparison of GTD Propagation Model Wide-Band Path Loss Simulation with Measurements," *IEEE Trans. Antennas and Propagation,* Vol. 37, No. 4, April 1989, pp. 499–505.

HYBRID METHODS

[62] G.A. Thiele and T.H. Newhouse, "A Hybrid Technique for Combining Moment Methods with the Geometrical Theory of Diffraction," *IEEE Trans. Antennas and Propagation,* Vol. AP-23, No. 1, January 1975, pp. 62–69.

[63] W.D. Burnside, C.L. Yu, and R.J. Marhefka, "A Technique to Combine the Geometrical Theory of Diffraction and the Moment Method," *IEEE Trans. Antennas and Propagation,* Vol. AP-23, No. 4, July 1975, pp. 551–558.

[64] G.A. Thiele and G.K. Chan, "Application of the Hybrid Technique to Time Domain Problems," *IEEE Trans. Antennas and Propagation,* Vol. AP-26, No. 1, January 1978, pp. 151–155.

[65] C.W. Chuang and W.D. Burnside, "A Diffraction Coefficient for a Cylindrically Truncated Planar Surface," *IEEE Trans. Antennas and Propagation,* Vol. AP-28, No. 2, March 1980, pp. 177–182.

Appendix F
Computer Subroutine Listings

"But when it comes to problems that arise in the real world around us, the mere existence of an algorithm is by no means the end of the matter . . . For the kinds of problem that are of concern to businessmen and applied scientists, the important issue is the existence of an efficient algorithm."

K. Devlin,
Mathematics: The New Golden Age (Penguin Books, Baltimore, 1988)

This appendix presents FORTRAN computer subroutines that can be used to calculate Fresnel integrals, the transition function, first-order and slope wedge diffraction coefficients, and several special functions related to curved-surface diffraction. These are defined in Appendix B. Except for the routine *FRESNEL,* all of the routines were developed at the Electroscience Laboratory of the Ohio State University [1] and are reprinted here with permission.

F.1 FRESNEL INTEGRALS

The Fresnel cosine integral $C(x)$ and sine integral $S(x)$, are defined by equations (B.1) and (B.2), respectively. They may be computed using subroutine *FRESNEL (X, C, S),* which was developed by J.W. Odendaal of the University of Pretoria. The routine is based on the algorithm presented in [2].

F.2 TRANSITION FUNCTION

The transition function $F(x)$ is defined by equation (B.3). It is computed using the routine *FFCT(XF)* [1]. The small argument form, interpolatory scheme for intermediate values of the argument, and large argument form used in the routine are described in Section B.2.

F.3 WEDGE-DIFFRACTION COEFFICIENT

The subroutine *WD* [1] can be used to calculate the first-order wedge diffraction coefficient, given in equation (6.20). The parameters are as follows:

D — Complex 4 × 1 matrix, where D(1) corresponds to D_1 in (6.20), and similarly for D(2), D(3), and D(4). Note that once D has been calculated using *WD*, it must be multiplied by $\sqrt{\lambda}$ to obtain the correct value of the diffraction coefficient.

R — Distance parameter, normalized to a wavelength. The three-dimensional parameters are given by (6.25) and (6.28); whereas the two-dimensional parameters are given by (4.60)–(4.61). Note that D(1), D(2), D(3), and D(4) all are calculated with the same R. In the general case where L^i, L^{ro} and L^{rn} are different, *WD* needs to be called three times. During the first call $R = L^i/\lambda$ so that D_1 = D(1) and D_2 = D(2). We then set $R = L^{rn}/\lambda$, call *WD*, and set D_3 = D(3). Finally, $R = L^{ro}/\lambda$, call *WD*, and set D_4 = D(4).

PHR — Angle ϕ in radians

PHPR — Angle ϕ' in radians

SBO — $\sin\beta_0$ (In the two-dimensional case SBO = 1)

FN — Parameter n given in (4.26).

The total first-order diffraction coefficient thus is given by

$$D_{s,h} = \sqrt{\lambda}\,(\mathrm{D}(1) + \mathrm{D}(2) \mp [\mathrm{D}(3) + \mathrm{D}(4)]) \tag{F.1}$$

The minus or plus sign is incorporated by the user, depending on the polarization required.

F.4 WEDGE-SLOPE–DIFFRACTION COEFFICIENT

The subroutine WDP [1] can be used to calculate the wedge-slope–diffraction coefficient, given in (4.226). The parameters are the same as those of *WD*. As in *WD*, D has to be multiplied by $\sqrt{\lambda}$ to obtain the correct value of the diffraction coefficient.

The total slope diffraction coefficient is given by

$$\frac{\partial D_{s,h}}{\partial \phi'} = \sqrt{\lambda}\,(\mathrm{D}(1) - \mathrm{D}(2) \pm [\mathrm{D}(3) - \mathrm{D}(4)]) \tag{F.2}$$

The minus or plus sign is incorporated by the user depending on the polarization required.

F.5 FOCK SCATTERING FUNCTIONS

The routines *PFUN(X)* and *QFUN(X)* [1] can be used to compute the soft and hard Fock scattering functions defined by equations (B.25) and (B.26), respectively.

F.6 UNIVERSAL FOCK RADIATION FUNCTIONS

The routines *G(X)* and *GB(X)* are altered versions of the original routines of the same name given in [1] and can be used to compute the hard and soft universal Fock radiation functions, $G(x)$ and $\tilde{G}(x)$, respectively, as defined by equations (B.37) and (B.38), respectively. The alteration just mentioned simply is a conjugation operation that is performed in the routines before the function value is returned to make them directly applicable for the time-dependence $e^{j\omega t}$ used in this text.

F.7 FOCK COUPLING FUNCTIONS

The routines *FU(X, UU)* and *FV(X, VV)* [1] can be used to compute the soft and hard Fock coupling functions $u(x)$ and $v(x)$ defined by equations (B.43) and (B.44), respectively.

REFERENCES

[1] W.D. Burnside, R.J. Marhefka, and N. Wang, "Computer Programs, Subroutines and Functions for the Short Course on the Modern Geometrical Theory of Diffraction" in "Notes for the Short Course on the Modern Geometrical Theory of Diffraction," Ohio State University.

[2] M.A. Heald, "Rational Approximations for the Fresnel Integrals," *Math. Comp.*, Vol. 44, No. 170, April 1985, pp. 459–461.

Index